12/91

The Reference Manual of Woody Plant Propagation

From Seed to Tissue Culture

D1264956

BY MICHAEL A. DIRR
AND CHARLES W. HEUSER, JR.

St. Louis Community College
at Meramec
Library

THE REFERENCE MANUAL OF WOODY PLANT PROPAGATION: FROM SEED TO TISSUE CULTURE

A practical working guide
to the propagation of over 1100 species,
varieties and cultivars.

BY

Michael A. *Dirr*
UNIVERSITY OF GEORGIA

AND

Charles W. *Heuser*, *Jr.*
THE PENNSYLVANIA STATE UNIVERSITY

Varsity Press, Inc.

P.O. BOX 6301
ATHENS, GEORGIA 30604

Library of Congress Cataloging in Publication Data

Dirr, Michael A.
Heuser, Charles W., Jr.
The Reference Manual of Woody Plant Propagation:
From Seed to Tissue Culture

Summary: A reference guide to the propagation of over 1100 woody plant species, varieties, and cultivars. Includes Index.

1. Plant Propagation. 2. Encyclopedia of Specific Propagation Requirements of Woody Plants. 3. Propagation practices of Seed, Cutting, Grafting/Budding, and Tissue Culture. Includes bibliographies and index.

ISBN 0-942375-00-9

Copyright © 1987 by Michael A. Dirr and Charles W. Heuser, Jr.

All rights reserved. No part of this book may be reproduced in any form or by any electronic or mechanical means including storage and retrieval systems without permission in writing from the publisher, except by a reviewer who may quote brief passages in review.

Printing: 4

Disclaimer of Liabilities: Due care has been taken in the preparation of this book to insure its effectiveness. The authors and publisher make no warrant, express or implied, with respect to the propagation procedures of this book. In no event will the authors or publisher be liable for direct, indirect, incidental, or consequential damages in connection with or arising from the furnishing, performance, or use of this book.

Illustrations by Bonnie Dirr. *Photography by* Michael Dirr.

DEDICATIONS

Many people have provided inspiration and encouragement during my educational journey. Professor Joseph McDaniel, University of Illinois, took me under his wing and shared years of plant wisdom and love with a young faculty member. Professor Emeritus Clarence E. Lewis, Michigan State University, a gentleman and a scholar, taught me that every plant possesses a certain intrinsic beauty. Mr. Alfred Fordham, former propagator, Arnold Arboretum, unraveled the secrets of woody plant propagation and, during my sabbatical, provided friendship and consul. Dr. L.C. Chadwick, Professor Emeritus, The Ohio State University, provided the inspiration in my first college plant materials course.

I learned from each but, most important, all taught me to teach myself. They exuded enthusiasm, knowledge, love of the subject matter and were willing to share it with a *student*.

Michael A. Dirr

For my father, Charles W. Heuser, Sr., and my mother, Kathleen R. Heuser.

Charles W. Heuser, Jr.

CONTENTS

CHAPTER ONE
Seed Propagation 11

A. INTRODUCTION . . . 11
B. SEED PROVENANCE . . . 11
C. SEED SOURCES . . . 13
D. SEED COLLECTION AND HANDLING . . . 13
1. Handling after collection . . . 13
2. Fleshy fruits and seeds . . . 13
3. Dried fruits and seeds . . . 15
E. SEED STORAGE . . . 15
F. SEED VIABILITY . . . 15
G. TREATMENTS TO OVERCOME SEED DORMANCY . . . 17
1. Pregermination treatments . . . 17
2. Hard seed coats . . . 17
3. Embryo dormancy . . . 17
4. Double dormancy . . . 19
H. POSTSCRIPT . . . 21

CHAPTER TWO
Cutting Propagation 23

A. INTRODUCTION . . . 23
B. WHY CUTTINGS? . . . 23
1. Integrity of characteristics . . . 23
2. Economics . . . 23
3. Avoidance of bud/graft incompatibilities . . . 23
C. TYPES OF CUTTINGS . . . 23
1. Leaf . . . 23
2. Stem . . . 25
a. Softwood . . . 25
b. Semi-hardwood (greenwood) . . . 25
c. Hardwood: Deciduous and broadleaf . . . 27
d. Hardwood: Needle evergreens . . . 27
3. Root cuttings . . . 27
D. COLLECTION AND HANDLING . . . 29
E. FACTORS AFFECTING ROOTING OF CUTTINGS . . . 29
1. Nutrition/carbohydrates/nitrogen . . . 29
2. Juvenility . . . 31
3. Timing . . . 31
4. Condition and type of cutting wood . . . 31
a. Clonal material . . . 31
b. Position of cutting on the plant . . . 33
5. Wounding and girdling . . . 33
a. Wounding . . . 33
b. Girdling . . . 33
c. Etiolation . . . 33
6. Hormones . . . 33
a. Introduction . . . 33
b. New chemicals . . . 35
c. Commercial root formulations . . . 35
d. Comparative study of root promoting chemicals and formulations . . . 35
e. Concentrations . . . 37
f. Chemical properties and costs of IBA and NAA . . . 37
g. Solvents for root promoting chemicals . . . 37
h. Quick dips compared to talcs . . . 39
i. Mixing quick dip and talc preparations . . . 39
F. MEDIA, FERTILIZERS . . . 39
1. The media components . . . 39
a. Sand . . . 39
b. Perlite . . . 41
c. Vermiculite . . . 41
d. Scoria and pumice . . . 41
e. Bark . . . 41
f. Peat . . . 41
2. Fertilizers . . . 41
G. ROOTING STRUCTURES, CONTAINERS, BOTTOM HEAT . . . 43
1. Rooting structures . . . 43
2. Containers for rooting . . . 43
3. Bottom heat . . . 43
H. MIST SYSTEMS . . . 45
1. Components . . . 45
a. Nozzles . . . 45
b. Controls-Timers . . . 45
c. Screen balance . . . 45
d. Electronic leaf . . . 45
e. Thermostat and timer . . . 45
f. Photoelectric cells . . . 45
I. FOG SYSTEMS . . . 45
J. LIGHT IN THE PROPAGATION HOUSE . . . 47
1. For propagation . . . 47
2. Inducing a flush of growth . . . 47
K. SANITATION . . . 47
L. AFTERCARE, OVERWINTERING, STORAGE . . . 47

CHAPTER THREE
Grafting and Budding 55

A. GRAFTING 55
B. LIMITATIONS OF GRAFTING 55
C. INCOMPATIBILITY 55
D. TIME MEASUREMENTS 57
E. PHYSIOLOGY OF GRAFTING 57
F. TOOLS AND ACCESSORIES 57
  1. Knives 57
  2. Tying and wrapping materials 57
G. GRAFTING METHODS 57
  1. Splice graft 57
  2. Whip-and-tongue graft 57
  3. Veneer and side grafts 57
  4. Saddle graft 59
  5. Cleft graft 59
  6. Cutting grafts 59
  7. Double working 59
  8. Root grafting 59
  9. Bare root grafting 59
H. CARE OF GRAFTS 59
  1. Closed case 59
  2. Open bench 61
  3. Poly bag chamber 61
  4. Hot-callus pipe 61
I. BUDDING METHODS 61
  1. Collection and storage of budsticks 61
  2. Time of budding 61
  3. T-budding 61
  4. Inverted T-budding 63
  5. Patch budding 63
  6. Chip budding 63
  7. Change purse budding 63
  8. Double shield budding 64
  9. Stages and time intervals in bud healing 64

CHAPTER FOUR
Tissue Culture 65

A. INTRODUCTION 65
B. APPLICATIONS OF TISSUE CULTURE 65
C. STAGES OF TISSUE CULTURE 67
  1. Stage I: Establishment 67
  2. Stage II: Multiplication 67
  3. Stage III: Preparation for re-establishment outside culture 67
D. PROCEDURES IN MEDIA PREPARATION 67
  1. Preparation of stock solutions 67
  2. Preparation of Murashige and Skoog medium 69
  3. Sterilization of media 69
E. EXPLANT PREPARATION AND STERILIZATION 69
  1. Surface sterilization 69
  2. Oxidative browning 71
  3. Internal contamination 71
F. HEALTH HAZARDS FROM STERILANTS AND RELATED PRODUCTS 71
G. SELECTED TISSUE CULTURE METHODOLOGY AND TYPES OF REGENERATION 71
  1. Shoot apex culture 71
  2. Axillary shoot proliferation 73
  3. Adventitious shoot initiation 73
  4. Organogenesis 73
  5. Embryogenesis 73
H. REGULATION OF EXPLANT GROWTH IN TISSUE CULTURE 75
I. ROOTING MICROCUTTINGS AND ACCLIMATION 75
J. LABORATORY REQUIREMENT AND DESIGN 75
K. SPECIFIC INFORMATION USEFUL IN TISSUE CULTURE 77

Encyclopedia 79
Scientific Name Index 225
Common Name Index 233

PREFACE

The idea for this book lay dormant for the past 15 years in the authors' minds. Over the past four years, we collected information and attempted to distill it to a workable quotient for the practitioner. This is not a book on the theories or principles of woody plant propagation. It is a grass roots effort to present information on seed, cutting, grafting and tissue culture that can be effectively utilized by the propagator. References are provided for those who require greater information. In most instances, we attempted to provide *specific* information relative to hormones, concentrations, media, condition of cutting wood, propagation systems, ad infinitum. For some plants the details are almost laborious; for others sketchy. We feel the latter provides at least a starting point and practitioners can refine the process.

The *Manual* does not have all the answers, and we eagerly solicit comments from readers that will allow us to improve future editions.

This book would not exist if researchers did not publish their results. To them, we offer thanks. During sabbatical, the senior author was allowed to peruse the propagation files at the Arnold Arboretum which Al Fordham, Jack Alexander and others built up over the years. To these individuals, a debt of gratitude is owed. Much of what is presented represents the authors' propagation research, observations and opinions.

The senior author wishes to thank his wife, Bonnie, for illustrating the *Manual*; Ms. Alice Richards, Ms. Beth Brinson and Mrs. Betty Johnson for proof-reading and organizational help; Mr. Tim Blalock for design and layout; and Guest Printing, Athens, Georgia for quality printing.

We consider this book a starting point and plan to improve and enlarge the material in future editions.

Michael A. Dirr, Athens, Georgia
Charles W. Heuser, Jr., University Park, Pennsylvania

HOW TO USE THE MANUAL

Plants are listed alphabetically by scientific name. If only the common name is known, please refer to the index.

Under each species four categories may be presented: SEED, CUTTINGS, GRAFTING, TISSUE CULTURE. Since many plants are propagated by only one or two methods, all possible categories will not be present.

Under SEED the optimum stratification treatments or other manipulations are described. Numerous factors affect germination, and readers should peruse the SEED PROPAGATION chapter. It provides background information that will make the ENCYCLOPEDIA more useful.

Remember that very few things are absolute in the biological world. The seeds of *Calycanthus floridus*, sweet shrub, require three months cold stratification. No doubt two months or four months will also work. Ample leeway is available for success. Precision is ideal; but seeds, like cuttings, can be quite forgiving.

Under CUTTINGS, for example, different hormone concentrations may be listed. These numbers are based on those presented in the literature and should serve only as guides. Generally, if one is not sure about the best starting point, use a commercial or hand-mixed concentration that approximates the published value. For example, a 4000 ppm IBA treatment will probably give the same results as 3000 or 5000 ppm treatments. In general, ball park figures are *more important* than values down to the last ppm.

For ready conversion of ppm to percent the following chart is provided. Sometimes in the text, percent and ppm are used together and this may present confusion for the reader.

Converison chart for percent (%) and parts per million (ppm)

%	ppm	grams of hormone/per liter of solvent to make the concentrations listed in the two left hand columns
0.1	1,000	1.0
0.25	2,500	2.5
0.5	5,000	5.0
1.0	10,000	10.
2.0	20,000	20.

The GRAFTING information is, perhaps, the least detailed. The chapter provides ample information for the common grafting practices. If more information is required the reader is encouraged to use Hartmann and Kester's *Plant Propagation: Principles and Practices* and/or Garner's *The Grafting Handbook*.

Information in TISSUE CULTURE is complete for some species; sketchy for others. Most nurserymen and propagators will benefit more from the SEED, CUTTING, and GRAFTING information than the TISSUE CULTURE. There are over 300 nurserymen in Georgia but none are propagating by tissue culture. The TISSUE CULTURE chapter provides a good overview of the process and provides background information for those interested in pursuing tissue culture propagation.

CHAPTER ONE

Seed Propagation

A. INTRODUCTION

Nature, in her special way, has provided a wonderful mechanism that permits the perpetuation of plant species. The reproductive structure is called a SEED and represents the end product of **gametasporagenesis, pollination** and **fertilization.** A seed, in simplest form, is a fertilized ovule which, when mature, contains an embryo composed of a root radicle and shoot primordium. The seed carries all the genetic information that, when properly translated, results in plants with characteristics similar to the parent. Seed propagation has been the basis for the perpetuation of approximately 250,000 angiosperm species and 700 gymnosperm species. Nature, for the most part, has provided for her own but man is inextricably bound with nature and, consequently, has manipulated seed propagation to his advantage and disadvantage.

Many nurserymen derive their livelihood from growing seedlings. The modern practices of seedling production combine art and science. If nurserymen depended solely on nature's whims for annual seed crops, they would soon be out of business (35, 44, 45, 46). Seed orchards (34), seed collecting equipment, storage facilities (42), dormancy treatments, fungicides, herbicides (31), seed bed preparation (43), harvesting equipment, postharvest storage facilities, and packaging equipment are integrated into modern day production (fig. 1).

At one time seedling production was the lifeblood of the nursery industry (15). Today the consumer is treated to red maple, *Acer rubrum*, with names like 'October Glory', 'Red Sunset' (fig. 2) and 'Northwood'. Trees that were selected for a specific trait...in this case fall color. To maintain this characteristic these trees must be vegetatively propagated by cuttings, budding (fig. 3), or tissue culture. Seedlings grown from seeds collected from these named clones are different from the parent. Most forest species (*Abies* spp., *Picea* spp., *Pinus* spp., *Pseudotsuga* spp., *Tsuga* spp.) are grown from seed (2, 22) but there is considerable interest in vegetative propagation of superior trees. No doubt, the future will offer more forest and ornamental trees produced through vegetative means, especially tissue culture.

For the nurseryman, the primary obstacle to seedling tree production is the variability of the progeny. The authors have never observed uniform, straight-trunked *Zelkova serrata* seedlings (fig. 4). On the other hand 'Village Green' and 'Green Vase' are essentially 100% uniform and, thus, salable. Green ash, *Fraxinus pennsylvanica*, when seed grown, yields a population of male and female (undesirable because of seed set) trees of variable growth habit and foliage quality. 'Marshall's Seedless', 'Summit', and 'Patmore' are male, straight-trunked, dark green foliaged, and superior to bulk run seedlings. Seedlings are necessary for large scale plantings, reforestation, breeding, maintaining genetic diversity and as rootstocks for budding or grafting superior types.

Many deciduous and evergreen shrubs are commonly produced from seed and little difference is observed in the progeny. *Lonicera maackii*, Amur honeysuckle, is a large, cumbersome shrub with white flowers and red fruits. Personally, the authors have not noticed sufficient differences among seed-grown populations to justify selection and naming of cultivars. Many references (2, 3, 20, 22, 28, 29, 33, 37, 38, 51) noted species that are most logically produced from seed.

Logically, if it were not for the wonderful variation that occurs in the process of gametasporagenesis (embryo sac development), pollination and eventual fertilization, the number of cultivars would be severely limited (fig. 5). For a short second, study figure 5, *Chamaecyparis obtusa* 'Nana Gracilis', and the varied seedlings resulting from a single batch of seed. All plants are the same age. In general foliage traits, they appear to belong to *C. obtusa*, otherwise they differ.

B. SEED PROVENANCE

The geographic source of seed is extremely important because seeds of a particular species do not perform uniformly under similiar conditions (16, 44, 45). Foresters pay strict attention to the provenance of a species. Provenance refers to the geographic area from which seed is collected. *Acer rubrum* ranges from Minnesota to central Florida. Minnesota trees, although recognizable as *A. rubrum*, do not respond the same when grown in Florida. The reciprocal is also true. A particular red maple planting on the Georgia campus flowers in mid to late January while the native trees do not flower until late February. The campus trees were purchased from a central Florida nursery that grew them from a local seed source. The Florida trees obviously require less chilling to break dormancy. These same trees do not harden sufficiently early in the fall and have suffered tremendous shoot dieback when early fall and winter freezes occurred.

Taking the red maple story a step further, all the named selections such as 'October Glory', 'Red Sunset', 'Autumn Flame', etc., are not cold-hardy at the University of Minnesota Landscape Arboretum where temperatures often reach −30°F. An introduction ('Northwood') selected from a native stand of *A. rubrum* near Excelsior, Minnesota (local provenance) proved reliably hardy and offered good orange-red fall color.

Flint and Alexander (18) present a case study of *Fraxinus americana*, white ash, and the provenance factors. Trees of northernmost origin (northern Michigan) withstood −40°F when tested in January; southern trees (southern Mississippi) withstood −22°F. In early October when the trees were again tested, northern trees withstood 2°F, southern trees only 14°F. Flint (17) and

Figure 1. Well managed seedbeds with irrigation and excellent weed control. The seedbeds are raised and covered with rotted sawdust. In spite of the "perfect" conditions the seeds of *Tilia americana* (right) germinated poorly while those of *Tilia cordata* (left) germinated in high percentages. A 50% germination rate on *Tilia americana* is considered good.

Figure 2. *Acer rubrum* 'Red Sunset' with excellent uniformity. This block of trees was produced from cuttings. There is no way to obtain this type of uniformity from seedling red maples.

Figure 3. A budded block of *Gleditsia triacanthos* var. *inermis*. The buds were placed on the understocks in August of the previous year. The trees in the photograph are one-year-old budded whips.

Figure 4. The variability in seed-grown *Zelkova serrata*. This tremendous variation is the primary reason nurserymen have turned to vegetative propagation.

Figure 5. The variation in seedlings in *Chamaecyparis obtusa* 'Nana Gracilis.' All plants are the same age yet note the differences in size and shape.

coworkers have studied *Cercis canadensis* (13), *Cornus florida* (figure 6a and b), *Pinus strobus* and noted differences between northern and southern sources. What this suggests to nurserymen is to grow seedlings from the immediate area if possible. Steavenson (45) described the importance of provenance for growing hardy seedlings of *Acer rubrum*, *A. saccharinum*, *Carya illinoensis*, *Cercis canadensis*, *Liquidambar styraciflua*, *Cornus florida*, *Liriodendron tulipifera* and other plants in Missouri.

Provenance not only affects hardiness but many factors, a most important one being growth rate. In Minnesota studies, *Cornus sericea* clones from the entire range showed tremendous variation in growth rates and rates of cold acclimation in the fall. Plants from mild climates hardened much more slowly than those from colder habitats, and so were prone to damage in early winter.

When buying seed from commercial sources always insist on the location where the seed was collected. If it came from Georgia, and your nursery is located in Caribou, Maine, do not waste the money or time. Most seed houses keep (or should) records on seed source and should be happy to supply such information.

C. SEED SOURCES

The nursery industry grows many different species and, in general, it is not practical to maintain seed orchards. However, it is logical to collect seed from healthy, vigorous trees that offer uniformity of habit (for a given species), foliage, and fall color. The authors have observed classic examples of seed source selection in the production of red maple. Trees grown from run-of-the-mill seed trees had crooked trunks, variable growth rates, and light green to dark green foliage. Trees grown from seeds collected from *Acer rubrum* 'Red Sunset', a superior cultivar, were straighter trunked, more uniform in growth rate and foliage characteristics. The two blocks of trees were growing in close proximity and a casual visitor could easily see the differences.

As discussed, always select from superior trees and, if cross pollination can be problematic, make sure seed trees are sufficiently isolated. Red maple in the vicinity of *Acer saccharinum*, silver maple, virtually ensures cross pollinated seed and subsequent red-silver seedlings which fall under the grex name *Acer* × *freemanii*. Two named clones of *Acer* × *freemanii* include 'Marmo' and 'Armstrong', of which the latter, although listed as an *Acer rubrum* form, is a hybrid.

Quercus (Oak) presents a tremendous problem, because most species within either the white (27) or red great groupings will cross with each other. Recently, the senior author designed a study to determine the feasibility of rooting a wide range of oak species. Seedlings were purchased from several commercial sources and the variation was disheartening. With *Quercus lyrata*, overcup oak, came *Quercus alba*, white oak, and *Quercus rubra*, red oak. Other species were similarly confused. Nurserymen buy seeds from collectors, seed houses and trade with other nurserymen. The correct identification of oaks is difficult and the chances for error are compounded when the collector does not know one from the other. Ask yourself how to separate pin, red, black, scarlet, shumard and the realization sets in that mistakes are bound to occur.

Many seeds are collected in arboreta and botanical gardens with variable success. The opportunities for cross pollination are numerous and the chance for new hybrids tremendous. Senior author grew *Acer triflorum* seed from a certain botanic garden and ended up with *Acer cissifolium* seedlings, obviously an error in identification. For many unusual plants, an arboretum or botanic garden is the only hope for procuring seeds.

Seed houses offer another outlet for seed and may/may not provide variety and quality. Table 1 provides a partial list of seed suppliers (39). Senior author has had variable success with "store-bought" seed. Know your supplier and request that he, in turn, supply information relative to age of seed, where collected, viability, etc.

D. SEED COLLECTION AND HANDLING

Possibly the most difficult aspect of seed collection is knowing when the seed is ripe. The tree or shrub does not hang out a sign saying "tree-ripened fruit". Years of close observation have taught the authors that every plant is inherently different. A "collecting calendar" much like a "cutting calendar" is a worthwhile undertaking. In Athens, Georgia, fruits of *Elaeagnus pungens*, thorny elaeagnus, ripen in April; *Chimonanthus praecox*, fragrant wintersweet, May-June; *Prunus* × *yedoensis*, yoshino cherry, May-June; *Calycanthus floridus*, common sweetshrub, August; *Ulmus parvifolia*, Chinese elm, late September; *Acer saccharum*, sugar maple, October; and *Cercidiphyllum japonicum*, Katsuratree, November. It is obvious from these examples that seeds are "ready" at different periods during the year.

In general seeds are ripe when there is no increase in fresh (or dry) weight. Knowing the normal size and maturation time of the fruits/seeds to be collected provide a good index as to ripeness. Some seeds which appear ripe (size, color) suffer from immature embryos (*Ilex*, *Magnolia*) and require a considerable after-ripening period (usually in warm temperatures) before the seed will germinate. No amount of cold stratification does any good until the embryo is physiologically ready to receive the stimulus.

Seeds of certain species benefit from early collection and immediate planting (33). For *Acer palmatum*, Japanese maple, *Viburnum* spp. (8), *Carpinus* spp., *Ostrya* spp., *Tilia* spp. (3), this approach may reduce germination time by one year. If seeds become dry and wrinkled it takes two years and longer for germination (49).

1. Handling after collection

Seeds occur in specialized structures called fruits and in angiosperm species are enclosed in one or more carpels (leaf-like structure). Fruits may be fleshy or dry, dehiscent or indehiscent and require special procedures during collection and thereafter. Most gymnosperms bear seeds in cones (figure 7a) which need to be collected before natural seed dispersal occurs. Several pine species require soaking and/or drying techniques to effectively induce cone opening. The specifics for the various cone-bearing gymnosperms are provided in the ENCYCLOPEDIA. Several gymnosperms [*Cephalotaxus* (fig. 7b), *Podocarpus*, *Taxus* (Fig. 7c), *Torreya*] bear naked seeds with a fleshy seed coat (aril). These seeds should be handled like those described under 2. Fleshy fruits and seeds.

2. Fleshy fruits and seeds

Included are those fruits or seeds with a mealy or fleshy covering such as *Malus*, *Pyrus*, *Prunus*, *Cornus*, *Cotoneaster*, *Magnolia* (fig. 8a, b), and *Euonymus*.

Studies (14, 40) have demonstrated the presence of inhibitors in these fleshy coverings. *Liriope muscari*, with the mealy covering intact, germinated in low percentages (25 to 38%) but when removed germination averaged 90% and above. Similar results

Fig. 6.a.

Fig. 6.b.

Figure 6. The importance of seed source selection on *Cornus florida*: a) a cold hardy form possibly from a northern provenance; b) injury on a cold-sensitive form.

Fig. 7.a.

Fig. 7.b.

Fig. 7.c.

Figure 7. Reproductive structures of gymnosperms: a) cone of *Pinus taeda*, loblolly pine; b) plum-like seed of *Cephalotaxus*; and c) fleshy seed of *Taxus*, yew.

were obtained with *Magnolia grandiflora* (Table 2). From a storage standpoint, removal of fleshy fruit or seed coats reduces the opportunity for significant fungal and bacterial growth. Cleaned seeds can be dried and dusted with a suitable fungicide and stored until planting or stratification.

Cleaning fleshy fruits is a messy but necessary task. Growers who handle hundreds (thousands) of pounds of fruit use professional cleaning machines. The fruits and water are placed into the machine and the rotating wheel slowly separates the flesh from the seed and is floated off. In the authors' experience, it is wise to soak the fruits or seed for 6 to 24 hours. In the case of extremely mealy/large fleshy fruits place them in plastic bags and allow them to soften naturally before attempting to separate seed from flesh. Most fleshy fruits with hard coated seeds (*Cotoneaster, Chionanthus, Cornus*) are easily cleaned, however, seeds of *Magnolia* and *Euonymus* are easily broken in the cleaning process. The presoaking treatment permits easier separation of seed and flesh. For *Magnolia* it may be necessary to clean the seeds by hand.

Senior author uses a food blender with the blades wrapped with electrical tape (thick) or covered with tygon tubing or some type of plastic. The container is filled 1/3 with water and an equivalent volume of seeds is added. Short bursts at low speed are best. The seeds, if solid, usually sink to the bottom and the pulp (macerate) can be floated off. Seeds are decanted into a screen (depends on size of seed), washed to remove debris, air-dried for 24 to 48 hours, and stored.

3. Dried fruits and seeds

Many woody plant species present a dilemma for the collector. If dehiscent fruits (fig. 9) are not collected before they open (shatter), then the seeds are lost. *Hamamelis* species offer good examples. The fruit is a two-valved, dehiscent capsule, containing two football-shaped, shiny black seeds. The seeds are actually ejected from the capsule and almost impossible to locate in grass and weeds. Fruits should be collected when yellow-brown, placed in closed grocery bags or cardboard boxes. The fruits, as they air-dry, expel the seeds with such force that they "pop" against the inside of the container.

In Georgia work, it is standard practice to collect leguminous fruits like redbud (fig. 10), coffeetree, honeylocust, and yellowwood before the pods split naturally. Pods are air-dried on paper or in large grocery bags; the material (pod & seed) broken and shaken, and the seeds screened. The dried pods can also be passed through a hammermill. The seeds are dried and stored in plastic bags at 34 to 38°F. This approach has been extremely successful and even the small seeds of katsura tree have been stored for 3 years and germinated in high percentages.

E. SEED STORAGE

Seeds of Woody Plants in the United States (22) provides in-depth information relative to longevity of woody plant seed in storage and should be consulted. Longevity information, when available, has been incorporated into the ENCYCLOPEDIA. Some of the best early work on seed storage was conducted by Barton and coworkers (4, 5, 6, 7, 10) and is discussed briefly here. The most important criterion is sound seed of high viability free of diseases and insects. Seed moisture content for most species should average 5 to 12%. *Betula papyrifera* seeds with 0.6% moisture have been stored successfully as have oaks with 30%. Large seeds like those of oak, buckeye, and horsechestnut are usually difficult to store for long periods. If possible, immediate sowing after collection is possibly the best procedure. Buckeye and horsechestnut may average 50% moisture.

Senior author has allowed yellow and red buckeye fruits to dry for several days until the capsules split and the seeds could be removed. During this time the seeds developed a wrinkled constitution indicating that the seeds were losing moisture. On the other hand, buckeye seeds (fig. 11) collected immediately after falling to the ground and placed in polyethylene bags under refrigeration (40°F), maintained their original constitution for 3 months. Each seed type must be handled differently and the ENCYCLOPEDIA describes the best method for each genus. Seed longevity can be improved by cold storage. Naturally dispersed seeds of pine and spruce may last one year. Seeds of the same genera stored at subfreezing temperatures remained viable for 6 to 10 years. Seeds of *Ulmus americana*, American elm, remain viable for several weeks after dispersal in spring. Seeds from the same species, dried to 3% moisture content and stored at 25°F, were viable after 15 years (22). *Quercus* seeds (fig. 12) as mentioned, lose viability quickly under dry conditions. Seeds stored in polyethylene bags at 39°F remained viable for 2½ years.

In senior author's work, polyethylene bags and glass jars with plastic, screw-type caps are used for storing seeds. Always store the seed after drying and close the bag or container tightly. Periodic cut tests or tetrazolium tests provide an index of seed viability over time.

Probably the most important considerations in successful seed storage involve (1) maintaining constant temperature and moisture conditions during storage. Fluctuations in the above reduce seed viability; and (2) maintaining as low a temperature and moisture level as possible. At low moisture content and temperature, the adverse effects of insects and diseases are effectively slowed or stopped.

Seed storage temperatures may range from 32 to 50°F with temperatures under 41°F best. Subfreezing temperatures (25°F) are acceptable. Some tree and shrub seeds have been frozen in liquid nitrogen (−320°F) and maintained their viability. Remember, what comes out of storage is only as good as what goes in.

F. SEED VIABILITY

Possibly the most overlooked aspect of seedage is the determination of viability. Although sound seeds supposedly sink and hollow do not, this is not always foolproof. Tetrazolium tests provide an indication of whether the seed is alive. Hartmann and Kester (28) provide a good discussion of the tetrazolium test and how to apply same. The seeds (entire or cut) are soaked in the chemical (0.1 to 1% solution) for 2 to 24 hours. The length of time depends on the seed. The chemical forms a red, insoluble compound in contact with living tissue. This indicates the seeds are alive. The chemical (2, 3, 5-Triphenyltetrazolium chloride) can be purchased from Sigma Chemical Co., P.O. Box 14508, St. Louis, MO 63178 (800-325-3010) or Research Organics, Inc., 4353 East 49th Street, Cleveland, OH 44125-1083 (800-321-0570). Embryo excision can be used to determine viability but is tedious. X-rays are used to determine seed quality (1). The easiest way to determine seed soundness (and probably viability) is a cut test. Simply cut the seed and check for whitish/solid tissue. This, too, is not foolproof but is about the best approach for nurserymen.

Authors feel that too many seed germination studies are con-

Fig. 8.a.

Fig. 8.b.

Figure 8. *Magnolia grandiflora*, southern magnolia: a) fruits are often placed on a bench until seeds start to dehisce; b) fleshy coated seeds emerging from the fruit can be easily collected and cleaned.

Figure 9. The 5-valved, dehiscent capsule of *Stewartia monadelpha*, tall stewartia. Wind and other elements effect the removal of the seeds. Ideally, the capsules should be collected just as they split at the tip, and then dried inside.

Figure 10. *Cercis canadensis*, redbud, fruits with hard-coated seeds. Collect the pods when brown but before natural opening takes place.

Figure 11. The large, milky, moist seeds of yellow buckeye, *Aesculus octandra*, lose moisture rapidly and must be planted or stored immediately after collection.

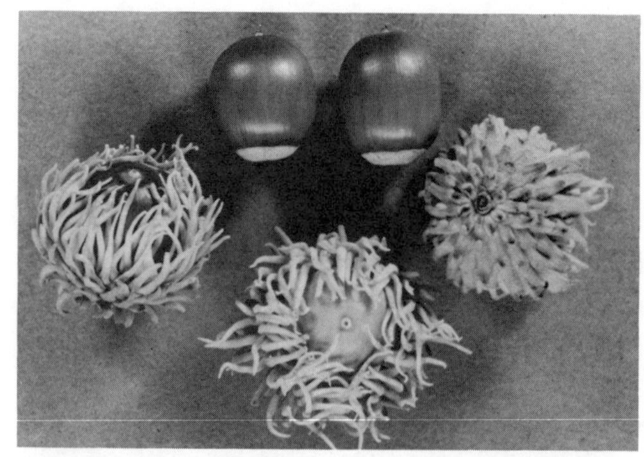

Figure 12. *Quercus acutissima*, sawtooth oak: the nuts (top two) are released from the cup early in fall and can be easily picked from the ground. Nuts should be checked for weevils.

ducted without determining the initial percentage of sound seed. In one particular study (23) with hard seeded woody legumes (*Cercis, Gymnocladus, Cladrastis*), the seeds were screened before running the study and after cut tests had 92, 98, and 98% sound seed, respectively. The actual final germination percentages were based on sound seeds and not total seeds. In a study (32) with Japanese maple, the author reported that the best stratification treatment (1 month warm/2 months cold) gave 48% germination. Unfortunately, the author never determined the percent sound seed and the 48% germination may have represented 100%, i.e., all sound seeds germinated.

Nurserymen need to assess viability, germination percentage, and natural attrition to determine rates of sowing to achieve a certain final density (# seedlings/square foot). A west coast seedling producer aims for 10 to 11 seedlings per square foot of bed space. This density allows for reasonable height and caliper increase. The grower noted that every species needs slightly different space allotments. The best density for *Sophora japonica*, Japanese pagodatree, is 6/square foot; *Cornus florida*, flowering dogwood - 18; *Ginkgo biloba*, ginkgo - 50; and *Acer rubrum*, red maple - 15.

This same grower mentioned certain interesting facts concerning aspects of seedling production on 10 acres. All seed is cleaned, no herbicides are used, mushroom compost is used to cover the seeds, a low rate of nitrogen (20 pounds actual N/acre) is used on the seed beds because the soil is naturally fertile, and total expenditure for seed alone averages $30,000. Table 3 provides a selected list of woody ornamental species and the relative seed costs. There are about 5000 seeds in a pound of *Acer griseum* and seldom is the seed soundness close to 10%. Figuring 10% sound seed times 5000 seeds per pound, there are 500 sound seeds at a cost of $85 per pound. This works out to $0.17 per good (sound) seed. This in part explains why a seedling *Acer griseum* will sell for 5 to 8 times more than a seedling *Acer rubrum*.

G. TREATMENTS TO OVERCOME SEED DORMANCY

Assuming that precautions have been made to procure and store viable seed there is still no guarantee that germination will proceed. A seed is a long way from a salable seedling and the road to germination is full of potholes. Senior author has watched many seed experiments ruined by diseases, insects, chipmunks and fate. Assume that something can go wrong and make preparations to fend off what often seems like the inevitable. Many researchers and practitioners (9, 19, 21, 30, 33, 36, 41, 47, 48, 50) presented worthwhile and often different approaches to solving dormancy and seed production problems.

1. Pregermination treatments

NONE: Many woody seeds possess no dormancy and can be sown as soon as possible after cleaning. For large batches of seed, it is wise to store seed and sow at a time in the greenhouse or outside that approximates the lengthening days of spring. Seedlings can then follow normal growth and dormancy cycles. For many growers, late spring frosts ruin many seedling crops. Planting later in spring to avoid the frosts rather than fall planting is practiced and should be considered by growers who suffer from the vagaries of spring weather. Senior author has germinated many seeds in the fall and winter months but had less than satisfactory growth because of short days, reduced sunlight, and generally lower temperatures (even in greenhouse). Additionally, watering (too much) and disease control may be more problematic.

In the ENCYCLOPEDIA under each species, the pregermination treatment is described. When "Direct Sow" or "No Pretreatment" are mentioned then the seed possesses no seed coat or internal dormancy and can be sown with good germination to follow.

2. Hard seed coats

Seeds may have hard, bony coats which are impervious to water. These must be degraded or broken to allow imbibition and the initiation of germination processes. One of the first prerequisites for seed germination is imbibition of water. Seeds of *Albizia*, silk tree; *Gleditsia triacanthos* (fig. 13), honeylocust; *Gymnocladus dioicus*, Kentucky coffeetree; *Cytisus*, broom; *Cladrastis lutea*, American yellowwood; and *Robinia*, locust, require only a seedcoat scarification or degradation to facilitate germination. Table 4 shows germination percentages of coffeetree after scarification in concentrated sulfuric acid. Note that without acid treatment germination was only 7 percent.

Hard seed coats can be broken down by: (1) Fall plant and let nature (microbial, physical and chemical weathering) take its course. (2) Mechanical means such as a file or abrasive wheel. Machines are available for large quantities of seed. (3) Hot water soaks can be used. Water is heated to 190°F and poured over the seeds which steep for 6 to 12 (24) hours. After this time the water is decanted and seeds are then planted. The degree of imbibition can be determined by swelling of the seeds. (4) Chemical (acid) treatment involves digestion of the seed coat with concentrated sulfuric acid or nitric acid (leathery coats like witch-hazel species). Extreme care must be exercised since these acids are caustic. Senior author uses the following procedure: (a) Wear goggles and protective clothing. If acid is spilled on skin wash immediately. (b) Use large glass vessel. (c) Place seeds inside *dry* glass vessel and decant about twice volume of acid over seeds. Stir with a glass rod for prescribed time mentioned in ENCYCLOPEDIA. Repeated stirring is necessary. (d) Always check seed for coat thickness by extracting a few, rinsing and cutting in half with pruners. Different seed lots of the same species may have thinner or thicker coats. (e) Decant acid and seeds through a screening device, wash under cold water for 5 to 10 minutes to remove acid. (f) Spread seeds on paper, separate to make sure they do not glump together and allow to dry at room temperature. (g) Seeds are now ready to sow as in the case of examples listed or can be placed in cold-moist or warm-moist stratification treatments depending on the dormancy involved. The scarification may completely or partially satisfy the warm stratification of doubly dormant seeds.

3. Embryo dormancy

This type of dormancy is related to conditions within the embryo which prevent germination even if environmental conditions are ideal. The embryo requires a period of **after-ripening**, usually induced by cold, moist stratification, after which biochemical processes occur that permit germination to proceed. This cold requirement varies according to species. A one month period is often sufficient for *Ulmus parvifolia*, Chinese elm; one to two months for *Prunus laurocerasus*, common cherrylaurel, and *Prunus* × *yedoensis*, Yoshino Cherry; and 3 months for *Cornus florida*, flowering dogwood, and *Nyssa sylvatica*, black tupelo. For some species, the radicle (primary root) emerges in cold stratification which provides a good indication that the seeds should be sown.

Figure 13. A fine stand of honeylocust seedlings; the seeds were acid scarified and spring planted.

Figure 14. *Acer palmatum*, Japanese maple, seedlings. Seeds are collected early, placed in sand:soil medium in flats, allowed to stratify outside and then moved to the greenhouse for germination and growth.

Table 1. Commercial sources of tree and shrub seeds.

NORTH AMERICAN SOURCES

1. Callahan Seeds
   6045 Foley Lane
   Central Point, OR 97502
2. Herbst Seed
   108-80 Candlewood Isle
   New Fairfield, CT 06812
3. Lawyer Nursery, Inc.
   950 Hwy. 200 West
   Plains, MT 59859
4. Maver Nursery
   Rt. 2, Box 265B
   Asheville, NC 28805
5. Plants of the Southwest
   1570 Pacheso St.
   Santa Fe, NM 87501
6. F.W. Schumacher & Co.
   36 Springhill Rd.
   Sandwich, MA
   02563-1023
7. Silvaseed Company
   P.O. Box 118
   Roy, WA 98580
8. Thompson & Morgan, Inc.
   P.O. Box 100
   Farmingdale, NJ 07727
9. Van Pines, Inc.
   West Olive, MI 49460
10. VBM Seeds
    4607 Wendover Blvd.
    Alexandria, LA 71301
11. Wild Seed
    2021 South Forest Ave.
    Tempe, AZ 85282
12. World Seed Service
    P.O. Box 1058
    Redwood City, CA
    94064

EUROPEAN SOURCES

1. Barilli and Biagi
   I-40 100 Bologna
   Casella Postale 1645-AD
   ITALY
2. B.V. 'Boomwekerij Udenhout'
   Schoorstraat 21,
   Postbus 31
   Udenhout,
   HOLLAND
3. Chiltern Seed
   Bortree Stile
   Ulverston
   Cumbria LA127PB
   ENGLAND
4. Establissement Versepuy
   Le Puy - 43000
   Haute Loire
   FRANCE
5. Florsilva Ansaloni
   I-40 100 Bologna
   Casella Postale 2100-EL
   ITALY
6. Forestry Commission
   Seed Branch
   Alice Holt Lodge
   Wrecclesham
   Farnham
   Surrey
   ENGLAND
7. A. J. Frost
   7080 Borkop
   DENMARK
8. Greenfingers Tree Seeds
   Indigo Rd.
   Stoneycroft
   Liverpool L1365H
   ENGLAND
9. Spren Lavinsen
   Kollerod Bygade 25
   3450 - Allerod
   DENMARK
10. Mosbacher Geholz-und Waldsamen
    Gammelsbach
    Postfach 1123
    D-6124 Bearfelden
    WEST GERMANY
11. The Old Farm Nurseries
    H. Den Ouden & Zoon B. V.
    P.O. Box 1
    2770 AA Boskoop
    HOLLAND
12. Paul Raeyamaekers
    Turnhoutsebaan 143
    Mol B-2400
    BELGIUM
13. Renz Nachf. GmbH & Co., K.G.
    727 Nagold-Emmingen
    WEST GERMANY
14. Samlesbury Tree Seeds
    The Boat House, Potter's Lane
    Samlesbury
    Preston PR50UE
    ENGLAND
15. Seeds of Bamber Bridge, Ltd.
    Lower Seed Lee Farm
    Brindle Road
    Bamber Bridge
    Preston PR56AP
    ENGLAND
16. Vilmorin-Andrieux
    Service Graine d'Arbres
    La Menitre
    49250
    Beaufort-en-Vallee
    Maine et Loire
    FRANCE

AUSTRALIAN SOURCES

1. Flamingo Enterprises
   P.O. Box 1037
   East Nowra
   N.S.W. 2540
   AUSTRALIA
2. H.G. Kershaw
   P.O. Box 88
   Mona Vale
   N.S.W. 2103
   AUSTRALIA
3. Nindethana Seed Service
   Narrikup
   Western Australia 6326
   AUSTRALIA

INDIAN SOURCES

1. Chandra
   Upper Cart Rd.
   P.O. Kalimpong 734301
   INDIA
2. P. Kohli & Son
   Park Rd.
   Srinagar
   Kashmir
   INDIA

Seeds in the process of stratification are undergoing chemical changes that will eventually allow germination. This process is termed after-ripening. For stratification to be effective, the seed must have: (1) imbibed water, (2) adequate aeration (oxygen), (3) proper temperature, and (4) time of exposure. The imbibition of water initiates the enzymatic processes that result in hydrolysis (breakdown) of stored compounds and mobilization to various sites within the seed. Once in stratification, seeds should not be allowed to dry, since a reversal of the after-ripening process may occur.

Oxygen is necessary for seed metabolic processes. If, for example, a pure carbon dioxide atmosphere is superimposed on nondormant seeds they develop dormancy and do not respond to stratification. **Avoid excess moisture** in the stratification medium since this reduces oxygen content. Plastic bags are excellent since they hold moisture but allow diffusion of gases. One grower uses large garbage bags to stratify seeds.

Temperature for cold stratification is important and the after-ripening process takes place from about freezing to around 55 to 60°F and ceases above this. In some cases the process is reversed at higher temperatures. For most seeds, 41°F appears to be an ideal temperature. At higher temperatures, 50°F for example, the after-ripening process takes longer to complete.

For warm stratification a range of 68 to 86°F appears optimum and even a more or less constant temperature (70 to 75°F) fosters the process.

After the stratification process is completed, seed germination proceeds best between 70 and 80°F.

The **time** of stratification is important in (a) releasing the seeds from dormancy and (b) unifying and hastening germination. Time is closely related to temperature. The shortest time for after-ripening occurs at optimum cold stratification temperatures. Within a given seed lot, individual seeds have different stratification requirements. Fordham (20) noted that *Cedrus deodara* seeds, if not given cold stratification, germinated sporadically over a two month period. A two month cold stratification produced even germination in 4 to 7 days. The same type of response is true for most species. The closer the temperature and stratification period are to optimum, the more uniform and rapid the germination.

Numerous approaches to satisfying embryo dormancy exist but the easiest involves sowing the seed in outdoor beds or flats and allowing nature to provide the proper conditions. Many large seedling producers do it this way with excellent success. Some growers place the seeds in sand in flats and cover with boards or suitable material and allow stratification to take place under natural conditions. For many species this is a simple and effective approach. The keys are good seed bed preparation (43), protection from birds and animals, proper seed density, attention to planting depth, moisture, fertilization, insect and disease prevention, and weed control (31).

Many growers provide 'artificial' stratification by placing seeds in a moist medium such as sand, vermiculite, peat, soil, etc., and mixing the seeds with several volumes of media. The mixture is then placed in refrigerated storage (34° to 41°F) for the proper time period and planted in spring (or designated time) to coincide with the normal growing season. For small batches of seed, the authors use sphagnum moss that is moist but not sloppy, plastic bags and good labels. The volume of stratification medium to seed is usually 4 to 1. The bags are placed in the refrigerator and a **seed calendar** is kept to note starting date and ending date. A weekly glance at the calendar is sufficient to indicate which seeds must be removed from storage and planted. Generally, the entire contents of the bag or container are placed on top of the medium and covered 1 to 3 times the diameter of the seed.

4. Double dormancy

Many seeds, especially those of woody plants, have hard, impermeable seed coats and embryo dormancy. The seed coat or fruit wall must be degraded so water can be imbibed. Then a cold-moist period will satisfy embryo dormancy. A classic example of the two step process is evident from Table 5. Seeds of redbud that were not acid scarified but given 1 and 2 months cold stratification germinated 0 and 7%, respectively. Seeds that were scarified first for 15, 30 and 60 minutes and then given 2 months cold stratification germinated about 87%. Scarified seeds that were not cold stratified germinated 1%. This is a classic example of the importance of breaking down the seed coat to allow water imbibition followed by a cold-moist stratification to satisfy embryo dormancy. Either treatment alone does not satisfy the dormancy. Periods of warm-moist (68 to 86°F) stratification followed by cold moist (33 to 41°F) stratification ususally satisfy the double dormancy and allow germination. The warm period permits microbial and chemical weathering of the seed coat which, in effect, simulates the acid scarification. Seeds with hard coats and embryo dormancy can be planted in late summer or early fall if the seeds are available. This early planting, while soil temperatures are still warm, permits seed coat degradation to occur and the subsequent winter cold satisfies embryo dormancy.

Kester (30) distinguishes plants that require special treatments in order to facilitate germination. One has a dormant root or immature embryo and must be conditioned by warm stratification before receiving the cold treatment. *Viburnum* and *Chionanthus* are two notable examples. The viburnums, in particular, have been well researched by Barton (8) and may require constant or fluctuating (68 to 86°F) temperatures from 12 to 17 months to induce root production. The treatment for shoot production (41°F) averages 3 months.

Again seeds can be sequenced through warm/cold periods under indoor conditions and then outplanted. Seeds of doubly dormant plants sown in fall usually germinate the second spring after planting. If traditionally double dormant (33) seeds (fig. 14) are collected early in late summer-fall, planted immediately, they will often germinate the following spring. The early planting when soil temperatures are warm allows microbial and chemical decomposition of the seed coat as well as after-ripening. Certain species like *Acer palmatum*, *Acer griseum*, *Acer triflorum*, *Chionanthus virginicus*, *Halesia carolina* (26) and *Stewartia pseudocamellia*, especially when dry, show erratic germination over a 2 to 5 year period. Senior author has worked with seeds of *Stewartia pseudocamellia* and *S. monadelpha*. After providing a 5 month warm/3 month cold period and sowing, seed germinated over a 5 month period in flats maintained in the greenhouse. The same type of response has occurred with *Halesia diptera*, *Halesia carolina* and *Hamamelis × intermedia*.

The doubly dormant seeds are not the easiest thing to gauge as to the exact duration of warm/cold. A Tennessee nurseryman told authors he plants seeds and waits, for there is no substitute for nature's approach. Mr. Jack Alexander, propagator, Arnold Arboretum, watches for radicle emergence from *Viburnum* seeds.

Table 2. *Magnolia grandiflora* seed germination.

Treatment	Percent germination	Days to germination from sowing date
Aril intact	29^z	125
Aril removed	73	125 (2 seeds up after 60 days)
Aril intact, 24 hour water soak	21	125 (1 seed up after 60 days)
Aril removed, 24 hour soak	76	125
Aril intact, 3 months cold stratification	None	-
Aril removed, 3 months cold stratification	General germination	35

^{z}Germination percentages are based on 5 replicates of 20 seeds per replicate (100 seeds). The seeds not subjected to cold stratification were treated and sown on Nov. 11. The seeds subjected to cold stratification were sown Feb. 11.

Table 3. Comparative cost and number per pound of seeds of selected woody species.

Species	Cost/pound in dollars	Number of seeds per pound
Acer griseum	85	5,000
Acer palmatum	48	16,000
Acer platanoides	5	3,000
Magnolia grandiflora	23	6,000
Pinus taeda	26	18,000
Quercus alba	3	120
Stewartia pseudocamellia	784	—
Tsuga canadensis	79	187,000
Ulmus parvifolia	114	120,000

Table 4. Germination of *Gymnocladus dioicus* seeds after acid scarification.

Time in sulfuric acid (hours)	Germination percentage
0	7a^z
2	93bc
4	100bc
8	95bc
16	82d
32	87bd

^{z}Numbers not followed by same letters significantly different at 5% level, Fisher's LSD test.

Table 5. The effect of scarification and cold stratification on the seed germination of *Cercis canadensis*, eastern redbud.

Scarification (minutes in acid)	Stratification days @ 41°F: 0	30	60
	Germination %		
0	1a^z	0a	7a
15	1a	69b	85c
30	1a	69b	87c
60	1a	81c	88c

^{z}Mean separation by Fisher's test. Means not followed by the same letter are signifcantly different at the 5% level.

This indicates that phase one of the dormancy breaking process is complete. The seeds are then placed in cold to satisfy embryo dormancy. Unfortunately all double dormant seeds do not send up flags like viburnum. If the seed is sound (viable) and has imbibed water it is in the perfect frame of reference to be subjected to the various types of stratification mentioned above. Seeds can be stored dry in refrigerated storage for years and will not germinate.

H. POSTSCRIPT

Commercial seed propagation requires comprehensive knowledge of numerous biological and cultural practices. In discussions with knowledgeable and successful seed producers, they acknowledge that success with a certain species from year-to-year is often elusive. Probably the most essential ingredient is sound seed, knowledge of how to treat the seed to induce germination, and once germinated, proper cultural practices to produce a salable seedling in one or two years.

In recent years, the use of bottomless milk cartons or plastic containers and the air-pruning technique have received significant research (11, 24, 25) and commercial attention. This method is more costly in the short run than bed-grown seedlings. However, for difficult-to-transplant species and/or those with large tap roots, especially *Carya*, *Fagus*, *Juglans*, *Quercus* and others, it offers great promise. Seeds, depending on their dormancy requirements, can be pretreated or stored and sown in spring into the containers which are located on wire benches above the ground. The roots grow to the bottom of the container and are root pruned (the root tips die). This process is repeated numerous times as the secondary and tertiary roots, ad infinitum, reach the bottom until a dense root mass is formed. The resultant seedling has a well branched root system and is ready for planting to the field in fall or overwintered and spring planted. Davis and Whitcomb (11) showed that air-pruned root growth of *Pinus thunbergiana*, *Sapindus drummondii* and *Pistacia chinensis* were greater in 2½″ diameter by 9 or 12″ deep containers. For nurserymen who grow oaks and other tap-rooted or difficult to transplant species, this method would be worthwhile investigating.

No one book offers all the answers and, in this reference, the emphasis is on the proper manipulations (pretreatments) to induce good germination. Several books every seed producer should have in the library include numbers 12, 20, 22, 28, and 38 which are fully cited in the back of the chapter. Also, the *Proceedings of the International Plant Propagators Society* contain invaluable information.

LITERATURE CITED

1. Allison, C. J. 1980. X-ray determination of horticultural seed quality. Proc. Int. Plant Prop. Soc. 30:78-86.
2. Barton, L. V. 1930. Hastening the germination of some coniferous seeds. Amer. J. Bot. 17:88-115.
3. Barton, L. V. 1934. Dormancy in *Tilia* seeds. Contrib. Boyce Thompson Inst. 6:69-89.
4. Barton, L. V. 1935. Storage of some coniferous seeds. Contrib. Boyce Thompson Inst. 7:379-404.
5. Barton, L. V. 1939. Storage of elm seeds. Contrib. Boyce Thompson Inst. 10:221-233.
6. Barton, L. V. 1941. Relation of certain air temperatures and humidities to viability of seeds. Contrib. Boyce Thomp. Inst. 12:85-102.
7. Barton, L. V. 1943. Effect of certain moisture fluctuations on the viability of seeds in storage. Contrib. Boyce Thompson Inst. 13:35-45.
8. Barton, L. V. 1958. Germination and seedling production of species of viburnum. Proc. Int. Plant Prop. Soc. 8:126-134.
9. Biggs, T. 1984. Recent developments in the propagation and establishment of plants from seed. Proc. Int. Plant Prop. Soc. 34:61-67
10. Crocker, W. and L. V. Barton. 1931. After-ripening, germination and storage of certain rosaceous seeds. Contrib. Boyce Thompson Inst. 3:385-404.
11. Davis, R. E. and C. E. Whitcomb. 1975. Effects of propagation container size on development of high quality tree seedlings. Proc. Int. Plant Prop. Soc. 25:448-453.
12. Dirr, M. A. 1983. Manual of woody landscape plants. Stipes Publ. Co., Champaign, IL.
13. Donselman, H. M. and H. L. Flint. 1982. Genecology of eastern redbud (*Cercis canadensis*). Ecology 63:962-971.
14. Fagan, A. E., M. A. Dirr, and F. A. Pokorny. 1981. Effects of depulping, stratification, and growth regulators on seed germination of *Liriope muscari*. HortScience 16:208-209.
15. Fillmore, R. 1951. A general review of woody plant propagation. Proc. Int. Plant Prop. Soc. 34:61-67.
16. Flint, H. L. 1970. Importance of seed source to propagation. Proc. Int. Plant Prop. Soc. 20:171-178.
17. Flint, H. L. 1974. Phenology and genecology of woody plants. Chapter 2.5 in Phenology and seasonality modeling, pp. 83-97. Spring-Verlag, N.Y.
18. Flint, H. L. and N. L. Alexander. 1982. Hardiness of white ash depends on seed source. The Plant Propagator 28(2):9-10.
19. Fordham, A. J. 1973. Dormancy in seeds of temperate zone woody plants. Proc. Int. Plant Prop. Soc. 23:262-266.
20. Fordham, A. J. and L. S. Spraker. 1977. Propagation manual of selected gymnosperms. Arnoldia 37:1-88.
21. Fordham, D. 1976. Propagation of plants from seed. Proc. Int. Plant Prop. Soc. 26:139-145.
22. Forest Service, USDA. 1974. Seeds of woody plants in the United States. Superintendent of Documents, Washington, D.C.
23. Frett, J. J. and M. A. Dirr. 1979. Scarification and stratification requirements for seed of *Cercis canadensis* L. (Redbud), *Cladrastis lutea* (Michx F.) C. Koch. (Yellowwood), and *Gymnocladus dioicus* (L) C. Koch. (Kentucky Coffeetree). The Plant Propagator 25(2):4-6.
24. Frolich, E. F. 1971. The use of screen bottom flats for seedling production. Proc. Int. Plant Prop. Soc. 21:79-80.
25. Gibson, J. D. and C. E. Whitcomb. 1980. Producing tree seedlings in square bottomless containers. Ornamentals South 3:12-15.
26. Giersbach, J. and L. V. Barton. 1932. Germination of seeds of the silverbell, *Halesia carolina*. Contrib. Boyce Thompson Inst. 4:27-37.
27. Hardin, J. W. 1975. Hybridization and introgression in *Quercus alba*. J. Arnold Arb. 56:336-363.
28. Hartmann, H. and D. E. Kester. 1983. Plant propagation: Principles and practices. Prentice-Hall, Inc., Englewood Cliffs, NJ.
29. Hess, C. W. M., Jr. 1973. Seedling propagation of difficult species. Proc. Int. Plant Prop. Soc. 23:284-287.
30. Kester, D. E. 1960. Seed dormancy and its relationship to nursery practices. Proc. Int. Plant Prop. Soc. 10:256-267.
31. Kuhns, L. J. 1983. Herbicides for conifer seedbeds. Proc. Int. Plant Prop. Soc. 33:439-444.
32. Laiche, A. J., Jr. 1983. Seed germination and seedling growth of Japanese maple. Proc. Southern Nurs. Res. Assoc. 28:217.
33. Leiss, Jorge. 1985. Seed treatments to enhance germination. Proc. Int. Plant Prop. Soc. 35:495-499.
34. Lovelace, R. W. 1982. Establishing and maintaining a seed orchard. Proc. Int. Plant Prop. Soc. 32:495-497.
35. Lovelace, W. 1981. Woody tree and shrub seedling production. Proc. Int. Plant Prop. Soc. 31:537-542.
36. Lovelace, W. 1984. Seedling production using companion grass crops. Proc. Int. Plant Prop. Soc. 34:595-598.
37. McMillan Browse, P.D.A. 1974. Raising hardwoods from seeds. Proc. Int. Plant Prop. Soc. 24:158-173.
38. McMillan Browse, P.D.A. 1979. Hardy woody plants from seed. Grower Books, London.
39. McMillan Browse, P.D.A. 1979. Discussion group report — Obtaining and treating seeds of hardy woody plants. Proc. Int. Plant Prop. Soc. 29:265-267.
40. Norton, C. R. 1980. Deleterious metabolic and morphological changes resulting from seed soaking prior to sowing. Proc. Int. Plant Prop. Soc. 30:132-134.
41. Pellett, H. 1973. Seed stratification. Proc. Int. Plant Prop. Soc. 30:132-134.
42. Petrie, R. W. 1974. Seed storage. Proc. Int. Plant Prop. Soc. 24:230-234.
43. Pinney, T. S., Jr. 1973. Seedbed management. Proc. Int. Plant Prop. Soc. 23:276-281.
44. Shugert, R. 1981. Seedling production in the eastern USA. Proc. Int. Plant Prop. Soc. 31:78-83.
45. Steavenson, H. 1973. Seedling propagation — solving the seed source problem. Proc. Int. Plant Prop. Soc. 23:281-284.
46. Steavenson, H. 1979. Maximizing seedling growth under midwest conditions. Proc. Int. Plant Prop. Soc. 29:66-71.
47. St. John, S. 1979. Successes and failures in starting a tree seedling nursery. Proc. Int. Plant Prop. Soc. 29:205-210.
48. Thompson, P. A. 1971. Research into seed dormancy and germination. Proc. Int. Plant Prop. Soc. 21:211-228.
49. Vertrees, J. D. 1978. Notes on propagation of certain Acers. Proc. Int. Plant Prop. Soc. 28:93-97.
50. Watkins, R. 1974. Aspects of seed supply and germination problems, such as dormancy, and their treatments. Proc. Int. Plant Prop. Soc. 24:304-308.
51. Wells, J. S. 1985. Plant propagation practices. American Nurseryman Publishing Co., Chicago, IL.

CHAPTER TWO

Cutting Propagation

A. INTRODUCTION

Cutting propagation is perhaps the most fascinating as well as frustrating area of plant propagation. Cuttings that rooted in high percentages last year may not have fared as well this year. The variables involved in successful cutting propagation are numerous and success is not necessarily guaranteed from year to year. Attention-to-detail, thorough knowledge of the factors that influence and, at times, the "correct phase of the moon" contribute to success. A colleague once said that if one cutting of a species or cultivar roots then the potential to root all of them exists. This philosophy keeps nurserymen, propagators and researchers trying even with taxa that resist every effort. *Chionanthus virginicus*, white fringetree, is virtually impossible to root, but its close relative *Chionanthus retusus*, Chinese fringetree, has been successfully rooted (93, 124).

Magnolia grandiflora, southern magnolia, has often refused to root in quantities approaching commercial acceptability (32). In Georgia studies, timing, juvenility, hormones, media, water management, bottom heat, polytent and mist were manipulated with little success. However, each study yielded new information that was applied to the next attempt. Success (figure 1) was obtained using coarse perlite in 4" deep flats, mist (2½ sec/5 min), polytent, 53% saran shade cloth, no bottom heat, raised greenhouse bench, July-August cuttings (Georgia) with leaves mature and terminal bud set (semi-hardwood), 4 to 6" long with 2 to 3 leaves at end, terminal bud intact, 5 second dip in 1% NAA in 50% ethanol (27).

This example serves to illustrate how each element in the propagation process may affect ultimate success (31). Know the plant(s) you work with. Be flexible and systematic in your thinking. If a particular cutting does not root the first time, do not give up. **Keep records** so that back tracking is possible. Relying on memory yields hazy recollections.

B. WHY CUTTINGS?

1. Integrity of characteristics

Cuttings are used to maintain specific characteristics. This reproduction process is termed CLONING since each individual has all the traits of the plant from which it was collected. *Calycanthus floridus* 'Athens' ('Katherine'), Athens sweetshrub, is a deliciously fragrant, yellow-flowered form, whose true characteristics can only be maintained by cuttings (28). Plants grown from seeds of 'Athens' may come partially yellow-flowered but are predominantly strawberry-brown. The adult form of English ivy, with its unique leaf morphology (figure 2), must be propagated by cuttings for seeds of the adult form produce the typical juvenile foliage forms (26). Numerous examples can be cited and any propagator/nurseryman who has grown seedlings from named clones understands full well the term variation. A plant propagated by cuttings will be true-to-type although occasional branch sports or chimeras are found. *Spiraea x bumalda* 'Anthony Waterer' is notorious for producing yellow-variegated shoots (figure 3). On any given plant, shoots may be all yellow or various combination of yellow and green. Many sports (mutations) have arisen on *Juniperus squamata* 'Meyeri', Meyer's singleseed juniper; *Juniperus chinensis* 'Pfitzerana', Pfitzer Chinese juniper; *Picea glauca* 'Conica', dwarf Alberta spruce (figure 4a); and *Picea mariana* 'Nana', dwarf black spruce (figure 4b), that have resulted in superior landscape plants (24,117). Branch sports are not necessarily bad or good, but their occurrence under cultivation is something the propagator should understand.

2. Economics

Another prime reason for cutting propagation is cost. It is considerably cheaper to root a cutting than graft or bud a particular clone. A break even point for cuttings is 50 to 60% compared to 90% or greater for grafting. This comment (at the time of publication) applies also to a comparison with tissue cultured plants. As tissue culture laboratories become more efficient, production costs should be reduced. Southern growers estimated that it costs $0.08 to 0.15 to produce a salable cutting if efficiency is high. Obviously this varies depending on species/cultivar.

3. Avoidance of bud/graft incompatibilities

Most shade and ornamental tree cultivars (clones) have been budded or grafted. Certain cultivars, especially *Acer rubrum* selections, display a high proportion of incompatibilities (figure 5a, b). This phenomenon occurs over a number of years and is frustrating to the grower and consumer. *Acer rubrum* 'October Glory' and 'Red Sunset' average 4 to 5% incompatibility per year. Over a 4-year period while the tree is being grown to salable size (3 to 4" caliper), the losses total 16 to 20%, and may represent the profit margin the grower hoped to reap. Because of this problem, many growers have changed to own-root (cutting) propagation, instead of grafting and budding (88). The quality of the trees is just as good and the incompatibility is avoided. Unfortunately, it usually takes another year in the production cycle to produce an own-root salable whip. See *Acer rubrum* in the ENCYCLOPEDIA for a step-by-step discussion of the cutting production process.

C. TYPES OF CUTTINGS

1. Leaf

Any vegetative portion of a plant constitutes a CUTTING. Leaves

Fig. 1.a.

Fig. 1.b.

Fig. 1.c.

Fig. 1.d.

Fig. 1.e.

Fig. 1.f.

Figure 1. (a) Variation in seedling progeny of *Magnolia grandiflora*; (b) a superb selection that must be maintained by vegetative propagation; (c) "typical" cutting success with M. *grandiflora*; (d) NAA concentration effects (0, 0.5, 1.0, 2.0%); (e) successful cutting propagation with 1.0% NAA (Note strong root systems); (f) potted rooted cutting 9 months after sticking.

(figure 6a) from *Saintpaulia* spp., African violet; *Sansevieria spp.*, snake plant; *Peperomia* spp.; and many houseplants will regenerate shoots and roots. However, **leaf cuttings** in woody plant propagation are of little value. **Leaf bud cuttings** consist of the leaf blade, petiole and a short piece of stem with the axillary bud (figure 6b). Rhododendron and camellia have been propagated from leaf bud cuttings but the method is not commercially popular. If cutting material is in short supply, leaf bud cuttings facilitate greater numbers. Leaf bud cuttings are akin to single node cuttings and are used with *Clematis*, red maple (43, 91) and other species.

2. Stem

STEM cuttings are usually divided into **softwood, semi-hardwood (greenwood)** and **hardwood** categories (figure 7). At the end of each category and the beginning of the next there is some overlap but for the most part they are readily distinguishable. **Hardwood cuttings** can be further divided into **deciduous, narrowleaf evergreen** and **broadleaf evergreen.**

There are several ways to assess the status of the cutting. Color of leaves and stems, firmness of wood, and time the end bud develops are suitable indices for different plants. The status of *Photinia x fraseri* 'Birmingham', red-tip photinia, can be assessed by color of leaves and stems. The emerging leaves are bright red and the color lasts for 3 to 6 weeks. This coincides with the **softwood stage.** When the leaves become green the wood is in a **semi-hardwood** condition. There is another later growth flush that follows the same pattern. In late fall the stems become gray-brown toward the base which indicates a **hardwood condition.** All stages root readily if a suitable indolebutyric acid (IBA) concentration (0.5 to 1.0%) is used (7, 8, 36) (Table 1). In fact, the plant can be rooted year-round in high percentages except for the first 2 to 4 weeks after the bright red shoots have emerged. Diametrically opposed to *Photinia* is *Myrica cerifera*, southern waxmyrtle, which roots in high percentages only from softwood cuttings (6, 97) taken in May, July, August (figure 8). December, January and February cuttings rooted approximately 20 percent. Hormone treatments did not greatly improve rooting over non-IBA treated cuttings.

a. Softwood

The emerging shoots of trees, shrubs and evergreens can be classified as **softwood.** The tissue is easily bruised with a nail; there is a gradation in leaf size with the end leaves small and undeveloped, older leaves more or less "full size"; shoots can be easily snapped and, when removed, their propensity to wilt is tremendous. For the latter reason, it is necessary to keep them cool and moist. Some species maintain a softwood condition into late summer/early fall. *Weigela*, *Lagerstroemia*, *Kerria* and *Spiraea* qualify here. Others like *Chionanthus virginicus* and *Euonymus alatus*, winged euonymus, produce a terminal bud and essentially cease the elongation process. The stem assumes a semi-hardwood or hardwood condition often by July-August. The softwood condition for most woody plants ranges from 2 to 8 weeks and coincides with May, June, to early July depending on latitudes. A softwood cutting rooted in June has a much better chance of survival than a August "softwood cutting". In the authors' opinion, an August softwood is physiologically not the same as a May-June softwood.

The "window" for successful rooting of softwood cuttings (4) can be relatively small and timing must be carefully monitored from year to year. Timing should not be based on a chronological but a physiological calendar. Spring may be 3 weeks early or 3 weeks late from the norm; a shift of 6 weeks. June 1 from year to year **is not** the same as far as the condition of the cutting.

Flowering crabapple, *Malus* spp., provides a good example of the narrow window effects. Softwood cuttings (11) taken in May (Illinois) and early July (Michigan, 13) rooted in high percentages. Later the percentages dropped dramatically. *Acer campestre*, hedge maple; *Acer platanoides*, Norway maple; and *Acer saccharum*, sugar maple, also rooted in a very narrow window (12, 13).

Softwood cuttings should average 2 to 5" long with several nodes. Upon collection, the cutting should not "flag" (wilt) like lettuce. This condition probably means the stem tissue is too soft. Extremely soft cuttings require extra care to prevent drying and are not as easy to stick as firmer-wooded cuttings. Extremely soft cuttings often rot in the rooting bench. Red maple, sugar maple, and Norway maple rot quickly if collected too soft (13). The rooting time for a softwood cutting may vary from 7 to 10 days for *Callicarpa*, beautyberry; *Lagerstroemia*, crape myrtle; and *Buddleia*, butterfly-bush; to 3 to 5 weeks for *Malus*, crabapple. A simple tug test (pull the cutting upward with a gentle pressure) indicates the cutting has rooted. At this time, it is wise to reduce or turn off the mist since softwood cuttings may decline rapidly. Softwood cuttings respond to hormones and even easy to root species like crape myrtle root more uniformly, rapidly, and have more profuse root systems than untreated cuttings (figure 9). A low concentration in the range of 1000 to 3000 ppm hormone in talc or solution is normally safe.

The rooted cuttings can be overwintered in ground beds with proper covering, poly-houses, or sold for fall planting. Growers often go directly to the container or field with softwood cuttings. Successful transplanting to the field depends on proper irrigation. If fall planted, winter damage can be particularly devastating on softwood cuttings.

b. Semi-hardwood (greenwood)

This category is applied to broadleaf evergreens such as rhododendron, photinia, osmanthus, euonymus, holly, magnolia, camellia and cherrylaurel taken in summer. This corresponds with mid-July to possibly early September. The growth flush is completed, the wood is reasonably firm and leaves have essentially matured. If growth is "soft" at the stem end, it is often best removed. The young leaves and stem tips of *Magnolia grandiflora* die in the rooting bed.

A large rhododendron producer in Oregon takes cuttings in July and finishes in early September. With experience, he has learned that certain species and cultivars root better in July; others in late August. The common denominator with all cuttings is the cessation of growth, and the semi-hardwood condition.

Deciduous trees and shrubs may also be rooted from semi-hardwood cuttings and taxa like *Magnolia x soulangiana*, saucer magnolia; *Chionanthus retusus*, Chinese fringetree; and *Hamamelis x intermedia* 'Arnold Promise' have been rooted from summer semi-hardwoods.

Cuttings are often 3 to 6" long with the basal one-half of the leaves removed. Some growers reduce leaf area to increase the number of cuttings that can be stuck and prevent excessive transpiration losses. There is no clear cut evidence that this is of great benefit. In a Georgia study, 0, 2, 4, or 8 leaves were left on *Euonymus alatus* 'Compactus' cuttings. Eight-leaf cuttings had more profuse root systems than 4 > 2 > 0 indicating benefit from maintaining adequate leaf tissue. In a mini-study with *Acer*

Figure 2. Juvenile (left) and adult (right) *Hedera helix*, English ivy, foliage. Seeds collected from the adult form yield juvenile foliage. To maintain foliage characteristics the adult form must be propagated by cuttings.

Figure 3. A branch sport of *Spiraea* × *bumalda* 'Anthony Waterer'. This type of bud mutation, if stable, can result in a new plant.

Fig. 4.a.

Fig. 4.b.

Figure 4. Branch reversion on (a) *Picea glauca* 'Conica' and (b) *Picea mariana* 'Nana'. Branch reversions are common on dwarf conifers. Many have been propagated and introduced as new plants.

rubrum 'Red Sunset', larger cuttings produced larger root systems than smaller cuttings (figure 10). Wounding may prove beneficial and from our experiences a hormone from 0.5% to often as high as 1 to 2% may be necessary. Rooting time varies from 4 to 6 weeks for *Photinia x fraseri* to 2 to 3 months for *Rhododendron* cultivars. *Rhododendron catawbiense* cultivars such as 'Nova Zembla', 'English Roseum', and 'Blue Ensign' have been rooted in high percentages in August with a 1% IBA 5-second dip (Table 2).

Needle evergreens, especially junipers, can be rooted as semi-hardwood cuttings. Rooting time for *Juniperus conferta*, shore juniper, may be only 4 weeks. Semi-hardwood cuttings are not as sensitive to desiccation as softwood cuttings. They also are generally easier to handle and stick. Most needle evergreens benefit from exposure to cold temperatures and generally root better when collected from September into winter. They can be stuck at a time that might coincide with a lull in the normal activities of the nursery business. Mist is used to keep the cuttings turgid during the rooting process.

c. Hardwood: Deciduous and broadleaf

Hardwood cuttings encompass deciduous plants that have lost their leaves in October-November and carry through to late winter (early spring before growth ensues); needle evergreens such as yew, juniper and arborvitae; and broadleaf evergreens such as holly, cherrylaurel and photinia. Some would argue that the last category does not belong here but having rooted holly, cherrylaurel, aucuba and others from January, February and March cuttings, we feel otherwise.

Hardwood cuttings of deciduous species can be propagated inexpensively. Generally, last season's growth is collected in fall through winter. Hardwood cuttings (figure 11) are manipulated in different ways. The stems are often cut with a ban saw 6 to 20" long (possibly longer), treated with a hormone (60), and either stuck directly in the field (fall or late winter) or callused/rooted by various methods and then out-planted. The methods include:

Bottom heat: The East Malling Experiment Station (72), England, uses heated bins (Garner) with the bottom of the cutting exposed to 70°F and the tops maintained at cool temperatures. Cuttings are treated with 2500 to 5000 ppm hormone and left for 4 weeks. After callusing the cuttings are transplanted before the buds break into growth.

Plastic bag: Hardwood cuttings are prepared, dipped into a hormone (2000 to 4000 ppm), sealed in poly-bags and placed in the dark at 50°F. After callusing/rooting they are planted out.

Winter treatment: Cuttings of uniform size are bundled together, buried outside or in a suitable structure with the tops down and ends several inches below the sand/sawdust/soil level. Bud development is delayed while callus and root development at the basal ends occur. In spring, bundles are removed and planted right side up. This approach has been used for crape myrtles.

Warm temperature: Cuttings are taken in fall, treated with hormone, stored under moist conditions around 70°F for 3 to 5 weeks. After this, cuttings may be out-planted in mild climates or held in cold storage until spring and then out-planted.

Outdoor ground beds: Work in Georgia (35) with bottom-heated (70°F) outdoor ground beds indicated that hardwood cuttings of evergreens including photinia, aucuba, holly, and juniper could be rooted in high percentages when collected in January, February and March, treated with 1% IBA in 50% alcohol, and placed in poly-covered frames. Non-bottom heated beds were largely unsuccessful. Deciduous species such as forsythia and crape myrtle rooted well, but only from January cuttings. Cuttings collected in February and March broke bud in the poly-covered beds and did not root well.

Everyone seems to approach hardwood cuttings in a different manner (14). The essence for success is good cutting wood of moderate size with ample "food" reserves. Tip cuttings are usually discarded. A hormone treatment in the range of 2000 to 5000 ppm appears beneficial and in Georgia studies 1.0% IBA did not result in basal burn. Too much moisture during the callusing/storage phase diminishes success. Keeping the tops cool prevents bud break and a carbohydrate drain (materials shunted to emerging leaves rather than the basal portion of stem where rooting is taking place). Also, if leaves emerge before roots, they die from a lack of water.

Cornus alba, tatarian dogwood, and C. *sericea*, redosier dogwood, are often propagated by hardwood cuttings. Long slender dormant stems are collected in January-February, placed in bundles, sawed 8 to 10" long, and stuck directly in the field with just an inch or two of the end of the stem showing. Crape myrtle can be handled this way (figure 12). The new shoots that emerge are extremely tender and the underlying root systems often sparse and fragile. Some nursing (water) is necessary in the early stages of development. Stoutemyer (106) presented a list of plants that could be effectively propagated from hardwood cuttings.

d. Hardwood: Needle evergreens

The needle evergreens are often rooted in late fall and winter under greenhouse conditions. *Thuja*, arborvitae; *Chamaecyparis*, falsecypress; *Juniperus*, juniper (particularly low growing types); and *Taxus*, yew, can be easily rooted although rooting times may take as long as 3 months. *Pinus*, pine; *Abies*, fir; *Picea*, spruce, and others are more difficult although some dwarf conifer selections are rooted in commercial quantities (49). Four to 8" (can be shorter) long terminal cuttings are stripped of basal needles, perhaps wounded, treated with a hormone and placed in peat:perlite, sand or bark. After rooting takes place the cuttings can be potted or placed in containers for growing on. Some growers use very small tip cuttings, probably no bigger than 2 to 3" long. These will root but it takes them longer to grow off and develop into a salable plant.

Cuttings are usually placed under mist but an enclosed poly-tent with or without bottom heat has proven useful. × *Cupressocyparis leylandii*, Leyland cypress, is most successfully rooted from February-March cuttings (34, 120). Over a 3 year period in Georgia studies, 90% plus rooting was achieved using 4 to 6" long cuttings with brown wood at base, removing tips, 3000 to 8000ppm IBA in 50% ethanol, and either mist or a poly-tent. Bottom heat was beneficial under mist but in the poly-tent did not seem to help. The cuttings root in 10 to 12 weeks and can be transplanted, fertilized and grown to 10 to 12" size the same season. *Thuja* and *Juniperus* have also been successfully rooted with this approach.

3. Root cuttings

Very few woody plant producers use root cuttings simply because of the time and cost involved. Root cuttings are taken in December through March depending on weather conditions. Roots from younger plants and those collected close to the main stems/trunk display a greater propensity to form shoots (figure

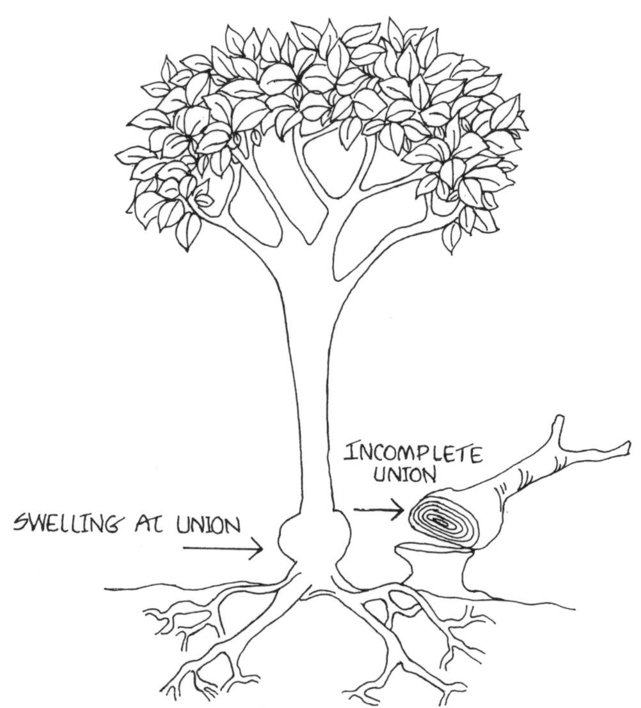

Fig. 5.a.

Fig. 5.b.

Figure 5. Diagrammatic representation (a) of the symptoms of graft incompatibility on *Acer rubrum*. Symptoms may include swelling at the graft union, premature fall color, and/or decrease in growth. Figure (b) is a 3″ diameter red maple that simply broke off at the graft union.

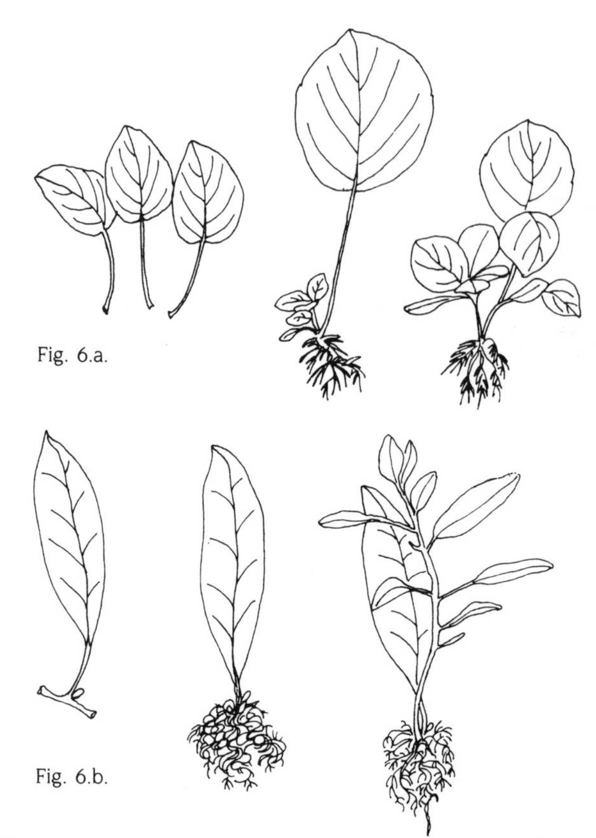

Fig. 6.a.

Fig. 6.b.

Figure 6. Leaf cuttings (a) are uncommon in woody plant propagation, however, (b) are used for rhododendron, magnolia, red maple, etc. Many cuttings can be made from a branch using leaf bud (single node) cuttings.

Figure 7. Stem cuttings (soft, semi-hardwood and hardwood) are usually prepared by removing the leaves from the bottom one-half.

13). Almost any plant that suckers in nature is a good candidate for production by root cuttings. *Rhus* spp., sumac; *Sassafras albidum*, sassafras; *Aralia spinosa*, devil's-walkingstick; *Elliottia racemosa*, Georgia plume; and *Albizia julibrissin*, silk-tree, are just a few. In fact, the latter two are effectively propagated only through root cuttings (48). Creech (21), Donovan (37), Donovan and Johnstone (38), Flemer (46), Orndorff (90) and Stoutemyer (105) provide lists of plants that can be propagated by root cuttings. Heuser (67) and MacMillan Browse (80) discuss the physiology and practicality of the process.

Roots can be dug, cleaned, fungicided and stored in late fall/early winter (57) or dug when ground permits in late winter. Polarity must be maintained and the root end closest to the stem should be planted upright. It is wise to cut the proximal end (closest to stem) with a cross-cut and the distal (farthest) with a slant. Root cuttings should vary in length depending on the diameter of the root piece. A 1/16 to 1/4" diameter root should measure perhaps 3 to 4", while a 3/8 to 1/2" diameter root will function at 1½ to 3".

Root cuttings are either stuck vertically (with proximal end up and level with or slightly above soil line) in containers containing a loose medium or placed horizontally in flats and covered with ½" of medium (figure 14a, b). The cuttings should be watered and maintained with even moisture either by covering the flats with plastic or placing in a polytent. In Georgia, bottom heat had a positive effect on Bradford pear, *Pyrus calleryana* 'Bradford', root cuttings (figure 15). As shoots emerge they may be severed and treated like softwood cuttings or the entire shoot/root transplanted (48).

Large root pieces may be dug, prepared, packed in damp medium and stored for a time and planted in well prepared nursery soil. A well known Minnesota nursery has used this method to produce *Rhus typhina* 'Laciniata', cutleaf staghorn sumac (22).

D. COLLECTION AND HANDLING

Textbooks on plant propagation are often unrealistic in prescribing that cuttings be taken in the early morning or late evening hours. Ideally, early morning is probably the best time because the cuttings are fully turgid and air temperatures cooler. Plastic bags with suitable moisture (ice is excellent), wet burlap, ice chests, etc., can be used to hold the cuttings. We collected 6,000 *Magnolia grandiflora* cuttings on a 100°F day, carefully iced them down, kept the bags in the shade, used a covered truck for transport and did not injure a single cutting. If in doubt as to the rate of heat buildup in a 15 to 20 gallon milky or black plastic garbage sack, simply stick your hand into the bag a few minutes after collection.

The cuttings, regardless of whether softwood, semi-hardwood or hardwood, should be brought to the preparation area and made ready for sticking as soon as feasible. Cuttings are often placed on wire mesh benches which are rigged with misting nozzles that can be turned on to keep cuttings moist. Cuttings of many species can be placed in cold storage (33-41°F) for several days without appreciable loss of rooting potential. Evidence exists that cuttings can be stored for extended periods and possibly shipped long distances in low pressure storage (41, 42). The cuttings are placed in sealed chambers and a vacuum is pulled preventing the buildup of ethylene and other gases.

In the process of collecting cuttings use true-to-name plants. Collecting in parks, from commercial plantings, from back yards, etc., almost guarantees some confusion. Was *Cotoneaster dammeri* 'Skogholm' or *C. d.* 'Royal Beauty' collected? Was that *Berberis candidula* or *B. verruculosa*, *Hamamelis vernalis* or *H. virginiana*, *Acer ginnala* or *A. tataricum*, *Prunus × cistena* or *P. cerasifera* 'Atropurpurea'...? The list of confusing plants is almost endless. Work with the local arboretum and botanic garden. The donation of a few plants for the new landscape planting is often repaid manyfold with cuttings of true-to-name species and cultivars.

Cuttings are also collected from container and field nursery stock and this is preferable to the above approach. The plants have usually been fertilized and sprayed so the chances of success are better. Nursery grown plants are often heavily fertilized and the cuttings may be quite soft. This may present a problem if not handled properly.

The best source of cuttings is a separate stock block that is properly labeled, fertilized and sprayed. The plants are often kept in a juvenile condition (thus facilitating greater rooting propensity) by frequent pruning.

As a precaution, cuttings should be soaked or dipped in benomyl or Captan either before or after the cuttings are stuck. As a matter of routine, in Georgia work, the flats or beds are drenched with benomyl after the cuttings are stuck. Several rooting powders (HormoRoot and Rootone) have thiram (a fungicide) in the formulation. Benomyl and Captan wettable powders can be mixed with IBA talc formulations (See Hormones).

E. FACTORS AFFECTING ROOTING OF CUTTINGS

1. Nutrition/carbohydrates/nitrogen

Ideally, the stock plant should not be water stressed, should have an adequate carbohydrate level and reasonable nutritional content. The age-old concept that a high carbohydrate/nitrogen ratio is beneficial to rooting and the reverse is detrimental can be looked on somewhat skeptically (1, 109). All essential elements/compounds are important in the rooting process and to ascribe a more significant role to one or two is probably misleading. Carbohydrates supply energy and carbon skeletons and probably dictate the quality of rooting response more than any other endogenous biochemical component. Nitrogen is a component of amino acids, proteins, and nucleic acids. In rooting studies with *Photinia × fraseri*, stockier cuttings produced more roots than thin-stemmed cuttings. However, both rooted in equivalent percentages and had more than adequate root systems to facilitate successful transplanting. The actual lower limits of nutrition and subsequent successful/unsuccessful rooting have not been established. An actively growing terminal cutting would be lower in carbohydrates than a cutting further down the same shoot, however, they may root the same or differently and the exact cause can seldom be determined. Cuttings taken from adequately maintained stock plants root in high percentages.

Micronutrient deficiency may affect rooting and Coorts (17) showed that cutting of *Juniperus chinensis* 'San Jose' and *Ilex crenata* collected from plants grown for two years under severe boron, iron, manganese, and zinc deficiency generally rooted in lower percentages than cuttings taken from plants grown with complete nutrient solutions. However, with San Jose, the number of roots per rooted cuttings increased in the zinc and boron deficient treatments compared to the complete nutrient plants. These results must be looked on somewhat skeptically from a

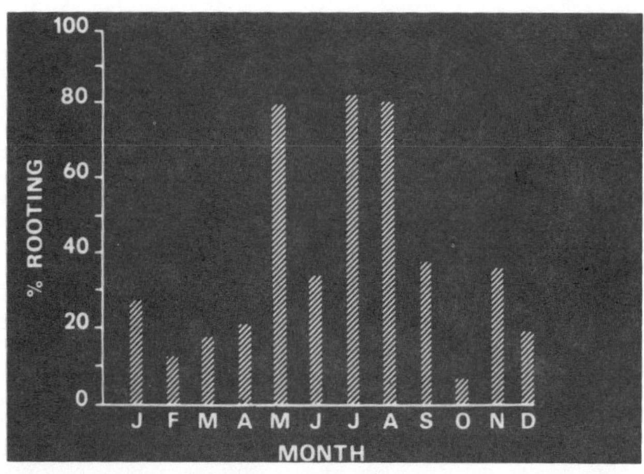

Figure 8. Seasonality of rooting in *Myrica cerifera*, southern wax myrtle. Note that May, July and August were peak months with rooting around 80% or above. All other months less than 40%.

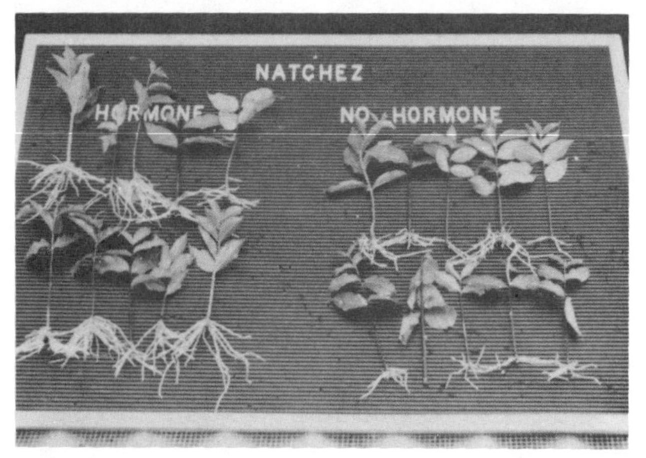

Figure 9. The effect of 1000 ppm IBA-quick dip on the rooting of *Lagerstroemia* 'Natchez'. Root systems are more profuse on cuttings treated with hormone. These cuttings transplant and grow off better than a cutting with one or two stringy roots.

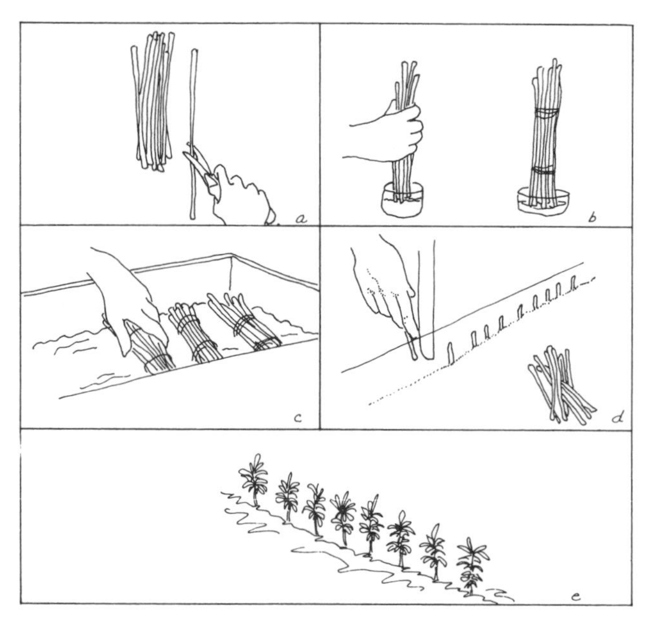

Figure 12. A schematic representation for hardwood cuttings; (a) prepared, (b) dipped in hormone, (c) callused in boxes, (d) placed in rows, (e) new growth evident in spring.

Figure 10. The effect of cutting size (single node) compared to multiple node on root quality of *Acer rubrum* 'Red Sunset'. The larger cutting with more leaf area produces more carbohydrates which are used in the formation of new roots.

Figure 11. Hardwood cuttings of *Lagerstroemia*, crape myrtle, are 6 to 8″ long, 1/4 to 3/8″ thick (greater storage reserves), and stuck in late fall or when the ground is workable in late winter.

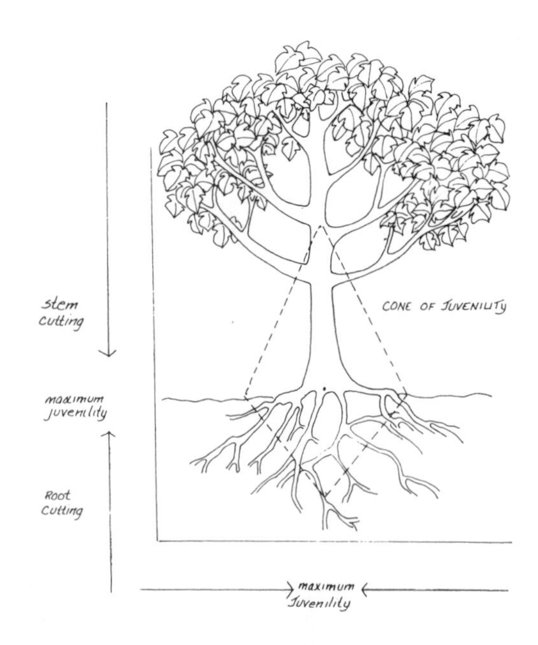

Figure 13. Juvenility, in practical terms, can be gauged by proximity to trunk/root interface. For stem cuttings the best area to collect material is within the cone of juvenility (above ground). This same effect applies to root cuttings. Generally, the closer the root cuttings are to the crown the better the regeneration potential.

practical standpoint since the stock plants were grown in water culture system for almost two years simply to induce sufficient deficiency to conduct the experiements. A nursery stock plant under moderate nutrition will not be affected by severe micronutrient deficiencies.

2. Juvenility

This factor may be the most overlooked element affecting the rooting of cuttings. Cuttings taken from young cutting-produced or seedling plants are much more amenable to rooting than cuttings from older plants (figure 13). *Cercidiphyllum japonicum*, Katsuratree, cuttings taken from one-year-old seedlings rooted 100%; cuttings from a 15 to 20-year-old tree did not root (26). Studies with Leyland cypress illustrate graphically this juvenility phenomenon (56). Cuttings from 5-year-old trees rooted 94%; 34% from 20-year-old trees, and 5% from 50-year-old trees. In this case, a juvenile-like condition is maintained by using young stock plants even though the stock plants did not arise from seedlings. This is an example of a partial return to a juvenile state. Schreiber and Kawase (102) rooted cuttings from 12-year-old *Ulmus americana* by collecting from the top of the tree, sprouts produced from 6 to 7.5′ high stumps and 1′ high stumps. Thirty-eight, 64% and 83% rooting resulted, respectively, emphasizing the strong influence juvenility plays in successful rooting.

Juvenility is attributed to many causes but the real reasons are largely deep, dark secrets. As plants age, it has been postulated that rooting inhibitors increase. Work with *Eucalyptus* has shown this to be true (92). Phenolics decrease in some plants with age and may affect rooting response since they supposedly function as cofactors with auxin in the rooting process (63, 64, 65). Cytokinins and gibberellins have been sprayed on various species and have introduced a measure of juvenility into the cuttings that resulted in improved rooting. Juvenility preconditioning is not practiced on a commercial scale.

The most logical approach for propagators/nurserymen who work with traditionally difficult-to-root plants is to maintain them in a juvenile condition by frequent pruning. In general, the further removed the cuttings are from the root system, the more difficult they are to root. Dirr was unable to root cuttings from a mature honeylocust, *Gleditsia triacanthos* var. *inermis*, but cuttings collected from root sprouts rooted readily. True juvenility is associated with seed-grown plants and represents a transition phase from germination to flowering and fruiting. Often, the mature condition is correlated with reproductive processes. Oak, beech, and other Fagaceae members hold their leaves throughout winter in a loose arrangement termed the "cone of juvenility" (figure 13).

Southern magnolia is extremely difficult to root, especially from cuttings collected from old trees. The procedure is to root a few and keep these as future stock plants while maintaining the reintroduced juvenility. The next generation is rooted from the first rooted cuttings. Nurserymen (23) have emphatically emphasized this approach for *Magnolia grandiflora* and other difficult-to-root species and cultivars.

Root cuttings are an excellent way to produce juvenile shoots that can be rooted easily. See *Albizia julibrissin* and *Elliottia racemosa* in the ENCYCLOPEDIA for the procedure. These juvenile shoots, once rooted, can be maintained as such by proper pruning and nutrition.

Rooting studies with *Chionanthus virginicus*, white fringetree, have proved "fruitless". Even cuttings taken from 3-year-old seedlings, which should be juvenile, failed to root (26,44). On the other hand, cuttings from 40-year-old *Halesia carolina*, Carolina silverbell, rooted readily (figure 16). Every species/cultivar is inherently different relative to juvenility/maturity. The only way to determine how striking the effect is to collect cuttings from different aged stock.

Perhaps the classic example is *Acer griseum*, papberbark maple. Cuttings from mature trees root poorly, if at all (47, 75). However, cuttings from seedlings root in reasonable percentages. The literature and the senior author's personal frustration attempting to root *Acer griseum* provide cause for reflection concerning juvenility/maturity. The relationship should be considered when rooting cuttings, especially difficult-to-root taxa. Hoogendoorn (69) maintains a seedling stock block and successfully (60%) roots *Acer griseum* from June cuttings (figure 17).

3. Timing

A book could be written on this subject as it relates to rooting cuttings. Certain plants can be rooted almost year-round, i.e., willow, red-tip photinia, forsythia, while others show definite seasonal trends. Timing was discussed under types of cuttings and will be briefly expanded upon here. Keeping records relative to time of taking cuttings as well as condition of the tissue are important. Many plants, like lilac (83, 116), southern wax myrtle (97), native azaleas (89) and numerous evergreens (87) have a narrow window of rootability. If timing is missed, a production year may be lost. Some nurserymen keep range charts (figure 18) for species and cultivars listing optimum and marginal times. The rooting response graph of southern wax myrtle (97) provides convincing evidence for proper timing relative to successful rooting. The effects of timing are largely a reflection of environmental conditions which, in turn, influence physiological conditions. Softwood cuttings correspond to the spring and early summer conditions; hardwood the onset of dormancy induced by decreasing photoperiod and low temperatures. However, concomitant with successful timing is the increase/decrease of root promoters/inhibitors; carbohydrate and nutritional balances/imbalances; cutting softness/hardness/physical barriers, etc. Doubtfully will the scientific aspect of cutting propagation evolve to the degree that cuttings can be analyzed for the above factors and be pronounced fit to root. Trial and error will continue to determine rooting response.

4. Condition and type of cutting wood

a. Clonal material

The particular stock plant, especially if seed grown, may exert a pronounced effect on subsequent rooting. Considerable variation exists in rootability among seedlings and clones (cultivars) of the same species. Tremendous variation in rooting response of cuttings collected from seedling Chinese elm, *Ulmus parvifolia*, was reported (66, 123). Of 13 clones, 3 did not root, 5 rooted less than 25% and 5 rooted greater than 25% at 68, 73, 81, 86 and 96 percent. Of these, only two grew well and developed good tree forms. The University of Minnesota Landscape Arboretum, before introducing a yellow flowered "Lights" series azalea, tested several yellow-flowered clones for propagability in tissue culture. Nurserymen should think along these lines before introducing any and every tree that shows a slightly different trait. Whether it is easily propagated often determines wholesale acceptance.

Cuttings should be collected from non-flowering shoots. Several studies (59) have shown that stem cuttings regenerate better

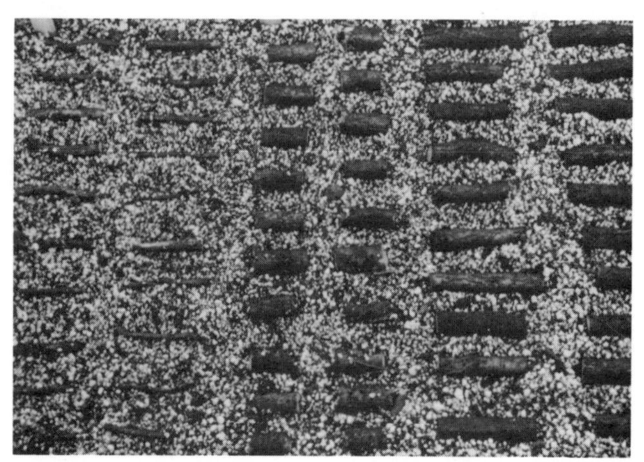

Fig. 14.a.

Fig. 14.b.

Figure 14. (a) Root cuttings of *Pyrus calleryana* 'Bradford' before being covered with soil. Cuttings can be placed horizontally or vertically (not shown) with the proximal end (closest to crown) sticking about 1/2" above the soil line. (b) Regenerated plants from root cuttings of *Gymnocladus dioicus*, Kentucky coffeetree. Each row represents a different clone (source of root cuttings). It is interesting to note the difference in growth rate.

Figure 15. Shoot regeneration from a root piece of Bradford pear. The greatest number of shoots occurs at the end of the root piece closest to the trunk.

Figure 16. Rooting of stem cuttings of *Halesia carolina*, Carolina silverbell. Cuttings were collected from a 40-year-old tree. The ideal IBA concentration was 2500 ppm-quick dip. This photo also points out that each species/cultivar has an optimum IBA concentration or range for maximum rooting.

Figure 17. A juvenile stock block of *Acer griseum*, paperbark maple. The plants are pruned hard every year to promote the long shoots which are then harvested for cuttings in June.

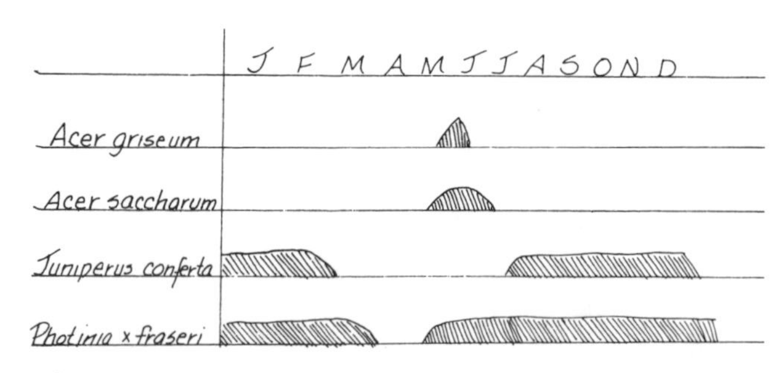

Figure 18. A timing chart for the cutting propagation of selected species. This provides a good record of when to take cuttings from year to year.

when taken before or after, rather than during the flowering stage. Hormone relationships are different and carbohydrates and nitrogen may be shunted to developing flowers and fruits. Hardwood cuttings of crape myrtle taken from flowering/fruiting stems did not root as well as those taken from strictly vegetative shoots.

The English literature refers to heel, mallet and normal (straight) cuttings. The first has some old wood attached, the second reminds of a croquet mallet with a small piece of attached older wood, and the third is the traditional straight cutting (figure 19). *Magnolia grandiflora* (23) roots betters if mallet cuttings are used. Leyland cypress seems to root better if the cuttings have developed the brown wood at their base (34). This is also true for *Taxus*, yew and needle evergreens. Obviously, the mechanical aspects of collecting, making and sticking heel and mallet cuttings are staggering compared to straight cuttings. If straight cuttings do not root, the other approaches might be considered.

b. Position of cutting on the plant

Everyone who has taken a college plant propagation course learned that cuttings taken from lateral branches of *Taxus cuspidata* var. *capitata*, upright Japanese yew, produced spreading plants, while vertical cuttings yielded an upright version similar in habit to the parent (figure 20). *Picea pungens* var. *glauca* 'Moerheimii', propagated from lower branches, forms a prostrate, rich silver-blue ground cover unless staked. *Topophysis* is the term applied to different growth characteristics when different parts of the plant are used as cuttings. *Orthotropic* refers to cuttings that grow vertically; *plagiotropic* to those that grow laterally. Topophytic variation is not a significant problem for most plants that are commonly propagated by cuttings but is a phenomenon worth considering.

There is evidence that cuttings taken from lateral shoots root better than cuttings from terminal shoots. Some of this response might be attributed to the juvenility factor and auxin relationships.

The **position** of the cuttings may affect the rooting response. Long shoots can be cut into many cuttings with or without a gradation in rooting response from terminal cuttings to subterminal cuttings (36, 91). There is no clear rule of thumb here and the only way to determine which cutting position works best is by testing.

5. Wounding and girdling

a. Wounding

For most species and cultivars wounding is not necessary. However, selected species of *Magnolia*, *Rhododendron*, *Ilex*, *Juniperus*, *Thuja* and others may benefit. The wound is about ½ to 1″ long and can be light or heavy, single or double-sided. The wounding induces internal hormonal changes that may improve rooting (40, 118), exposes active cells to hormone preparation, and alleviates physical barriers to root penetration. December cuttings of *Ilex opaca* 'Jersey Knight' and 'Jersey Princess' showed a definite response to wounding with roots emerging only from the wounded side (26).

Semi-hardwood and hardwood cuttings may benefit from wounding. Wounding is labor intensive and should be avoided unless significant improvement in rooting can be realized. A thin piece of bark can be removed, the stem can be scored with razor blades, or special tools can be used to facilitate the process. For extremely difficult to root species and cultivars, a comparative wound/no wound study is warranted. Often just the simple act of wounding may be the key to successful rooting.

Many needle evergreens are wounded in the process of stripping the needles. With Leyland cypress, callus and root production occur around the wounded areas. Juniper, arborvitae, yew, and false cypress are sufficiently wounded when the needles are stripped.

b. Girdling

This is a modified type of wound formed by ringing or cutting a circular strip of bark from the stem (figure 21). Difficult-to-root trees (oak, walnut, pine) have been preconditioned by girdling and the shoots rooted successfully on the tree or removed from the tree and rooted under mist (58, 61). The process is slow and not conducive to large numbers but warrants consideration if budding/grafting facilities are not available, for maintaining a particular clone on its own roots, and introducing a measure of juvenility into the tree (see Juvenility section). Hayes Regional Arboretum, Richmond, IN, has applied this technique to propagation of native Indiana trees (61). Superior plants were located in the wild, girdled and treated according to the method described by Hare (58). The girdling cut is made about 1/4″ to 1/2″ wide to the cambium, a powdered slurry consisting of 1% IBA, 1% PPZ (1-phenyl-3-methyl-5-pyrazolone), 20% sucrose, and 5% Captan in talc is applied followed by moist peat and an aluminum foil wrap. Hare has had great success with loblolly pine, *Pinus taeda*, and water oak, *Quercus nigra*. This girdling treatment period lasted 6 weeks and the cuttings, when removed, rooted in high percentages.

Obviously, girdling can be successful without applying the concoction described above. The girdling blocks the downward translocation of carbohydrates, hormones, and possibly rooting cofactors and concentrates them in a confined area. In a sense, the cutting has been primed to root by the pretreatment (girdle). The use of black tape or aluminum foil around the girdled area may also improve the rooting response.

c. Etiolation

Bassuk et al. (2) have combined etiolation and banding to induce rooting of *Fagus sylvatica*, European beech; *Carpinus betulus*, European hornbeam, *Syringa vulgaris*, common lilac and other species. The process is a modification of Gardner (50) and Howard's (71) earlier work. In brief, dormant plants are covered with black plastic (99% light reduction) and when new growth is 2 to 3″ long the north side of the covering is removed to begin weaning the soft yellow green shoots. At this time black adhesive tape is banded at base of shoots (keeping bases etiolated) while allowing leaves to green up. Cuttings are then collected at intervals after the black plastic is removed and treated with IBA in the conventional manner and placed on the rooting bench. Rooting of beech averaged 69% compared to 5% for no shading or banding pretreatment. Hornbeam 43% compared to 15%. Obviously, for extremely difficult-to-root plants this method has possibilities. It should be mentioned that the stock plants were either established in-ground hedges or container-grown plants.

6. Hormones

a. Introduction

A review of the pertinent literature shows that numerous chemical compounds have been tested for root-promoting activity. It is estimated that well over 10,000 chemicals show positive formative effects. The science of chemical plant pro-

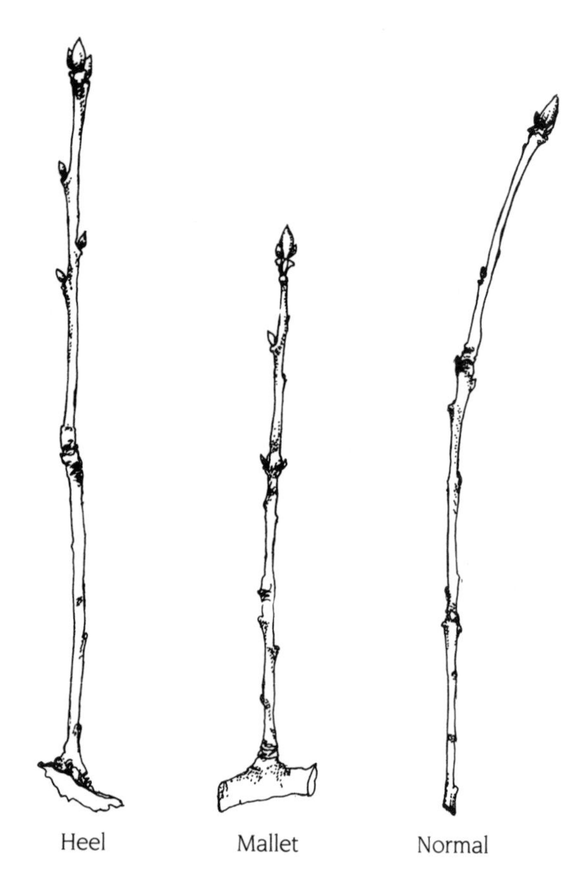

Figure 19. Heel, mallet and normal cutting. For certain plants, especially those with large pith, a heel or mallet may improve rooting response. A normal cutting is fine for the majority of species.

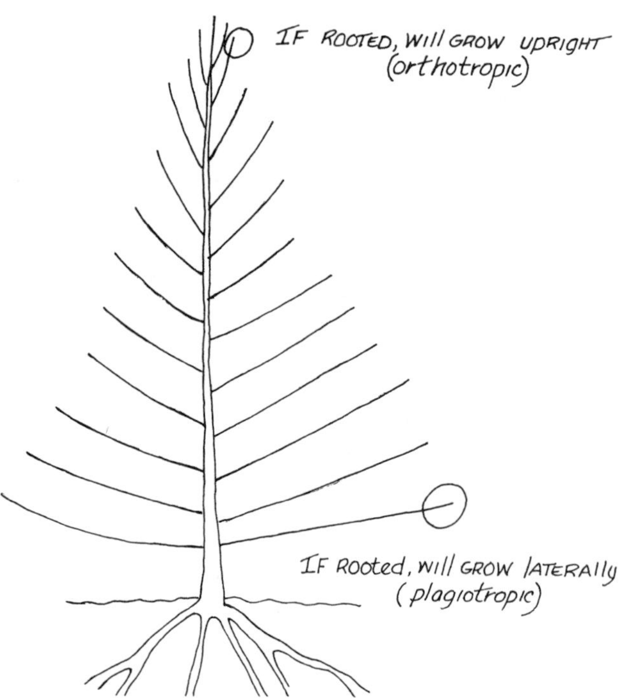

Figure 20. The position from which a cutting is taken can affect subsequent growth habit. The lateral cutting will develop into a spreading (plagiotropic) plant; the upright into a vertical (orthotropic) plant.

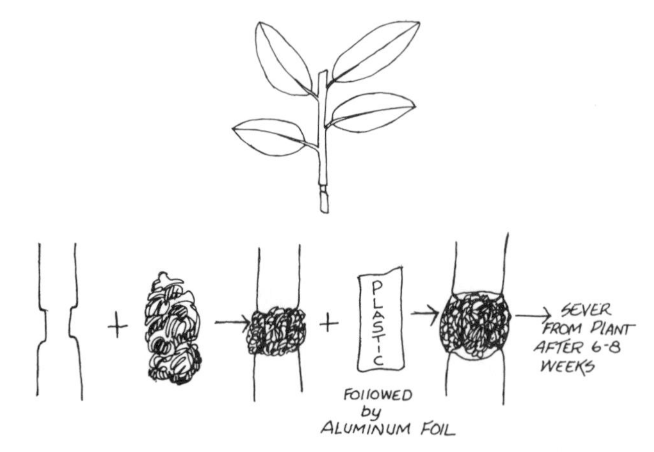

Figure 21. The process of girdling and preconditioning a shoot. A ring of bark is removed, hormone is applied, moist peat packed around area with plastic wrapping and finally aluminum foil. Entire shoot is severed below the treated area and handled like a conventional cutting.

Fig. 22.a.

Fig. 22.b.

Fig. 22.c.

Figure 22. The effect of (a) Hormodin #2 (3000 ppm IBA-talc), (b) 3000 ppm IBA-50% ethanol quick dip, and (c) Dip 'N Grow (2000 ppm IBA + 1000 ppm NAA in alcohol solvent) on rooting of *Photinia* × *fraseri*.

pagation began in 1934 with the discovery of a naturally occurring auxin, indole-3-acetic acid (IAA). The demonstration that two synthetic (do not occur in higher plants) auxins, indole-3-butyric acid (IBA) and naphthaleneacetic acid (NAA), induce a greater rooting response was shown by Zimmerman and Wilcoxon in 1935 (127).

Modern plant propagation revolves around the use of IBA, NAA, and their derivatives. Both 2,4-dichlorophenoxyacetic acid (2,4-D) and 2,4,5-trichlorophenoxypropionic acid (2,4,5-TP) have been used for rooting cuttings. They show potent root promoting activity but are readily translocated throughout the cutting and may delay bud break or induce other adverse effects.

b. New chemicals

Recently several new synthetic aryl ester formulations of IAA and IBA have been touted as being superior to acid formulations of IAA and IBA (54,55). Phenyl indole-3-acetate (P-IAA) resulted in 95 to 154% more roots in bean compared to IAA treatment. Substantial increases also occurred when cuttings were treated with 3-hydroxyphenyl indole-3-acetate (3HP-IAA). On a molar basis both were 10 times more effective than IAA. Phenyl indole-3-butyrate (P-IBA) increased the percentage of rooted *Pinus banksiana* cuttings (taken from seedlings) over IBA—treated cuttings but did not affect the number or length of roots. The aryl esters of IAA had no effect on Jack pine. Other compounds [indole-3-butyramide (NP-IBA) and phenyl indole-3-thiobutyrate (P-ITB)] have also appeared from the same source.

Dirr (30) tested P-IBA, NP-IBA and P-ITB against IBA using *Photinia × fraseri*. P-ITB produced greater rooting than IBA and subsequent studies with 21 different species and cultivars have shown it to be as effective as IBA. P-ITB has a considerably safer toxological profile that IBA with an LD_{50} of 5 g per kilogram compared to 0.1 g per kilogram for IBA.

Kawase (73,74) identified a willow rooting substance (WRS) that has improved the rooting of certain plants. Results have not always been reproducible but success with *Betula alleghaniensis*, yellow birch, was impressive. When IBA was used in conjunction with WRS excellent rooting occurred compared to IBA—treated cuttings. The basic format for making the WRS involves taking current year stems (leaves removed), cut into small pieces, pack into a container, and cover with water (20). This mixture is allowed to steep for 24 hours and is then drained off. The resultant extract is used to treat the cuttings. Cuttings should be placed upright in the willow extract and allowed to absorb for 24 hours and then stuck. The extract which is quite stable has been refrigerated for six years and retained its effectiveness. Apparently most willow species work with equal effectiveness.

c. Commercial root formulations

Some commercial preparations that are/have been available to nurserymen are presented in table 3. Many are no longer available, and the current crop of favorites includes Hormodin, Hormo-Root, Hormex, Rootone, Dip 'N Grow and Wood's. The first four are talc (powder) formulations that contain IBA and other hormones at various concentrations. The chemicals and inert materials (carriers) of various formulations along with the addresses of the manufacturers are presented in table 4.

d. Comparative study of root promoting chemicals and formulations

Tables 5 and 6 and figure 22 show the formulations and results of a rooting study comparing the effectiveness of various root promoting compounds on *Photinia × fraseri* (25). Photinia is an excellent test plant because it does not root without a relatively high exogenous supply of hormone (7,8,36). A relatively low level of IBA, NAA or derivatives (0.3% hormone total for each treatment) was used to determine if the carrier or formulation made a difference.

Rooting was dramatically affected by the treatments from a low of 0 percent in the water control to 100% with Dip 'N Grow. The three control treatments (water, 50% ethanol and water plus boron) did not stimulate rooting. This agrees with previous work which showed that a rooting hormone was essential for rooting. Chloromone-treated cuttings rooted only 17%. Obviously, it does not contain appreciable quantities of either IBA or NAA, two compounds that promoted reasonably good rooting in *Photinia × fraseri*.

Naphthaleneacetamide is a component of Rootone hormone powders and resulted in 67% rooting and reasonable root number and length. In the mung bean bioassay, it results in many small roots that do not elongate to any degree.

Perhaps the most striking result occurred with Hormodin #2 (0.3% IBA in talc). Rooting was only 3% with an average length of one centimeter. In all cases the liquid preparations, whether in water, alcohol or dimethylformamide, were superior to the talc formulation. The difference in response can be explained by the low solubility of IBA. In a talc formulation the IBA must first go into solution before being absorbed into the cutting. Rapid absorption of IBA did not occur from the talc source compared to the liquid formulations.

NAA was more effective than KNAA (potassium salt). NAA was dissolved in 50% ethanol while the KNAA was dissolved in water. The alcohol acts as an effective carrier and penetrant thus facilitating increased movement of the hormone into the cuttings. The same trend was observed with IBA and KIBA. Interestingly, NAA and KNAA were superior to IBA and KIBA in promoting rooting of *Photinia*.

KIBA plus boron was particularly effective in stimulating rooting. This response occurs with other plants. It is suspected that boron serves as a carrier or at least facilitates transport of molecules. The boron would hasten the movement of IBA into the cutting. Boron, when included with exogneously applied growth regulators, increases the translocation of these compounds.

The IBA + NAA treatment resulted in 83% rooting while the KIBA + KNAA treatment rooted 23%. The 50% alcohol solvent apparently facilitated auxin movement into the stem tissue. The K-salts of IBA and NAA are as effective as the acids. The limiting factor may be the rate of absorption into the stem tissue.

Cuttings treated with Wood's and Dip 'N Grow rooted 97 and 100%, respectively, and had the greatest root numbers and root lengths compared to other treatments. Both were diluted 1:5 which resulted in 0.2% IBA and 0.1% NAA in the treatment solution. The IBA + NAA and KIBA + KNAA treatments contained the same amount of active ingredients (auxins). The only difference was the solvent system. Logically it must be concluded that the solvent system (carrier) can have a pronounced effect on the effectiveness of a rooting compound. The carrier facilitates rapid absorption of the rooting compound.

Previous work has shown that *Photinia × fraseri* roots maximally when treated with 0.5 to 1.0% IBA applied as a concentrated dip. In this study, the 0.3% IBA in 50% alcohol or 0.2% IBA + 0.1% NAA in 50% alcohol were not sufficient to induce 95 to

Figure 23. Rooting of *Acer rubrum* 'Bowhall' as influenced by IBA-quick dips. Note the tremendous root systems at 0.5, 1.0 and 2.0%. Figure 16 also shows the effect of high concentrations of IBA on *Halesia carolina*.

Fig. 24.a.

Fig. 24.b.

Fig. 24.c.

Fig. 24.d.

Figure 24. (a) Several propagation media (left to right, top to bottom — vermiculite, peat:perlite, sand, bark, perlite), (b) *Magnolia grandiflora* in sand (left) and perlite (right)...all cuttings in sand died, (c) commercial production in outdoor ground beds in sandy soil with a 1″ layer of sand, (d) rooted cuttings in these same beds.

Fig. 25.a.

Fig. 25.b.

Figure 25. (a) Polycovered propagation houses have become standard in the industry. (b) Ground beds with a single mist line through the center of each bed.

100% rooting. However, these same levels in a different solvent system (Woods, Dip 'N Grow) did result in 97 to 100% rooting. By using an appropriate solvent the effect of the hormone is enhanced. This means that lower levels of the rooting compounds can be used.

e. Concentrations

Every species/cultivar responds to hormone treatment in different ways (33, 39, 76, 77, 112, 123). The optimum IBA concentration for one species may not be effective in promoting rooting in another. Ideally, every plant has an optimum hormone concentration range. Note the term range was stated and not a single concentration. *Halesia carolina* (figure 16) rooted within a narrow range and a 2500 ppm IBA-quick dip proved optimum. A range of concentrations from 1000 to 5000 would possibly be as effective as the 2500 ppm treatment. Obviously the control, 10,000, 20,000 and 30,000 ppm treatments were too low or too high. Bowhall red maple (figure 23) rooted over a much wider concentration range than *Halesia carolina*. Theoretically, a rooting hormone concentration curve could be determined for every species and cultivar. On a practical basis, if one is unsure of what levels to use follow the rates described in the ENCYCLOPEDIA. The idea is to hit the best average for the greatest number of plants being propagated. Based on work with many different plants, the authors believe a 2500 to 5000 ppm IBA-quick dip is a good starting point.

An idea of the latitude in hormone rates is afforded by a study the senior author conducted with *Ilex glabra* 'Leucocarpa'. IBA solutions from 1000 to 10,000 ppm with 1000 ppm increments plus a control (0 IBA) were used. For the 11 solutions (0 to 10,000 ppm) rooting was 33, 77, 87, 97, 93, 93, 97, 77, 90, 93, 80, respectively. Cuttings did not suffer basal necrosis and all appeared more or less equal in root quality. Table 1 presents data from a study with *Photinia × fraseri*. Essentially there was no difference in rooting percentage or quality with 0.5, 1.0 and 2.0% IBA, again indicating the tremendous "latitude of rootability". However, at 0 and 4.0% IBA, rooting was reduced.

If a propagator is interested in determining the "ideal" rate for a particular plant simply use a range of concentrations from 0, 2500, 5000, 10,000, 20,000 ppm IBA (usually lower for NAA). Excessively high concentrations produce basal burn (figure 16). For some plants, high concentration quick dips have been used with good success (10, 15). Usually rooting takes place above the "burned" end of the cuttings.

f. Chemical properties and costs of IBA and NAA

The essence of all the commercial preparations centers around IBA, NAA and their derivatives. IAA, although naturally occurring, is seldom used as a rooting compound because it is broken down by a naturally occurring enzyme (IAA oxidase) system, and is destroyed by light and a bacterium (*Acetobacter* sp.) that is widely distributed. This same organism has no effect on IBA or NAA.

Robbins (Kansas State University) determined quantitatively the stability of IBA solutions. Both 1000 and 5000ppm IBA-solutions were placed in clear or amber bottles at 32, 43 or 87°F for 4 or 6 months. There was no change in IBA concentration or rooting activity after 6 months at any temperature. The solutions changed color (became darker) and this was attributed to temperature (higher) and not light. After 19 months at room temperature storage, only 26% of the original IBA remained. Based on Robbins work, a nurseryman can keep a solution for 6 months and probably longer and should always label the solutions as to the date they were prepared.

NAA seems to be entirely light stable and is probably similar to IBA in stability over time. It should be mentioned that when pure chemicals are purchased the label prescribes storing IBA at 32 to 41°F, IAA at 32°F and NAA at room temperature which serves as an indication of their relative heat stability.

IBA and NAA are designated in chemical catalogs and the literature as alpha, beta or gamma forms. In short, the gamma form of IBA is the most effective while the alpha form of NAA is 100 times more effective than the beta form in promoting rooting of cuttings. There is a significant cost difference with the beta form of NAA selling for 14 times more than the alpha form. As a rule, IBA offers much more lattitude than NAA for rooting cuttings. Cuttings of a particular species or cultivar will root over a wide range of IBA concentrations. NAA is more toxic to cuttings than IBA and may produce "burning". One large nursery firm utilizes NAA exclusively for cutting propagation. When a nurseryman compares the cost of NAA to IBA, there is good reason to at least run comparative effectiveness studies between the two chemicals. 1986 catalog prices show IBA selling for 10 to 20 times as high as NAA. For example, one supplier (Research Organics) sells 25 grams of IBA for $18.75 and the same amount of NAA for $2.00.

Combinations of IBA and NAA are often used; the idea being to derive the best effects of both in a single treatment. The literature is full of testimonials to the combination, but there are many studies that show no superiority of the combination over IBA alone. The combination approach is a type of insurance for if the IBA doesn't elicit a response then the NAA will. In an extensive study with several *Malus* taxa, cuttings treated with IBA rooted in higher percentages, had greater number of roots and root length than NAA or IBA plus NAA treated cuttings (11).

g. Solvents for root promoting chemicals

Many nurserymen/propagators use the concentrated quick dip method. This involves dissolving the acid form of IBA or NAA in an organic solvent. IBA is soluble in ethanol, while NAA is soluble at the ratio of 1 to 30 parts alcohol (3.3%). This presents no problem to the nurseryman for it is doubtful he would need to exceed the solubility limits of NAA. The standard solvent is usually a 50% alcohol/water mixture, but any concentration is acceptable as long as the pure chemical dissolves. Isopropyl alcohol (50 or 70%) is a suitable solvent and is available from the pharmacist. Polyethylene glycol (PEG) (Carbowax) is frequently used as a solvent and carrier. DMSO, dimethyl sulfoxide, is also used; but care should be exercised. It penetrates skin as well as cuttings, and rubber gloves should be worn. The basis for utilizing DMSO is that it "carries" the IBA into the tissue and, therefore, elicits a more uniform and possibly rapid rooting response. In humans, it causes bad breath, something approximating garlic. DMF, dimethylformamide, is also an excellent solvent/penetrant but much more toxic than DMSO. Tables 7 and 8 list the relative solubility and toxicity of IBA and NAA, plus suitable solvents. Care should always be exercised when mixing the pure chemicals or making dilutions from formulations like Wood's and Dip 'N Grow.

In addition to the pure acids, various salts of these acids have been formulated and are available. They are sold as the potassium or sodium salt. In general, they cost more than the acid forms but their free solubility in water makes them "easier" to handle. Technically, they are as effective and stable as IBA

Fig. 26.a.

Fig. 26.b.

Fig. 26.c.

Figure 26. Various degrees of sophistication in the construction of ground bed propagation structures. (a) Bottom heated bed in a polycovered house, (b) and (c) commercial production outside under low wire frame, polycovered tunnels or hoop, polycovered tunnels. Mist lines are used during propagation in (b) and (c). The photographs were taken after cuttings had rooted.

Figure 27. Raised bench propagation with bottom heat (note white PVC pipe at bottom left). In this case flats are placed on top of the PVC tubes.

Figure 28. A greenhouse bench with bottom heat, wire frame and polycover. Excellent for winter propagation of needle and some broadleaf evergreens. Needs attention to prevent excessive heat buildup during sunny days.

Figure 29. Outdoor lath area for propagating and hardening off tender rooted cuttings and grafts.

and NAA (126). However, they may not be as effective in rooting certain plants because the carrier (water) does not function as effectively as a penetrant compared to alcohol, DMSO, DMF or PEG.

Certain plants are sensitive to alcohol and other solvents and may drop leaves soon after treatment. For *Berberis thunbergii* var. *atropurpurea* 'Crimson Pygmy', *Calycanthus floridus*, *Chionanthus virginicus* and others the K or Na salts in water or the talc formulations are more suitable. Mr. Jack Alexander, propagator, Arnold Arboretum, noted that roses in general are extremely sensitive to alcohol dips and best success is achieved with a talc formulation. Carrier toxicity response can only be determined by trial and error.

h. Quick dips compared to talcs

The relative effectiveness of talc formulations compared to quick dips has been the subject of numerous articles (52, 62, 70, 82, 108). There is ample evidence to indicate the superiority of the quick dip over the powders. Meahl and Lanphear (82) reported that a quick dip was equal or superior to powder in the promotion of rooting. They also noted that, on an equivalent basis, 8000 ppm IBA powder was not as good as 8000 ppm IBA solution. Supposedly, one pound of the powder treats 35,000 cuttings. Our guess is that with the waste involved with powders something like 25,000 cuttings could be treated. Sixteen ounces of solution does not go as far as 16 ounces of talc. Never return the powder to the can or solution to the stock bottle. Use a small vessel for the powder or the solution. Never stick the cuttings in the original talc can or in the stock solution.

The general superiority of quick dips is probably related to the uniformity of coverage and perhaps the more rapid absorption of IBA. It is reasonable to assume that IBA or NAA in solution will be more rapidly absorbed by the cutting than that applied in a powder form which has to be solubilized. The question has been raised many times as to how the 5-second dip became standard. Earlier work showed that a 5-second dip was as effective as a 160-second dip in promoting rooting (82). A 320-second dip decreased rooting. It was determined that the decrease was caused by the 50% alcohol and not the IBA. This may be one of the few occasions when "haste does not make waste". It should be mentioned that extremely concentrated quick dips of 20,000 to 40,000 ppm IBA or NAA will often "burn" the base of the cutting. Rooting may occur in the untreated region just above the "burn" or the cutting may rot and die.

Another technique that has been used is a dilute IBA or NAA solution and a longer soaking period. The solution may range from 20 to 200 ppm and the soak period from 6 to 24 hours. This technique appears to be effective but involves a time lag that is not inherent in the quick dip method. Howard (70) reported that similar levels of rooting were obtained with dipping times of 5000 ppm IBA for 5 seconds, 500 ppm for 30 seconds, 50 ppm 18 minutes. The need for 24-hour soaks in aqueous solutions is due to the low solubility of IBA in water. Howard has also shown that best rooting occurred when cuttings were dipped as shallow as possible. Most nurserymen dip the cutting about one inch, and this is perfectly acceptable. Blazich (5) dipped 5″ long terminal cuttings of *Ilex crenata* 'Helleri' 0, 0.5″, 1″, 1.5″ and 2.0″ deep in 5000 ppm IBA in 50% propanol. Cuttings were dried 20 minutes before sticking 2″ deep into 1 peat:1 vermiculite medium under mist. After 8 weeks cuttings were evaluated and as dipping depth increased, the number of roots per cutting and distribution of roots on basal stem increased. Increased depth increased necrosis. A 1″ dip produced rooting equal to the 2″ dip but with less necrosis.

i. Mixing quick dip and talc preparations

An easy method of mixing stock solutions has been described by Berry of Flowerwood Nursery, Alabama (Table 9) (3) and is amended here. Clean glassware, reagent grade chemicals, appropriate solvents [isopropyl alcohol (70%) can be purchased at drug store] and a reasonable beam balance (cost $100 to 125) are the prerequisites. The local pharmacist will also weigh the chemicals.

Note well that in the process of dilution the use of water alone with the acid formulations of IBA and NAA may result in the precipitation of the hormone out of solution. As a precaution always perform the dilutions carefully. Pour the diluting solvents gently into the stocks and provide gentle agitation. For the higher concentrations like 2500 and 5000 ppm it is always wise to use isopropyl alcohol (or other suitable alcohol) as the diluting solvent. If isopropyl alcohol (70%) is diluted with 1 part water, the alcohol content becomes 35%. A 1:9 dilution reduces the alcohol to 7% and if the IBA or NAA is 2500ppm or higher it may come out of solution. Slow heating of the precipitated solution with gentle stirring will bring the precipitated material back into solution. The potassium or sodium salts of IBA and NAA are made in the same manner except the solvent will be water. Make the stock solutions and store them in the refrigerator.

New formulations and chemicals appear on the market periodically but none have competed effectively with IBA, NAA and their various derivatives. For over 50 years, these chemicals have been the backbone of cutting propagation. When integrated properly with the other rooting factors discussed in this chapter they are especially effective.

F. MEDIA, FERTILIZERS

Cuttings can be rooted in any substrate providing good air/water relationships are maintained (25 to 40% air space is perhaps ideal) (figure 24). Over the years, everything from bark to sand to pumice to soil has been used. For the commercial propagator, cost, ease of handling, availability, and reproducibility of results must be considered. Peat and perlite make excellent rooting components when mixed in various ratios, but are expensive. Sand is inexpensive but at something approaching 70 to 100 pounds per cubic foot is a back-breaker and not always consistent. Many nurserymen in the southeast use straight pine bark or pine bark mixed with peat, sand and/or perlite. Some Tennessee nurserymen root directly outdoors in sandy soil while Oregon nurserymen use pumice (a type of volcanic rock).

Numerous studies (16, 19, 96, 101, 113) have been conducted to determine the best medium. The answer that continually surfaces is that one does not exist. If we could pick one ubiquitous medium for **most** plants it would be 2 coarse perlite:1 sphagnum peat by volume. However, *Magnolia grandiflora* had remained excessively wet in this medium and died under Georgia experimental conditions. Straight perlite worked perfectly. One learns quickly that there are almost as many media as propagators and plants.

1. The media components

a. Sand

Sharp builders sand with a particle diameter of 0.5 to 2mm is recommended. The mixing of sand with another component may

Figure 30. Plastic containers (2¼" by 2¼" by 5") can be used in any type of propagation system. Here containers are lined in outdoor propagation frames.

Figure 31. Oasis-type cube with roots of *Acer rubrum* 'October Glory' evident after four weeks. The "cubes" lend themselves to easy handling and mechanical planting.

Figure 32. A raised bench in a greenhouse outfitted with bottom heat. Warm water is circulated through the PVC pipe. The idea is to maintain the temperature of the rooting medium at 70 to 75°F.

Figure 33. Mist propagation in a polycovered structure. The idea is to maintain an even film of moisture on the leaf surface to prevent desiccation until the roots have formed.

Fig. 34.a.

Fig. 34.b.

Figure 34. (a) Deflector-type nozzle that provides a flat, circular pattern of water. Too often mist patterns are irregular and coverage is not complete. (b) Mist nozzle is welded to steel rod and can be moved around in the propagation bed to cover dry spots.

actually decrease aeration and drainage. Fine sands (0.05 to 0.25mm) are virtually worthless for propagation. Plants rooted in sand may have a rather coarse root structure. Sand is exceedingly heavy (70-100 lb/cubic foot), low in nutrients, and moisture holding capacity. Sand should be sterilized before use. The pH of the sand (depending on parent material) differs and can change with the pH of the water since sand has no buffering capacity. If water is alkaline, rooting of ericaceous plants can be affected (45). The real advantages lie in low cost and wide-spread availability. Sand has been mixed with peat, perlite and other amendments to produce suitable media.

b. Perlite

Perlite is produced from a crushed, aluminum-silica volcanic rock. When heated rapidly to a temperature of 1800°F, the rock fragments expand like popcorn to form white, light-weight particles with sealed internal air spaces. It is sterile, chemically inert, with no cation exchange capacity and has a pH of 7.0 to 7.5. Horticultural (#2) or coarse grades are the most effective for making propagation media and are usually sold in 4 cubic foot bags. Perlite weighs about 8 lb/cubic foot, does not decay, and is resistant to abrasion. It is more expensive than sand but comparable to vermiculite. Dry perlite is dusty and may induce coughing. Use a breathing mask and wet the perlite before use. It can be sterilized and used again. Root growth is often coarse in straight perlite and, since it contains no nutrients, cuttings should be transplanted as soon as rooted.

c. Vermiculite

Vermiculite is a clay mineral containing high amounts of potassium and magnesium that is produced by heating to 1400°F. The expanded vermiculite particles are composed of a series of plate-like layers with a high capacity for water absorption and nutrient retention. Vermiculite is sterile and very light in weight (6 lb/cubic foot). When moistened it is easily compressed and with time breaks down and shrinks in volume. As a pure rooting medium it can be used for quick rooting plants but has no value for long term propagation although *Prunus persica* (18), peach, and *Magnolia grandiflora* (9), southern magnolia, have been successfully rooted. The pH ranges between 7.0 and 7.5. Vermiculite is available in several particle sizes and the coarser particles are best for propagation. Vermiculite is a good seed starting medium and has been mixed with peat to produce the peat-like mixes that are popular for seed germination. Vermiculite is usually sold in 4 and 6 cubic foot bags.

d. Scoria and pumice

Scoria is a naturally occurring volcanic rock that has been crushed and screened for size. It is porous and weighs about 30 lb/cubic foot. *Pumice* is a white, natural glass that was formed when red-hot lava foam flowed from a volcano and cooled to a solid so quickly that crystals did not form. Pumice is relatively light and can be substituted for perlite. It is primarily silicon dioxide and aluminum oxide, with small quantities of iron, calcium, magnesium, and sodium in the oxide form.

e. Bark

In recent years, hardwood and pine barks have become the cornerstone of the container plant industry. Plants are grown in milled bark or bark mixed with sand or other amendments. Pine bark is generally inexpensive, abundant and effective. It has been used alone or mixed in 3 bark:2 peat:2 perlite ratios and other combinations (94, 95). Bark used for propagation is milled and should have 70 to 80% of the particles in the 1/40 to 3/8″ diameter range and 20 to 30% less than 1/40 inch. Pine bark has a cation exchange capacity range of 30 to 57 me/100 g., pH range of 3.5 to 5.0, and suppresses various *Pythium*, *Phytophthora* and *Fusarium* species (98). Bark is relatively light weight (19 lb/cubic foot air dried) and inexpensive compared to peat and perlite (Tables 10 and 11). Comparative costs of propagation media are presented in Table 11.

Softwood cutting of *Vaccinium ashei*, rabbiteye blueberry, cultivars rooted best in milled pine bark or bark:perlite medium, compared to sphagnum peat or peat:perlite (96). Bark:perlite was generally superior to bark alone. Bark:perlite had 26% air filled pore space compared to 17% for bark alone. Water filled pore space was about 60% for each medium. Bark will break down with time and lose air space. Commercially, bark is receiving widespread acceptance in those parts of the country where available.

f. Peat

Peat is an elusive term and encompasses sphagnum moss peat, hypnaceous moss, reed and sedge peats, and humus or muck peat. Sphagnum is important in propagation and represents the dehydrated or living portions of acid-bog plants in the genus *Sphagnum*. It is light in weight (when dry), relatively sterile, can absorb 10 to 20 times its weight in water, has a pH of 3.5 to 4.0, and has certain fungicidal properties. It can be purchased in coarse grade (good for air layering, etc.), in a shredded condition (common form used in propagation), and a fine milled state (used for germinating seeds, particularly ericaceous plants). Sphagnum peat contains small amounts of nutrients but plants kept for any time exhibit deficiency symptoms. It is mixed with sand and perlite to create excellent propagation media.

2. Fertilizers

In general there is no benefit to incorporating, broadcasting, misting, or watering fertilizer into the rooting medium until the cuttings have rooted (31, 122, 128). The roots serve as the absorptive organs for nutrients. Very limited uptake occurs through the stem of an unrooted cutting. However, if fertilizer is pre-incorporated there is a reservoir of nutrients available when the first roots emerge. Tukey (114) and Wott and Tukey (125) studied the effects of nutrient mist on the rooting of cuttings. In general there was no significant enhancement of rooting other than darker green leaves on the nutrient misted cuttings. Most researchers (109, 128) agree that the benefit to the cutting occurs after rooting because fertilizer-treated cuttings have more profuse root systems and are nutritionally primed to grow off better when transplanted (122).

Osmocote (18-6-12) (Sierra Chemical Co., Milipitas, CA 95035) is a particularly effective formulation for incorporation or top-dressing when used at rates of 2 to 6 lb/cubic yard. Gouin (51) surface applied 18-6-12 Osmocote to azaleas directly after sticking and 22 days later at 1/4 to 1/2 ounce/square foot. Osmocote treated cuttings were darker green, held leaves better and produced more breaks the following spring. Rooting percentages were not affected. Urea-based formulations (sulfur coated urea, urea formaldehyde and IBDU) are unsuccessful. These fertilizers release ammonia which can be toxic. Whitcomb (122) presented evidence for Osmocote incorporation especially for hollies. *Ilex crenata* 'Hetzii' receiving no Osmocote had a root grade of 5.1 and 18 branches after one growing season; 2 lb of Osmocote produced a 7.1 root grade and 27 branches; 4 lb - 7.6 and 31; and 6 lb - 8.1 and 32. The latter two rates are not significantly different from the 2 lb rate.

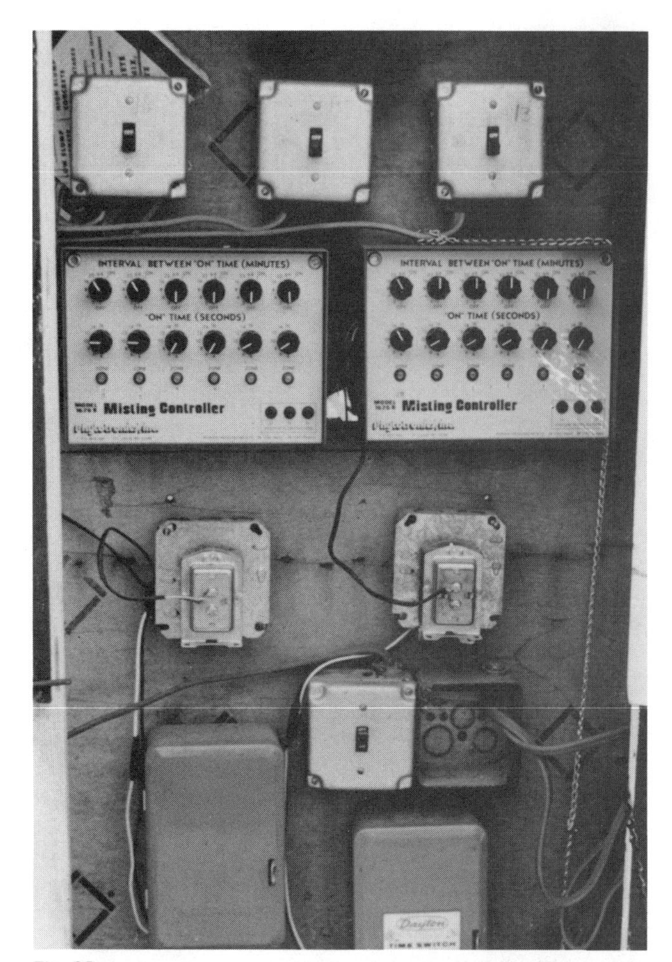

Fig. 35.a.

Fig. 35.b.

Figure 35. (a) Mist control unit with time clocks to regulate the time mist is turned on in the morning and off at night each day (a day clock), an internal timer that turns on the mist every so many minutes and a control unit that turns on different lines at different times. (b) The solenoid receives an electrical signal from the main unit and a valve is opened allowing water into the mist line.

Figure 36. A screen balance mist controller where the screen acts like (simulates) a leaf. The weight of water holds the screen down. As it evaporates the screen raises and the mist is turned on.

Figure 37. A high humidity "fog-like" system that produces smaller water droplets than mist and may result in improved rooting because of less moisture in the medium.

In Georgia work, rooted cuttings are often transplanted immediately after rooting and top dressed with Osmocote 18-6-12. The cuttings green up and grow off better than non-fertilized plants. More important, certain plants like *Disanthus cercidifolius*, *Malus* spp., *Acer* spp., *Viburnum* spp., *Betula* spp., *Hydrangea* spp., when they come out of the rooting bench in August are essentially dormant and will not grow unless fertilized. Osmocote applications induce buds of these often troublesome-to-overwinter species and cultivars to break and elongate. It is almost amazing to observe the transformation induced by surface-applied Osmocote or liquid fertilizers.

G. ROOTING STRUCTURES, CONTAINERS, BOTTOM HEAT

1. Rooting structures

Although cuttings can be rooted under a variety of environmental conditions, there is no practical substitute for quality rooting structures. Although they take many forms, their primary functions are to provide (1) an atmosphere conducive to low water loss from the cuttings, (2) protection from the elements, (3) ample but not excessive light, (4) proper rooting temperature [70 to 75°F (80°F) for most taxa], efficiency and reproducibility of results.

Glass greenhouses were once the standard for propagating and growing, but cost has precluded extensive new construction. **Plastic covered greenhouses** (figure 25) now account for about three times the area covered by glass. Many suitable books (79) and pamphlets are available on greenhouse construction and should be consulted for detailed instructions.

Many commercial propagators use plastic covered hoop or quonset houses because of low cost, ease-of-construction and effectiveness. If ground beds (figure 26a, b, c) are used for rooting, the houses should be sited to maximize drainage. Too many ground beds end up becoming the great dismal swamp. Raised benches of treated wood, metal pipe frame or other suitable materials can be constructed (figure 27). The benches may be fitted with sides to hold the rooting medium. Possibly the best approach is a raised bench and the use of flats to hold the medium. This permits easy ingress/egress and probably makes sanitation easier.

For certain plants that require a long rooting period over the winter months a polyethylene covered bench in the greenhouse makes an inexpensive and excellent rooting device (49) (figure 28). These are often outfitted with heating cables to insure a 65 to 75°F medium temperature. These frames must be ventilated on bright, sunny days and require a certain amount of management to function properly. These same high humidity chambers are excellent for germinating seeds.

In recent years many nurserymen have developed low cost, plastic covered hoop frames fashioned from treated wood, wire, plastic, etc. In Georgia work (35), 2″ by 6″, 8″ or 10″ treated lumber is used for the sides. Eight foot lengths (can be longer) are cut in half and used for the ends. Pieces are nailed together to produce a 4′ by 8′ unit that will hold 1500 to 2000 cuttings. A 1″ wood drill is used to bore 2″ deep holes into the sides, 4 to 5 per 8′ length, these are fitted with 7 to 8′ lengths of 3/4″ diameter black irrigation tubing to form the hoops. Clear or milky (preferred) plastic is used to cover the houses. A mist line is run under the frame with 30″ risers spaced at 24 to 30″ distances. The beds may be placed on top of the ground or sunk (to take advantage of ground heat). Railroad ties, landscape timbers, telephone poles, etc., have been used for sides. Several propagation nurseries use this type of system entirely and other combinations and permutations. A producer must always think about cost effectiveness and there is certainly no one best system for every firm.

Numerous other structures such as hot beds, cold frames, wet tents (121) and lathhouses can be used (figure 29). In south Georgia, cuttings of some species (photinia, crape myrtle) are stuck directly in one gallon containers and rooted in the field, under mist. This approach obviates additional handling since the cuttings are grown to salable size in the same container.

2. Containers for rooting

Although many cuttings are rooted in ground beds (bed-grown liners), large flats, etc., there is a move to compartmentalized (small rooting container) production. Major container manufacturers make the flats and containers (figure 30). Obviously there is additional expense up front for the producer, but the cost of handling, shipping, etc., may outweight initial cost. Many growers feel transplanting losses are minimized with rooted cuttings produced in a container since a small amount of medium prevents drying of the tender roots. The rooting container should be at least 3½ to 4″ deep to allow for reasonable drainage. Peat, fiber, sponge, oasis-type rooting pots and cubes (figure 31) (Smither's Oasis, P.O. Box 118, Kent, OH 44240) are also popular. The latter have been used for years to root poinsettias and mums. Red maple and crape myrtle have been successfully propagated in these "cubes" and transplanted to the field. The planting of the entire cube minimizes root damage and facilitates successful transplanting. The obvious advantage of rooting in containers/cubes is that root systems are not disturbed and successful overwintering/transplanting is facilitated. Several studies (47, 103) have demonstrated the advantage of not disturbing root systems of *Acer* spp., *Hamamelis* spp., *Stewartia* spp., etc., until the plants have been overwintered and new shoot growth has ensued the following spring. The "cubes" might also be impregnated with hormone to promote rooting and fertilizer to stimulate shoot and root growth. This is a fertile area for propagation research since "cubism" lends itself to mechanization and ease of handling/transplanting.

A relatively new rooting cube is offered by Grow-Tech, Inc., 400 Casserly Road, Watsonville, CA 95076. A sponge-like material, affectionately referred to as "rubbert dirt," has been used in the florist industry for a number of years. Tapered plugs of various diameters are placed in premolded styrofoam blocks. The plugs are predrilled and cuttings are easily stuck. The roots grow down, do not spiral and are well branched. For mechanical transplanting and perhaps long distance shipping the system has merit. Square, ribbed, or tapered containers reduce root spiraling and subsequent transplanting loss. If possible, a mesh bottom is advantageous since the cuttings (or seedlings) are effectively air pruned (122). This results in a more fibrous, well branched root system. Horticultural supply houses offer a wide selection of rooting containers. Each grower must decide what works best for his operation. For many growers, the best system is never realized and each new product that appears on the marketplace is probably worth testing.

3. Bottom heat

For certain plants, bottom heat is definitely beneficial, especially with cuttings taken in the winter months. Needle evergreens, holly, magnolia and others (59) benefit. However, the vast ma-

Fig. 38.a.

Fig. 38.b.

Fig. 38.c.

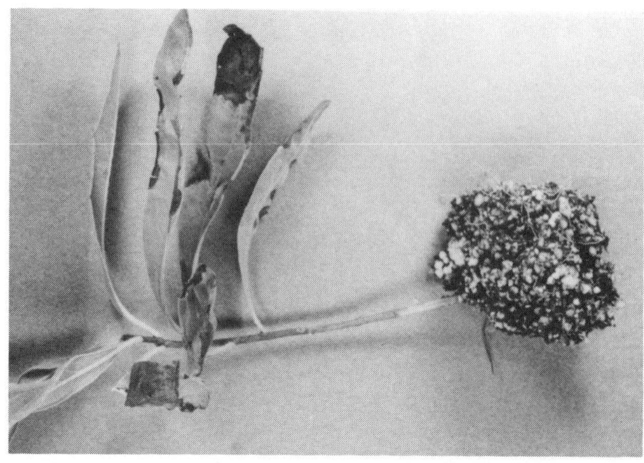

Fig. 38.d.

Figure 38. Lighting after rooting is essential to induce a growth flush on certain difficult-to-overwinter species. Note (a) lights that are used to induce a growth flush, (b) rhododendrons induced into a flush of growth, (c) Japanese maple with a growth flush and (d) well rooted deciduous azalea, *Rhododendron canescens*, that will not survive the overwintering period unless induced to grow.

Figure 39. A chamber of propagation horrors unit that was used to root plants. Common sense dictates a clean propagation house.

Figure 40. Deep pit house for overwintering cuttings and grafts. The floor is 6 feet below ground level and the temperature at ground level is maintained at 28°F.

jority of plants, especially summer softwood and semi-hardwood cuttings, do not require bottom heat. Plants propagated in outdoor beds during winter benefitted from bottom heat (35). On a recent nursery tour to Oregon, it appeared about one half of the growers used bottom heat of one kind or another (29). Bottom heat may be supplied by lead or plastic coated electric cables (30, 60, 120′ lengths manufactured by General Electric), electric heated mats, or various hot water circulated systems. Biotherm (Biotherm Engineering, 611 Mountain View Avenue, Petaluma, CA 94952) is a commercially available unit that uses flexible EDPM tubing, a hot water heater, and pumping system to circulate the water. Many growers rig systems to suit their needs. PVC pipe is placed at 6″ spacing the length of the bench and covered with sand or other material to evenly dissipate the heat (figure 32). The cuttings may be rooted in the beds or flats placed on the surface. The essence of any bottom heat system is to provide sufficient energy to maintain medium temperatures in the range of 65 to 75°F (80°F).

H. MIST SYSTEMS

Leaves of softwood and semi-hardwood cuttings wilt and die if not kept moist in some fashion. Intermittent mist as pioneered by Templeton (110, 111) has become the standard for modern cutting propagation. This electrically timed system provides a film of water over the leaves which lowers leaf temperature, increases humidity, reduces transpiration and respiration (figure 33). Mist systems can be set up in any of the structures mentioned previously. The development of the intermittent mist system has made the propagation of softwood and normally difficult to root species much easier. In fact, the two most significant advances in cutting propagation are the development of IBA/NAA and mist.

Intermittent mist reduces the volume of water applied to the cuttings and may alleviate excess moisture accumulation in the media. However, intermittent mist is not fail-safe and like any electrical/mechanical system must be checked on a routine basis. Nozzles clog, time clock motors burn out, lightning or electrical problems short circuit the clocks, etc.

1. Components

a. Nozzles

Deflection (figure 34), oil-burner, and whirling action are the principal types available for constructing the mist lines. Deflection types are the most widely used and come in a variety of designs and prices. They provide a flat circle of fog-like, mist spray and depending on the orifice deliver 4 to 14 gallons of water per hour. One of the common deflector nozzles costs $0.80 to $1.00. Forty or 80 grade PVC pipe in 1/2 to 3/4″ diameter is used for building the lines. Generally mist deflector nozzles perform best when water pressure average 35-40 psi or more. Nursery supply catalogs sell all the components for building mist systems.

b. Controls-Timers

The electronic time clock is probably the most foolproof method. Two timers; a 24 hour and the other of various times (6, 12, 30, 60 minutes) act in series. The first turns the system on and off in the morning and evening; the second turns on the mist depending on frequency desired. Controllers (timers) are available that control multiple solenoids (and thus multiple mist lines at the same time). A typical mist system might be designed like that presented in figure 35a, b.

c. Screen balance

A stainless steel screen simulates an actual leaf (not very well in many cases) and is attached to a lever actuating a mercury switch (figure 36). When the actual leaf surface is dry (stainless steel leaf also), the lever arm raises and kicks on the switch which turns on the mist. As water accumulates on the screen the lever returns to a down position and the mist is switched off. The idea is reasonably sound since the stainless steel leaf "reads" environmental conditions (cloudy versus sunny days) and acts accordingly. Mist systems on timer turn on and off at defined intervals (intermittent) regardless of the weather conditions. However, the screen balance accumulates mineral deposits and dirt, become heavier, and is slower to react to environmental changes. Spiders have webbed the balance (level) arm to the housing and prevented the arm from moving. The screen balance is far from fail-safe and needs more attention than timed mist systems. The screen needs to be cleaned on a regular basis.

d. Electronic leaf

A small piece of plastic containing two terminals is placed under the mist along with the cuttings. Alternate wetting and drying of the terminals make and break the electric circuit which in turn controls the solenoid valve. Mineral deposit buildups may cause the leaf to malfunction (operate continuously) so cleaning is a matter of course.

e. Thermostat and timer

A thermostat is placed with the cuttings. When temperature at cutting level reaches a certain point, the solenoid is turned on and mist applied. The cooling effect of the mist reduces temperature and the mist is turned off.

f. Photoelectric cells

Photoelectric cells work on the premise of the relationship between light intensity and transpiration. The cell conducts current in proportion to light intensity, actuates a magnetic counter, and opens the solenoid. The higher the light intensity; the more frequently mist is applied.

The last three control systems are not used to any extent. The screen leaf has proponents, but can be unreliable. A commercial unit (Mist-A-Matic) is available from E. C. Geiger Co., Harleysville, PA. For best results, timed mist is the most logical approach. The nurseryman/propagator must remember that all systems are fallible. Diligence and vigilance insure success. Water management is possibly the most important ingredient in successful propagation. The outstanding propagators know when to turn the water on and off.

Water quality should be good for high levels of salts may prove troublesome. Nozzles can be clogged from dirty, sandy water and filters may have to be installed in the supply lines. Nozzle orifices and strainers often need to be cleaned.

All tap water (city water) has a measure of Ca, Mg, Na, K, etc., and low levels present no problem. Table 12 lists levels of common elements found in tap water. Water high in sodium or potassium carbonates, bicarbonates, or hydroxides can be detrimental, especially when coupled with low levels of calcium salts.

I. FOG SYSTEMS

Fogging systems have been around for years (107, 115) but have become commercially feasible in recent years because of new

Table 1. Effect of indolebutyric acid on the rooting of *Photinia* × *fraseri*.

Hormone treatment (IBA)	Rooting percent	Number	Length (cm)
Control (0)	8	0.3	2
0.5%	93	10	107
1.0%	95	17	156
2.0%	83	17	140
4.0%	55	11	60

Table 2. Rooting of large leaf *Rhododendron* cultivars.

Cultivar	Rooting percentage
Anna Rose Whitney	95z/
Blue Ensign	85
Cheer	85
English Roseum	80
Ignatius Sargent	75
Nova Zembla	95
President Lincoln	75

z/Cuttings taken 8-13; evaluated 12-7; treated with 1% IBA solution, wound, peat: perlite, mist.

Table 3. Commercial rooting preparations encountered in plant propagation literature.*

Auxan	Proliferol
Auxilan	Ree Root
C-Mone	Rhizopan
Chloromone**	Rootagen
Dip 'N Grow**	Rootone**
Hormex 1, 3, 8, 16, 30, 45**	Seradix
Hormodin 1, 2, 3**	Stim Root
HormoRoot A, B, C, 1, 2, 3, 4**	Synergol
Hormovita	
Jiffy Grow	Wood's Rooting Compound**

*This list is not exhaustive.
**Preparations that are commonly used in the United States.

Table 4. Commercial root promoting formulations, ingredients, and sources.

Hormex No. 8

Active ingredient: Indole-3-butyric acid	0.8%
Inert ingredient: Talc	99.2%

Hormex	1 = 0.1% IBA
	2 = 0.3% IBA
	16 = 1.6% IBA
	30 = 3.0% IBA
	45 = 4.5% IBA

Brooker Chemical Corp.
P.O. Box 9335
No. Hollywood, CA 91609

Table 4 (cont.). Commercial root promoting formulations, ingredients and sources.

Hormodin 3

Active ingredient: Indole-3-butyric acid	0.8%
Inert ingredient: Talc	99.2%

Hormodin	1 = 0.1% IBA
	2 = 0.3% IBA

MSDAGVET
Division of Merck & Co., Inc.
Rahway, NJ 07065

Hormo-Root "C"

Active ingredient: Thiram (tetramethylthiuram disulfide)	14.00%
Indole-3-butyric acid	0.80%
Inert ingredient: Talc	84.20%

HormoRoot	A = 0.1% IBA
	B = 0.4% IBA
	1 = 1.0% IBA
	2 = 2.0% IBA
	3 = 3.0% IBA
	4 = 4.0% IBA

Hortus Products Co.
P.O. Box 275
Newfoundland, NJ 07435

Rootone

Active ingredient: 1-naphthaleneacetamide	0.067%
2-methyl-1-naphthaleneacetic acid	0.033%
2-methyl-1-naphthaleneacetamide	0.013%
Indole-3-butyric acid	0.057%
Active as fungicide:	
Thiram (tetramethylthiuram disulfide)	4.00%
Inert ingredient: Talc	95.830%

Union Carbide Agricultural Products Co., Inc.
P.O. Box 12014
T.W. Alexander Dr.
Research Triangle Park, NC 27709

Dip 'N Grow

Active ingredient: Indole-3-butyric acid	1.0%
1-naphthaleneacetic acid	0.5%
Inert ingredient: Alcohol	98.5%

Alpkem Corporation
Clackamas, OR 97015

Synergol

Active ingredient: K salts of: Indol-3-yl butyric acid (IBA) and 1-naphthylacetic acid (NAA) plus fungicide and synergistic additives...10,000 ppm total (5000 ppm of each chemical)

Silvaperl Products Ltd.
P.O. Box 8, Dept. 71
Harrogate
North Yorkshire, HG2BJW
(Tel: 0423-070370)

equipment (84, 85, 86, 100). Fogging produces less moisture than intermittent mist systems and facilitates "better" rooting of cuttings that traditionally fare miserably under "wetter" intermittent mist conditions. Fogging differs from mist by producing smaller size water particles that stay suspended in air longer thus producing high humidity. The AgriTech (figure 37) system uses an oscillating fan and nozzle to produce a 90° pattern that extends for 30 to 40'. Growers have tried the system with moderate success. The mechanical problems/aspects have outweighed the advantages over traditional timed mist. Recently, a fogging machine (Mee Industries, Inc., 1629 South Del Mar Avenue, San Gabriel, CA 91776) has been described (100). There are advantages to high humidity (fog) propagation but its long-term performance and particularly reliability need to be determined before nurserymen change from the traditional systems.

J. LIGHT IN THE PROPAGATION HOUSE

1. For propagation

The question is often asked 'How much light is required for cuttings in the rooting bench'? In general, cuttings do not require high light levels until they root. Photosynthesis may drop almost to the compensation point (no net gain in dry matter production) until rooting takes place. Hormones produced in the roots provide the signal to the leaves that engages the photosynthetic mechanisms. In essence, a cutting is in a state of suspended animation and for survival the first priority is roots. Notice how few species or cultivars make shoot growth while in the rooting bench indicating a distribution of carbohydrates for root production.

From a stress standpoint, it is worthwhile shading cuttings to reduce water loss and temperatures. Many species (*Acer palmatum*, *Acer rubrum* and *Betula nigra*) will die within hours if the leaf surface is allowed to dry. Reducing the heat load with shade cloth is advantageous. As a matter of reference, on a bright summer day total available foot candles average 10,000. Even if photosynthesis was operating at 100% efficiency, no more than 1000 would be utilized. The cutting that is barely photosynthesizing is utilizing approximately 100 foot candles.

2. Inducing a flush of growth

For selected species it is often necessary to induce a flush of growth after rooting which in turn results (through photosynthesis) in increased carbohydrate reserves. These reserves are largely concentrated in the roots and serve as an energy source during the overwintering period and during budbreak in the spring. Deciduous azalea (119) species and cultivars, Japanese maple, red maple (78), witch-hazel, and others (103) must be induced to break bud, for overwintering losses can be staggering. Supplemental lighting, either continuous or interrupted from 10 p.m. to 2 a.m., may promote new growth.

A 60 to 75 watt incandescent bulb placed 3 feet apart and 3 feet above the top of the cuttings is sufficient to promote growth (figure 38). Usually a line of light bulbs is placed over each bench in the rooting house. The lights should be turned on after rooting has taken place and continued for 6 to 8 weeks. Smalley et al. (104) showed that rooted cuttings of *Acer rubrum* 'October Glory' that broke bud and grew overwintered 100%, those that did not overwintered 58%. Cuttings that overwintered 100% averaged root carbohydrates of 16%; the 58% group less than 10%.

K. SANITATION

Cleanliness is a necessary prerequisite for successful cutting propagation (99). Clean benches, flats, sterilized media, good water and a preventative fungicidal program are a must. Too often, in our travels, we have observed dilapidated propagation houses (figure 39) that on the inside look like the great dismal swamp. At times, it appeared that the origin of life may have occurred in the slime and ooze of a propagation bench.

Fungal problems are minimized if the propagator starts with clean, healthy stock plants from which to collect cuttings. A soak of Captan or Benlate before cuttings are prepared is advisable. Once cuttings are placed in flats or benches a weekly drench of Benlate and perhaps Subdue (for damping off and root rot caused by Phycomycetes) is warranted. Several papers (53, 81) reported beneficial effects of Captan and Benlate on the rooting of cuttings. One propagator noted that benomyl (Benlate) not only increased the percent rooting but also the quality. With the use of any chemical there is always the chance of injuring a particular species or cultivar. Propagators must keep good records relative to chemical phytotoxicity.

An excellent publication that could benefit every propagator is "Diseases of Woody Ornamental Plants and Their Control in Nurseries". It is available from Publications Office, Department of Agricultural Communications, North Carolina State University, Raleigh, NC 27695 ($5.00). It contains excellent tables on modern fungicides and sterilants and their effectiveness in the propagation house.

L. AFTERCARE, OVERWINTERING, STORAGE

The ideal situation for cutting producers is to sell all plants as soon as they root. Since this seldom happens, nurserymen must carefully manage the newly rooted plants. In our experience as soon as the cutting has rooted the mist should be reduced and flats removed, the rooting bed ventilated (if enclosed in plastic) and the cuttings hardened gradually. After weaning, softwood cuttings can be left in place and overwintered, transplanted, or sold. Some growers leave rooted softwood cuttings in the bed for two years. The cuttings may die when transplanted immediately after rooting. Vibrunum, maple, stewartia, witch-hazel, and others are better left in place until the following spring when new growth emerges. Carbohydrate and hormonal levels affect the success of these newly transplanted rooted cuttings.

Rooted cuttings of some species have been stored for as long as 5 months in polyethylene bags at 35 to 40°F. Considerable reasearch needs to be conducted in this area. Every plant displays a different response to cold storage which must be determined by trial and error. In general, an ideal temperature for the root system is between 34-41°F. The soil often freezes but not to a depth that injures the young roots.

Most nurserymen overwinter rooted cuttings in the structures in which they were rooted. Milky plastic is a better covering since light penetration is reduced and temperatures do not fluctuate as widely. Microfoam is also used to cover outdoor rooting ground beds. Deep pit storage (68) has been used which good success for traditionally difficult-to-overwinter species. The house is sunk in the ground and takes advantage of ground heat (figure 40).

Table 4 (cont.). Commercial root promoting formulations, ingredients, and sources.

Wood's Rooting Compound

Active ingredient:	Indole-3-butyric acid	1.03%
	1-naphthaleneacetic acid	0.51%
Inert ingredient:	Ethanol SD 3A	78.46%
	Dimethylformamide	20.0%

Earth Science Products Corp.
P.O. Box 327
Wilsonville, OR 97070

Chloromone

Active ingredient:	Indole-naphthylacetamine	0.1%

Chloromone Co., Inc.
Upper Montclair, NJ 07043

Pure indolebutyric acid and naphthaleneacetic acid as well as K-salts

Aldrich Chemical Co.
940 East Saint Paul Avenue
Milwaukee, WI 53233
(414) 273-3850

Baker, J.T., Chemical Co.
222 Red School Lane
Phillipsburg, NJ 08865
(201) 859-5411

ICN Pharmaceuticals, Inc.
K&K Labs Division
121 Express Street
Plainview, NY 11803
(516) 433-6262

Pfaltz and Bauer, Inc.
375 Fairfield Avenue
Stamford, CT 06902
(203) 357-8700

Research Organics, Inc.*
4353 East 49th Street
Cleveland, OH 44125-1083
1-800-321-0570

Sigma Chemical Co.
P.O. Box 14508
Saint Louis, MO 63178
(800) 325-3010

United States Biochemical Corporation
P.O. Box 22400
Cleveland, OH 44122
(800) 321-9322

Phenyl indole-3-thiobutyrate

GRO/TECH Inc.
P.O. Box 347
Rapid City, SD 57709
(605) 394-6400

*Excellent supplier.

Table 5. Composition of treatments used for the *Photinia* × *fraseri* rooting study.

Treatment	Growth regulator and concentration	Solvent
Water		Water (Distilled deionized in all cases)
Ethanol		Water (50% ethanol)
Water + 50 ppm B	50 ppm B from H_3BO_4	Water
Chloromone	(5:1 dilution)	Water
Naphthaleneacetamide	Naphthaleneacetamide (0.3%)	50% ethanol
Hormodin #2	Indolebutryic acid (0.3%)	Talc
NAA	a-Naphthaleneacetic acid (0.3%)	50% ethanol
KNAA	Potassium salt of a-Naphthaleneacetic acid (0.3%)	Water
IBA	Indolebutyric acid (0.3%)	50% ethanol
KIBA	Potassium salt of indolebutyric acid (0.3%)	Water
KIBA + 50 ppm B	Potassium salt of indolebutyric acid (0.3%) + 50 ppm B from H_3BO_4	Water
IBA + NAA	Indolebutyric acid (0.2%) Naphthaleneacetic acid (0.1%)	50% ethanol
KIBA + NAA	Potassium salt of indolebutyric acid (0.2%), potassium salt of naphthaleneacetic acid (0.1%)	Water
Woods	Diluted 4:1 Indolebutyric acid (0.2%) Naphthaleneacetic acid (0.1%), 4% Dimethylformamide	Water
Dip 'N Grow	Dilute 4:1 Indolebutyric acid (0.2%) Naphthaleneacetic acid (0.1%)	Water

Table 6. The effects of selected rooting compounds on the rooting percentage, root number and length of *Photinia* × *fraseri* stem cuttings.

Treatment	Root parameters		
	Percent	Number	Length (cm)
Water	0	0	0
Ethanol (50%)	7	0	0
Water + 50 ppm B	3	0	0
Chloromone	17	1	5
Naphthaleneacetamide	67	6	28
Hormodin #2 (0.3% IBA)	3	0	1
NAA	87	11	59
KNAA	53	5	30
IBA	63	7	46
KIBA	33	2	8
KIBA + 50 ppm B	83	8	47
IBA + NAA	83	8	47
KIBA + KNAA	23	2	10
Woods	97	26	116
Dip 'N Grow	100	24	151

Table 7. Chemicals used in rooting formulations: their molecular weights, solubilities and toxicities.

Chemical	Molecular weight	Solubility	Toxicity
Indoleacetic acid (Indole-3-acetic acid)	175.18	Sparingly soluble in H_2O or chloroform, freely soluble in alcohol, soluble in acetone, ether.	Unknown
Indolebutryic acid (Indole-3-butyric acid)	203.23	Practically insoluble in H_2O or chloroform, soluble in alcohol, ether, acetone.	LD 50 intraperitoneal in mice: 100 mg/kg
1-Naphthaleneacetic acid (a-Naphthaleneacetic acid)	186.20	Soluble in 30 parts alcohol. Freely soluble in acetone, ether, chloroform. Soluble in H_2O at 17°C (63°F)-0.38 g/l.	LD 50 orally in rats: 1.0 g/kg
1-Naphthaleneacetamide	185.20	Soluble in alcohol.	Unknown
2,4-Dichlorophenoxyacetic acid (Sodium salt)	221.04	Almost insoluble in H_2O, soluble in organic solvents, soluble in oil	LD 50 orally in rats: 100 mg/kg
2,4,5-Trichlorophenoxyacetic acid (Isopropyl ester)	255.49	Almost insoluble in H_2O, soluble in alcohol, soluble in oil	LD 50 orally in rats: 300 mg/kg
2-(2,4,5-Trichlorophenoxy) propionic acid (Silvex)	269.53	(A water soluble salt with triethanolamine is called silvex-amine). Soluble at 25°C (77°F) in H_2O: 0.014%, acetone: 15.2%, methanol: 10.5%, ether: 7.13%	LD 50 orally in rats: 650 mg/kg

Table 8. Solvents that can be used for dissolving the acid formulations of IBA, NAA and other root promoting chemicals.

Solvent	Molecular weight	Solubility	Toxicity
Acetone (2-Propanone)	58.08	Miscible with H_2O, alcohol, DMF, chloroform, ether, most oils.	LD 50 orally in rats: 10.7 ml/kg
Glycerol (1,2,3-Propanetriol; glycerin)	92.09	Miscible with H_2O, alcohol. 1 part dissolves in 11 parts ethyl acetate. 1 part dissolves in 500 parts ethyl ether. Insoluble in benzene, chloroform, carbon tetrachloride, carbon disulfide, petroether, oils.	LD 50 orally in rats: 31.5 g/kg; intravenous (i.v.): 7.56 g/kg
N,N-Dimethylformamide** (DMF)	73.09	Miscible with H_2O & most common organic solvents.	LD 50 in mice: 1122 mg/kg orally; 1.19 ml/kg intraperitoneal. Vapor harmful
Dimethyl Sulfoxide** (DMSO)	78.13	Soluble in H_2O, ethanol, acetone, ether, benzene, chloroform.	LD 50 orally in rats: 20 g/kg. Skin contact causes 1° irritation, redness, itching, scaling
Polyethylene Glycol 600* (Carbowax 6000)	570-630	Soluble in H_2O, many organic solvents. Readily soluble in aromatic hydrocarbons.	Low toxicity; LD 50 40 g/kg orally
Ethanol, Ethyl Alcohol*	46.07	Miscible with H_2O & with many organic liquids	LD 50 orally in rats: 13.7 g/kg
Isopropyl Alcohol, Isopropanol* (2-Propanol)	60.09	Miscible with H_2O, alcohol, ether, chloroform. Insoluble in salt solutions.	LD 50 orally in rats: 5.8 g/kg; 100 ml ingestion can be fatal.
Methanol, Methyl Alcohol	32.04	Miscible with H_2O, ethanol, ether, benzene, ketones & most other organic solvents. Usually is a better solvent than ethanol, dissolves many inorganic salts.	Poisoning may occur from ingestion, inhalation or percutaneous absorption Death from 30 ml has been reported. Usual fatal dose 100-250 ml.
n-Butyl Alcohol (1-Butanol)	74.12	Solubility at 77°F, 9.1 ml/100 ml H_2O, miscible with alcohol, ether & many other organic solvents.	LD 50 orally in rats: 4.36 g/kg

*Indicates the safest and most commonly used solvents.

**Effective solvents and penetrants but caution should be exercised, especially with dimethylformamide.

Table 9. Procedure for making concentrated rooting solutions.

SOLUTIONS

Basic formulations (approximate values)

1. 10,000 ppm (1.0% IBA, NAA or whatever hormone solution (STOCK) (Double for 2% solution): Dissolve 5 g hormone in 1 pint (16 ounces) isopropyl alcohol (70%).
2. 5000 ppm (Mix 1 part STOCK: 1 part isopropyl alcohol).
3. 2500 ppm (Mix 1 part STOCK: 3 parts isopropyl alcohol).
4. 1000 ppm (Mix 1 part STOCK: 9 parts isopropyl alcohol).

For combination treatments such as IBA and NAA simply mix the various STOCK solutions together. In combination treatments, NAA is usually one-third to one-half the concentration of IBA because of its greater toxicity.

Example: 1. To prepare a solution containing 5,000 ppm IBA and 5,000 ppm NAA mix 1 part of 10,000 ppm IBA STOCK with 1 part 10,000 ppm NAA STOCK

2. For a 5000 ppm IBA and 2500 ppm NAA mix 2 parts of IBA STOCK + 1 part of NAA STOCK + 1 part isopropyl alcohol

TALCS

Hormone-fungicide-talc formulation

1% IBA or 1% NAA talc

Mix 1 g IBA or NAA + 9g Benlate + 90 g talc

A 2% formulation can be made by adding 2 g IBA or NAA.

Table 10. Comparative weight and pH of selected rooting media.

Medium	Air-dry weight (lb./cu.ft.)	Field-capacity weight (lb./cu. ft.)	pH
½ peat - ½ sand	47.0	60.8	4.47
½ peat - ½ perlite	9.7	24.9	4.58
100 percent sand	64.5	76.7	6.62
100 percent milled pine bark	19.1	27.6	4.90
½ milled pine bark - ½ sand	47.3	60.8	5.15
½ milled pine bark - ½ perlite	13.6	22.9	5.23

Table 11. Comparative cost of media used for rooting cuttings of woody plants.

Medium	Cost of medium per cu. ft.	Cost of medium per sq. ft. of bed area 4″ in depth
Vermiculite (Horticultural grade)	1.20	0.40
Perlite (Horticultural grade)	1.40	0.47
Canadian peat moss	1.75	0.58
Sand	0.25	0.08
½ peat - ½ sand	1.00	0.33
½ peat - ½ perlite	1.58	0.53
Milled pine bark	0.40	0.13
½ milled pine bark - ½ sand	0.33	0.11
½ milled pine bark - ½ perlite	0.90	0.30

Table 12. Elemental content of water used to mist cuttings.

	ppm										
	P	K	Ca	Mg	Mn	Fe	B	Cu	Zn	Na	Mo
Mist*	0.076	2.50	15.8	2.55	n.d.**	0.004	0.019	0.013	0.044	3.82	0.005

*City water, Athens, GA

**Not detected.

LITERATURE CITED

1. Ali, N. and Westwood, M.N. 1966. Rooting of pear cuttings as related to carbohydrate, nitrogen and rest period. Proc. Amer. Soc. Hort. Sci. 88:145-150.
2. Bassuk, N., D. Miske, and B. Maynard. 1984. Stock plant etiolation for improved rooting of cuttings. Proc. Int. Plant Prop. Soc. 34:543-550.
3. Berry, J. 1985. Techniques in propagation at Flowerwood Nurseries Inc. Georgia Nursery Notes, July-August 1985, p. 6-9.
4. Bicket, J. and T. E. Bilderback. 1982. The impact of seven treatments on rooting softwood cuttings. Amer. Nurseryman 156(11):85-86.
5. Blazich, F. 1979. Effect of auxin dipping depth on rooting of 'Helleri' holly cuttings. Proc. Southern Nurs. Assoc. Res. Conf. 24:225-227.
6. Blazich, F.A. and V.P. Bonaminio. 1984. Propagation of southern wax myrtle by stem cuttings. Proc. Southern Nurs. Assoc. Res. Conf. 29:250-209.
7. Bonaminio, V.P. and F.A. Blazich. 1982. Response of red tips photinia to rooting compounds. Proc. Southern Nurs. Assoc. Res. Conf. 27:243-246.
8. Bonaminio, V.P. and F.A. Blazich. 1983. Response of Fraser's photinia stem cuttings to selected rooting compounds. J. Environ. Hort. 1:9-11.
9. Brailsford, W.M. 1983. Asexual *Magnolia grandiflora* propagation at Shady Grove Nursery. Proc. Int. Plant Prop. Soc. 33:622-624.
10. Brown, B.F. and M.A. Dirr. 1976. Cutting propagation of selected flowering crabapple taxa. The Plant Propagator 22(4):4-5.
11. Burd, S.M. and M.A. Dirr. 1977. Propagation of selected *Malus* taxa from softwood cuttings. Proc. Int. Plant Prop. Soc. 27:427-431.
12. Chapman, D.J. 1979. Propagation of *Acer campestre*, *A. platanoides*, *A. rubrum*, and *A. ginnala* by cuttings. Proc. Int. Plant Prop. Soc. 29:345-347.
13. Chapman, D.J. and S. Hoover. 1981. Propagation of shade trees by softwood cuttings. Proc. Int. Plant Prop. Soc. 31:507-511.
14. Cheffins, N.J. 1975. Nursery practice in relation to the carbohydrate resources of leafless hardwood cuttings. Proc. Int. Plant Prop. Soc. 25:190-193.
15. Chong, C. 1981. Influence of higher IBA concentrations on rooting. Proc. Int. Plant Prop. Soc. 31:453-460.
16. Cook, C.D. and B.L. Dunsby. 1978. Perlite for propagation. Proc. Int. Plant Prop. Soc. 28:224-228.
17. Coorts, G.D. 1969. The effect of minor element deficiency on rooting woody ornamentals. The Plant Propagator 15(3):15-16.
18. Couvillon, G.A. 1985. Propagation and performance of inexpensive peach trees from cuttings. Acta Horticulturae 173:271-282.
19. Cowan, J.M. 1973. Peat/sawdust mixture as a propagating medium. Proc. Int. Plant Prop. Soc. 23:384-387.
20. Cox, Jeff. 1981. Organic discoveries. Organic Gardening 29(9):130-132.
21. Creech, J.L. 1954. Propagating plants by root cuttings. Proc. Int. Plant Prop. Soc. 4:164-167.
22. Cross, R.E. 1981. Propagation and production of *Rhus typhina* 'Laciniata', cutleaf staghorn sumac. Proc. Int. Plant Prop. Soc. 31:524-527.
23. Curtis, W.J. 1965. Rooting of *Magnolia grandiflora*. Proc. Int. Plant Prop. Soc. 15:142-143.
24. Den Ouden, P. and B.K. Boom. 1965. Manual of cultivated conifers. M. Nijhoff, The Hague, Netherlands.
25. Dirr, M.A. 1983. Comparative effects of selected rooting compounds on the rooting of *Photinia* × *fraseri*. Proc. Int. Plant Prop. Soc. 33:536-540.
26. Dirr, M.A. 1983. Manual of woody landscape plants. Stipes Publ. Co., Champaign, IL.
27. Dirr, M.A. 1986. Cutting propagation of *Magnolia grandiflora*. Proc. Southern Nurs. Assoc. Res. Conf. 31:200-203.
28. Dirr. M.A. 1986. Field Notes — *Calycanthus floridus* 'Athens'. Amer. Nurseryman 163(8):136.
29. Dirr, M.A. 1986. Flourishing fields add new ideas to tour. Nursery Manager 2:28-45.
30. Dirr, M.A. 1986. New root promoting chemicals and formulations. Proc. Southern Nurs. Assoc. Res. Conf. 31:204-209.
31. Dirr, M.A. 1986. The nuts and bolts of cutting propagation. Amer. Nurseryman 163(7): 54-64.
32. Dirr, M.A. and B. Brinson. 1985. *Magnolia grandiflora*: A propagation guide. Amer. Nurseryman 162(9):38-50.
33. Dirr, M.A. and J.J. Frett. 1983. Rooting chinese elm and Japanese zelkova. The Plant Propagator 26(2):10-11.
34. Dirr, M.A. and J.J. Frett. 1983. Rooting of Leyland cypress as affected by indolebutyric acid and boron treatment. HortScience 18:204-205.
35. Dirr, M.A., W. Raughton, Jr., and J.J. Frett. 1983. Outdoor ground bed propagation of broadleaf deciduous, needle and broadleaf evergreen cuttings. Proc. Southern Nurs. Assoc. Res. Conf. 28:229-232.
36. Dirr, M.A., W. Raughton, Jr., and J.J. Frett. 1983. Rooting of *Photinia* × *fraseri* using terminal and subterminal cuttings and indolebutyric acid dips. Proc. Southern Nurs. Assoc. Res. Conf. 28:233-235.
37. Donovan, D.M. 1976. A list of plants regenerating from root cuttings. The Plant Propagator 22(1):7-8.
38. Donovan, D.M. and R. Johnstone. 1977. A supplementary list of plants propagated by root cuttings. The Plant Propagator 23(2):14-15.
39. Doran, W.L. 1957. Propagation of woody plants by cuttings. Univ. Mass. Expt. Sta. Bull. No. 491.
40. Edwards, R.A. and M.B. Thomas. 1979. Influence of wounding and IBA treatments on the rooting of several woody perennial species. The Plant Propagator 24(4):9-12.
41. Eisenberg, B.A., G.L. Staby and T.A. Fretz. 1978. Low pressure and refrigerated storage of rooted and unrooted ornamental cuttings. Proc. Int. Plant Prop. Soc. 28:576-587.
42. Eisenberg, G.A., G.L. Staby, T.A. Fretz and T.R. Erwin. 1976. Low pressure storage of rooted and unrooted geraniums. Proc. Int. Plant Prop. Soc. 26:215-219.
43. English, J.A. 1981. Rooting *Acer rubrum* cultivars using single node cuttings. Proc. Int. Plant Prop. Soc. 31:147-148.
44. Fagan, A.E. and M.A. Dirr. 1980. Fringe trees — ready to be propagated. Amer. Nurseryman 152(7):14-15, 114-117.
45. Falkenstrom, K. and M.A. Dirr. 1976. Factors affecting the rooting of *Rhododendron* P.J.M. cuttings. The Plant Propagator 22(1):6-7.
46. Flemer, W., III. 1961. Propagating woody plants by root cuttings. Proc. Int. Plant Prop. Soc. 11:42-47.
47. Fordham, A.J. 1969. *Acer griseum* and its propagation. Proc. Int. Plant Prop. Soc. 19:346-349.

48. Fordham, A.J. 1969. Production of juvenile shoots from root pieces. Proc. Int. Plant Prop. Soc. 19:286-287.
49. Fordham, A.J. and L.J. Spraker. 1977. Propagation manual of selected gymnosperms. Arnoldia 37:1-88.
50. Gardner, F.E. 1937. Etiolation as a method of rooting apple variety stem cuttings. Proc. Amer. Soc. Hort. Sci. 34:323-329.
51. Gouin, F.R. 1974. Osmocote in the propagation house. Proc. Int. Plant Prop. Soc. 245:337-341.
52. Gray, H. 1959. The quick dip alcohol solution as an aid to rooting cuttings. Proc. Int. Plant Prop. Soc. 4:208-214.
53. Grigsby, H.C. 1965. Captan aids rooting of loblolly pine cuttings. Proc. Int. Plant Prop. Soc. 15:147-151.
54. Haissig B.E. 1979. Influence of aryl esters of indole-3-acetic and indole-3-butyric acid on adventitious root primordium initiation and development. Physiol. Plantarum 47:29-33.
55. Haissig, B.E. 1983. N-phenyl indole-3-thiolohydrate enhances adventitious root primordium development. Physiol. Plantarum 57:424-440.
56. Halliwell, B. 1970. Selection of material when propagating Leyland cypress. Proc. Int. Plant Prop. Soc. 29:338-339.
57. Hamilton, W.W. 1974. Container production of sweet fern. Proc. Int. Plant Prop. Soc. 24:364-366.
58. Hare, R.C. 1979. Modular air-layering and chemical treatments improve rooting of loblolly pine. Proc. Int. Plant Prop. Soc. 29:446-454.
59. Hartmann, H.T. and D.E. Kester. 1983. Plant propagation — Principles and practices. 4th Edition, Prentice Hall, Englewood Cliffs, NJ.
60. Hassan, H., G.A. Couvillon and F.A. Pokorny. 1984. Hardwood cutting propagation of peaches — influence of IBA concentrations on rooting. Proc. Southern Nurs. Assoc. Res. Conf. 29:201-204.
61. Hendricks, D.R. 1984. Air layering of native woody plants. Proc. Int. Plant Prop. Soc. 34:528-531.
62. Hess, C.E. 1959. A comparison between quick dip methods of growth substance application to cuttings. Proc. Int. Plant Prop. Soc. 9:41-45.
63. Hess, C.E. 1961. Characteriziation of rooting cofactors extracted from *Hedera helix* L. and *Hibiscus rosa — sinensis* L. Proc. Int. Plant Prop. Soc. 11:51-57.
64. Hess, C.E. 1962. A physiological analysis of root initiation in easy and difficult to root cuttings. Proc. 16th Int. Hort. Cong. 375-381.
65. Hess, C.E. 1963. Why certain cuttings are hard to root. Proc. Int. Plant Prop. Soc. 13:63-70.
66. Hickman, G.G. and C.E. Whitcomb. 1983. Propagating lacebark elm from cuttings. Proc. Southern Nurs. Assoc. Res. Conf. 28:221-222.
67. Heuser, C.W. 1977. Factors controlling regeneration from root cuttings. Proc. Int. Plant Prop. Soc. 27:398-402.
68. Hoogendoorn, C. 1977. Deep pit storage of newly propagated plants. Proc. Int. Plant Prop. Soc. 27:485-486.
69. Hoogendoorn, D.P. 1984. Propagating *Acer griseum* from cuttings. Proc. Int. Plant Prop. Soc. 34:570-573.
70. Howard, B.H. 1974. Factors which affect the response of cuttings to hormone treatments. Proc. Int. Plant Prop. Soc. 24:142-143.
71. Howard, B.E. 1980. Etiolation of leafy summer cuttings. Rept. East Malling Res. Station for 1979, pp. 70-71.
72. Howard, B.N. 1981. Propagation by leafless winter cuttings. The Plantsman 3:99-107.
73. Kawase, M. 1970. Root-promoting substances in *Salix alba*. Physiol. Plantarum 23:159-170.
74. Kawase, M. 1971. Diffusible rooting substances in woody ornamentals. J. Amer. Soc. Hort. Sci. 96:116-119.
75. Kling, G.J. and M.M. Meyer, Jr. 1983. Effects of phenolic compounds and indoleacetic acid on adventitious root initiation in cuttings of *Phaseolus aureus*, *Acer saccharinum* and *Acer griseum*. HortScience 18:352-354.
76. Knowles, J.W., W.A. Dozier, Jr., and C.H. Gilliam. 1984. Rooting of softwood Bradford pear cuttings using different rates and formulations of IBA. Proc. Southern Nurs. Assoc. Res. Conf. 29:222-223.
77. Lamb, J.G.D., J.C. Kelly, and P. Bowbrick. 1985. Nursery stock manual. Grower Books, London.
78. Lane, Bryce H. and S.M. Still. 1981. Influence of extended photoperiod and fertilization on rooting *Acer rubrum* L. 'Red Sunset' cuttings. Proc. Int. Plant Prop. Soc. 31:571-577.
79. Langhans, R.W. 1980. Greenhouse management. Halycon Press, Ithaca, NY.
80. MacMillan Browse, P.D.A. 1980. The propagation of plants from root cuttings. The Plantsman 2:54-62.
81. McGuire, J.J. and V.H. Vallone. 1971. Interaction of 3-indolebutyric acid and benomyl in promoting root initiation in stem cuttings of woody ornamental plants. Proc. Int. Plant Prop. Soc. 21:374-380.
82. Meahl, R.P. and F.O. Lanphear. 1967. Evaluation of the quick dip method of treating stem cuttings with rooting hormones. The Plant Propagator 13(2):13-15.
83. Mezitt, E.V. 1978. Propagation by cuttings of lilacs and other hard-to-root species by the subirrigation method. Proc. Int. Plant Prop. Soc. 28:494-496.
84. Milbocker, D.C. 1980. Reducing energy requirements with ventilated high humidity propagation. Proc. Int. Plant Prop. Soc. 30:306-307.
85. Milbocker, D.C. 1980. Ventilated high humidity propagation. Proc. Int. Plant Prop. Soc. 30:480-482.
86. Milbocker, D.C. 1983. Ventilated high humidity propagation. Proc. Int. Plant Prop. Soc. 33:384-385.
87. Mitsch, J. 1975. Propagation of dwarf conifers. Proc. Int. Plant Prop. Soc. 25:81-84.
88. Moller, G.A. 1985. How one Oregon grower produces trees from softwood cuttings. Amer. Nurseryman 162:68-69.
89. Nienhuys, H.C. 1980. Propagation of deciduous azaleas. Proc. Int. Plant Prop. Soc. 30:457-459.
90. Orndorff, C. 1977. Propagation of woody plants by root cuttings. Proc. Int. Plant Prop. Soc. 27:402-406.
91. Orton, E.R., Jr. 1978. Single node cuttings: A simple method for the rapid propagation of plants of selected clones of *Acer rubrum* L. The Plant Propagator 24(3):12-15.
92. Paton, D.M., R.R. Willing, W. Nichols, and L.D. Pryor. 1970. Rooting of stem cuttings of eucalyptus: A rooting inhibitor in adult tissue. Australian J. Bot. 28:175-183.
93. Perry, F.B. and S.S. Harper. 1983. Cutting propagation of *Chionanthus retusus* (Lindl. and Paxt.), Chinese fringetree. Proc. Southern Nurs. Assoc. Res. Conf. 28:239-241.
94. Pokorny, F.A. 1979. Pine bark container media — An overview. Proc. Int. Plant Prop. Soc. 29:484-498.
95. Pokorny, F.A. 1982. Pine bark as a soil amendment. Proc. Southern Nurs. Assoc. Res. Conf. 27:131-139.
96. Pokorny, F.A. and H.F. Perkins. 1966. Utilization of milled pine bark for propagating woody ornamental plants. Forest Prod. J. 17:43-48.

97. Pokorny, F.A. and M.E. Austin. 1982. Propagation of blueberry by softwood terminal cuttings in pine bark and peat media. HortScience 17:640-642.
98. Pokorny, F.A. and M.G. Dunavent. 1984. Seasonal influence and IBA concentration effects on rooting terminal cuttings of southern wax myrtle. Proc. Southern Nurs. Assoc. Res. Conf. 29:209-214.
99. Powell, C.C., Jr. 1977. Control of disease problems as it relates to plant propagation. Proc. Int. Plant Prop. Soc. 27:477-479.
100. Press, T.F. 1983. Propagation: Fog not mist. Proc. Int. Plant Prop. Soc. 33:100-109.
101. Robinson, E.H. 1967. Peat-perlite as a rooting medium. Proc. Int. Plant Prop. Soc. 17:363-364.
102. Schreiber, L.R. and M. Kawase. 1975. Rooting of cuttings from tops and stumps of American elm. HortScience 10:615.
103. Smalley, T.J. and M.A. Dirr. 1986. The overwinter survival problems of rooted cuttings. The Plant Propagator 22(3):10-14.
104. Smalley, T.J., M.A. Dirr, and G.G. Dull. 1986. The effect of extended photoperiod on budbreak, overwinter survival rates and tissue carbohydrate levels of *Acer rubrum* 'October Glory' rooted cuttings. J. Amer. Soc. Hort. Sci. 112: In press.
105. Stoutemyer, V.T. 1968. Root cuttings. The Plant Propagator 14(3):4-6.
106. Stoutemyer, V.T. 1969. Hardwood cuttings. The Plant Propagator 15(3):10-14.
107. Stroombeek, E. 1958. The propagation of softwood cuttings in the foghouse. Proc. Int. Plant Prop. Soc. 8:47-53.
108. Stroombeek, E. 1959. Hormone application by the quick dip method. Proc. Int. Plant Prop. Soc. 9:51-54.
109. Struve, D.K. 1980. The relationship between carbohydrates, nitrogen and rooting of stem cuttings. The Plant Propagator 27(2):6-7.
110. Templeton, H.M. 1953. The electronic leaf. Proc. Int. Plant Prop. Soc. 3:131-133.
111. Templeton, H.M. 1953. The phytotector method of rooting cuttings. Proc. Int. Plant Prop. Soc. 3:51-54.
112. Thiman, K.V. and J. Behnke-Rogers. 1950. The use of auxins in the rooting of woody cuttings. Harvard Forest, Petersham, MA.
113. Tilt, K. and T.E. Bilderback. 1984. Effects of physical properties of propagation media on the rooting response of woody ornamentals. Proc. Southern Nurs. Assoc. Res. Conf. 29:216-221.
114. Tukey, H.B., Jr. 1962. Leaching of metabolites from above ground plant parts, with special reference to cuttings used for propagation. Proc. Int. Plant Prop. Soc. 12:63-70.
115. Warner, Z.P. 1966. Fogging machines versus intermittent mist. Proc. Int. Plant Prop. Soc. 16:167-169.
116. Wedge, D. 1977. Propagation of hybrid lilacs. Proc. Int. Plant Prop. Soc. 22:432-436.
117. Welch, H.J. 1979. Manual of dwarf conifers. Theophrastus. Little Compton, R.I.
118. Wells, J.S. 1962. Wounding as a commercial practice. Proc. Int. Plant Prop. Soc. 12:47-55.
119. Wells, P. 1970. Overwintering of deciduous azalea cuttings. Proc. Int. Plant Prop. Soc. 20:366.
120. Whalley, D.N. 1979. Leyland cypress — rooting and early growth of selected clones. Proc. Int. Plant Prop. Soc. 29:190-202.
121. Whitcomb, C.E. 1982. Rooting cuttings under a wet tent. Proc. Int. Plant Prop. Soc. 32:450-451.
122. Whitcomb, C.E. 1984. Plant production in containers. Lacebark Publ., Stillwater, OK.
123. Whitcomb, C.E. 1984. Propagating trees from cuttings. Proc. Southern Nurs. Assoc. Res. Conf. 29:200-201.
124. Witte, W.T. 1984. Rooting summer cuttings of Chinese fringetree. Proc. Southern Nurs. Assoc. Res. Conf. 29:215.
125. Wott, J.W. and H.B. Tukey, Jr. 1967. Influence of nutrient mist on propagation of cuttings. Proc. Amer. Soc. Hort. Sci. 90:454-461.
126. Zimmerman, P.W. and A.E. Hitchcock. 1937. Comparative effectiveness of acids, esters, and salts as growth substances and methods of evaluating them. Contrib. Boyce Thompson Inst. 8:337-350.
127. Zimmerman, P.W. and F. Wilcoxon. 1935. Several chemical growth substances which cause initiation of roots and other responses in plants. Contrib. Boyce Thompson Inst. 7:209-229.
128. Zimmerman, R.H. 1958. Effects of liquid fertilizers on rooting cuttings. Proc. Int. Plant Prop. Soc. 8:162-164.

Grafting and Budding

A. GRAFTING

Grafting is one of the oldest known methods of plant propagation and dates back 2000 years or more (22). It is defined as the process of joining two plants or plant parts together in such a manner that they will unite and continue their growth as one. The main reason for grafting is to propagate plants that are difficult or impossible by other vegetative methods. Another important reason is to grow plants on roots other than their own for such purposes as size or disease control. It is not the intent of this chapter to discuss in detail the subject of grafting, but to cover the important methods. Several excellent books on grafting and budding include R.J. Garner, *The Grafter's Handbook*, Oxford University Press; H. T. Hartmann and D. E. Kester, *Plant Propagation: Principles and Practices*, Prentice-Hall, Inc..

A graft consists of two parts, the scion and understock. The scion consists of a short stem piece with two or more buds and is that part of the graft combination which develops into the top (shoot) of the plant. If the scion is reduced to a single bud with a thin slice of wood, the technique is called bud grafting or budding.

The understock, also referred to as rootstock or stock, is the lower part that becomes the root system. Understocks can be seedling or clonal in origin. Seedling understocks are more widely used, however, they are not genetically uniform. Asexually propagated clonal understocks provide genetic uniformity, disease resistance, growth modification and prevent incompatibility problems.

Although costly [Bassuk (1) reported a total cost of $0.80 per budded liner], grafting is still a valuable method of propagation. Major reasons for the continued use of grafting include (2): 1) Propagating a plant that is not easily or conveniently increased by other methods, 2) Obtaining benefits of a particular rootstock for reasons such as disease resistance, soil tolerance, or special growth forms, 3) Changing the cultivar of established plants, 4) Repairing damaged plants, and 5) Indexing for viruses.

B. LIMITATIONS OF GRAFTING

There are limitations to the successful grafting of two different plants. The botanical relationship between two plants is not a guarantee of success; long term observational experience provides the best guide. When two species cannot be successfully grafted together they are said to be incompatible. As a general rule, the closer two plants are taxonomically related to each other, the greater the chance of forming a successful union. From a commercial nursery point of view, grafting is limited to plants that have a continuous cambium. Among woody plants, no successful long term grafts are reported between different families, although there are short lived examples with herbaceous plants such as *Melilotus alba*, white sweet clover (Leguminosae), and *Helianthus annuus*, sunflower (Compositae) (14). In short, it is not possible to successfully graft an oak (Fagaceae) and maple (Aceraceae).

Grafting between genera within the same family is possible, however, the number of cases is limited. *Poncirus trifoliata* is used as a dwarfing understock for *Citrus* (orange); *Pyrus* (pear) is grafted on *Cydonia* (quince) as a dwarfing understock; *Hamamelis* is grafted to *Sycopsis* and *Distylium*; and a number of *Cotoneaster* species have been grafted to *Pyrus* (pear) and *Crataegus* (hawthorn).

Different species within the same genus are compatible in some cases but incompatible in others. As noted by Hartmann and Kester (6), grafting within *Citrus* is successful and often practiced. Similarly, *Prunus dulcis* (almond), *P. armeniaca* (apricot), *P. domestica* (European plum), and *P. salicina* (Japanese plum) are compatible on *P. persica* (peach). However, some cultivars of almond are incompatible on *P. cerasifera* × *P. munsoniana* 'Marianna', while others are not. The complexity of grafting between species within the same genus is further demonstrated by reciprocal grafts. 'Marianna' plum is compatible as a scion on peach, but the reverse graft, peach on 'Marianna' plum, is not.

Grafting between clones and seedlings of the same species for all practical purposes is successful, however, incompatibility problems do exist. *Pseudotsuga menziesii* (Douglas fir) clone, *Quercus palustris* 'Sovereign', and *Acer rubrum* cultivar graft failures have been reported. The incompatibility problem between *A. rubrum* cultivars and *A. rubrum* seedling rootstocks has led to the development of cutting propagation methods for those cultivars (See *Acer rubrum* in the ENCYCLOPEDIA).

C. INCOMPATIBILITY

Mahlstede and Haber (11) noted that incompatibility may be expressed in a number of ways including: 1) Combinations which never form a successful union, 2) Combinations in which only a small number of unions form, 3) Types in which the union is successful initially, but the plant eventually dies, 4) Combinations that produce deficiency symptoms or nutritional disorders, 5) Combinations that result in dwarf trees, 6) Types that produce differential growth at, or close to the union, 7) Combinations causing degeneration of tissue systems, abnormal distribution of stored food reserves, and premature defoliation.

Although the causes of graft incompatibility are not completely understood, genetic differences and diseases (virus and mycoplasma causal agents) are important factors. Because incompatibility problems can be related to virus and mycoplasma causal agents, understock and scion material free of known diseases should be used.

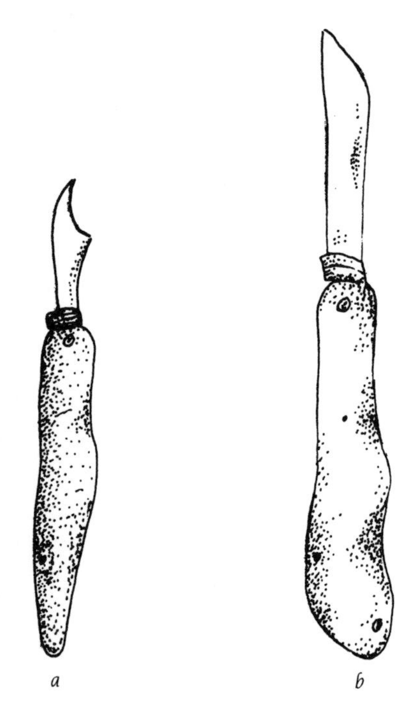

Figure 1. Budding (a) and grafting (b) knives. A budding knife has a curved cutting edge. A grafting knife has a straight edge.

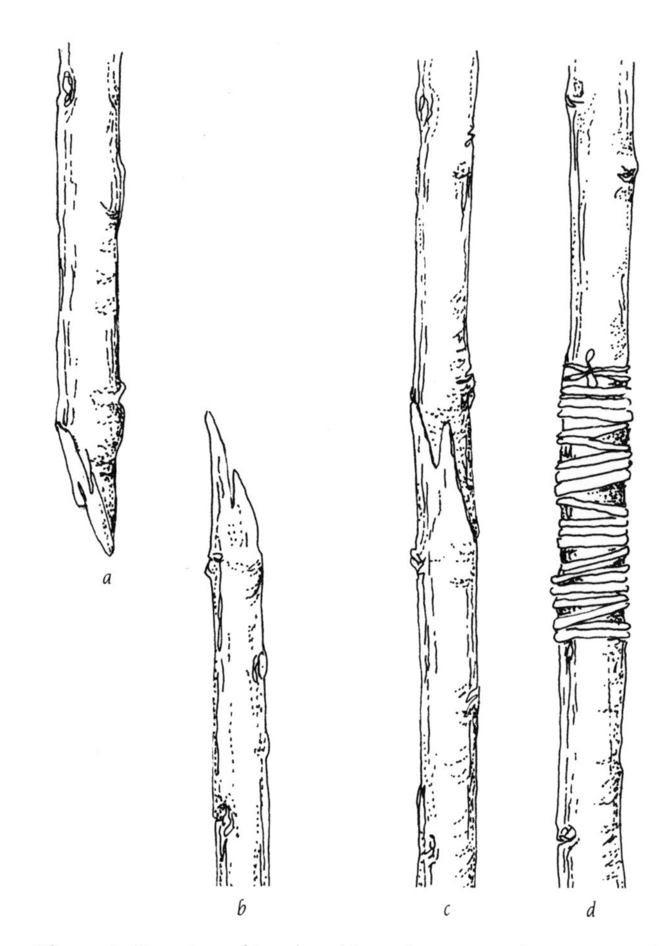

Figure 3. Steps in making the whip and tongue graft: a) prepared scion; b) prepared understock; c) matched graft; d) graft tied.

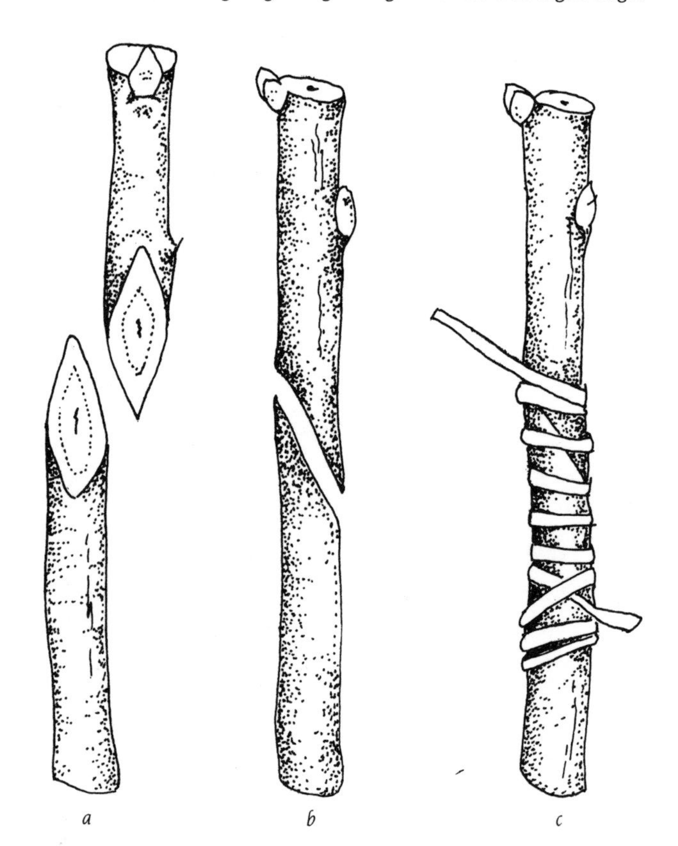

Figure 2. Steps in making the splice graft: a) single slanting cuts are made on the scion and understock; b) the parts are fitted to match cambium layers; and c) the graft is tied.

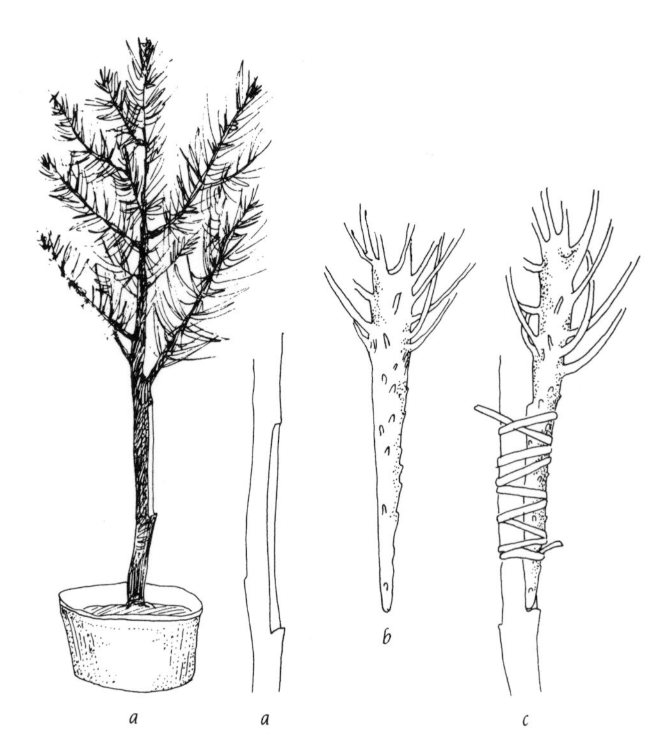

Figure 4. Steps in making the veneer graft with a narrow-leaf (needle) evergreen: a) position of the cut on the understock; b) the scion prepared for insertion; c) the completed graft.

D. TIME MEASUREMENTS

Time measurement studies on the grafting of different plants are limited. In a study by Gaggini (5), *Fagus, Prunus, Clematis, Picea, Hibiscus*, and *Cupressus* were examined in detail. The results are presented below; however, the times should be used only as a guide. *Clematis* cultivars - split leaf-bud root graft, tied with very thin raffia. Average time for grafting, 0.69 minutes. *Cupressus macrocarpa* 'Donard's Gold' - side veneer graft, tied with raffia, stocks cut back to 1′ high. Average time for grafting, 0.95 minutes. *Fagus sylvatica* 'Riversii' - side tongue graft, tied with raffia. Average time for grafting, 1.92 minutes. *Hibiscus syriacus* cultivars - veneer graft, tied with 2-ply fillis. Average time for grafting, 0.67 minutes. *Picea pungens* 'Glauca Pendula' - side veneer graft, tied with raffia, removal of needles from base of scion. Average time for grafting, 1.41 minutes. *Prunus* × *hillieri* 'Spire' - Whip and tongue, root wiped with rag, raffia-tied, graft painted with cold wax. Average time for grafting, 2.12 minutes. *Prunus mume* 'Beni-shi-don' - whip and tongue, on washed roots, tied with mutscene twine. Average time for grafting, 1.10 minutes.

E. PHYSIOLOGY OF GRAFTING

Aeration, temperature, humidity, inherent capacity to form callus et. al., affect the success of the graft. For example in a developmental study with different ploidy levels in *Juniperus* (4), the sequence of events was similar at all ploidy levels, but the timing of each stage of development was markedly different. When 'Fountain' (2n=22) was a component of the graft, callus xylem bridging occurred early, whereas with 'Pfitzeriana Kallay' (4n=44) new tissue formation was delayed. Temperature, aeration and moisture are primarily responsible for success or failure in most grafted plants.

Temperature has the strongest effect. Warm temperatures promote callus formation; cool temperatures inhibit or reduce it, while extremely high temperatures are detrimental. In a comparison of temperature effects on apple (16) and walnut (17), apple was found to be less temperature sensitive. Apple grafts showed little callus growth below 50°F or above 95°F. In contrast, little callus growth occurred below 70°F or above 100°F with walnut. Callus growth increased linearly between 50 and 75°F for apple and 70 and 85°F for walnut. Walnut, therefore, has a narrower temperature range for success. High humidity is essential for callus formation and that is the reason grafts are waxed and/or tied, and often placed in high humidity chambers.

F. TOOLS AND ACCESSORIES

1. Knives

Of the tools employed for grafting, knives are the most important. A large number of excellent knives are available, however, the success of the grafting operation depends more on the sharpness of the knife than on the type. The best understock, scion wood and equipment cannot compensate for dull knives that produce ragged wounds. For propagation work, the two general types of knives used are the budding knife and grafting knife (Fig. 1). The knives have either a fixed or folding blade, with the fixed blade stronger and longer lasting. A grafting knife should be of good quality and have a straight-edged blade flat on one side and sloped on the other. The purpose of sharpening the blade on one side only is to provide a flat backing on the blade so that it will make a flat cut into the plant. With the budding knife, a cutting edge that curves away to the tip is usually preferred because it makes it easier to separate the bark flaps during T-budding. Also, a sharp pointed knife will tend to enter the wood of the rootstock. Table 1 presents a list of supply houses that offer grafting and budding supplies.

2. Tying and wrapping materials

Tying and wrapping materials hold the scion and stock closely together and prevent the callus from forcing the pieces apart. Control of moisture loss also occurs with some materials, such as parafilm (American Can Co.). Almost any tying material can be used to hold the scion and understock until union formation occurs, but some are better than others. Widely used materials include: adhesive tape (like surgical tape), parafilm, plastic, polyethylene tapes, raffia, rubber budding strips, and twine (waxed or nonwaxed).

G. GRAFTING METHODS

Numerous grafting methods exist, however, only a few are of major importance to nurserymen. Techniques such as approach grafting, inarching, and bridge grafting are seldom used in propagation nurseries. When the terms bench or pot grafting appear in the ENCYCLOPEDIA, they refer to indoor (winter) grafting activity using potted (established) and, in certain cases, bareroot understock.

1. Splice graft

This is one of the simplest grafts. Long slanting cuts are made on the scion and rootstock (Fig. 2a). The cut surfaces are joined together so that the cambial zones are in contact and tied (Fig. 2b and c). This graft is limited to plants that heal rapidly because the method of joining adds no strength to the union except for the tying material. This method is generally limited to indoor bench grafting.

2. Whip-and-tongue graft

The whip-and-tongue graft, which is a modified splice graft, does not suffer from the splice graft limitations. The whip-and-tongue graft (Fig. 3) is widely used for joining together comparatively small scion and rootstock parts, usually not more than one inch in diameter. Best results are obtained when the stock and scion are the same size. The graft is easy to make because the interlocking edges form a strong union before tying. The initial long slanting cuts, about 1 to 1½″, are made like the splice graft. On each of the cut surfaces a second cut or tongue is made (Fig. 3a and b). It is started downward at a point about one half the distance between the pith and the tip of the outer edge of the bark. The second cut is one half the length of the initial slanting cut. The scion and rootstock are then inserted into each other (Fig. 3c), with the tongue interlocking and the cambial zones in contact, and tied (Fig. 3d). If the union is to be waxed, grafting twine often is used.

3. Veneer and side grafts

The veneer (Fig. 4) and side graft are similar and widely used to propagate shrubs and conifers. The understock stem is cleared of leaves in the region of the graft and preferably cut first. This sequence eliminates laying the scion down and possible contamination. A cut about 1¼″ long is made on a straight portion of the stock that is free of side branches, and as close to the soil as possible (Fig. 4a). The cut is made through the bark and slightly into the wood. At the bottom of the first cut a second cut is made downward and inward about 3/16 to 1/4″ and through the flap. This will leave a short lip of bark and wood at the base. The scion is prepared by making a cut of equal

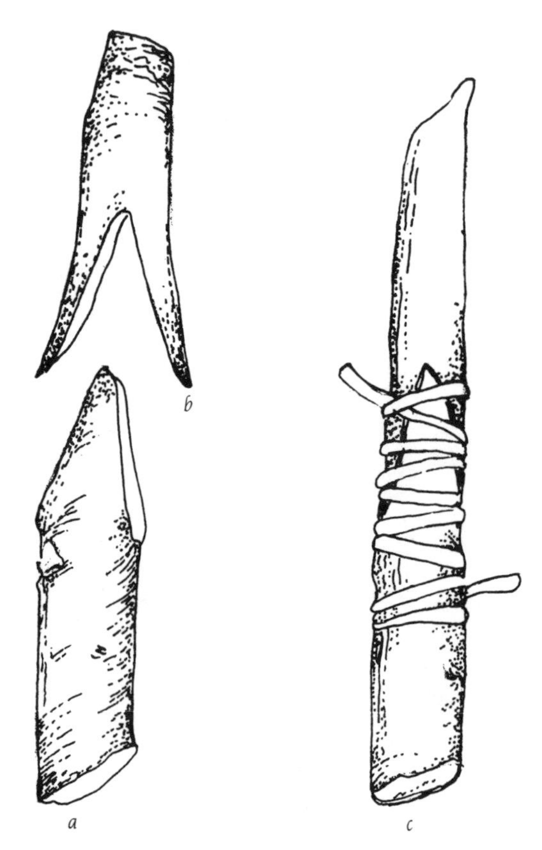

Figure 5. Steps in making the saddle graft: a) two slanting cuts on opposite sides of the understock to form a blunt point; b) a section at the base of the scion is removed to match the understock; c) the graft is tied after the cambium layers are matched.

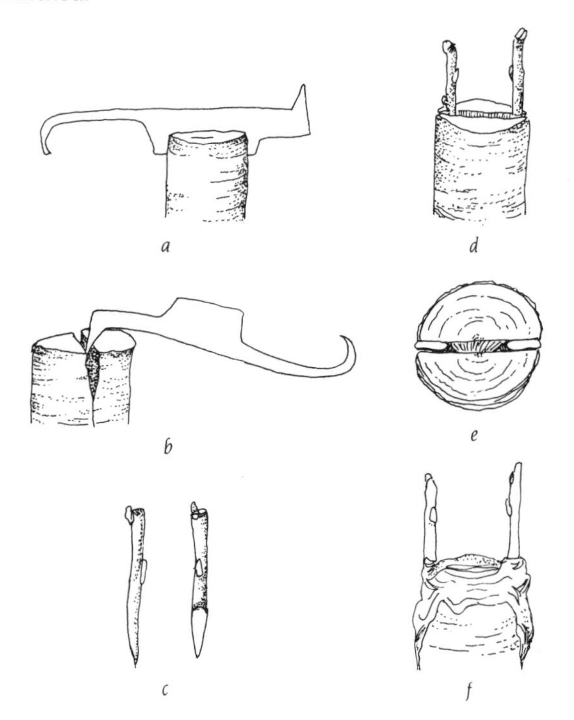

Figure 6. Steps in making the cleft graft: a) using the cleft iron to make the cut; b) spreading the cut to insert the scions; c) the basal slant cuts on the scions; d) placement of scions in cleft; e) cambium region of scion in contact with cambium zone of cleft graft; f) graft waxed.

Figure 7. Steps in making the piece root graft using a whip graft: a) prepared root piece; b) prepared scion; c) joined parts; d) tied graft.

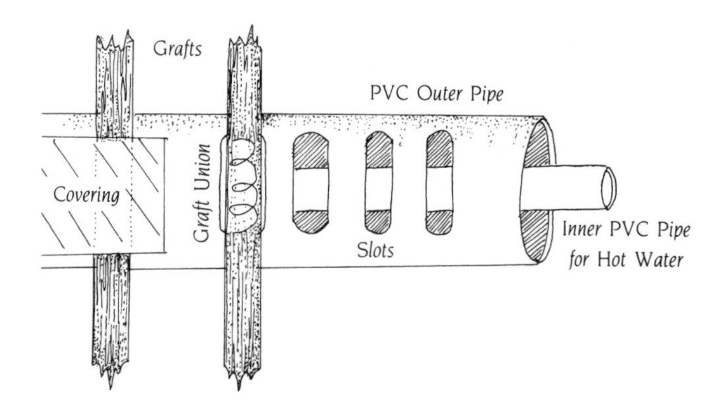

Figure 8. A cross section of the hot-callusing device with a graft union in place.

length from top to bottom on the straightest side and deep enough to expose the wood. A second cut is made across the base from top to bottom and slanting downward at the same angle as the cut lip on the understock (Fig. 4b). The scion and understock are joined, the cambium layers matched and tied (Fig. 4c). If the stock and scion are not the same size then the cambium layers should be matched along at least one side (20).

The side graft is similar to the veneer, except that the second cut is not made to the understock, thus leaving a long flap. The scion is given three cuts, with the first about 2″ long, the second not quite as long on the opposite side and the third slanting across the base. The scion is then fitted to the understock with the longer side against the stock, the flap brought up and against the shorter cut side and tied.

Wells (21) lists plants commonly propagated by these two methods. Plants listed as being grafted with the veneer type include: *Abies* taxa, *Cedrus atlantica*, *Clematis*, *Cupressus* taxa, and *Rhododendron indica*. The side graft is mentioned as being used with *Acer palmatum* and *A. japonicum*, *Camellia* taxa, *Chamaecyparis* taxa, *Cornus florida*, *Fagus* taxa, *Ginkgo*, *Hamamelis mollis*, *Ilex opaca*, *Magnolia grandiflora*, *Rhododendron* taxa, and *Viburnum carlesii*. Plants using either type include: *Cryptomeria japonica*, *Juniperus* taxa, *Picea pungens*, and *Thuja* taxa.

4. Saddle graft

The saddle graft (Fig. 5) is not used as often as some other types because it is more time consuming to make. However, it is valuable in certain situations for rhododendron grafting and plants with fairly large stems. The scion and rootstock should be the same size. The top of the rootstock is removed close to the soil (1 to 1½″) and trimmed on two opposite sides to form a wedge (Fig. 5a). The base of the scion is cut to remove a wedge (Fig. 5b). Be sure the bark at the base of the scion does not separate from the wood or the scion will not take. Rootstock and scion are joined together and securely tied (Fig. 5c). All cut surfaces are waxed if the grafts are not placed in a grafting case.

5. Cleft graft

The cleft graft (Fig. 6) is often used for field grafting, especially when the stock is larger than the scion. Branches or tree trunks from 1 to 3″ in diameter are best. Preferred grafting time is late winter or early spring when the buds of the rootstock are beginning to swell but not actively growing. The understock is cut off at a right angle to the main axis where the union is desired. A cleft (split) is then made in the end of the stub with special cleft grafting tool (Fig. 6a) or large knife, such as a butcher knife. The cleft is held open by the wedge-shaped prong on the cleft grafting tool or by the use of a wooden or metal wedge (Fig. 6b). The scion is made from the previous season's growth and includes 2 to 3 buds (Fig. 6c). The lower end is cut to a wedge shape about 1 to 1¼″ long with the outer edge slightly thicker for better cambium contact. Larger stock (tree trunk) receives 2 scions while smaller stock receives one scion (Fig. 6d and e). If both scions grow, one is removed after the first year. Generally the union is not tied, the pressure of the understock stub being strong enough to hold the scion in close contact. After setting the scion, all the cut surfaces including the tip of the scion are covered with grafting wax (Fig. 6f).

6. Cutting grafts

A cutting graft (also called twig graft) is made by grafting a leafy scion to an unrooted leafy cutting. This technique, although not widely practiced, has been utilized with citrus and rose. The method presented is that used with citrus (3). Scion and rootstock cuttings are taken from spring growth that has hardened off. The scion and rootstock are splice grafted. A veneer graft is also used with this technique. After tying the graft is handled as a cutting. The cuttings are treated with 8000 ppm IBA, inserted into the rooting medium, and placed in a fog chamber that maintains high humidity. High humidity is required because the scion has no contact with the rooting medium. Bottom heat (70 to 72°F) is supplied to speed rooting. After rooting and union healing the grafts are hardened off.

7. Double working

A double worked graft contains an intermediate piece of stem between the scion and rootstock. Double shield budding is similar, with the intermediate piece reduced to a thin sliver of wood and bark. The major reasons for double working are to produce dwarf trees and overcome incompatibility problems. Several different methods are possible to produce double worked grafts (6). 1) All three pieces — the scion, intermediate piece and rootstock — can be bench grafted at the same time. After union formation the grafts are field planted. 2) Rootstocks can be spring planted and then fall budded with the intermediate stock. The intermediate stock is grown for one year and then fall budded with the desired scion cultivar. 3) The intermediate stock can be bench grafted in the winter to the rootstock and, after callusing, field planted. These are then fall budded with the desired cultivar. 4) The desired scion can be bench grafted to the intermediate piece in winter and, after callus formation, field grafted to the understock.

8. Root grafting

In root grafting, the roots of seedling or clonal rootstocks serve as understock (Fig. 7). If the whole root is used it is termed *whole root grafting* and if a piece is used it is *piece root grafting*. A whip-and-tongue graft is the method of choice and grafting is carried out from December to March. The initial long slanting cuts for the scion and understock (about 1 to 1½″ long) are made like the splice graft. On each of the cut surfaces a second cut or tongue are made (Fig. 7a and b). The cut starts downward at a point about one half the distance between the center of the stem or root and the tip of the outer edge of the bark. The second cut is 1/2 the length of the initial slanting cut. The scion and rootstock are then joined (Fig. 6c) with the tongues interlocking and the cambial zones in contact, and tied (Fig. 7d). After tying, the grafts are stored at about 40°F for callus development. When weather conditions permit, the grafts are planted out.

9. Bare root grafting

This type of grafting is carried out in the winter with fall dug understock that is cold stored until used. Plants such as *Malus* (apple), *Cornus* (dogwood), *Magnolia* (magnolia), and *Acer* (maple) can be grafted by this technique. After tying, the grafts are dipped in wax and boxed in alternating layers of moist material such as peat moss. After callusing, the grafts can be stored at close to freezing until planted out.

H. CARE OF GRAFTS

1. Closed case

For graft healing, especially with leafy cuttings, the grafts are placed under double glass,a closed case or polytent in a heated greenhouse. Moist peat is placed in the base of the case to provide humidity. Under high humidity conditions, drying of the

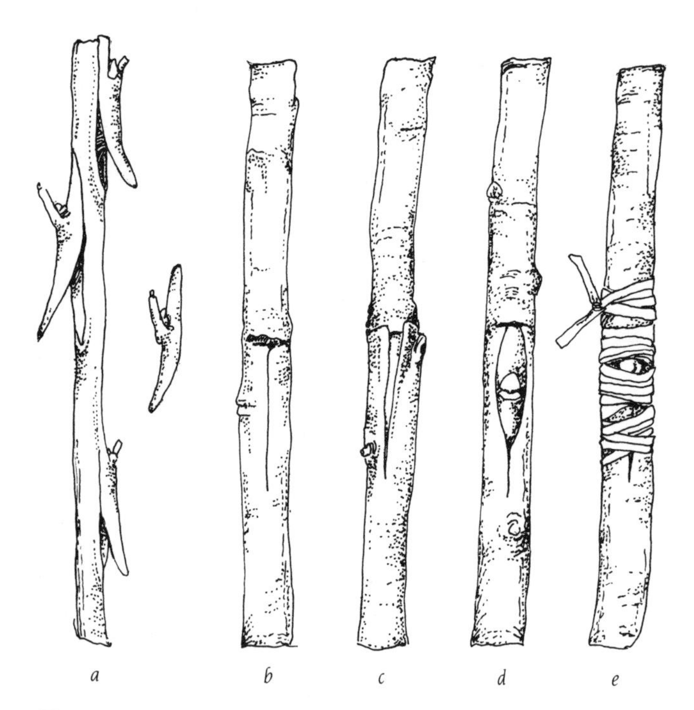

Figure 9. Steps in making the shield bud graft: a) bud stick and shield; b) T-shaped cut through the bark of the rootstock; c) bark raised to admit the shield; d) bud inserted; e) bud tied.

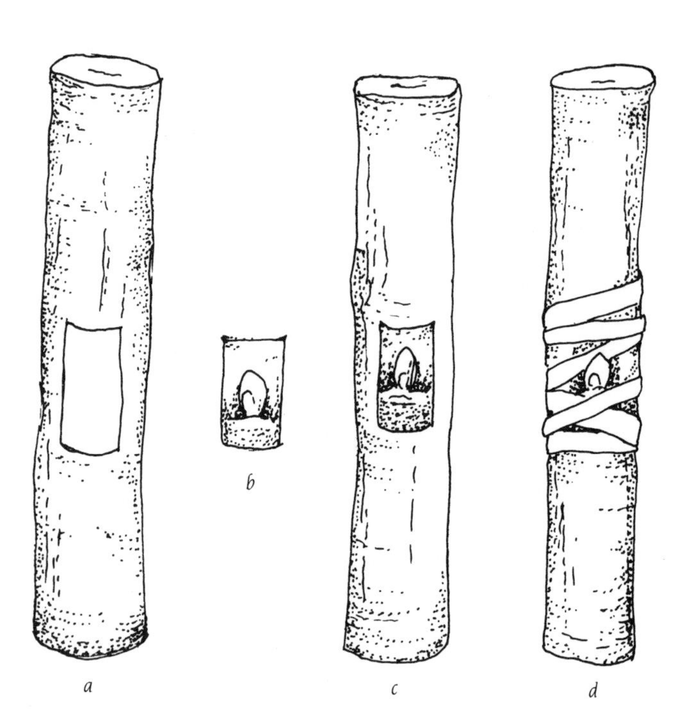

Figure 11. Steps in making the patch bud: a) the position of the removed patch on the understock; b) the scion patch bud; c) the completed graft before tying; d) the union is wrapped so that all cuts are covered.

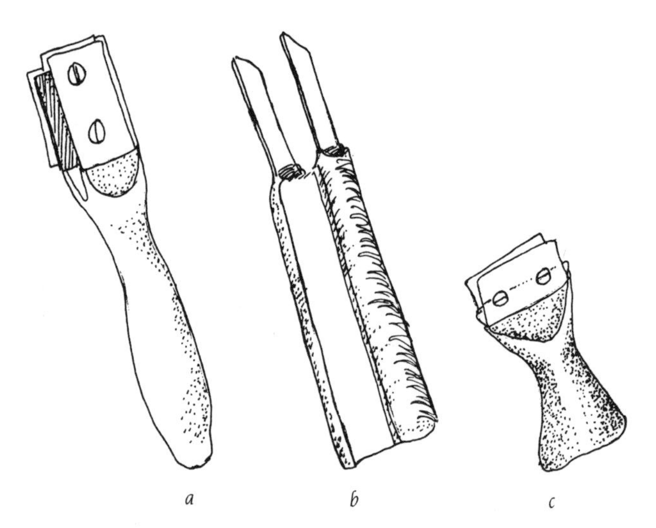

Figure 10. Patch budding tools: a) Double-bladed knife made with razor blades; b) double-bladed knife; c) tool consisting of four rectangular blades.

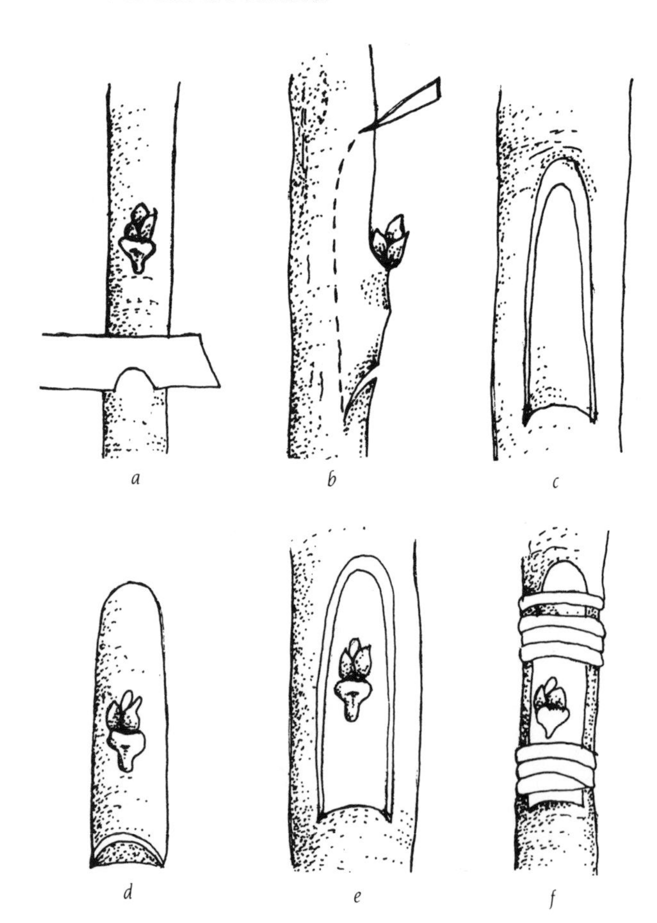

Figure 12. Steps in making the chip bud: a) the first cut is at an angle of about 20° to form a lip; b) the second one is made 1½" above the first; c) appearance of understock ready for budding; d) bud ready for insertion; e) bud inserted; f) bud wrapped.

union is not a problem and the unions are usually just tied. The grafts are not watered until extensive callus formation is visible; at this time airing becomes necessary. Heavy shading should be used to lower temperatures and prevent scion stress. Once the union is established, the grafts are gradually hardened-off (8).

2. Open bench

Completed grafts can be placed on an open greenhouse bench with pots plunged in peat, perlite, sand or any combination at a depth sufficient to cover the union. The purpose of burying the union is to prevent drying out. The time period will depend on the plant type and temperature. Initially the medium temperature should be kept between 65 and 75°F for 4 to 6 weeks. A heat source under the bench allows the maintenance of a cooler air temperature (50 to 60°F). The tops are supplied with adequate moisture by syringing, shading, or covering with polyethylene. When polyethylene is used, the covering should be lifted daily to dry the grafts for disease control.

After callusing is evident, the grafts are ready for hardening-off. The understock is pruned back about half-way, and the grafts are placed on the medium for an additional 4 to 6 weeks. At the end of this period the remaining top is removed (20).

3. Poly bag chamber

This technique is useful when space is limited or for grafting specialized forms such as standards. After the graft is made and tied, a ball of wet sphagnum moss, the size of a lemon, is tied to the rootstock one inch below the union. A plastic bag is then inflated and put over the scion and tied with a rubber strip below the sphagnum moss. Grafts are then placed in a greenhouse. No other care, except possibly watering the containerized standards, is needed (15).

4. Hot-callus pipe

The hot-callus pipe was first introduced by Lagerstedt (9) for grafting difficult plants. The method presented is a modification by Strametz (19). Strametz constructed the first large scale hot callusing pipe with 1200′ of 2″ PVC pipe that contained 9600, 1/2 and 5/8″ slots cut perpendicular to the length of pipe (Fig. 8). Hot water is circulated through a ½″ PVC pipe inside the 2″ slotted pipe. A 40-gallon hot water heater, in concert with a 1/4 H.P., 1725 RPM electric circulating pump, and a 20-gal. expansion tank placed 6′ above the pump are used. This system performs best between 20 to 30 psi at the pump which is placed on the cool side of the water heater. An exit temperature of 81 to 82°F is maintained. The top of the slots are first covered with 1/8″ closed cell foam and then a layer of 6 mil black poly. A slit is cut in the center of each slot in the pipe. Moist sawdust between the pipes is used to cover bare roots. The best location for a hot-callus sytem is where temperatures can be maintained cold to prevent premature bud break on the scion while the unions are healing. The system has worked well with *Acer*, *Cedrus*, *Cercidiphyllum*, *Corylus*, *Fagus*, *Malus*, *Prunus* and *Sequoia*. Spruce union formation is not favored by the high temperature.

I. BUDDING METHODS

Budding or bud grafting is a type of grafting in which the scion is reduced to a single bud with a thin shield of bark and wood. Budwood should be true to type and free of known virus or other diseases. Budding is a popular method of propagation because it is fast and makes maximum use of scion wood. It is widely used with fruit trees and many ornamental plants such as *Fraxinus* (ash), *Malus* (crabapple), *Gleditsia* (honeylocust), *Tilia* (linden), *Acer* (maple) and *Rosa* (rose). The bud is inserted into the understock and later the understock above the inserted bud is removed and a new top develops. T-budding, the most popular budding method, is limited to the time of the year when the bark is slipping, however, chip budding can be used when the bark is not slipping.

1. Collection and storage of budsticks

Although propagators generally collect only enough budsticks for a particular day's needs, the collection and storage of budsticks can bring about increased operating flexibility including: 1) Earlier budding which extends the growing season of the budded plants, 2) Increased production from the same number of budders, 3) Budding wood can be cut in its prime condition, 4) The need for stock blocks can be reduced.

The method presented is used for rose budsticks and also serves as a model for other plants (10). Ideally, budsticks for overwinter storage are collected during late September and early October from mature stems of medium caliper. The top of the stem is removed above the first 5-leaflet leaf and all leaves are stripped. Stems are bundled in groups of 50 and placed in polyethylene lined butcher's paper, then in wet newspaper, then a polyethylene bag and sealed. *Botrytis* mold will not be a problem if as much air as possible is removed from the bag, the wood is dry, and no leaf material adheres to the budsticks.

Successful storage depends on keeping the temperature at 28 to 31°F. In the spring the packages are removed from storage a day before use and held at 40°F until needed by the budders. This technique has worked satisfactorily for *Malus* (apple and crabapple), *Pyrus* (pear) and *Prunus* (plum).

Fraxinus (ash), *Catalpa* (catalpa), *Ulmus* (elm), *Gleditsia* (honeylocust) and *Tilia* (linden) have been handled differently. The budsticks are placed in plastic bags with a small amount of moist sphagnum moss and held at 34 to 36°F (7).

2. Time of budding

Budding, which usually refers to T-budding, is accomplished when the bark readily separates from the wood and the new buds on the young shoots are mature. This means that budding can extend from June to September.

"June budding" refers to a budded deciduous tree that is grown in a single season (13). This results from early budding and rapid forcing techniques. June budding is commonly used with the *Prunus* species such as almond, apricot, nectarine, peach and plum. Budding is accomplished as early as the scion and rootstock will permit. Roostocks are fall or early spring seeded and are ready for budding in California as early as May 15. The budsticks are collected from current season's wood. Most nurserymen use T-budding. The bud is knitted to the understock in about 4 days and the understock is headed back by half at that time. Ten days later the remaining understock is removed and the grafting rubber strip is cut away if still present. Within 2 to 3 weeks the bud has grown 8 to 12″ in height. During this growth period one to several suckers will have grown and they must be nipped back but not removed because the leaf area on the suckers is important for success. After scion growth has reach 12″ or more, the suckers are completely removed.

3. T-budding

T-budding or shield budding is the most common method of budding. The T-name comes from the shape of the incision while shield is derived from the shape of the scion. A special sharp

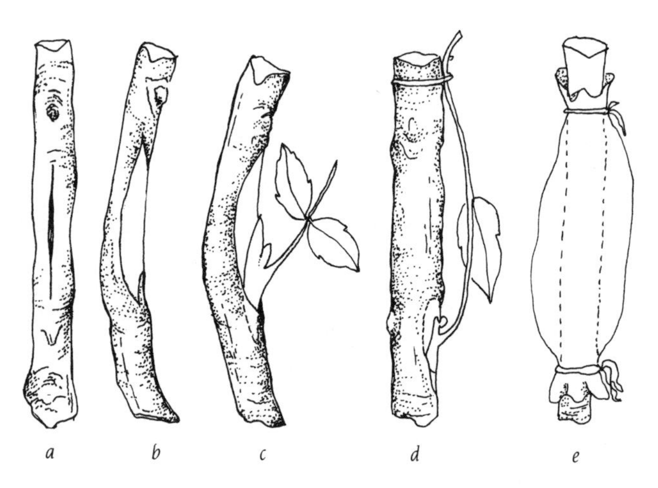

Figure 13. Steps in making the change-purse bud: a) single cut 3 to 5″ long; b) stem is flexed and flaps are opened; c) bud shield is pushed downward until covered by flaps; d) bud shield is pushed upward, the stock is released, the bud is tied in, and leaf tied to stem; e) entire bud union and leaf covered with plastic film and tied at top and bottom.

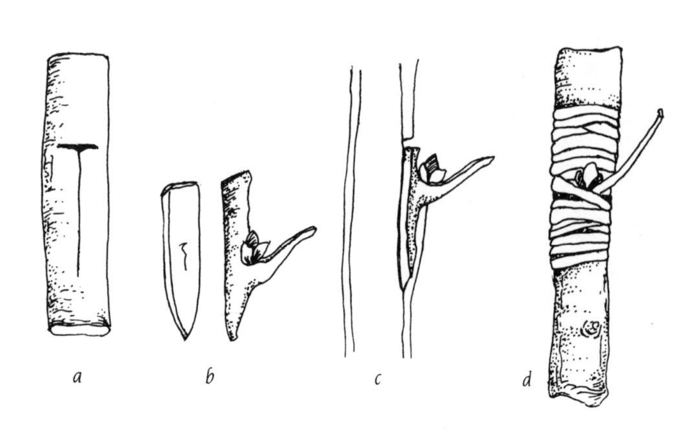

Figure 14. Steps in making the double shield bud: a) normal T cut in understock; b) intershield and shield bud; c) intershield and shield bud inserted into understock; d) bud tied.

Table 1. A partial list of vendors supplying grafting and budding supplies.

American Can Co.
Greenwich, CT 06830

A.H. Hummert Seed Co.
2746 Chouteau Avenue
St. Louis, MO 63103

A.M. Leonard, Inc.
6665 Spiker Road
Piqua, OH 45356

Ben Meadows Co.
3589 Broad St.
P.O. Box 80549
Atlanta, GA 30366

Brighton By-Products Co., Inc.
P.O. Box 23
New Brighton, PA 15066

D&L Growers Suppliers, Inc.
546B East 28th
Division Highway
Lititz, PA 17543

E.C. Geiger
Box 285, Rt. 63
Harleysville, PA 19438-0332

Forestry Supplies, Inc.
205 W. Rankin St.
Jackson, MS 39204-9987

Geo. K. Groff, Inc.
224 Maple Ave.
Bird-in-Hand, PA 17505

Good-Prod Sales, Inc.
825 Fairfield Ave.
Kenilworth, NJ 07033

MacKenzie Nursery Supply, Inc.
P.O. Box 372
3891 Shepard Rd.
Perry, OH 44081

Penn State Seed Co.
906 Wyoming Ave.
Forty Fort, PA 18704 budding knife is essential. Generally, it is useless to bud if the bark is so tight that it must be forcibly lifted. T-budding is limited to understocks that are pencil thick to 1″ in diameter. Larger stock often has bark that is too thick.

The bud is removed from the bud stick in the form of a shield by making a thin cut starting about ½″ below the bud and continuing under the bud to about ½″ above the bud (Fig. 9a). A second horizontal cut is made above the bud to remove the bud. The knife should be allowed to penetrate the wood slightly. There are two methods of preparing the bud — with the wood piece in or out. Many propagators remove the sliver of wood from the bud shield (wooding the bud), although leaving the wood in rarely decreases success. If the wood is left in the shield, the shield is cut thinner; thicker if it will be removed. The sliver of wood is removed by sliding it sideways so as not to remove the small piece of wood, called the bud trace, from the bud. Removal of the bud trace reduces the chance of success.

A T-shaped cut is made through the other tissues down to the wood, but no deeper, by first making a vertical 1″ cut and then a transverse cut at the top to form the T-shaped incision (Fig. 9b). The incision can be made on any smooth side of the understock, however, some propagators place it on the windward side or on the north side. The cross cut is made with a rolling movement of the knife which lifts the corners where the two cuts cross (Fig. 9c). If the bark does not lift easily to admit the bud as described above, it is raised with the point of the blade or the flat end of a special budding knife handle.

The shield is next inserted by pushing it under the flaps of the bark of the understock (Fig. 9d). The shield piece is well covered by the bark flaps with the bud exposed and the union is tied together until the healing is complete (Fig. 9e). Almost any tying material can serve to hold the bud in place, however, plastic strips, raffia, rubber budding strips, rubber patches and waxed cotton or twine are most often used.

4. Inverted T-budding

The inverted T-bud is used in localities having particularly wet conditions during budding. Water that runs down the stem and into the incision can prevent healing of the union. The inverted T technique prevents water from entering the incision. The inverted T is widely used with citrus. The technique is the same as for T-budding except that the horizontal cut is made at the bottom of the vertical cut and the shield cut is started above the bud.

5. Patch budding

This method is used for grafting thick barked plants, such as *Juglans* (walnut) and *Carya* (pecan). The bark must slip on the rootstock and scion. A distinguishing characteristic of this method is the presence of a rectangular bud patch that is placed in the same sized rectangular patch opening on the understock. Special patch budding knives have been developed (Fig. 10).

Scionwood and understock bark of similar size are best. The patch is first cut into the understock by initially making the two horizontal cuts and then the vertical cuts (Fig. 11a). The patch is removed from the understock and then the same size patch is cut in the scion wood to obtain the bud (Fig. 11b). After placing into the open patch of the understock (Fig. 11c), the bud is tied (Fig. 11d). If conditions favor desiccation, the bud should be waxed. The ties are removed after callus formation.

6. Chip budding

Chip budding appears to have greatest relevance as a replacement for conventional T-budding when the period of cambial activity is limited and the bark is not slipping. Although in theory chip budding can be conducted any time of the year, successful budding is confined to the period when the stock is growing actively.

The rootstock is pruned to provide a clean stem. The first cut is made about 1/8″ deep at an angle of about 20° into the stem to make an acute lip (Fig. 12a). A second cut is made 1½″ above the first, entering the stem and cutting down to meet the first cut (Fig. 12b). The second cut should be U-shaped and not A-shaped (Fig. 12c).

The bud chip is prepared by similar cuts (Fig. 12d) and placed into the prepared cut (Fig. 12e). The bud chip should be sized so that the cambiums of the stock and scion match. An essential requirement is that the union be completely covered with a polyethylene film (Fig. 12f), such as parafilm (American Can Co.), to protect the chip during the healing process. This is essential because unlike the T-bud the shield is not protected by the bark flaps. The length of time the wrap is left on will depend on the growing conditions.

7. Change-purse budding

Change-purse budding (Fig. 13) involves a single cut on the understock that is opened like a change-purse (18). The technique is used with *Juglans nigra* (black walnut), *J. regia* (English walnut), *J. cinerea* (butternut), and *Carya illinoensis* (pecan). In Kentucky, the technique is used from June 1 to July 20 and is governed by the availability of buds in the proper condition. The best shield bud for change-purse budding has light colored pith and a large plump bud. As growth of the new shoots proceeds, the color of the pith changes to a dark color in nut trees. Once the pith has become dark, the buds should not be used. Reduce the leaf over the potential shield to 4 to 6 leaflets and leave a ½ to ¾″ piece of rachis above the upper pair of leaflets. The shield is cut to 1¼ to 2″ long. While holding the shield bud by the petiole, remove a thin strip of tissue along both sides of the upper half of the shield to increase cambial contact.

The understock cut is made on the south or southwest side of the understock. A single 3 to 5″ vertical cut is made on a smooth area of the stem (Fig. 13a). The cut is only deep enough to penetrate the bark because wood injury can cause bleeding that may lead to failure. After the cut is complete, the stem is slightly flexed to open the bark flaps which can be separated with a thumbnail (Fig. 13b).

To insert the bud, bend the stock to force open the slit, push the bud beneath the flaps and then push the shield upward until it meets resistance (Fig. 13c). Release the stock and the flaps will pull tightly against the shield. The shield is tied with an 8 × 3/8″ budding strip and then the leaf is pushed up and lightly tied to the understock (Fig. 13d). A piece of clear plastic about 10 × 12″ is used to enclose the bud until healed (Fig. 13e). About 1/3 of the stock top is removed at this time.

After 7 to 10 days the buds are examined and any with yellow or abscised leaflets are opened and the leaflets removed and retied. Between 14 and 20 days, the plastic is opened at the base and after 20 to 30 days the plastic wrap is removed. The stock at this time is cut back to 4 to 6″ above the bud and when the bud has made 4 to 8″ of growth the stub is cut off.

8. Double shield budding

This technique (Fig. 14) was developed to overcome graft incompatibility between pear and quince combinations. A normal T-cut is made in the understock (Fig. 14a). Two bud sticks are required. One to produce the scion, and the other (intershield) a type compatible with both the scion and understock. A bud shield is cut from the intermediate and discarded. A second shield, approximately 1/16″ in thickness, is cut from the location of the first shield to form the intershield (Fig. 14b). The shield from the desired scion wood is removed, the cambiums matched with the intershield and the two segments inserted into the T incision on the understock (Fig. 14c). Tying completes the operation (Fig. 14d).

9. Stages and time intervals in bud healing

Detailed studies on the healing stages and time intervals have been made with a number of different plants. The developmental sequence of events is similar, however, the timing will vary depending on plant type, temperature, and time of the year. Stages and times for the healing of fall budded citrus serve as examples (12):

Stage of development	Approximate time after budding
First cell division	24 hours
First callus bridges	5 days
Differentiation:	
a. in the callus of the bark flaps	10 days
b. in the callus of the shield	15 days
First occurrence of meristematic layers in the callus between shield and bark flaps	15 days
First occurrence of vessels:	
a. in the callus of the bark flaps	15 days
b. in the callus of the shield	20 days
Lignification of the callus completed:	
a. in the bark flaps	25-30 days
b. under the shield	30-45 days

Grafting and budding play a significant role in the propagation of woody plants. However, over the past 10 to 15 years, their importance has decreased, primarily because of the continued development of cutting propagation techniques and the astronomical advances in tissue culture. The extra manipulations necessary to graft or bud a plant raise the cost to levels even grater than that for tissue culture. For certain plants grafting and budding are the only reliable methods of propagation. The grower should consider all factors before proceeding full speed with any *single* vegetative propagation method.

LITERATURE CITED

1. Bassuk, N., B. Maynard, and J. Creedon. 1986. Stock plant etiolation and banding for softwood cutting propagation — working toward commercial application. Proc. Int. Plant Prop. Soc. 36: In press.
2. Browse, P.M. 1977. Why graft? Proc. Int. Plant Prop. Soc. 27:56-57.
3. Dillion, D.F. 1981. Propagating dwarf citrus with hydronic radiant heated benches. Proc. Int. Plant Prop. Soc. 31:527-532.
4. Evans, G.E. and H.P. Rasmussen. 1972. Anatomical changes in developing graft unions of *Juniperus*. J. Amer. Soc. Hort. Sci. 97:228.
5. Gaggini, J.B. 1971. A time measurement study: bench grafting of woody plants under glass. Proc. Int. Plant Prop. Soc. 21:275-292.
6. Hartmann, H.T. and D. Kester. 1983. Plant propagation: Principles and practices. Prentice-Hall, Inc., Englewood Cliffs, N.J.
7. Holmes. K.D. 1966. Storage of rooted cuttings, unrooted cuttings, scions and budwood. Proc. Int. Plant Prop. Soc. 16:251-254.
8. Humphrey, B. 1978. Propagating by grafting under glass at Hilliers Nursery. Proc. Int. Plant Prop. Soc. 28:482-490.
9. Lagerstedt, H.B. 1981. A device for hot callusing graft unions of fruit and nut trees. Proc. Int. Plant Prop. Soc. 31:151-159.
10. Mackay, I. 1962. The collection, storage and use of budwood. Proc. Int. Plant Prop. Soc. 12:142-144.
11. Mahlstede, J.P. and E.S. Haber. 1957. Plant propagation. John Wiley & Sons, Inc. New York.
12. Mendel, K. 1936. The anatomy and histology of the bud union in citrus. Palest. Jour. Bot. 1:13-56.
13. Mertz, W. 1964. Deciduous June bud fruit trees. Proc. Int. Plant Prop. Soc. 14:255-259.
14. Nickell, L.G. 1946. Heteroplastic grafts. Science. 108:389.
15. Savella, L. 1983. Saddle grafting. Proc. Int. Plant Prop. Soc. 33:425-426.
16. Shippy, W.B. 1930. Influence of environment on the callusing of apple cuttings and grafts. Amer. J. Bot. 17:290-327.
17. Sitton, B.G. 1931. Vegetative propagation of black walnut. Mich. Agr. Expt. Sta. Tech. Bull. No. 119.
18. Stoltz, L. 1982. "Change purse" budding of nut trees. Proc. Int. Plant Prop. Soc. 32:616-619.
19. Strametz, J.R. 1983. Hot-callus grafting of filbert trees. Proc. Int. Plant Prop. Soc. 33:52-54.
20. Vermeulen, J.P. 1983. Side veneer grafting. Proc. Int. Plant Prop. Soc. 33:422-425.
21. Wells, J.S. 1955. Plant propagation practices. The Macmillan Co., New York.
22. Young, M.J. 1983. Review of grafting. Proc. Int. Plant Prop. Soc. 33:417-421.

CHAPTER FOUR

Tissue Culture

A. INTRODUCTION

Fifteen years past, many scoffed at the idea that tissue culture might become a competitive force for propagating nursery crops. Today the process is employed worldwide for propagation of pitcher plants, Kiwi fruit, red maple, ad infinitum. Millions of cultivated plants are now produced through tissue culture. In the United States, over 100 commercial laboratories are functioning and a partial list is presented in Table 1. Twyford estimates that its California facility will produce 25 million plants per year.

The propagators and nurserymen who are considering starting a tissue culture laboratory should examine, actually scrutinize, the cost. Simple basement type laboratories can be built for $2000 to $5000 but are really only suitable for the tinkerer and hobbyist. Stoltz (27) described a home-made tissue culture laboratory that could be constructed for under $1000. Large tissue culture laboratories consume hundreds of thousands to millions in establishment costs. Senior author knows one nurseryman who invested $250,000 establishing a tissue culture laboratory. Along the establishment path, contamination wiped out most of the cultures and very few plants had actually been derived from the laboratory. When queried whether he would do it again, the answer was a rather definite "I doubt it."

Tissue culture requires attention-to-detail and unless practiced as art and science, the entire process is rather unforgiving. Establishment of pathogen-free cultures, hormone and media manipulations, economical multiplication rates, rooting, acclimatization, etc., are factors that must be carefully regulated. Entire culture rooms have become infested with mites that are carried on dust particles. Constant and knowledgeable vigilance is necessary for successful tissue culture.

If a propagator is considering establishing a tissue culture laboratory, then considerable spade work is a must. Read the available literature and visit several laboratories to fully understand the scope of the operation. A fully equipped tissue culture laboratory is one thing; an economically functioning laboratory another.

Several books that every tissue culture practitioner should own are listed in Table 2. The reference by George and Sherrington (7) is *outstanding*.

Perhaps the most logical way to benefit from tissue culture is to contract with a commercial laboratory to produce specific plants. For example, a South Carolina nurseryman discovered a weeping form of *Acer rubrum*, red maple, and wanted to increase stock as quickly as possible. He proceeded to make contact with a commercial laboratory. The commercial laboratory may require an upfront cash stipend to work with the plant and a contract that guarantees the nurseryman buy back so many plants (often 5,000 or more at a prescribed price). Another option includes an hourly laboratory fee with a maximum of 15 to 20 hours per month until the new plant is established in culture. Also, predetermined quantities of plants will be sold to the nurseryman at a specific unit price.

Be cautioned! Tissue culture laboratories do not always deliver the goods on time or in the numbers requested. Before contracting with a laboratory, always confer with other customers. Tissue cultured plants are not inexpensive. Senior author knows of one situation where tissue cultured southern magnolia about 2 to 3" high sells for $1.75. Tissue cultured plants run the gamut in price from $.20 to $.30 to $1.50 to $2.00 per plant. Table 3 presents prices of a select number of herbaceous and woody plants that are offered by commercial tissue culture firms in the U.S. See Kyte (12) for a discussion of production costs.

Tissue culture offers an alternative to conventional vegetative propagation. By no yardstick is it a panacea for the propagation of all woody plants. There is no consistent economic evidence that a tissue culture plant will outgrow rooted cuttings or budded trees. Significant variation is often evident in nursery blocks of tissue culture plants. In Oregon, senior author observed tissue culture clonal flowering crabapples in the field that showed sufficient variation to be mistaken for seedlings. The grower noted that the plants were grown for one year and then cut back to a single bud in late winter (like a budded tree) in order to produce a uniform straight-trunked liner.

B. APPLICATIONS OF TISSUE CULTURE

Tissue culture has four potential applications (20): 1) production of natural products, 2) genetic improvement of crops and germplasm storage, 3) production of disease free plants, and 4) rapid multiplication. The last application offers the greatest significance to the commercial propagator.

Mass micropropagation originiated with attempts to produce "virus free" orchids through meristem culture. Morel (18, 19) showed that shoot tips proliferated into protocorms which could be divided and recultured to produce additional plants. Since orchid propagation is slow by the normal division technique, the potential was quickly recognized for greatly increased production. Current interest in tissue culture as a propagation system can be traced to the success achieved with orchids.

A feature of mass propagation is the production of large numbers of clonal plants in a short time period. For example, starting with a single shoot explant and a multiplication rate of 5 every month, over one million plants can be produced in 10 months. Micropropagation on a commercial scale has been demonstrated for a number of woody species (7). McCown and Amos (15), working with *Betula* species, achieved propagation

Table 1. Commercial tissue culture laboratories in the United States (partial list).

Briggs Nursery Inc.
4407 Gleason Road
Olympia, WA 98501

California-Florida Plant Corp.
5600 Stevenson Blvd.
P.O. Box 1367
Freemont, CA 94538

Carter & Holmes Inc.
1 Mendenhall Road
P.O. Box 668
Newberry, SC 29108

Cedar Valley Nursery
3833 McElfresh Rd., SW
Centralia, WA 98531

Ceres 2000 Inc.
P.O. Box 2927
Winter Haven, FL 33880

Crown Zellerbach Corp.
Forestry Research
P.O. Box 368
Wilsonville, OR 97070

Daisy Farm Inc.
Plant Tissue Culture Division
10,000 SW 64th Street
Miami, FL 33173

Driscole Strawberry Ass. Inc.
404 San Juan Road
Watsonville, CA 95076

Paul Ecke Poinsettias
P.O. Box 483
Encinatas, CA 92024

Fennell Orchid Company
26715 SW 157th Avenue
Homestead, FL 33032

G and B Orchid Laboratory
3588 Merrimac St.
San Diego, CA 92117

G and S Laboratories
645 Stoddard Lane
Santa Barbara, CA 93108

Green Plant Research
P.O. Box 735
Kaaawa, Hawaii 96730

Hartman's Bromeliads, Ferns & Exotic Plants
P.O. Box 90
Palmdale, FL 33944

Hyclone, Inc.
P.O. Box 3190
Conroe, TX 77305

International Island Resources
P.O. Box 11693
Santa Rosa, CA 95406

International Paper Company
77 West 45th Street
New York, NY 10036

Jungle-Gems Inc.
300 Edgewood Road
Edgewood, Maryland 21040

K.M. Nursery Inc.
P.O. Box 847
Carpinteria, CA 93013

Knauss Plant Laboratory Inc.
P.O. Box 1605
Lake Worth, FL 33460

Knight Hollow Nursery
236 E. Sunset Ct.
Madison, WI 53705

Lindemann Laboratories Inc.
Cornell Industry Research Park
61 Brown Rd.
Ithaca, NY 14850

The Marie Selby Botanical Gardens
800 South Palm Avenue
P.O. Box 4155
Sarasota, FL 33578

Microplant Nurseries Inc.
13357 Portland Road NE
Gervais, OR 97026

Native Plants Inc.
University Research Park
360 Wakara Way
Salt Lake City, Utah 84103

The New York Botanical Garden
Cell & Tissue Culture Lab.
Harding Lab.
Bronx, NY 10458

Oakdell Nurseries
P.O. Box 1145
Opopka, FL 32703

Oglesby Nurseries Inc.
3714 SW 52nd Ave.
Hollywood, FL 33023

Oki Nursery Inc.
8649 Kiefer Blvd.
Box 7118
Sacramento, CA 95826

Phyton Technologies
7327 Oak Ridge Hwy.
Knoxville, TN 37931

Phyto Tech Laboratory
ABC Nursery
21822 S. Vermont Ave.
Torrance, CA 90502

Plant Tissue Culture Lab
1232 S. Bonnie Cove Ave.
Glendora, CA 91340

Sherman Plant Laboratory
Sherman Nursery Co.
1300 W. Grove St.
Charles City, IA 50616

Sunny Borders Nurseries Inc.
1709 Kensington Rd.
Kensington, CN 06037

T & Z Nursery Inc.
28 West S. 21 Roosevelt Rd.
Winfield, IL 60190

Transplant Nursery Inc.
3815 W. Doris Ave.
Oxnard, CA 93030

Twyford Plant Laboratories, Inc.
P.O. Box 664
15245 Telegraph Road
Santa Paula, CA 93060

Vita Leaf Laboratories Inc.
823 Woodmere Drive
Lafayette, IN 47904

Volkmann Bros. Greenhouses
2714 Minert St.
Dallas, TX 75219

Walters Gardens Inc.
96th Avenue & Business I-196
P.O. Box 137
Zeeland, MI 49464-0137

Weyerhaeuser Corp.
Tacoma, Washington

Yoder Brothers Inc.
P.O. Box 160
Fort Meyers, FL 33902 rates of 500,000 plantlets per year using 125 sq. ft. of culture space.

C. STAGES OF TISSUE CULTURE

Murashige (20) was the first to articulate the stages and the significance. The three stages are: 1) establishment of the aseptic culture, 2) multiplication of the plant part (explant), and 3) preparation (rooting) for re-establishment of plants in soil. Stage 3 is sometimes divided into two stages with the fourth stage being the acclimation of plantlets from the aseptic culture environment to the free-living environment.

1. Stage I: Establishment

The objective of stage I is the establishment of an aseptic (sterile) culture. Factors that affect success include explant source, elimination of contamination from explant, proper culture medium, and cultural conditions (temperature and light). When selecting the explant, consider the following: 1) organ serving as the tissue source, 2) physiological or ontogenetic (developmental) age of the organ, 3) season the explant is being obtained, 4) size of the explant, and 5) quality of the plant supplying the explant. Since explant sources from the same plant are not equal in response the choice of the explant is most critical. Cultures have been established from many organs or tissues including: shoot tip, lateral bud, bulb scale, leaf, root, stem, embryo, seedling, cotyledon, hypocotyl, rhizome tip, inflorescence, corm, ovary, flower bud, nucellus, and tuber. For most woody species, the actively growing shoot tip is the best explant source.

2. Stage II: Multiplication

The objective of this stage is the rapid increase in organs (primarily shoots) or other structures that will eventually give rise to a complete plant. Asexual embryogenesis (like seeds, only clonal) or adventitious shoot initiation (from callus) may provide faster rates of increase, but are less desirable than axillary shoot production because of possible increased production of genetically aberrant plants (not true-to-type).

3. Stage III: Preparation for re-establishment outside culture

Stage III's function is preparation of the plantlets for establishment in soil outside of the culture environment. For success, culture conditions should be distinct from the previous two stages. Stage III involves the rooting of microcuttings, hardening the plantlets to provide resistance against moisture stress, and weaning from a dependency to a self-sustaining state.

D. PROCEDURES IN MEDIA PREPARATION

Media formulations can be prepared from stock solutions made by weighing out the individual chemical components or premixed powder formulations. Premixed formulations are simpler to use but also more expensive than buying the individual chemicals in bulk.

For commercial tissue culture operations, it is probably sufficient to use commercially purchased distilled water without further purification. However, any commercial water source should be checked for possible toxic contaminants before use. Tap water should not be used. The following example shows the steps involved in the preparation of stock solutions from individual chemical components.

1. Preparation of stock solutions

Stock solutions are concentrated solutions of a single compound or groups of compounds, aliquots of which will be combined to prepare the media. Stocks are prepared ahead of time to reduce steps in media preparation and to enhance accuracy in weighing substances. Salt stock solutions are usually made at 100 times concentration. Often more than one substance can be combined in a stock solution. Never combine compounds that will form precipitates. Precipitates are insoluble compounds that form by chemical reaction and are no longer effective. Compounds containing calcium and phosphate or sulfate, and magnesium and phosphate should not be mixed together. The following materials can be combined without forming precipitates: 1) nitrates, 2) sulfates, 3) halides such as chlorides and iodides, and 4) borates, molybdates and phosphates. Iron is prepared separately as a chelated compound (in most cases with sodium EDTA). After preparation, stock solutions are stored under refrigeration, except for organic stocks, such as vitamins and hormones, which are most logically frozen. If either precipitates or microbial contamination occur, the solutions should be discarded.

The stock solutions are prepared with distilled or deionized water (double distilled water) and only chemicals of reagent grade purity are used.

For stock solutions 1 through 4, pour 750 ml of distilled water into a one liter flask and add each chemcial. After dissolving the chemicals, bring volume to 1000 ml. These 4 stock solutions are 100 times final media concentration. Use 10 ml of solution per liter for preparation of tissue culture media.

1) KNO_3	190.0 g/liter	3) $CaCl_2 \bullet 2H_2O$	44.0 g/liter
NH_4NO_3	165.0 g/liter	$CoCl \bullet 6H_2O$	2.5 mg/liter
2) $CuSO_4 \bullet 5H_2O$	2.5 mg/liter	KI	83.0 mg/liter
$MgSO_4 \bullet 7H_2O$	37.0 g/liter	4) H_3BO_3	0.62 g/liter
$MnSO_4 \bullet H_2O$	1.69 g/liter	KH_2PO_4	17.0 g/liter
$ZnSO_4 \bullet 7H_2O$	0.86 g/liter	$NaMoO_4 \bullet 2H_2O$	25.0 mg/liter

For stock solution 5, pour 750 ml of distilled water into a one liter flask and add each chemical. After dissolving, bring volume to 1000 ml and store in a dark bottle since the solution is not stable in light. This stock solution is 100 times concentration. Use 10 ml of stock solution per liter.

5) $FeSO_4 \bullet 7H_2O$	2.78 g/liter
Na_2EDTA	3.72 g/liter

For stock solution 6, pour 750 ml of distilled water into a one liter flask and add each chemical. After dissolving, bring volume to 1000 ml. This stock is 100 times concentration. Use 10 ml of stock solution per liter.

6) glycine	200.0 mg/liter
myo-inositol	10.0 g/liter
nicotinic acid	50.0 mg/liter
pyridoxine HCl	50.0 mg/liter
thiamine HCl	100.0 mg/liter

7) Cytokinin stock solution. Cytokinins are weak bases and can be dissolved in dilute acid solutions. Weigh 25 mg of N_6Benzyladenine (other cytokinins are mixed the same way) and place in a 250 ml volumetric flask. Add 5 ml of distilled water and with a dropper add 0.1 to 1 N HCl dropwise while stirring until dissolved. Bring to volume with distilled water. Stock contains 0.1 mg/ml N_6Benzyladenine. Dimethylsulfoxide (DMSO) can also be used as a solvent.

Table 2. Selected tissue culture reference books

Cell Culture and Somatic Cell Genetics of Plants. Volume 1. Laboratory Procedures and Their Applications. 1984.
I.K. Vasil (ed.)
Academic Press, Inc.
New York, New York

Experiments in Plant Tissue Culture. 1985.
J.H. Dodds and L.W. Roberts
Cambridge University Press
New York, New York

Handbook of Plant Cell Culture. Volume I. Techniques for Propagation and Breeding. 1983.
D.A. Evans, W.R. Sharp, P.V. Ammirato, Y. Yamada (eds.)
Macmillan Publishing Co.,
New York, New York

Handbook of Plant Cell Culture. Volume 2. Crop Species. 1984.
W.R. Sharp, D.A. Evans, P.V. Ammirato, Y. Yamada (eds.)
Macmillan Publishing Co.
New York, New York

Handbook of Plant Cell Culture. Volume 3. Crop Species. 1984.
P.V. Ammirato, D.A. Evans, W.R. Sharp, Y. Yamada (eds.)
Macmillan Publishing Co.
New York, New York

Introduction to In Vitro Propagation. 1982.
D.F. Wetherell
Avery Publishing Group, Inc.
Wayne, N.J.

Plant Cell Culture: A Practical Approach. 1985.
Dixon, R.A. (ed.)
IRL Press
Washington, D.C.

Plant Propagation by Tissue Culture. 1984.
E.F. George and P.D. Sherrington
Exegetics, Ltd.
Eversley, Basingstoke
Hants.
RG27 OQY England

Plant Propagation — Principles and Practices. 1983.
H.T. Hartmann and D.K. Kester
Prentice Hall, Inc.
Englewood Cliffs, NJ

Plants From Test Tubes. 1983.
Lydiane Kyte
Timber Press
P.O. Box 1631
Beaverton, Oregon 97075

Plant Tissue and Cell Culture. 1977.
H.E. Street (ed.)
University of California Press
Berkeley, California

Plant Tissue Culture — Methods and Applications in Agriculture. 1981.
T.A. Thorpe
Academic Press, NY

Plant Tissue Culture: Theory and Practice. 1983.
S.S. Bhojwani and M.K. Razdan
Elsevier
New York, New York

Proceedings of Conference on Nursery Production of Fruit Plants through Tissue Culture — Applications and Feasibility. 1980.
USDA-SEA. Ag. Res. Results.
ARR-NE11

Tissue Culture for Plant Propagators. 1976.
R.A. deFossard
University of New England
Armidale, New South Wales, Australia

Table 3. Current prices of selected woody and herbaceous plants available from commercial tissue culture laboratories.

Species	Price per rooted cutting
Acer platanoides 'Crimson King'	$0.59
Acer rubrum 'Autumn Flame'	0.59
Amelanchier alnifolia 'Regent'	0.60
Betula nigra 'Heritage'	0.55
Kalmia latifolia	0.65
Magnolia stellata 'Royal Star'	0.60
Nandina domestica 'Harbour Dwarf'	1.05
Nephrolepis exaltata 'Boston'	0.42
Rosa carolina	0.45
Spathiphyllum 'Mauna Loa'	0.50

Prices obtained from 1985-1986 price lists of Hyclone, Knight Hollow, and Microplant Nurseries.

Table 4. Surface sterilants used to treat explants prior to culture.

Sterilant	Concentration	Treatment time (minutes)
Antibiotics	4-50 mg/liter	30-60
Bromine water	1-2%	2-10
Calcium hypochlorite	9-10%	5-30
Ethanol/isopropanol	70%	quick dip
Hydrogen peroxide	10-12%	5-15
Mercuric chloride	0.1-1%	2-10
Sodium hypochlorite	10-20%[1]	5-30

[1] 10-20% (v/v) of a commercial bleach solution

8) Auxin stock solution. Auxins are weak acids and can be dissolved in dilute base solutions. Weigh 25 mg of indole-3-acetic acid (other auxins are mixed the same way) and place in a 250 ml volumetric flask. Add 5 ml of distilled water and with a dropper add 0.1 to 1N KOH dropwise while stirring until dissolved. Bring to volume with distilled water. Stock contains 0.1 mg/ml indole-3-acetic acid.

2. Preparation of Murashige and Skoog medium

The procedure outlined assumes that the stock solutions have been previously prepared. Similar procedures would be used for preparation of other media such as those listed in table 5.

1) Pour 700 ml of distilled or deionized water into a 2000 ml flask or beaker and place on a hot plate-stirrer. The container volume should be about 2 times the final media volume.

2) Gently place a magnetic stir bar into the container and turn on the stirrer and hot plate.

3) Add 10 ml each of stock solutions 1 through 6.

4) Add sucrose (30 grams/liter is standard with MS medium).

5) Add cytokinin stock at desired concentration. If 1 mg/liter final concentration is desired add 10 ml of stock (0.1 mg/ml x10ml=1mg actual cytokinin).

6) Add auxin stock at desired concentration. Follow same concentration procedure described under 5.

7) Bring volume to 1000 ml.

8) Adjust pH to desired level (generally between 4.5 and 6.0) with 0.1 to 1N KOH or HCl while stirring.

9) Add agar.

10) Heat to about 92°C until the solution has cleared. Do not allow solution to boil.

11) Dispense into containers and cap. Best to use a pipetting device that can deliver 10, 15, 20 ml of solution.

12) Sterilize in a pressure cooker or autoclave for 15 to 20 minutes at 15 pounds pressure.

13) Cool, label and store in a clean place until used.

14) Ideally, freshly prepared media should be used as soon as possible after preparation.

3. Sterilization of media

Media are sterilized by autoclaving at 15 pounds pressure and 250°F. The duration will vary with the volume of medium but 15 minutes is the usual length of time. However, the larger the volume the longer it takes for heat penetration; thus, the longer the time exposure required.

Some media constituents are destroyed by high temperatures and cannot be autoclaved. In cases of heat instability, the usual sterilization technique involves membrane filter (cold) sterilization of the unstable compound(s) and addition to the medium after heat sterilization. Membrane filtration equipment comes in various sizes. For large volumes, the funnel type apparatus is used, while with small quantities, a syringe type is preferred.

In practice the filtration equipment is pre-sterilized by autoclaving. Heat labile substances are prepared in concentrated solutions, pH adjusted, filtered and distributed to the medium that has been cooled to 95°F.

E. EXPLANT PREPARATION AND STERILIZATION

The principal objective of Stage I is the establishment of pathogen-free cultures. Plant parts free of algae, bacteria, fungi and other microorganisms are required. The most common contaminants are bacteria and fungi that are present on the surface of the plant material and the usual method of removal is surface sterilization. Microbial contaminants are undesirable because they often outgrow the explant, compete for nutrients, and produce toxic metabolic products.

All microorganisms must be removed by the sterilization methods. Explants should be taken from greenhouse or growth chamber grown plants because field grown plants have a higher number of microorganisms. There are a number of chemical agents for surface sterilization of plant material (Table 4) (29).

Household bleach (Clorox or Purex, containing 5.25% sodium hypochlorite) or ethanol are the most commonly used. Mercuric chloride can be very toxic to plant material and is undesirable because it contains mercury. The effectiveness of most disinfesting chemicals is enhanced if a small quantity of wetting agent (0.05%) like detergent or Tween 20 is added. The addition of a wetting agent reduces surface tension and allows better penetration. It is necessary to determine the optimal conditions for each tissue.

1. Surface sterilization

A typical surface sterilization procedure for shoot tissue includes the following steps:

1) A preliminary washing of the untrimmed explant in running water for about 10 minutes. Some labs use a weak bleach solution (1:100, v:v) + a detergent with agitation.

2) Trim explant. This and the next steps are conducted under a sterile transfer hood.

3) Place in bleach solution (1 part commercial bleach + 9 parts distilled water) with stirring for 15 minutes. Some labs use a vacuum pump or an ultrasonic cleaner for better penetration.

4) Rinse in 3 changes of sterile distilled water and allow a 5 minute soak in the last change. Some plant parts, such as buds in contact with soil, may require surface sterilization in several steps. Following the first treatment, superficial tissue is removed and the process repeated again.

5) Cut to desired implanting size and trim any damaged tissue before transferring to sterile media.

When the surface of the material is not part of the explant material, a more drastic treatment can be used. An example would be the use of seeds for the production of sterile explants.

1) Place seeds in cheesecloth, soak in 95% ethanol for about 10 seconds, and rinse in sterile distilled water (under transfer hood).

2) Place intact seeds in bleach solution (1 part commercial bleach + 9 parts distilled water) for 20-30 minutes.

3) Rinse in 3 changes of sterile distilled water with a 5 minute soak in the last change.

4) Place the intact seed on a sterile medium. Embryos may also be extracted from the sterilized seeds.

The objectives of the sterilization treatment are to kill all microorganisms and yield an explant capable of growing when placed on the sterile medium. If sterile explants are not pro-

Table 5. Formulations of selected plant tissue culture media.

Compound	Murashige Skoog (21)	Woody Plant Medium (14)	Anderson (1)	White's (30)	Gamborg B5 (6)	Heller (9)	Nitsch's (22)	Linsmaier and Skoog (13)
	concentration (mg/liter)							
NH_4NO_3	1650	400	400				720	1650
KNO_3	1900		480	80	2500		950	1900
$Ca(NO_3)_2 \bullet 4H_2O$		556		300				
$NaNO_3$						600		
$(NH_4)_2SO_4$					134			
$MgSO_4 \bullet 7H_2O$	370		370	720	250		185	
Na_2SO_4				200				
$Ca_3(PO_4)_2$								
$CaCl_2 \bullet 2H_2O$	440	96	440		150	75	166	440
KH_2PO_4	170	170	380			125	68	170
$NaH_2PO_4 \bullet H_2O$				16.5	150			
KCl				65		750		
K_2SO_4		990						
$MgSO_4 \bullet 7H_2O$		370				250		370
$FeSO_4 \bullet 7H_2O$	27.8	27.8	55.7		27.5		27.8	27.8
Na_2EDTA	37.3	37.3	74.5		37.3		37.3	37.3
$FeCl_3 \bullet 6H_2O$						1		
$Fe_2(C_4H_4O_6)_3 \bullet 2H_2O$								
$Fe_2(SO_4)_3$				2.5				
$MnSO_4 \bullet H_2O$		22.3	16.9		10		18.9	
$MnSO_4 \bullet H_2O$	22.3			7		0.01		22.3
$ZnSO_4 \bullet 7H_2O$	8.6	8.6	8.6	3	2.0	1	10	8.6
H_3BO_3	6.2	6.2	6.2	1.5	3.0	1	10	6.2
KI	0.83			0.75	0.75	0.01		0.83
$Na_2MoO_4 \bullet 2H_2O$	0.25	0.25	0.25		0.25		0.25	0.25
$CuSO_4 \bullet 5H_2O$	0.025	0.25	0.025		0.025	0.03	0.025	0.025
$CoCl \bullet 6H_2O$	0.025		0.025		0.025			0.025
$NiCl \bullet 6H_2O$						0.03		
$AlCl_3$						0.03		
myo-inositol	100	100	100		100		100	100
nicotinic acid	0.5	0.5		0.5	1.0		5.0	
pyridoxine HCl	0.5	0.5		0.1	1.0		0.5	
thiamine HCl	0.1	1.0		0.1	10.0	1.0	0.5	0.40
biotin							0.05	
folic acid							0.5	
glycine	2.0	2.0		3.0			2.0	
Ca D-pantothenic acid				1.0				

duced or the explants are killed by overzealous sterilization, modifications are required. It is necessary to determine the optimal conditions for each type of explant. The concentration of the bleach solution and time of soaking are variables that can be easily modified.

2. Oxidative browning

If the tissues brown after cutting, place in an antioxidant solution. Browning results from oxidation of polyphenolic compounds. Not only does the explant brown, but the medium will also become discolored. Explants that become discolored by oxidative browning often do not establish in culture. Oxidative browning can be controlled or reduced by: 1) reducing the amount of polyphenolic compounds and their oxidative products, 2) inhibiting the action of polyphenoloxidases, and 3) lowering the supply of oxygen.
Methods for accomplishing these include:

1) Minimize damage during dissection of the explant.

2) Place explant, after dissection but before disinfestation, into a solution of ascorbic acid (100 mg/liter) or citric acid (150 mg/liter), or a combined solution.

3) Place the disinfested explant into a filter sterilized solution of ascorbic acid or citric acid, or rinse with a sterile solution of cysteine hydrochloride (100 mg/liter) or PVP-40 (polyvinylpyrrolidone) (2.5 microMolar) before placing into culture.

4) Use liquid media instead of semi-solid media.

5) Do not agitate the liquid medium.

6) Use a filter paper bridge in a liquid medium.

7) Use reduced light or complete darkness during explant establishment.

8) Place the explant in fresh liquid media daily for about 7 days starting 12 hours after culture initiation.

9) Add activated charcoal to the culture medium at 600 mg/liter or place into the culture medium an antioxidant such as ascorbic acid, citric acid, polyvinylpyrrolidone (insoluble PVP).

3. Internal contamination

Clonally propagated plants frequently are infested with viruses, as well as bacteria and fungi. Often cultures may be infected internally with bacteria, as well as fungi, but the infection may not become apparent until after several subcultures. The unknown infections can lead to serious problems, and they should be removed as soon as possible — preferably at the end of the first transfer.

The recovery of pathogen free tissue involves the isolation of tissue free of known pathogens. Culture indexing is a useful procedure for eliminating contaminated plants. The first step in culture indexing is testing individual cuttings or plants for the presence of bacteria and fungi. The nutrient broth test provides a simple assay for detecting such contaminants. Nutrient broth is prepared by dissolving 8 grams Difco Bacto-nutrient broth mix in 1000 ml of distilled water. Broth is dispensed in 10-ml aliquots into 25 × 150-mm culture tubes, capped, and autoclaved. The nutrient broth tubes are inoculated with a small piece of tissue and placed into darkness at room temperature and observed after 24 hours and again after 7 days. The development of cloudiness in the broth indicates bacterial infection. Fungal infection is indicated by the presence of mycelium and may take longer to develop. The shoot apex culture technique can be used to clean up infected plants.

Once pathogenic bacteria and fungi have been removed from plants, they can then be put through a system to eliminate viruses. This part of the system involves thermotherapy and meristem-tip culture, followed by indexing for detectable viruses (25). During thermotherapy, plants are grown at 100°F during a 16-hour day and 95°F during an 8-hour night for 3 weeks to reduce the concentration of spherical viruses in the plant (10). After treatment, the meristem tips, 0.1-0.5 mm in height containing the meristem plus the first leaf primordia, are removed and grown in vitro. After elongation and rooting, the plants are grown in isolation and virus indexed to test for elimination.

F. HEALTH HAZARDS FROM STERILANTS AND RELATED PRODUCTS.

Many of the commonly used sterilants and equipment can pose health hazards. Sodium and calcium hypochlorite solutions should be used with care. Inhalation can cause bronchial irritation and skin contact can result in irritation. Hypochlorite solutions in the presence of ultraviolet light release chlorine gas which is a serious health hazard. Never pipette hypochlorite solutions with your mouth. Mercuric chloride, another disinfectant, is poisonous and probably should not be used in commercial labs.

There are a number of instrument sterilization techniques available, however, it is a general feeling that alcohol dips and open flames in a hood should be avoided because of the fire hazard.

Ultraviolet light poses health risks. One should never look at a live tube because of potential eye damage. Working under an ultraviolet light also poses a potential cancer risk. An ultraviolet light generates ozone, a powerful oxidizing agent.

G. SELECTED TISSUE CULTURE METHODOLOGY AND TYPES OF REGENERATION

There are five fundamental types of vegetative regeneration in tissue culture systems: 1) shoot apex culture, 2) axillary shoot production, 3) adventitious shoot initiation, 4) organogenesis, and 5) embryogenesis (8).

1. Shoot apex culture

Because of the wide application of shoot apex cultures in horticulture, there has been misuse of botanical nomenclature (25, 26). The terms "meristem culture", "meristemming", and "mericloning" are widely used when in reality propagators are culturing relatively large stem tips. The apical shoot meristem, by definition, is that portion of the shoot apex lying distal to the youngest leaf primordium. The shoot apex refers to the apical meristem plus a few subjacent leaf primordia. A true apical meristem is about 8-100 micrometers in height and is very difficult to culture. It is therefore not practical for the general propagation of plants. Shoot apex culture, however, has found application in the production of pathogen-free stock (the removal of bacteria, fungi and viruses). It is standard practice for the culture of a number of important horticultural crops including potato, chrysanthemum, sweet potato, and geranium.

The in vitro requirements for the growth and development of isolated shoot apices vary with size, intended use and plant genotype. An isolated shoot apex may: need a high concentration of potassium (23); require auxin (28,31) or gibberellin (23) for growth, or cytokinin for multiple shoot formation; or be in-

Table 6. Plant growth substances and the range of concentrations frequently used in tissue culture.

Substance	Concentration
Auxins	
2,4-Dichlorophenoxyacetic acid—2,4-D	0.01-10 mg/l
Indole-3-acetic acid—IAA	0.1-10 mg/l
Indole-3-butyric acid—IBA	0.1-10 mg/l
Naphthaleneacetic acid—NAA	0.1-10 mg/l
P-chlorophenoxyacetic acid	0.1-10 mg/l
Cytokinins	
6-(y,y-dimethylallylamino) purine (2iP)	0.1-30 mg/l
Kinetin	0.1-10 mg/l
N_6Benzyladenine-BA	0.1-10 mg/l
Thidiazuron	0.01-10 mg/l
Other Organic Substances	
Adenine sulfate	20-200 mg/l
Ascorbic acid	100 mg/l
Citric acid	150 mg/l
Sucrose	20-30 g/l
Undefined Ingredients	
Bacto malt extract	50-5000 mg/l
Bacto yeast extract	50-5000 mg/l
Casein hydrolysate	50-5000 mg/l
Coconut milk/water	100-150 ml/l
Orange juice	50-300 ml/l

Table 7. Atomic weight of elements used in tissue culture.

Element	Symbol	Atomic weight
Aluminum	Al	26.98
Boron	B	10.81
Calcium	Ca	40.08
Carbon	C	12.01
Chlorine	Cl	35.45
Cobolt	Co	58.93
Copper	Cu	63.54
Hydrogen	H	1.00
Iodine	I	126.90
Iron	Fe	55.84
Magnesium	Mg	24.31
Manganese	Mn	54.94
Molybdenum	Mo	95.94
Nickel	Ni	58.17
Nitrogen	N	14.00
Oxygen	O	15.99
Phosphorus	P	30.97
Potassium	K	39.10
Sodium	Na	22.98
Sulfur	S	32.06
Zinc	Zn	65.37

Table 8. Molecular weight of compounds used in tissue culture.

Compound	Molecular weight
Salts	
Aluminum chloride, hexahydrate-$AlCl_3 \bullet 6H_2O$	241.4
Ammonium nitrate-NH_4NO_3	80.1
Ammonium phosphate, dibasic-$(NH_4)_2HPO_4$	132.1
Ammonium phosphate, monobasic-$NH_4H_2PO_4$	115.0
Ammonium sulfate-$(NH_4)_2SO_4$	131.2
Boric acid-H_3BO_3	61.8
Calcium chloride, dihydrate-$CaCl_2 \bullet 2H_2O$	147.0
Calcium nitrate, tetrahydrate-$Ca(NO_3)_2 \bullet 4H_2O$	236.2
Calcium phosphate-$Ca_3(PO_4)_2$	310.1
Cobolt chloride, hexahydrate-$CoCl \bullet 6H_2O$	237.9
Copper sulfate, pentahydrate-$CuSO_4 \bullet 5H_2O$	249.7
Ferric chloride, hexahydrate-$FeCl_3 \bullet 6H_2O$	270.3
Ferric sulfate-$FeSO_4 \bullet 7H_2O$	277.8
Ferric tartrate-$Fe_2(C_4H_4O_6)_3 \bullet H_2O$	573.9
Ferrous sulfate-$Fe_2(SO_4)_3$	399.9
Magnesium sulfate, heptahydrate-$MgSO_4 \bullet 7H_2O$	246.5
Manganese chloride-$MnCl_2 \bullet 4H_2O$	197.9
Manganese sulfate, monohydrate-$MnSO_4 \bullet H_2O$	169.0
Manganese sulfate, tetrahydrate-$MnSO_4 \bullet 4H_2O$	223.0
Nickel chloride, hexahydrate-$NiCl \bullet 6H_2O$	237.7
Potassium chloride-KCl	74.6
Potassium iodide-KI	166.0
Potassium nitrate-KNO_3	101.1
Potassium phosphate, dibasic-K_2HPO_4	174.2
Potassium phosphate, monobasic-KH_2PO_4	136.1
Potassium sulfate-K_2SO_4	174.3
Sodium molybdate, dihydrate-$NaMoO_4 \bullet 2H_2O$	242.0
Sodium nitrate-$NaNO_3$	85.0
Sodium phosphate dibasic heptahydrate	268.1
Sodium phosphate, monosodium monohydrate-$NaH_2PO_4 \bullet H_2O$	138.0
Sodium sulfate-Na_2SO_4	142.0
Zinc sulfate, heptahydrate-$ZnSO_4 \bullet 7H_2O$	287.5
Growth Regulators	
2,4-Dichlorophenoxyacetic acid-2,4-D	221.0
6-(y,y-dimethylallylamino) purine-2iP	203.2
Adenine sulfate-$AdSO_4$	404.4
Gibberellic acid (K-salt)-GA_3	384.5
Indole-3-acetic acid-IAA	175.2
Indole-3-butyric acid-IBA	203.2
Inositol-myo	180.2
Kinetin	215.2
N_6Benzyladenine-BA	225.3
Naphthaleneacetic acid-NAA	186.2
P-chlorophenoxyacetic acid	186.6
Thidiazuron	220.2
Zeatin	219.2
Other Growth Factors	
Arginine (L form)	174.2
Ascorbic acid	176.1
Asparagine (L form)	132.1
Biotin	244.3
D-pantothenic acid, Ca salt	476.5
Disodium ethylenediamine tetraacetic acid-Na_2EDTA	372.3
Folic acid	441.4
Glutamine (L form)	146.2
Glycine	75.1
Nicotinic acid	123.1
Pyridoxine HCl	205.7
Thiamine HCl	337.3
Carbon Sources	
Fructose	180.2
Glucose	180.2
Sucrose	342.3
Sorbitol (and similar sugar alcohols)	182.2

dependent of hormone requirements (23). Both agar and liquid media have been utilized for the culture of shoot apices. Liquid cultures use a filter paper bridge.

Procedure: The procedure outlined is that used for the culture of geranium shoot apices (11).

1) Add 10 ml of MS medium supplemented with 2 mg/liter IBA or 1 mg/liter NAA, 0.04 mg/liter kinetin, 2 mg/liter L-cysteine, 100 mg/liter m-inositol, 100 ml/liter of coconut milk, and agar (0.4 or 0.9%) at pH 5.5-5.7 to 25 × 150 mm test tubes. Cap and autoclave the medium.

2) Apical shoots about 1 cm in length are removed from the geranium plants and placed in sodium hypochlorite solution for about 15 minutes.

3) Provide three separate sterile water rinses. Place in a sterile Petri dish containing moistened filter paper.

4) Carefully remove the terminal 0.5 to 1.0 mm of the shoot apex. A dissecting microscope is necessary for this operation.

5) Transfer the shoot apex to the agar medium.

6) Culture in a growth chamber at 21°C (70°F) and light set at about 3000 lux with a 16 light/8 dark hour cycle. The geranium shoot apices develop shoot growth and roots in a few weeks. Other plants may take longer. The plantlets can be transferred to soil after shoot and root growth have occurred.

2. Axillary shoot proliferation

With axillary shoot proliferation, lateral meristems are stimulated to break and grow. This type, with repeated subculture, can produce an exponential type of growth and forms the basis for most commercial micropropagation.

Procedure: The procedure outlined is that used for the culture of *Kalmia latifolia*, mountain laurel (14).

1) Prepare 50 ml Erlenmeyer flasks containing 15-20 ml of Woody Plant Medium supplemented with 0.8 to 3.2 mg/l 2iP and autoclave.

2) Excise 2 to 3 cm long shoot tips from actively growing plants and remove leaves larger than 1 cm.

3) Dip in 70% ethanol and then treat with 10% household bleach with a wetting agent added (0.05% Tween-20) for 10-15 minutes.

4) Rinse in 3 changes of sterile distilled water, remove any injured tissue, and place in individual flasks.

5) The liquid medium is changed at 12 and 24 hours, and then daily for 1 week. After one week the shoots are transferred to stationary test tubes and liquid medium to ½ their height. The stationary medium is changed every 3 weeks. After 1 to 2 months axillary shoots are produced.

6) Shoots are removed when approximately 2 cm long and placed on solid medium (6% agar). The shoot-tip cultures are multiplied by removing several elongating shoots from the basal mass and subculturing. Optimal number of shoots is produced at 1.6 mg/l 2iP.

7) Cultures are grown in rooms with 24 hours cool white fluorescent lighting (100-300 foot candles) and a temperature of 28-30°C.

8) Shoots are rooted in 100% peat medium in a warm (30 to 35°C) high humidity chamber under 24 hour cool white fluorescent lights.

3. Adventitious shoot initiation

Excised plant parts of some plants can be induced to develop adventitious shoots directly. The method is useful for the propagation of a number of plants traditionally propagated by adventitious shoots, but also has been shown to occur with many plants during axillary shoot proliferation. For example, Zimmerman and Broome (31) reported that expanding leaves of subcultured shoots of *Vaccinium corymbosum*, blueberry, that touch the medium develop adventitious callus and adventitious shoots. Many other plants show a similar response.

Procedure: The procedure outlined is used for the culture of *Saintpaulia*, African violet (12).

1) Prepare culture tubes with 10 ml of MS medium (Murashige and Skoog) containing 2.5% sucrose, 1.0% agar, naphthaleneacetic acid at 0.1 mg/l, and benzyladenine at 0.01 mg/l.

2) Cut fully expanded leaves with petioles from an African violet plant. Wash in water to remove any soil, remove the blade and quickly dip in 95% ethanol before immersing in the commercial bleach solution (1 to 20 dilution) for 15 min. Wash in 3 changes of sterile double distilled water (DDH_2O) and place in a sterile Petri dish.

3) Trim off damaged cells, slice into 2 mm cross sections, and place in the culture tubes. Callus production and regenerating plantlets should occur in 4-6 weeks. Greater numbers of plantlets can be generated if, after the initial differentiation of plantlets, the petiole cross section is divided and transferred to fresh medium.

4. Organogenesis

During the early work on the chemical characterization of kinetin (a cytokinin), cultured tobacco pith tissue was used extensively. It was observed that if the proper balance of auxin and kinetin was present, pith tissue divided, enlarged and produced a loosely arranged mass of undifferentiated cells (callus). Skoog and Miller (24) further showed that when the kinetin and auxin ratio was altered to a relatively high concentration of cytokinin (higher than for callus growth), shoots were produced. Conversely, a relatively high auxin to kinetin ratio produced roots. These observations have lead to many of the widespread techniques currently used in micropropagation.

Procedure: The procedure outlined is that used for the differentiation of shoots from tobacco callus.

1) Prepare Murashige and Skoog medium and supplement with NAA at 0.1 mg/liter and benzyladenine at 1.0 mg/liter. Adjust the pH to 5.8, add 1 gram of agar. Dissolve the agar, pour 50 ml into 125 ml Erlenmeyer flasks, and autoclave.

2) With sterile forceps transfer a large piece of tobacco callus stock to a sterile petri dish. Cut the callus into small pieces and transfer to flasks.

3) Place the cultures in a growth room at about 27°C and constant light (40 foot candles) for a period of 5 to 8 weeks.

5. Embryogenesis

The capacity of higher plants to produce embryos is not limited to the development of the fertilized egg. Embryos can be induced to form in cultured plant tissues and these are referred to as somatic embryos. The phenomenon was first observed in suspension cultures of *Daucus carota* (carrot). Somatic embryogenesis is a general phenomenon in plants and has been reported in more than 30 plant families.

Table 9. Gelling agents frequently used in tissue culture.

Agent	Description
Bacto-agar	Purified for use in microbiological culture media; fine granular form; dissolves rapidly yielding clear solutions; available from Difco Laboratories, Detroit.
Gelrite	Derived from bacteria (*Pseudomonas* sp.) Clear gel, use less than agar; has been mixed with agar with success; available from Kelco (see table 14.)
'Noble' Agar Phytagar-CG Phytagar-I Sigma Agar TC Agar	Derived from extracts of red algae, polysaccharides made of various glucose linkages; variable in purity and cost; may contain impurities that affect results; for most tissue cultures used at 0.5% to 1.0%; lower concentrations result in a softer medium.

Table 10. Useful units of measurement.

Weight Measurements
1 kilogram (kg) = 1000 grams = 2.2 pounds (lb)
1 pound = 16 ounces (oz) = 453.4 grams
1 ounce (oz) = 28.4 grams
1 gram (g) = 1/453 of a pound
1 milligram (mg) = 1/1000 of a gram
1 microgram (ug) = 1/1000 of a milligram
1 mole = formula or molecular weight in grams of a compound
1 millimole = 1/1000 of mole
1 micromole = 1/1000 of a millimole

Volume Measurements
1 liter (l) = 1000 ml = 1000 cubic centimeters (cm) = 1.06 quart
1 milliliter = 1 cubic centimeter (ml or cm) = 1/1000 of a liter
1 microliter (ul) = 1/1000 of a milliliter
1 gallon (gal) = 4 quarts (qt)
1 quart (qt) = 2 pints (pt) = 946 milliliters

Length Measurements
1 meter (m) = 3.28 feet = 39.4 inches (in.)
1 meter = 100 centimeters (cm) = 1000 millimeters (mm)
1 centimeter = 1/1000 of a meter
1 centimeter = 10 millimeters
1 millimeter (mm) = 1/10 of a centimeter

Temperature Conversion

$$°C = \frac{(°F-32) \times 5}{9}$$

$$°F = \frac{°C \times 9}{5} + 32$$

Concentrations
parts per million (ppm) = milligrams/liter
molar (M) = 1 mole of a compound in 1 liter of solution
millimolar (mM) = 1/1000 of a mole of a substance in 1 liter of solution

Table 11. Common acids and their chemical and physical properties.

Acid	Molecular weight	Molarity	Weight percent	Density (g/cc)
Acetic	60.05	17.4	99.5	1.051
Hydrochloric	36.47	12.4	38.0	1.188
Phosphoric	98.00	14.7	85.0	1.689
Sulfuric	98.08	17.6	94.0	1.831

Table 12. The pH values of 0.1N solutions of selected acids and bases.

Acids	pH value	Bases	pH value
Hydrochloric acid	1.0	Sodium bicarbonate	8.4
Sulfuric acid	1.2	Sodium carbonate	11.6
Phosphoric acid	1.5	Potassium hydroxide	13.0
Acetic acid	2.9	Sodium hydroxide	13.0

Table 13. Instruments, supplies and chemicals used in tissue culture.

Instruments
automatic media pipetter
balances for both milligram and gram range
Bunsen burner/alcohol lamp
carts and trays for transporting cultures and supplies
cold sterilization apparatus and filters
dissecting microscope
instrument sterilizers
magnetic stirrer/hot plate
microwave
pH meter

Nonchemical Supplies
aluminum foil
beakers — assorted sizes
brushes — assorted sizes
carboys for water storage
clamps — assorted sizes
cork borers
culture tube racks for 25 mm diameter tubes
culture tubes — 25 × 100 and 25 × 150 mm size
Erlenmeyer flasks — 250, 500, 1000, 3000 ml
filter papers — assorted
forceps — large and micro tweezers
graduate cylinders — 10 to 4000 ml
gym rubber tubing
Petri dish box for sterilization
Petri plates — 100 × 15 mm
Pipettes — 0.1 to 10 ml or adjustable pipetters
razor blades — single edge
scalpel handles and blades
scissors
spatulas
syringes and needles
test tube closures
washing bottles

Chemicals
see table 8

Growth Regulators
see tables 6 and 8

Other Growth Factors
see tables 6 and 8

Carbon Sources
see table 8

Undefined Substances
see table 6

Gelling Agents
see table 9

Somatic embryoids can develop in vitro from three sources: 1) vegetative cells of mature plants, 2) reproductive tissues, and 3) seedling tissues such as cotyledons and hypocotyls. The somatic embryos have been produced in culture by two processes, direct and indirect embryogenesis. Direct embryogenesis uses tissues that produce embryos directly without the intervention of callus. These tissues are said to have pre-embryonically determined cells. The second method, indirect embryogenesis, utilizes an intermediate callus step and is referred to as having induced embryonically determined cells. Embryogenesis follows a two step process. The first step involves the development of a callus that is stimulated by the presence of a high auxin level (2,4-D has proven effective). The second step utilizes a change to a medium without auxin and high nitrogen.

Procedure: The procedure outlined is used for the differentiation of somatic embryos from *Daucus carota*, carrot (5). This procedure assumes that a carrot cell liquid culture is available. The cell suspension is growing in a MS salt medium supplemented with 2,4-D (1.0 mg/liter) and sucrose (3% w/v)

1) Prepare Petri dishes containing MS salts, zeatin (0.2 mg/liter), sucrose (2% w/v), and agar (1% w/v). No auxin is present.

2) Aliquots (2 ml) of suspension culture are pipetted to the surface of the agar medium and dispersed.

3) Seal dishes with Parafilm and incubate at 25°C (77°F) in the dark for 2-3 weeks.

H. REGULATION OF EXPLANT GROWTH IN TISSUE CULTURE

The cytokinins and auxins (17, 24) are possibly the most important components of the medium and regulate in large measure the performance of the explant. Cytokinins are more important than auxins, which are often omitted from stage I and stage II. One commercial tissue culture laboratory uses Woody Plant Medium supplemented with 1.0 to 2.0 mg/liter BA. If this approach does not work, then different concentrations or cytokinins are used. In general BA is the cytokinin of choice for most woody and herbaceous plants and should be tested. Large scale factorial type tests combining many different cytokinins and auxins become cumbersome and costly.

For ericaceous plants, Anderson's or Woody Plant Medium (WPM) with 2iP are recommended. 2iP is the cytokinin of choice for ericaceous plants. Both WPM and Anderson's are low salt media and seem to work better with many woody plants.

Murashige and Skoog (MS) media is used with a wide range of plants. Senior author uses it routinely for herbaceous plants. If explants elongate excessively or simply languish, the original MS salt concentration can be reduced by one-half. Again BA, 2iP, kinetin and other cytokinins and auxins can be combined with the basal MS medium.

Recently, a synthetic compound, thidiazuron, has been used to proliferate a *Acer rubrum* × *A. saccharinum* hybrid, apples and other plants. It is much more effective at lower concentrations than the normal cytokinins in promoting shoot development, and concentrations of 0.5 ppm or less are used.

There is no easy way to know where to start with a specific plant. Every clone is different in its absolute tissue culture requirements. The most logical place to start is with the literature. The ENCYCLOPEDIA provides specifics for many species and gives pertinent references where additional information can be obtained. George and Sherrington's (7) reference is particularly valuable since references are provided for most plants produced in tissue culture up to 1983-84.

I. ROOTING MICROCUTTINGS AND ACCLIMATION

Tissue culture plantlets are often difficult to establish outside of culture. They are very tender (like young seedlings) and must be hardened off with care. The microcuttings are prepared for transplanting from the artificial environment of the culture vessel to the free-living existence of a greenhouse or similar environment. Microcuttings can be handled in three ways for rooting (8).

1) Individual microcuttings can be recultured in a sterile medium for rooting. The medium may have an increased auxin content (often IBA or NAA), no cytokinin, and reduced mineral content. In other cases the cuttings can be dipped in auxin before culture, or left in the auxin containing medium for only a few days and then transferred to a non-auxin medium.

2) An elongation phase may be included. In this case the cultures are transferred to a medium without cytokinin (or very low cytokinin) which may contain gibberellic acid.

3) Microcuttings can be placed into rooting medium outside of culture.

Three systems can be used to root cuttings outside of culture (13). With all systems, humidity is the critical factor and it is absolutely necessary that the microcuttings *not* dry out.

1) Flats or trays of plugs can be put into individual chambers, either clear for controlled environment rooting or translucent for greenhouse production. This is useful if the production consists of a large number of different plant types. Capillary mats can be used to provide even watering.

2) The next step would be a sweat box (polyethylene covered structure) enclosing an entire bench.

3) The third alternative is a fog system in an entire greenhouse.

McCown (16) reported that the Techniculture plug system approaches the "ideal" system for rooting some woody plants. However, not all plants respond equally. With ericaceous plants, the pH of Techniculture plugs (6.3) is too high. The trays are soaked overnight prior to planting with a dilute acidifed fertilizer solution (0.4 g 20-20-20 + 0.42 ml 85% H_3PO_4/liter, pH 5.5). Some clones die at the base in the Techniculture trays, which indicates the presence of toxins in the plugs.

To discourage mold growth some growers wash plantlets coming out of culture to remove any medium adhering to the microcuttings. Apply appropriate fungicides such as Captan or Benlate if necessary during rooting and acclimation.

Whatever the system, the next step after rooting is a gradual reduction in humidity, increased lighting, and acclimation to greenhouse conditions.

Studies have evaluated the performance of *Betula*, *Ulmus*, *Amelanchier*, *Thuja*, *Solanum and Mentha* in field plantings that were established by direct planting of micropropagules established in Techniculture plugs (16). Survival was high (90% or greater) under proper conditions of field preparation and irrigation.

J. LABORATORY REQUIREMENT AND DESIGN

A tissue culture facility should contain the following:

1) An open lab area with provision for either independent or

Table 14. Suppliers of tissue culture equipment, supplies and chemicals.

American Scientific Products
1430 Waukegan Rd..
McGaw Park, IL 60085

Aldrich Chemical Co., Inc.
940 W. Saint Paul Ave.
Milwaukee, WI 53233

American Optical Corporation
Scientific Instrument Division
Sugar and Eggert Rd.
Buffalo, NY 14215

Bausch and Lomb
1400 North Goodman St.
Rochester, NY 14602

Becton Dickinson Labware
1950 Williams Dr.
Oxnard, CA 93030

Bel-Art Products
Pequannock, NJ 07440-1992

Bellco Glass, Inc.
340 Edrudo Road
Vineland, NJ 08360

Bio-Rad Laboratories
32nd and Griffin Ave.
Richmond, CA 94804

Cadillac Plastic and Chemical Co.
2427 6th S.
Seattle, WA 98108

Calbiochem
P.O. Box 12087
San Diego, CA 92112-4180

Carl Zeiss, Inc.
444 5th Ave.
New York, NY 11042

Carolina Biological Supply Co.
2700 York Rd.
Burlington, NC 27215
or
Box 187
Gladstone, OR 97027

Cole-Parmer Instrument Co.
7425 North Oak Park Avenue
Chicago, IL 60648

College Biological Supply Co.
21707 Bothell Way
Bothell, WA 98011

Corning Medical and Scientific
Corning Glass Works
Corning, NY 14831

Costar
205 Broadway
Cambridge, MA 02139

Curtin Matheson Scientific, Inc.
P.O. Box 1546
Houston, TX 77251

Difco Laboratories
P.O. Box 1058A
Detroit, MI 40432

Dynalab Corporation
P.O. Box 112
Rochester, NY 14601

E. Leitz
Link Drive
Rockleigh, NJ 07647

Eastman Kodak Co.
Eastman Organic Chemicals
343 State Street
Rochester, NY 14650

Fisher Scientific
711 Forbes Ave.
Pittsburgh, PA 15219

Flanders Filters, Inc.
P.O. Box 1219
Washington, North Carolina 27889

Flow Laboratories, Inc.
936 W. Hyde Park Blvd.
Inglewood, CA 90302
or
1710 Chapman Ave.
Rockville, MD 20852

FMC Corporation
Marine Colloids Division
BioProducts Department
Rockland, Me 04841

Fungi Perfecti
P.O. Box 7634
Olympia, WA 98507

Germfree Laboratories
7435 NW 41 Street
Miami, FL 33166

GIBCO
519 Aldo Ave.
Santa Clara, CA 95050
or
3175 Staley Rd.
Grand Island, NY 14072

Helena Plastics
632 Irwin St.
San Rafael, CA 94901

Integrated Air Systems, Inc.
3750 Cohasset St.
Burbank, CA 91504

KC Biological, Inc.
P.O. Box 5491
Lenexa, Kansas 66215

Kelco
8355 Aero Dr.
San Diego, CA 92123

Labconco Corporation
8811 Prospect
Kansas City, MO 64132

Magenta Corp.
4149 W. Montrose Ave.
Chicago, Ill. 60641

Millipore Corporation
Bedford, MA 01730

Nalge Co.
P.O. Box 365
Rochester, NY 14502

New Brunswick Scientific Co., Inc.
1130 Somerset St.
New Brunswick, NJ

Nikon, Inc.
623 Stewart Ave.
Garden City, NY 11530

Nuclepore Corporation
7035 Commerce Circle
Pleasanton, CA 94566

Olympus Corporation
2 Nevada Dr.
New Hyde Park, NY 11042

Oxoid USA Inc.
9017 Red Branch Rd.
Columbia, MD 21045

Satorius Filters, Inc.
26575 Corporate Ave.
Hayward, CA 94545

Sigma Chemical Co.
P.O. Box 14508
St. Louis, MO 63178

Scientific Supply and Equipment, Inc.
1818 E. Madison
Seattle, WA 98122

Harry Sharp & Son
420 8th Ave. North
Seattle, WA 98108

Thomas Scientific
99 High Hill Rd.
P.O. Box 99
Swedesboro, NJ 08085-0099

Water Distillers
730 Goodwin Ave.
Penngrove, CA 94951

Whatman Inc.
9 Bridewell Place
Clifton, NJ 07014

Wheaton Scientific
1000 North 10 Street
Millville, NJ 08332

United States Plastic Corp.
1390 Neubrecht Rd.
Lima, OH 45811

general working spaces. Some equipment and supplies will be necessary for all workers and should be readily available.

2) Culture rooms where cultures can be grown under controlled conditions of temperature and light.

3) Transfer areas for aseptic manipulations.

4) Equipment for cleaning glassware — sinks and washing machines.

5) Cabinet and shelf space for storage of glassware and chemicals.

6) Equipment for sterilizing media, instruments, etc.

7) Utilities such as electric, water and gas.

8) A supply of distilled and deionized water.

9) Various instruments, supplies, and chemicals (table 13).

Design of the laboratory is a function of current and projected production (3). Factors to be considered when designing a laboratory include methods: 1) to improve materials flow through the laboratory, 2) to strictly isolate clean areas from other areas of the facility, 3) to improve movement of personnel in and out of clean areas, and 4) to plan so that future expansion can be accommodated readily. In recent years there have been a number of publications on this subject that can be consulted for additional information (2, 4, 12, 29). The paper by Broome (3) is a particularly good reference.

The work stations in a laboratory facility are divided into clean and dirty areas and include the office, shipping and receiving, greenhouse, storage, preparation room, cleanup and sterilizing, transfer room, growth room(s), employee center and possibly cold storage (3). The size of each room will vary depending on the projected number of units and plant type propagated.

Dirty areas. The office is the main entrance to the building and provides controlled access to the clean areas of the laboratory. The shipping and receiving area provide room for a number of functions including the shipping and receiving of materials and plants, preparation of plants for shipping, water heating and purification, and maintenance. If a greenhouse is present it should exit from receiving and shipping, and provide storage for supplies used in that area. General storage provides room for supplies used in both dirty and clean areas. An employee center as a separate room is not required but provides an area for workers to relax.

Clean areas. The preparation room is that part of the facility used to prepare the media. Equipment and materials needed would include storage for chemicals and other materials for media preparation; equipment such as pH meter, refrigerator and freezer, balances, mixers, hot plates, microwave, automatic pipetters; and work space including sinks, bench surface area, and storage. An isolated washing and autoclaving room provides space for isolating heat and humidity generating functions of a laboratory. A fan can be used to discharge excess heat and humidity. In smaller labs commercial type pressure cookers can be used. The transfer room provides space for the transfer hoods, and cooling and storage of media. Growth rooms contain open shelves with lighting provided by cool and warm fluorescent bulbs. A light level of 35 to 40 micromole $s^{-1} m^{-2}$ at the top of the culture is adequate. Ballasts should be remotely wired outside of the culture room to provide better temperature control. Multiple growth rooms provide insurance in case of breakdown, allow for growing crops that require different environmental conditions, and provide space for rooting the microcuttings if a greenhouse is not available. A greenhouse is useful for rooting cuttings or the growth of the plantlets. Fog type systems are better for rooting the microcuttings than mist type systems.

K. SPECIFIC INFORMATION USEFUL IN TISSUE CULTURE

The following tables contain information that is useful in tissue culture propagation. The tables are listed here with their description for quick reference by the reader.

Table 1. Commercial tissue culture laboratories in the United States.

Table 2. Selected tissue culture reference books.

Table 3. Current prices of selected woody and herbaceous plants available from commercial tissue culture laboratories.

Table 4. Surface sterilants used to treat explants prior to culture.

Table 5. Formulations of selected plant tissue culture media.

Table 6. Plant growth substances and the range of concentrations frequently used in tissue culture.

Table 7. Atomic weight of elements used in tissue culture.

Table 8. Molecular weight of compounds used in tissue culture.

Table 9. Gelling agents frequently used in tissue culture.

Table 10. Useful units of measurement.

Table 11. Common acids and their chemical and physical properties.

Table 12. The pH values of 0.1 N solutions of selected acids and bases.

Table 13. Instruments, supplies and chemicals used in tissue culture.

Table 14. Suppliers of tissue culture equipment, supplies and chemicals.

LITERATURE CITED

1. Anderson, W.C. 1975. Propagation of rhododendrons by tissue culture: Part 1. Development of a culture medium for multiplication of shoots. Proc. Int. Plant Prop. Soc. 25:129-135.
2. Broome, O.C. 1986. Laboratory design. pp. 351-364. In: R.H. Zimmerman, R.J. Griesbach, F.A. Hammerschlag, and R.H. Lawson (eds). Tissue culture as a plant production system for horticultural crops. Martinus Nijhoff. Boston.
3. Broome, O.C. 1986. Summary of panel discussion on laboratory design. pp. 365-366. In: R.H. Zimmerman, R.J. Griesbach, F.A. Hammerschlag, and R.H. Lawson (eds). Tissue culture as a plant production system for horticultural crops. Martinus Nijhoff. Boston.
4. Brown, D.C.W. and T.A. Thorpe. 1984. Organization of a plant tissue culture laboratory, pp. 1-12. In: I.K. Vasil (ed). Cell culture and somatic cell genetics of plants, Volume 1. Academic Press, Orlando, FL.
5. Dodds, J.H. and L.W. Roberts. 1982. Somatic embryogenesis. pp. 89-97. In: Experiments in plant tissue culture. Cambridge University Press. Cambridge, England.
6. Gamborg, O.L., R.A. Miller and K. Ojima. 1968. Nutrient requirements of suspension cultures of soybean root cells. Exp. Cell Res. 50:148-151.
7. George, E.F. and P.D. Sherrington. 1984. Plant propagation by tissue culture. Exegetics, Ltd. Hants., England.
8. Hartmann, H.T. and D.E. Kester. 1983. Plant propagation: Principles and practices. Prentice-Hall, Inc., Englewood Cliffs, NJ.
9. Heller, R. 1953. Recherches sur la nutrition minerale des tissue vegetaux cultives in vitro. These Paris and Ann. Sci. Nat. Bot. Biol. Veg. 14:1-223.
10. Hollings, M. 1965. Disease control through virus-free stock. Ann. Rev. Phytopathology 3:367-396.
11. Horst, R.K., S.H. Smith, H.T. Horst, and W.A. Oglevee. 1976. In vitro regeneration of shoot and root growth from meristematic tips of *Pelargonium* × *hortorum* Bailey. Acta Horticulturae 59:131-141.
12. Kyte, L. 1983. Plants from test tubes: An introduction to micropropagation. Timber Press, Portland, OR.
13. Linsmaier, F.M. and F. Skoog. 1965. Organic growth factor requirements of tobacco tissue cultures. Physiol. Plant. 18:100-127.
14. Lloyd, G. and B.H. McCown. 1980. Commercially-feasible micropropagation of mountain laurel, *Kalmia latifolia*, by use of shoot-tip culture. Proc. Int. Plant Prop. Soc. 30:421-427.
15. McCown, B.H. and R. Amos. 1979. Initial trials of commercial micropropagation with birch. Proc. Int. Plant Prop. Soc. 29:387-393.
16. McCown, D.D. 1986. Plug systems from micropropagules. pp. 53-60. In: R.H. Zimmerman, R.J. Griesbach, F.A. Hammerschlag, and R.H. Lawson (eds). Tissue culture as a plant production system for horticultural crops. Martinus Nijhoff, Boston.
17. Miller, C.O., F. Skoog, M. Saltze, and F.M. Strong. 1955. Kinetin a cell division factor from deoxyribonucleic acid. J. Amer. Chem. Soc. 77:1329.
18. Morel, G.M. 1964. Tissue culture: A new means of clonal propagation of orchids. Amer. Orch. Soc. Bul. 33:473-478.
19. Morel, G.M. 1975. Meristem culture techniques for the long-term storage of cultivated plants. pp. 327-332. In: O.H. Frankel and J.G. Hawkes (eds). Crop genetic resources for today and tomorrow. Cambridge University Press, Cambridge, England.
20. Murashige, T. 1974. Plant propagation through tissue culture. Ann. Rev. Plant Physiol. 25:135-166.
21. Murashige, T. and F. Skoog. 1962. A revised medium for rapid growth and bioassays with tobacco tissue cultures. Physiol. Plant. 15:473-497.
22. Nitsch, J.P. and C. Nitsch. 1969. Haploid plants from pollen grains. Science 163:85-87.
23. Shabde-Moses, M. and T. Murashige. 1979. Organ culture. In *Nicotiana*. Procedures for experimental use. R.D. Durbin (ed). pp. 40-51. U.S. Department of Agriculture. Tech. Bull. No. 1586.
24. Skoog, F. and C.O. Miller. 1957. Chemical regulation of growth and organ formation in plant tissues cultured in vitro. Symp. Soc. Exp. Biol. 11:118-131.
25. Smith, R.H. and T. Murashige. 1970. In vitro development of the isolated shoot apical meristem of angiosperms. Amer. J. Bot. 57:562-568.
26. Smith, S.H. and W.A. Oglevee-O'Donovan. 1977. Meristem-tip culture from virus infected plant material and commercial implications. pp. 453-460. In W. R. Sharp, P.O. Larsen, E.F. Paddock, V. Raghavan (eds). Plant cell and tissue culture. The Ohio State University Press. Columbus, Ohio.
27. Stoltz, L.P. 1979. Getting started in tissue culture — equipment and costs. Proc. Int. Plant Prop. Soc. 29:375-381.
28. Street, H.E. (ed) 1977. Plant tissue and cell culture. Univ. of Calif. Press, Berkeley, CA.
29. Wetherell, D.F. 1982. Introduction to in vitro propagation. Avery Publishing Group, Inc., Wayne, NJ.
30. White, P.R. 1963. Cultivation of animal and plant cells. Ronald Press, N.Y.
31. Zimmerman, R.H. and O.C. Broome. 1980. Blueberry micropropagation. Proceedings of the conference on nursery production of fruit plants through tissue culture — Applications and feasibility. pp. 44-47. USDA — SEA Ag. Res. Results. ARR-NE-11.

Encyclopedia

Abelia × grandiflora Glossy Abelia

SEED: The fruit is a one-seeded leathery achene and should be collected and sown when ripe. Since glossy abelia flowers from June until frost, the fruits mature over a long period of time. Unless one is interested in breeding and/or selecting superior seedlings, seed production is not the logical choice.

CUTTINGS: June through August is the best time to take cuttings although fall and late winter hardwood cuttings will work. One nurseryman uses mid-July through October cuttings. A hormone treatment hastens, unifies, and improves the rooting response. A range of 1000 to 2000 ppm IBA liquid or talc proved optimum with good rooting occurring in 4 to 5 weeks. Rooting percentages will approach 90% to 100%. 1000 ppm NAA in talc also produced superior results. Hardwood cuttings root but percentages are generally not good. Ten to fifteen percent rooting was reported. Early June (GA), untreated or 5000 ppm IBA, rooted 60% in 6 weeks in perlite under mist. Cuttings doubled their size while in rooting bench.

Glossy abelia can be transplanted immediately after rooting with excellent success. The species is also photoperiod sensitive, and a night interruption period of two hours induces continued growth and flowering. *Abelia* 'Edward Goucher' is similar to A. *× grandiflora* in rooting requirements. Early July cuttings rooted 100% in 4 weeks under mist. Work in Hawaii indicated the need for a well-drained medium with perlite producing the best results [*The Plant Propagator* 24(3):5-6. 1978].

Abeliophyllum distichum Korean Abelialeaf

SEED: Fruit is a two-celled compressed capsule that is winged all around. Seeds should germinate immediately if collected and sown when ripe.

CUTTINGS: Softwood cuttings root well when treated with a hormone, preferably IBA in the range 1000 to 3000 ppm. Cuttings of half-ripened (greenwood) taken in July will root when treated with IBA. Bottom heat may be helpful. Senior author has not had good success with this species.

Abies Fir

Pyramidal, elegantly formal coniferous evergreens with bluish to dark green needles. The cones ripen in fall and the period for collecting is short...about one month. Cones disintegrate after seed dispersal. Cones are placed in sacks and air dried for several weeks or months in drying sheds. Cones can be kiln (85 to 100°F) or air dried for 1 to 3 weeks at 70 to 85°F. Seeds are separated from cones mechanically. Seed can be stored in sealed containers under refrigeration. Seed with 9 to 12% moisture content has been stored 5 years.

SEED: Interestingly, the literature offers conflicting information relative to the germination requirements. Seeds are described as requiring no cold stratification to 2 to 3 months. Best results are otained with fresh seed that is fall planted or stratified 1 to 3 months. One month cold stratification proves effective for most firs. Seed collection is not particularly easy since cones shatter at maturity. Collect cones in fall, allow cones to dry inside until seeds are released. Sow seed immediately outside or stratify.

CUTTINGS: Difficult and not practiced commercially except with some of the dwarf cultivars (*q.v.*). Cuttings display topophysis and lateral branches when rooted tend to grow horizontally (plagiotropic) while vertical leaders grow upright (orthotropic). Obviously this confounds the problem of finding sufficient cutting material since the upright leaders tend to be higher in the tree. Firs, like spruces, often develop one or two coarse brittle roots. Transplanting is seldom easy and, if successful, the rooted cutting displays no great propensity to grow. Take cuttings after several freezes, 8000 ppm IBA-talc, well drained medium, bottom heat, polytent. One researcher reported that cuttings of *Abies alba* took 9 months to root when handled as above; untreated cuttings did not root.

GRAFTING: Cultivars are side- or veneer-grafted on seedling understock in winter. *Abies alba*, silver fir, and *Abies balsamea*, balsam fir, are considered suitable understocks for all *Abies* clones. *Abies nordmanniana*, Nordmann fir, is compatible with short needle firs. *Abies concolor*, white fir, is compatible with long needle firs. *Abies grandis*, giant fir, is also used. General feeling that *Abies* species are compatible with each other and understock does not present a significant problem.

The following procedure produces good results. Understock in heeled in peat under glass keeping soil temperature at ±50°F in November; raised to 58°F in January; graft in January-February when new white roots are visible; grafts heeled in peat, 60°F bottom heat, polytent with union above peat line; grafts gradually hardened and placed outside in ground beds under about 50% shade. One east coast nurseryman insisted that shading was necessary. A long-term study of interest is presented in *The Plant Propagator* 31(2):6-8 (1985). In short, 6 clones of A. *amabilis* were grafted on rootstocks of A. *amabilis*, A. *grandis*, A. *lasiocarpa*, and A. *procera*. First year survival was acceptable for all species; after 3 years A. *procera* had significantly lower survival. After 3 years, height, growth and vigor of scions on A. *amabilis* understock were significantly better than on other species rootstocks.

At the Arnold Arboretum the following rootstock/scion combinations have worked. A. *balsamea* with A. *alba*, A. *amabilis*, A. *cilicica*, A. *fargesii*, A. *fraseri* 'Prostrata', A. *lasiocarpa* 'Compacta', A. *procera* 'Glauca', A. *veitchii*, A. *veitchii* var. *olivacea*; A. *concolor* with A. *cephalonica*, A. *concolor* 'Violacea'; A. *firma* with A. *alba*, A. *concolor* 'Conica' and 'Violacea', A. *firma*, A. *homolepis*, A. *koreana*, A. *lasiocarpa* 'Compacta'.

Abies grandis is used in England in place of A. *alba* because of disease problems with the latter. Based on the literature, almost any fir can serve as suitable understock for another species.

Abies balsamea Balsam Fir

SEED: See *Abies*.

CUTTINGS: See discussion under *Abies*. A few selected forms are propagated from cuttings most notably 'Nana'. Mid-December-January (ideal time in Oregon), 8000 ppm IBA-talc plus Benlate. 'Nana' — after frost (New York), current season's growth, 1½ to 2½" long, strip lower ½" of needles, 8000 ppm IBA-talc, sand, cool greenhouse, slow rooter but roots well, no disturbance until new growth occurs and hardens, then transplant. A. *pinsapo* 'Glauca' is also rooted this way but makes only 1 or 2 roots.

TISSUE CULTURE: Dormant bud (adult) explants produced adventitious shoots from pith area. Embryonic shoots yielded direct adventitious shoot formation. Refer to the following: *In Vitro* 12:333 Abst. 158 (1976), *In Vitro* 13:41-48 (1977). Applied and fundamental aspects of plant cell, tissue and organ culture, pp. 93-108 (1977), Reinert and Bajaj (eds).

Abies concolor White or Concolor Fir

SEED: See discussion under *Abies*.

CUTTINGS: Difficult; to authors' knowledge no cultivars of the species are produced from cuttings.

Abies fraseri Fraser Fir

SEED: Interesting study (*Proc.* SNA *Res. Conf.* 225-227, 1980) showed no seeds germinated at 95°F and few at 50°F. At constant temperatures, 59°F with a 45 minute light treatment, seeds germinated 35%. Best results (43%) were obtained at alternating temperatures of 86°F (8 hr)/68°F (16 hr) with light treatment.

CUTTINGS: Considerable interest in this evergreen because of value to Christmas tree industry and the fact it takes 5 years to produce a suitable transplant from seed. Early October (North Carolina), 5-year-old stock plants, cuttings chilled at 41°F for 4 to 10 weeks, wound, 5000 ppm IBA-solution in 50% isopropyl, peat:sand, mist, rooted 80 to 90% in 10 weeks. Rooting decreased as age of stock plant increased. 5000 ppm IBA-solution increased rooting of 12 and 22-year-old plants. Chilling of early sampled cuttings was required. Without this treatment, many cuttings died or partly defoliated in mist bed. IBA in combination with chilling was also necessary for high rooting percentages. Lateral cuttings maintained plagiotropic growth pattern, i.e., grew horizontally.

Cuttings were also taken on March 1 (N.C.) from 5, 12, and 22-year-old trees. The treatments were approximately the same and when treated with 5000 ppm IBA in 50% isopropyl alcohol plus wound rooted 92, 50, 29% depending on age of stock plant. Obviously cuttings from younger trees root in high percentages. See *Proc.* SNA *Res. Conf.* 25:222-225 (1980) and *HortScience* 15:96-97 (1980).

Softwood cuttings collected in late June, mid-July and early August have been rooted successfully with IBA treatments.

GRAFTING: Used as understock for A. *koreana* 'Prostrate Beauty', A. *lasiocarpa* 'Compacta', A. *procera* 'Glauca' and A. *pinsapo* 'Glauca'.

Abies koreana Korean Fir

CUTTINGS: Interesting fir in that it cones at a young age. One report noted that it was easy to root but resultant plants were prostrate. See *Abies* for explanation. 'Prostrata' or 'Prostrate Beauty', early to mid-June (Boston), 3000 to 8000 ppm IBA-talc, peat:perlite, mist, rooted 50 to 60% in 2 to 3 months.

Acanthopanax henryi Henry's Aralia

SEED: A lustrous black, 2 to 5-seeded, ⅓" diameter berry that ripens in September-October and should be collected at this time. Seeds should be extracted from pulpy matrix. A 6 month warm/3 month cold stratification period has yielded good results. Some root radicals developed during cold treatment. After sowing, germination was completed in seven days.

CUTTINGS: Softwood collected in mid to late June, treated with 3000 ppm IBA-talc, sand, mist, rooted 55%.

OTHER METHODS: Division is effective if just a few plants are desired. Divide with a sharp spade before growth initiates in spring. Transplant to permanent place in landscape, nursery row or container.

Acanthopanax sieboldianus Fiveleaf Aralia

SEED: Same as described under A. *henryi*.

CUTTINGS: Mid-August, 8000 ppm IBA-talc plus thiram, mist, rooted 95% in 30 days. There is a beautiful, delicate, creamy-variegated form that can be propagated from softwood cuttings.

Acer Maple

The maples are at the forefront of landscape plants. Their importance in the marketplace is significant. *Acer palmatum*, Japanese maple, *Acer platanoides*, Norway maple, *Acer rubrum*, red maple, *Acer saccharinum*, silver maple, and *Acer saccharum*, sugar maple, are the most important landscape species. Propagation techniques have shifted from budding to own-root cuttings and tissue culture. Most quality growers will not sell a budded red maple, for example, because of the incompatibility problems associated with the cultivars. See Tubesing, *The Plant Propagator* 21(3):11-13 (1975) for a bibliography of maple propagation. See: Propagation of Acers by cuttings, *The Plant Propagator* 12(1):4-6 (1966).

Acer barbatum (A. *floridanum*) Florida Maple

SEED: See *Acer saccharum*. The Florida maple has many characteristics common to sugar maple. Since it is found wild in the Coastal Plain, there is inherent heat tolerance. Fall color is akin to that of sugar maple and for that reason it is a valuable small tree where excessive summer heat limits successful culture of sugar maple.

Acer buergeranum Trident Maple

SEED: Exceedingly easy. Collect in October (when wings are yellow-brown) and sow outside or stratify for 2 to 3 months at 40°F. Some seeds generally germinate in bag. Seeds are virtually 100% sound and percentages will be high. Senior author's research has indicated this is an easy species to grow from seed. One report noted that extremely dry seed imported from Korea was soaked, cold stratified and germinated only 20%. However, the following spring, germination was very heavy.

CUTTINGS: Successful cutting propagation has been elusive. Cuttings taken from a mature tree in late June and treated with either 2.0% ppm IBA or 25% chloromone, fine sand, 60 and 36% rooting. Trees were at least 4 to 5 years old. Late June, 3 to 4 nodes, tip pair removed, wound, 8000 ppm IBA-talc, 9 perlite: 1 peat, mist, 70°F bottom heat, rooted 95%. Five cultivars rooted 75 to 95% with same treatment. No mention was made as to the age of stock trees. Senior author has had no success with cuttings from mature trees even with high IBA (1% solution). There are definitely superior selections that could be made.

GRAFTING: Cultivars can be pot grafted on seedling understock (see grafting section for details). Several cultivars exist and Vertrees' book serves as a good reference. Some cultivars have been rooted successfully (see cuttings above).

Acer campestre Hedge Maple

SEED: Benefit from early harvesting and should be collected when wings are yellowish brown. Seeds should be fall planted or stratified. If seeds are allowed to dry, germination will not occur until the second year. One month warm/3 to 6 months cold stratification is necessary to break dormancy. Also soaking dried seed in warm water, then 4 months cold stratification may be good. European work reported dormancy was associated with the pericarp (fruit wall) and testa (seed coat). Best germination occurred after one month cold/6 to 7 months warm.

CUTTINGS: Early June (Michigan), mature tree, 8000 ppm IBA-talc, mist, 75°F bottom heat, rooted 75%. Cuttings collected 2 weeks earlier or 2 and 4 weeks later rooted 40, 4 and 24%, respectively [PIPPS 29:345-348 (1979)]. 'Compactum' — Grafted in winter, grafted plant grew 2½′ high, 2′ wide in 9 years. Late June (Oregon), wound, 8000 ppm IBA-talc, perlite: peat, mist, bottom heat, rooted 25% in six weeks. 'Pulverulentum' — Rooted 90% under same conditions in Oregon. Sixty-eight percent, early August, 8000 ppm IBA-talc and thiram, mist.

GRAFTING: If cultivars are pot grafted, this should be accomplished on seedling understocks. 'Queen Elizabeth' is produced by budding on seedling understock in July-August (Oregon).

Acer capillipes

SEED: Collect in fall with a slight yellow-brown color to the samara and direct sow. Stratification for 3 or 4 months results in heavy germination. Seedlings are susceptible to damping off and a preventative fungicidal treatment is advisable. "Green" seed may germinate without treatment but best results are assured when above specifications are followed.

CUTTINGS: Collected from a 40-year-old tree in mid August, wound, 1% IBA, rooted 92% when evaluated 9 weeks later. Hormone treatment is essential, for cuttings wounded and not treated with hormone rooted only 4%. July (Ohio) cuttings, 2% IBA-talc, sand, mist, rooted 100% and all were overwintered successfully. Cuttings treated with 8000 ppm IBA-talc rooted only 68%.

NOTE: This is a snakebark maple with beautiful green bark striped with white. Seeds are often void based on authors' observations. This is a *monoecious* species and cross pollination may be necessary to assure good seed set.

Acer carpinifolium Hornbeam Maple

SEED: Collect in fall and sow immediately. Germination will occur in spring. Three months cold stratification will satisfy dormancy requirement.

CUTTINGS: Late July, 4 sand:1 peat, mist, 70°F bottom heat, rooted 73% in 9 weeks. Mid August (Boston), 8000 ppm IBA-talc, sand, mist, rooted 80%. Late June (Oregon), 3-4 nodes, tip pair removed, wound, 8000 ppm IBA-talc, 9 perlite:1 peat, mist, 70°F bottom heat, rooted 90% in 6 weeks.

Acer circinatum Oregon Vine Maple

SEED: Warm (1 to 2 months)/cold (3 to 6 months) stratification period is recommended. Seeds collected when in color transition (also still green) and planted should germinate the following spring. If allowed to dry, dormancy becomes more pronounced probably due to pericarp impermeability. Seed collected in September (Oregon), stratified for 4 months at 41°F, germinated in high percentages the first year. Seed collected in October after drying and given same treatment, germinated the second spring after planting.

CUTTINGS: Difficult to root. Layering has been successful but is extremely slow.

GRAFTING: 'Monroe', a dissected-leaf form, was grafted successfully on *Acer palmatum* understock (80%) but only 5% on *A. circinatum*.

Acer cissifolium Ivy-leaved Maple

SEED: Little sound seed is set, at least on trees the authors have observed. Three months cold stratification or fall planting will produce good germination. Senior author provided 3 months cold stratificaiton to seeds collected from a grove of trees (cross pollination insured) with excellent germination 3 weeks after planting.

CUTTINGS: This is one of the easiest maples to root. August, 35-year-old tree, wound, 1% IBA-talc, polytent, rooted 100%. Wounding definitely improved rooting response. Untreated cuttings rooted 80%. July cuttings (Ohio), 1% IBA-talc, sand, mist, rooted and overwintered 100%.

Acer davidii David Maple

SEED: Two to 3 months cold stratification results in excellent germination. Wise to harvest seed with a slight green color remaining in the samara. See A. *capillipes*.

CUTTINGS: Generally difficult to root.

Acer ginnala Amur Maple

SEED: Reports for seed treatments are variable but this depends on condition of seed at time of planting. Fresh seed can be directly sown and will germinate in spring. Three months cold stratification will also suffice. If seed becomes excessively dry, 1 to 2 months warm/3 to 4 month cold, or light scarification followed by 3 to 4 months cold is required to facilitate good germination.

CUTTINGS: Considerable research has been conducted with this species owing to its adaptability and popularity. Softwood cuttings, June, 1000 to 5000 ppm IBA-solution, peat: perlite, mist, rooted readily (90%). There is a significant decline in rooting as the tissue matures. The authors have found this true for the species and 'Compactum'.

Greenhouse forced shoots (2 to 3″ long) from 3 to 6-year-old plants, 8000 ppm IBA-talc, sand, polytent, stuck in January rooted 90% in 3 to 4 weeks. This forcing method worked well on *Acer palmatum* and *Acer triflorum* and may represent a method for rare and difficult-to-root plants although commercially it probably does not have much merit.

Late June (Michigan), 8000 ppm IBA-talc, mist, 75°F bottom heat, rooted 86%. Cuttings taken 2 and 4 weeks earlier rooted 45 and 35%, respectively.

'Durand Dwarf' and 'Compactum' can be rooted following above methodology.

After rooting in July the species and 'Compactum' can be induced to produce a growth flush under normal daylength by applying fertilizer.

Acer griseum Paperbark Maple

SEED: The biggest problem is poor seed quality. Over the years, seeds the senior author collected ranged from 1 to 8% sound (embryos present). An English source noted 2% sound seed, while cut tests over a number of years at the Arnold Arboretum showed 20% sound seed from a tree in 1962 and 80% in 1968 (June test). Seed production is extremely variable from year to year even on the same tree.

Seeds are doubly dormant and if fall planted require 2 years with some germinating the third year and beyond. The pericarp wall is extremely tough and dormancy is caused by a physical barrier as well as internal embryo conditions (*HortScience* 16:341). The senior author has cold stratified seed for 90 days, split the fruit wall, extracted the embryos, planted them in vermiculite and most grew. This method is not feasible for large scale propagation, but if a few trees are desired and one does not want to wait 2 to 3 years, it might be worth the effort.

CUTTINGS: Extremely difficult (perhaps impossible) to root by conventional techniques. Over the years, using many clones (10 to 15 years and older), times, hormones, etc., senior author had one cutting that rooted. There is evidence for clonal differences in rootability. Cuttings collected in late June (Ohio) from 8-year-old trees, wounded, 2% IBA-talc, sand, mist, rooted 17 to 80%. Fifty-five percent of cuttings of one clone were overwintered successfully. No information was provided for the others.

Late June cuttings from 1 and 2-year-old seedlings rooted 80 to 85% when treated with 8000 ppm IBA-talc, sand, mist. The cuttings (young or from more mature wood) should not be transplanted after rooting but overwintered in a cool greenhouse, pithouse or suitable structure where temperatures range from 33 to 40°F. Even cuttings from seedlings, if collected too late, root in low percentages. The same individual who achieved 80 to 85% rooting also had 30% rooting by taking the cuttings too late.

One-hundred percent rooting of seedling material was reported using a combination of catechol (4.5×10^{-3} M) and IAA (1.1×10^{-3} M) - 24 hour soak. Several combinations of the above stimulated the greatest number of roots [*HortScience* 18(3):352-354 1983]. Classic paper [PIPPS 34:570-573 (1984)] describes a commercial nurseryman's unbridled success. The basic recipe: Seedling stockblock, pruned in March (Rhode Island) to induce long shoots, 3rd week in June (timing is critical, wood can't be too hard or soft), sand, 8″ long cutting, tip removed with only 1 pair of leaves remaining, 8000 ppm IBA-talc, 3″ deep in medium, mist, Benlate and Captan applied regularly, rooting takes place in 8 to 10 weeks, lifted with a spading fork, average 60% rooting. Rooted cuttings are potted in 2½″ clay pots in soil: peat: sand and placed pot to pot in a greenhouse to reroot. Plants syringed and given bottom heat until mid October. Pots moved to deep pit house, covered with ½″ peat moss and watered in. Maintained at minimum 28°F during winter. When shoots emerge in spring (June) plants are planted in outdoor beds under 50% shade. After 3 years they are sold for lining out material or transplanted to the field.

Extended photoperiod to force the cuttings into growth after rooting has been tested but the cuttings did not respond.

Root cuttings do not work for this or any of the trifoliate maples.

GRAFTING: For best results, A *griseum* should be grafted on seedling A. *griseum*. The other trifoliate maples, according to some, do not serve as suitable understocks. Work in Holland indicated the grafts were successful for 2 to 3 years but eventually failed. Senior author and cohorts budded A. *griseum* on A. *saccharum* in August with 40% success. The budded stock was dug, brought into the greenhouse and grew 18 to 30″ the first three months. One commercial grower grafts A. *griseum* on A. *saccharum* but long term prospects are open to question. Senior author has two of these grafted trees in his garden. A conspicuous bulge is noticeable at union (A. *saccharum* is twice diameter of A. *griseum*).

Acer henryi Henry Maple

SEED: Should be planted in the fall with germination occurring in spring. Also 3 months cold stratification should facilitate good germination.

CUTTINGS: Late June cuttings from 4 to 5 year old tree, 2% IBA-talc, sand, mist, rooted 100%.

NOTE: This is a handsome maple similar to *Acer cissifolium* but larger. Based on available literature and authors' observation, its propagation requirements are similar to that species.

Acer japonicum Fullmoon Maple

SEED: Treatment should be similar to that for A. *palmatum*. If collected fresh and planted, reasonable to good germination will occur the following spring. Dried seed may be doubly dormant and take two years to germinate. Three months cold stratification is recommended for fresh seed. Three months warm/3 months cold is needed for dry seed.

CUTTINGS: Mid June (Boston), treated with 8000 ppm IBA-talc plus thiram, mist, rooted 66% and 100% in 6 to 7 weeks. Cuttings were moved to cold storage in early December and returned to greenhouse in March. After growth ensued (early April) the rooted cuttings were potted successfully.

'Aconitifolium' — Stock plants forced in greenhouse in March. Six to 10″ long, 1 to 4 node cuttings were taken late May - early June, wounded on one side, 2% IBA-talc, sand, mist, 70 to 75°F bottom heat. Cuttings rooted in 30 to 45 days, were potted in rose pots, returned to mist bed until roots reached bottom of container. After removal from mist, cuttings were maintained in greenhouse, placed under extended photoperiod from 8 p.m. to midnight, and later intermittent supplementary light. Plants grew until November. Cuttings that did not put on new growth after rooting did not survive the winter. 'Aureum' was treated as above but success was not as great.

GRAFTING: Most cultivars are grafted on A. *palmatum* seedlings. This is accomplished by a side, wedge, or top graft on seedling understock in January-February. Success ranges from 95 to 100%. Summer (late July to late August) grafting has been practiced in England with 80 to 95% success.

Acer leucoderme Chalkbark Maple

SEED: Treat like *Acer saccharum*. Sow in fall or provide 3 months cold stratification. High percentage of empty fruits.

Acer mandshuricum Manchurian Maple

SEED: Difficult to find sound seed. Apparently this species is not as difficult as the other trifoliate maples but still is doubly dormant. Excised embryo of A. *mandshuricum* germinated in the presence of light while the other trifoliate species require light and gibberellic acid.

CUTTINGS: See *Acer griseum*. Very difficult to root. Mid June cuttings, 8000 ppm IBA-talc and thiram, mist, rooted 50%. No notation on age of stock plants or whether they were successfully overwintered.

Acer miyabei Miyabe Maple

SEED: Fall planting or 3 months cold stratification will overcome dormancy. Percent sound seed is usually low. Seed should be collected as early as possible to prevent seed coat drying and a subsequent deeper dormancy. This species is very close to A. *campestre* and may have similar dormancy problems.

CUTTINGS: Mid-May, Chloromone, rooted 44% in 14 weeks. Late July from basal shoots, 8000 ppm IBA-talc plus thiram, rooted 60% after 9 weeks. Root systems were sparse. Late June cuttings from 25-year-old tree, sand, mist, 2% IBA-talc or 25% Chloromone, rooted 98 and 95%, respectively. Mid-June (Chicago), terminal cuttings with tip removed, 5-6" long, soaked in Benlate: water solution, peat:perlite, 70°F bottom heat, mist, rooting evaluated after 9 weeks. Cuttings from 55-year-old tree rooted 20, 37, and 23% with 0.5, 1.0 and 1.5% IBA-5 second dip; cuttings from 7-year-old hedge plants rooted 57, 70, 40 and 43% with 0.1, 0.5, 1.0, and 1.5% IBA-5 second dip, respectively.

GRAFTING: The Morton Arboretum, Lisle, IL has grafted the species on seedling understock or seedling *Acer campestre*. Generally, it is considered important to graft milky-sap species on milky-sap understock.

Acer negundo Boxelder

SEED: Dioecious with the fruits borne in pendent racemes. Usually great quantities are present. Three months cold stratification or fall planting are recommended. If the seed is extremely dry, either a hot water soak or mechanical/chemical rupture of the pericarp is necessary before stratification.

CUTTINGS: Surprisingly limited information is available on the species. Reports indicate that it can be rooted from cuttings. Softwood to greenwood transition would be best. Cuttings of 'Aureo-variegatum' taken in mid-September, 8000 ppm IBA-talc, rooted readily by mid-October.

GRAFTING: There are many cultivars and European nurserymen use side, whip and tongue grafts, or chip budding. Chip budding has been mentioned as being more effective for 'Elegantissima' than the other methods. Seedling understocks of the species are best.

TISSUE CULTURE: Immature embryo explants produced callus which led to root formation only. Z. *Pflanzenphysiol.* 99:191-198 (1980).

Acer nigrum Black Maple

SEED: Similar to A. *saccharum*, 3 months cold stratification or fall plant.

CUTTINGS: Early June, early July, mid-July cuttings rooted 60, 56, 52%, respectively, when treated with 8000 ppm IBA-talc plus 3.8% Benlate, peat: perlite, mist. See PIPPS 31:517 (1981).

GRAFTING: English literature indicates A. *nigrum* can be successfully grafted on A. *pseudoplatanus* or A. *coriaceum*. Several old forms: 'Ascendens' (upright form from Rochester parks) and A. x *senecaense* (A. *leucoderme* and A. *saccharum*) have been successfully grafted on A. *nigrum* and A. *saccharum* understocks. Newer forms ('Green Mountain', 'Green Column') are budded on sugar maple understock.

Acer nikoense Nikko Maple

SEED: See *Acer griseum* for specifics. Seed exhibits a double dormancy. Even synthetic warm/cold stratification periods do not seem to hasten germination. If fall planted, seed will germinate in two to three years.

CUTTINGS: Like the other trifoliate maples, extremely difficult to root unless treated like the seedling cuttings of *Acer griseum*.

GRAFTING: The best understock is seedling A. *nikoense*. Unfortunately, production of seedlings is confounded by lack of sound seed and double dormancy.

Acer palmatum Japanese Maple

SEED: Germination success is variable depending on source of seed, time of collection, and pretreatment. In general seed should be collected when green or red and before it dries on the tree. Plant it directly and some germination will take place the following spring. Dry seed should be soaked in water at 110°F for 2 days and then given 3 to 5 months cold stratification before sowing. Seed that is collected green/red, cleaned, dusted with fungicide, stratified in moist peat at 40°F for 3 to 5 months should germinate. However, dried seed from Japan that was sown in the fall germinated over a 5-year period.

Another key to success is sound seed. Nurserymen should conduct cut tests to determine the percent viable seed. Any seed purchased dry should be subject to a cut test. The warm water soak is also recommended followed by cold stratification before sowing. GA soaks have not proven beneficial in promoting germination.

CUTTINGS: Many individuals root Japanese maples from cuttings. The success ratio varies from 0 to 100%. In general, softwood cuttings should be 6 to 8" long (smaller in less vigorous cultivars), wounded, 1 to 2% IBA-talc or solution, well drained medium, mist. When rooted they should be left undisturbed until they have gone through a dormant period. Supplemental light can be used to induce the cuttings to produce a new flush of growth. This is accomplished immediately after rooting. Commercial growers use this approach as a matter of routine. The senior author has observed rooted cuttings of 'Bloodgood' under lights. The plants that produced a new flush of growth survived the overwintering period in greater percentages than those that did not. Normally rooting takes about 6 to 8 weeks.

Several nurserymen force plants in the greenhouse and utilize the soft shoots. This can be accomplished as early as February and March. The cuttings can be treated as described above and have a long growing season to accumulate carbohydrates and, thus, survive overwintering. See Wells, J.S., *Amer. Nurseryman* 151(9):14, 117-120 (1980).

Juvenility strongly influences rooting, for cuttings from a 50-year-old tree rooted 25%, while 80% rooting was obtained from 4-year-old seedlings that were forced for 30 days in a greenhouse.

Late June (Tennessee), 3 node, 2.5 to 4.5" long, lower leaves stripped, single ½" long wound, 1.0, 1.5 and 2.0% IBA-50% ethanol-5 second dip, peat: perlite, 77°F bottom heat, mist, evaluated after 81 days. For the 3 IBA concentrations 'Bloodgood' rooted 97, 80, 83%; A. *palmatum* seedling — 83, 90, 90%; 'Crimson Queen' 97, 100, 87%; 'Viridis' — 100, 90, 97% and 'Ever Red' — 37, 53, 57%; respectively. Unfortunately no mention was made of stock plant age or overwintering success, but the results indicate that, with attention-to-detail, *Acer palmatum* and cultivars can be rooted in high percentages.

Timing: 'Bloodgood' (May 20, New Jersey) plus or minus a week or two on either side. Cuttings can be rooted up to mid-June but this poses problems in overwintering.

Type: Select strong, thick, vigorous shoots of current season's growth. One specialty grower suggests that shoots should be the size of a pencil. The only feasible way to do this is the maintenance of stock plants that are cut back heavily (thus maintaining juvenility), fertilized and watered on a regular schedule.

Handling: Collect in morning from fully turgid plants and do not allow to dry out. Place cuttings in water or moist container.

Preparation: 6 to 8″ long, 2 to 3 nodes, soft tip should be removed, lower two leaves removed, 1 to 1½″ long heavy wound on one side.

Hormone: 2% IBA-talc plus 5% Benlate. This higher concentration is particularly effective for Japanese maples.

Medium: Best medium is peat: perlite but peat: sharp sand is also satisfactory. Cuttings should be stuck 2″ into the medium.

Mist: Essential to keep leaf surface moist from sunrise to sunset. Allowing the foliage to dry out for short periods of time can be detrimental. Timed mist is the safest approach for Japanese maples.

Bottom Heat: Optimum medium temperature is 70 to 75°F. Air temperature may reach 90°F as long as cuttings are kept moist.

Cuttings should root in 3 weeks. As soon as the roots reach 2 to 3″ in length, lift and pot. A loose, well drained medium is important and during transplanting care should be exercised so the roots are not damaged.

From the middle node (assuming a 3 node cutting was used), one leaf should be removed. The potted cuttings should be put back under the mist until root growth is evident at the periphery of the root ball. The plants should then be weaned from the mist (reduce misting cycle). This may take 3 weeks.

Supplementary Light: Once hardened the cuttings can be moved to a shaded greenhouse or growing area and immediately provided supplementary light. The lights are usually positioned about 3′ above the plants and a 60 to 75 watt bulb supplies sufficient illumination to trigger the response. The light may be supplied as an interrupted night treatment from 10 pm to 2 am, 9 pm to 4 am or a cyclic on/off during this period. One grower has the lights on a timer and runs them 5 minutes on, 5 minutes off. Shoots will emerge from the nodes where the single leaf was removed. Leaves serve a source of abscisic acid, a growth inhibitor, and when removed a bud is often released from the imposed dormancy and can grow if conditions are correct. The light provides the necessary stimulus. Other buds may also grow. The plants should be lightly fertilized with liquid fertilizer (20-20-20) or low levels of Osmocote 18-6-12 (¼ recommended rate) and maintained under these conditions throughout the summer to encourage maximum shoot growth. One researcher (*Scientia Hortic.* 27:341-347) says that no fertilizer should be applied during or after the rooting process because plants do not properly acclimate and may die during the overwintering period.

Overwintering: Supplementary light should be discontinued in October and the plants allowed to come into natural dormancy. Plants should be held at air temperatures of about 33°F throughout the winter. The plants go through their normal hardening/dormancy phases at these temperatures.

Post Overwintering: The young plants can be shifted to containers or lined out in the field. The vigorous growth induced by light treatment should be cut back one-half. Plants will break into new growth and produce a 12 to 18″ high, quality plant in the growing season.

Postscript: These methods work on most cultivars of *Acer palmatum*, especially the more vigorous clones. All cultivars do not root with equal facility so a range of rooting success is to be expected. The *dissectum* types are more difficult due to slow growth and small amount of cutting material. The above recipe follows the recommendations of Mr. James Wells who spent a lifetime working with this species.

GRAFTING: Grafting is practiced for many of the cultivars especially var. *dissectum*. They are side grafted on potted seedling (1 or 2-year-old) understock during the winter months. For the slow growing types this is the only feasible method of production. In England, summer (July-August) grafting is practiced with success. The procedure is the same for winter grafting except 4 to 8″ scions with 3 to 4 nodes are taken from plants that are no longer actively growing. Leaves are removed with a portion of petiole remaining. Take ranges from 80 to 95% [PIPPS 26:169-170 (1976)].

Acer pensylvanicum Striped Maple

SEED: Fall planting or 3 to 4 months cold stratification should suffice. Based on authors' observations there is a low percentage of sound seed. This is a dioecious species and lack of sound seed could relate to the absence of a pollinator or the fact that pollen is shed ahead of stigmatic receptance and vice-versa.

CUTTINGS: Rooting studies by the authors have met with failure although *A. capillipes* and *A. tegmentosum* have been rooted.

GRAFTING: The species has been successfully grafted on seedlings of *A. pseudoplatanus*.

Acer platanoides Norway Maple

SEED: Matures later than most maples. Most seed appears to be sound based on cut tests the authors have conducted. Either fall planting or 3 to 4 months cold stratification will satisfy the dormancy requirement.

CUTTINGS: 3″ long, 2 nodes, mid-June (Michigan), 8000 ppm IBA-talc, mist, 75°F bottom heat, rooted 85%. Same researcher [PIPPS 31:507-511 (1981)] reported 80% rooting from early and mid-July cuttings. Cuttings sampled two weeks earlier or later rooted 5% and 25%, respectively. This suggests that timing is critical. Interestingly, only one article out of forty in PIPPS dealt with cutting propagation. Several old references indicated cuttings from young trees rooted 50% when treated with auxin. Cuttings from a 60-year-old tree rooted less than 1%.

GRAFTING: Based on observations of Oregon shade tree producers, the bud take for Norway maple cultivars seldom approaches 100%. A block of 'Crimson King' or 'Royal Red' may have a significant number of green plants (buds that did not take) interspersed. The typical procedure is summer T-budding on seedling Norway understock. However, bench (pot) grafting in January-February is practiced in Europe with variable take. 'Crimson King' has been successfully chip-budded (70%) compared to T-budded (7%) in one test.

Several growers noted emphatically that budwood taken from shoots that are still elongating (terminal not set) is superior to hardened growth. Also larger buds are superior to small ones. Shield buds placed at nodes where lateral bud existed produced better results than bud placement at internode. A shield bud with a thin layer of wood was more successful than one without.

TISSUE CULTURE: 'Crimson Sentry', 'Crimson King', *et. al.*, have been successfully tissue cultured. The authors have seen plants being hardened for fall planting. In one study, stem explants produced callus with no mention of shoot production |*Plant Physiol.* 63(5):Abst. 758 (1979)|.

Acer pseudoplatanus Planetree or Sycamore Maple

SEED: Fall planting or 2 to 3 months cold stratification. One study noted heavy germination (root radicle emergence) in the bag after 3 months at 41°F.

CUTTINGS: Some work has been conducted in England with softwood cuttings taken from seedlings but no final report was given. Another report noted it was difficult to root.

GRAFTING: *Acer pseudoplatanus* is a good understock for many maples and is used extensively in England. A. *p.* 'Atropurpureum' (purple on underside of leaf) may be even more universal since A. *platanoides* can be successfully grafted on it.

The cultivars are not used to any degree in the United States but throughout Europe are extremely popular. Cultivars such as 'Brilliantissimum', 'Leopoldii', 'Worleei', etc., are bench (pot) grafted in January-February on seedlings of the species. Take approximates 85% for a clone such as 'Brilliantissimum'. Budding is also practiced but special precautions are followed based on work in England (PIPPS 19:206).

Acer pseudosieboldianum Purplebloom Maple

SEED: Treat similar to *Acer palmatum*. One report stated that 3 months cold stratification produced good germination.

CUTTINGS: Late June (Ohio), 4 to 8″ long, basal leaves removed, wound, 2% IBA-talc, sand, mist, rooted 61%. Also mid-June cuttings (Boston), 1% IBA, mist, rooted 40%. No results were provided on overwintering success but this species is very similar to A. *palmatum* and should be handled similarly.

Acer rubrum Red Maple

SEED: Ripen in spring (May) and have no dormancy. They can be planted immediately and, like silver maple, will germinate. Soaking seeds in cold running water may leach inhibitors and hasten germination. There is some indication that if pericarp becomes dry a short cold stratification period may be required. Provenance is extremely important for southern seed sources are not reliably hardy in the north. A 1 month cold period may prove beneficial.

CUTTINGS: Production practices have changed dramatically in the last 5 to 8 years and now most cultivars are rooted from cuttings to avoid the problem of incompatability. New cultivars are being introduced yearly and one of the most important criteria for selection should be the ease of rooting from cuttings.

There are many ways to successfully root *Acer rubrum*. Orton |*The Plant Propagator* 24 (3): 12-15 (1978)| described the single node method for 'October Glory'. In his work, early August 'October Glory' single node cuttings from 4-year-old budded trees, 8000 ppm IBA: Benlate (10:1 mixture), 4 sand: 1 peat, mist, 76°F bottom heat, night lighting from 10:00 pm-2:00 am resulted in 98% rooting; June-July cuttings, 3000 to 6000 ppm IBA solution, well drained medium, uniform moisture on leaves (leaves scorch if not evenly moist), with or without bottom heat, rooting takes place in 21 to 28 days. Cuttings should not be distrubed after rooting. Many growers root in 2¼ by 2¼ by 5″ rose pots to eliminate transplanting. Timing is critical for cuttings taken in August and later root in lower percentages and often fail to overwinter.

Interrupted or extended photoperiod results in budbreak and growth (24 to 30″) after rooting. Plants on natural photoperiod may break but usually grow no more than 6″. Mid-August cuttings given an extended photoperiod did not respond indicating that bud dormancy was too deep seated (due to build up of inhibitors). Also, plants that put on new growth after rooting overwinter in significantly higher percentages (100%) compared to those that do not break bud (58%).

The following recipe is being used at a large West Coast nursery with outstanding success. Nine inch long cuttings, 2 to 4 nodes, June and July, 2¼ by 2¼ by 5″ pots, 1000 ppm IBA and 500 ppm NAA solution, peat:perlite, mist, Osmocote, Micromax, root in three weeks. Two weeks later, cuttings have another 6″ growth, are hardened off outside, stored in cooler, spring planted. Three batches are produced per growing season. The quality is outstanding!

PRODUCTION: Rooted red maple cuttings require different handling to produce a salable whip compared to traditional budded trees. Cuttings are lined out in spring and grown for a year. They are cut back to a node with good buds in winter. If two breaks occur one is removed and by the end of the second season, a straight 5 to 6′ tree with a quality root system is the product.

GRAFTING: Previously most cultivars were summer budded on seedlings of the species. This practice has now been largely discontinued due to the incompatability problems. The literature also mentions the use of A. *saccharinum*, silver maple, as a successful understock. However, many of the incompatability problems could be related to the use of A. *saccharinum* as an understock.

TISSUE CULTURE: Red maple is being successfully produced and outplanted in commercial nurseries. An early report indicated a problem with shoot multiplication but apparently this has been overcome.

Other studies are as follows: Shoot tip explants led to axillary shoot proliferation |PIPPS 29:382-386 (1979), *HortScience* 17:533 Abst. 431 (1982)|. Shoot tip explants from greenhouse-grown plants were established on a modified MS medium containing IBA and BA. Shoot proliferation of an *Acer rubrum* × A. *saccharinum* hybrid was achieved on a medium containing 0.1-0.5 microM Thidiazuron |*HortScience* 20(3):593 Abst. 506|.

Acer rufinerve Redvein Maple

SEED: See *Acer pensylvanicum*; 3 months cold stratification resulted in germination.

Acer saccharinum Silver Maple

SEED: Ripens in spring and germinates immediately. Has no dormancy. Provenance is important, for southern seed sources are not reliably hardy in the north and northern seed sources tend to grow slower in southern locales.

CUTTINGS: Root readily and the procedure described for 9″

cuttings under A. *rubrum* has been extremely successful. To the authors' knowledge the single node concept has not been tested with silver maple. It will, no doubt, work. Cuttings taken in November rooted with 84% efficiency. Also, the remarks under A. *rubrum* about photoperiod will probably apply to A. *saccharinum*. Catechol stimulated root initiation on this species. 'Wieri' responded well to 1.0% IBA throughout the summer.

Acer saccharum Sugar Maple

SEED: Fall planting or 2 to 3 months cold stratification is sufficient. Worthwhile to conduct cut tests since A. *saccharum* does not form consistently sound seed. In one study, 20, 40, and 50 days cold stratification resulted in 15, 38, and 65% germination, respectively.

Seedlings do not grow as fast as red or silver maple and in the commercial arena it may be difficult to produce a one-year seedling that is suitable for budding. Seed source plays a significant role and nurserymen should collect/buy seed from local sources.

CUTTINGS: Difficult to root compared to A. *rubrum* and A. *saccharinum* but some success has been reported. Mid-June (Ontario) juvenile cuttings rooted 65 to 89%. Bottom leaves were fully elongated, apical meristem still growing. Medium was 2 soil: 1 vermiculite:1 peat with mist. Larger diameter cuttings rooted better than smaller ones. Hormones did not improve rooting nor did wounding. 75% of cuttings overwintered when mulched with oak leaves. Vermont work [*The Plant Propagator* 22(1):3-6 (1976)] described 0 to 83% rooting from 4 different difficult or easy to root ortets. Untreated cuttings rooted better than those treated with hormone. One report noted that 0 to 4% IBA-solutions enhanced rooting of this species. Forty-five percent rooting with 5000 ppm.

GRAFTING: Almost all the cultivars, and there are many, are budded on seedling understock. Sugar maple is difficult to bud owing to the small size of the understock and toughness of the wood. Percentage take is variable and not particularly high.

Interestingly, several reports have indicated that A. *saccharum* can serve as an understock for A. *nikoense* and A. *griseum*. Long term prospects for the success of these grafts is somewhat suspect. See comments under A. *griseum*.

Acer sieboldianum Siebold Maple

SEED: Should be handled like A. *palmatum*. If collected and planted in fall before it has dried, reasonable germination can be expected the following spring. Three months cold stratification has induced 32% germination. Produces seed at an early age, often when 6 to 10-years-old. Plant in fall after harvest and germination should ensue the following spring.

CUTTINGS: Difficult, for late July cuttings collected by senior author, 8000 ppm IBA talc plus thiram, peat: perlite, mist, did not root and showed no inclination of doing so. Cuttings were collected from a mature tree.

GRAFTING: *Acer palmatum* should serve as a successful understock.

Acer spicatum Mountain Maple

SEED: Fall plant or 3 to 4 months cold stratification.

CUTTINGS: July cuttings, 4 to 8″ long, wound, 2% IBA-talc, sand, mist, rooted 25% but none successfully overwintered.

Acer tataricum Tatarian Maple

In many morphological characteristics this species is similar to A. *ginnala* and its propagation requirements are also similar.

SEED: Fall plant or 3 to 4 months cold stratification result in excellent germination.

CUTTINGS: Early May cuttings (too soft) did not root particularly well as most died; however, those treated with 1% IBA-talc rooted 40%. Mid August cuttings rooted 56% in 8 weeks when treated with 8000 ppm IBA-talc. Same batch treated with 1% IBA-talc rooted 74%.

Acer tegmentosum Manchu-striped Maple

SEED: Fresh seed can be fall sown or given 4 months cold stratification. If seed is allowed to dry it should be soaked or given a warm stratification period prior to fall planting.

CUTTINGS: Mid August (Boston) from a 31-year-old tree, 8000 ppm IBA-talc, coarse sand, polytent, rooted 89% in 5 weeks. Late July (Ontario), 6″ long, seedling trees, ½ Captan:½ Seradix 3 (8000 ppm IBA-talc), 4 sand:1 peat, mist, 70°F bottom heat, rooted 59% in 9 weeks. July cuttings, 4 to 8″ long, wound, 1.5% or 8000 ppm IBA-talc, sand, mist, rooted 75 and 70%, respectively. Cuttings overwintered 100%. There is some indication that 2-year-old wood is better than 1-year wood.

Acer triflorum Threeflower Maple

SEED: Doubly dormant and when fall planted will germinate the second spring and sporadically thereafter. Seed, unfortunately, is often not sound and cut tests have shown about 5% sound seed. Nine months warm followed by 3 months cold gave reasonable germination. If seed is received dry it may be prestratified for 6 months and then sown. Germination is less than 1% the first year but is very good the second.

CUTTINGS: Sinilar to *Acer griseum*. Late June cuttings (Ohio), 4 to 8″ long, wound, 2% IBA-talc or 25% Chloromone (solution), sand, mist, rooted 58 and 17%, respectively. Trees used as source of cutting material were probably 5 to 8 years old. Cuttings taken from a 30-year-old tree did not root. Cuttings taken from 2-year-old seedlings that were forced in the greenhouse (February) rooted 56 to 75%. High IBA (1%) killed some of the cuttings and best rooting was obtained without hormone. These cuttings did not grow rapidly and no indication of survivability was given. All cuttings were rooted in coarse sand in a polytent.

GRAFTING: Seedling A. *triflorum* is the best understock although other trifoliate maples might be tested.

Acer truncatum Purpleblow Maple

Acer truncatum subspecies ***mono*** Painted Maple

SEED: Germination is enhanced when seed is harvested with a slight green color to the samara. If fall planted or provided 3 to 4 months cold stratification, germination will be high.

One problem the senior author has noticed is lack of sound seed, especially on subsp. *mono*. It is a monoecious species but apparently male and female flowers do not open at the same time and, hence, most seeds are void of embryos.

CUTTINGS: Difficult. Work with the species and subspecies *mono* resulted in no rooting.

NOTE: The A. *truncatum* complex including subsp. *mono* and *mayrii* are beautiful and exceedingly cold hardy landscape plants. Rich dark green leaves, excellent yellow-pumpkin fall color, small stature and apparent insect and disease resistance are sufficient reasons for extended use.

Actinidia

A group of vines that produce fragrant flowers and edible fruit. A. *arguta* is extremely cold hardy while A. *chinensis* (Kiwi) is suited only for Zone 8 (+10 to +20°F) and warmer climates. In general, flowers are polygamo-dioecious or dioecious so male and female plants are necessary. The fruit is a fleshy berry and seed must be removed from the flesh by maceration. A. *kolomikta* is grown for the white and pink tipped leaves.

Tissue culture has proven easy and A. *arguta* and A. *kolomikta* have been proliferated in culture at the Arnold Arboretum. In fact, these species have been used as model systems to study cytokinin biochemistry.

Actinidia arguta Bower Actinidia

SEED: Fall plant or 3 months cold stratification proves optimum. It is best to use cutting material to be sure of the sex.

CUTTINGS: Softwood/greenwood cuttings root readily and the early literature reported 80 to 90% rooting using IBA or NAA. Untreated cuttings, early summer, rooted 88% under high humidity (tent). September cuttings, IBA-soak, sand, rooted 86% in 4 weeks, only 42% without hormone.

Actinidia chinensis Kiwi

SEED: If extracted fresh and planted will germinate readily. Fruits can be stored at 41°F until planting time, then seed is extracted. If seed is allowed to dry, it develops a dormancy which can be broken with 3 to 4 weeks cold stratification or alternating cold/warm stratification temperatures for 2 to 3 weeks. Seedlings grow rapidly but should be provided a preventative treatment for damping off. Seedlings appear 2 to 3 weeks after sowing.

CUTTINGS: Rooted from summer softwoods with varying degrees of success. One, two, and three node cuttings are used with a single leaf attached to upper node (leaf may be reduced in size), 6000 to 8000 ppm IBA-talc or solution, mist, and rooting occurs in 3 to 4 weeks. Cuttings can be potted, overwintered, and lined out in spring. Overwintering success is often variable. Reports indicate abundant callus may occur. This can often be overcome with higher hormone treatments. New Zealand growers feel that rooted cuttings are inferior to grafted plants. Others challenge this contention.

Apparently the cultivars show variable propensities to root. One male is suggested as a pollinator for 7 females. Work from Monrovia Nursery indicated 3000 ppm IBA and 3000 ppm NAA-solution was superior to other treatments in rooting the female 'Chico Hayward' (72%) and 'Chico Male' (33%).

Hormone is essential and hardwood cuttings of 'Bruno' taken in early winter rooted 39% without IBA and 99% when treated with 5000 ppm IBA-dip. Mist and bottom heat (80°F) were used. Rooting occurred in 6 weeks. Hardwood cuttings have also been rooted successfully using 1000 ppm NAA. For several cultivars, 'Hayward' and 'Abbott', 5000 ppm IBA in 50% alcohol provided optimum. Work in California with hardwood cuttings taken in late December-early January, 2 bud cuttings, 60% perlite: 40% sand, IBA in talc or 3333 ppm IBA plus 1666 ppm NAA quick dip showed best rooting occurred with liquid formulation at 73%. 1.6%, 3.0% and 4.5% IBA in talc produced 55%, 62%, and 68%, respectively.

Root cuttings in late winter-early spring ('Abbott'), 2″ long and 1/5 to 3/5″ wide, horizontal in sand, mist, 74°F bottom heat, greenhouse, produced shoots in 60 days. Early March hardwood cuttings from 6-year-old plants, treated with 1000 ppm NAA-10 sec dip, vermiculite, maintained with 70 to 90% humidity, rooted in high percentages and were planted into nursery in mid-May. By late fall, plants averaged 3′ high and survival rate was 95%.

GRAFTING: Top grafted by whip or cleft graft in winter on seedling understock that was started 12 months earlier. The grafted plant is sold 18 months after the seed was planted. T-budding is also practiced in September and October (California).

TISSUE CULTURE: Many reports in literature of tissue culture propagation [PIPPS 34:236-243 (1984)]. Stem and root segment explants led to direct shoot formation, or through callus culture, adventitious shoots and embryogenesis [J. *Hort. Sci.* 50:81-83 (1975)]. Stem segment explants through callus culture produced adventitious shoots [*Acta Bot. Sin.* 21:339-344 (1979)]. Shoot tip and nodal explants through shoot tip culture gave rise to shoot proliferation which rooted. Plant tissue culture 1982. Proc. 5th Int. Cong. Plant Tiss. Cell Culture, Japan. Jap. Assn. Plant Tissue Cult. Tokoyo Fujiwara (ed) pp. 737-738 (1982).

Actinidia kolomikta Kolomikta Actinidia

SEED: Apparently, the seed is doubly dormant and 3 months warm/3 months cold stratification resulted in good germination 23 days after sowing. One report noted good germination after 3 months cold stratification and seed treated with 500 ppm GA_3 plus 50 ppm kinetin germinated in reasonable percentages. Another mentioned 5 months was best.

CUTTINGS: Senior author's work indicated this was not particularly easy to root. June cuttings, 8000 ppm IBA, sand: perlite, mist, rooted 40%. The newly rooted plants do not show the characteristic white and pink leaf splotching. This develops as plants mature.

Actinidia polygama Silver-vine

SEED: Fall planting or 3 months cold stratification. Seeds germinated in 20 days after 3 months cold stratification.

CUTTINGS: Softwoods root easily.

Aesculus Buckeye, Horsechestnut

A group of trees and shrubs characterized by large, compound-palmate leaves, showy flowers of white to almost red, and interesting capsular fruits that dehisce in late summer/fall to expose glistening rich brown seeds. The seeds can be collected from the ground or the entire fruit collected when brown, placed in a dry room and allowed to dehisce. This latter approach is good for shrubby species such as A. *pavia*, A. *parviflora* and A. *sylvatica*. It always seems to be a race between the squirrels and the seed collector...too often the squirrels win. Seeds of all species are composed of fats and lipids and degenerate quickly after collection. Seeds allowed to dry in a warm room show ridges and furrows in a short time indicating drying and respiration are proceeding exceedingly fast. Cold storage of seed (34 to 38°F) at high humidity extends viability for about 12 months.

Germination drops from 85 to 65% with time. Seeds should be sown immediately after cleaning or cold stratified. Seeds should be covered about once their diameter. Since seeds within a particular species are variable in size, some growers grade to a uniform size (weight) in order to produce a reasonably uniform crop of seedlings. See McMillan Browse, *Plantsman* 4(3):150-164 (1982).

Aesculus arguta Texas Buckeye

SEED: Fall plant or 3 months cold stratification.

Aesculus californica California Buckeye

SEED: Requires no stratification and fresh seed will germinate without pretreatment.

Aesculus × carnea Red Horsechestnut

SEED: Fall plant or 3 to 4 months cold stratification. Interesting hybrid species and will come relatively true-to-type.

GRAFTING: Bench (pot) grafting using a side or veneer graft in January-February followed by placement in polytent with heat until graft knits.

One English nursery inserts the scion (not bud) on the side of the understock into a T-cut in summer in the field. Grafts are waxed and understock is defoliated. 'Briotii' is also done this way and more successfully than traditional budding. Seedling understock is A. *hippocastanum*.

A. × *carnea* or 'Briotii' can also be budded on A. *glabra*. Buds of A. × *carnea* are placed about 5 to 6″ above ground in early June (Cincinnati). After 3 weeks, the entire top of the understock is removed. The new bud (A. × *carnea*) makes 12 to 18″ growth before fall. The idea is to produce a larger caliper, salable tree sooner.

Aesculus glabra Ohio Buckeye

SEED: Fall plant or 3 to 4 months of cold stratification.

Aesculus hippocastanum Common Horsechestnut

SEED: Fall plant or 3 to 4 months cold stratification. Seed stored dry at 30°F germinated 60% after one year in storage. Seed from same collection stored dry at 75°F failed to germinate and after 2 months had completely deteriorated.

CUTTINGS: May-August (Michigan), Chlorox dip, 8000 ppm IBA-talc plus 3.8% Benlate, peat: perlite, mist, evaluated 6 weeks after sticking. Late May (64%), early June (72%), late June (44%), early July (72%), late July (48%) and August (0%). Unfortunately the age of stock plant was not mentioned or survival percentage. See PIPPS 31:507 (1981).

GRAFTING: As described under A. × *carnea*, the preferred understock is A. *hippocastanum*. The most common form is 'Baumannii', a double-flowered, sterile form that is certainly superior to the species. Apparently, seedling A. *hippocastanum* is a good understock for most *Aesculus* species. A. *parviflora*, a shrubby, suckering species has been successfully grafted upon it.

Aesculus indica Indian Horsechestnut

SEED: Fall plant or 3 to 4 months cold stratification. To the authors' knowledge the tree is not used in the east, midwest, or south. It is a handsome tree that offers large whitish pink flowers after A. *hippocastanum* (mid to late June in London). 'Sydney Pearce' is a deeper pink form of great beauty.

GRAFTING: As described under A. × *carnea* but preferably with seedling understock of A. *indica*.

Aesculus octandra (*flava*) Yellow Buckeye

SEED: Fall plant or 3 to 4 months cold stratification. Abundant seed is set on mature trees. Capsules remind of A. *pavia*, i.e., somewhat pear-shaped, light brown and smooth.

Aesculus parviflora Bottlebrush Buckeye

SEED: No pretreatment is necessary and seed will germinate upon planting. Usually a large root develops with only limited top growth (1 to 2 inches). The senior author has grown A. *parviflora* in flats in the greenhouse but lost most of them due to overwatering and subsequent "rot". Be sure to plant in deep containers or outdoor seed beds since the root system needs room to develop.

CUTTINGS: Early April (Urbana, IL), taken from sucker growth under a mature shrub, 0 or 1000 ppm IBA-solution, peat: perlite, mist, rooted 70 and 80% in 4 weeks, respectively, and overwintered 100%. Cuttings collected later required 2500 to 5000 ppm IBA for best rooting (50 to 60%). These cuttings were juvenile, hence, the high rooting percentages [*The Plant Propagator* 23 (4):6-7 (1977)]. Cuttings taken from June 15-July 15 (Boston) can be rooted.

Root cuttings, February-March, placed in vermiculite in greenhouse, produced 50 to 60% plants. December root cuttings, 3″ long, 3/16″ wide, placed in pots with 1/4″ showing produced shoots. The shoots root readily owing to the juvenility factor.

Cuttings are worth considering for the production of this plant, since superior forms exist. 'Roger's', with long inflorescences, and var. *serotina*, later flowering, might be candidates for vegetative propagation.

GRAFTING: There is no reason to graft this species except to avoid the suckering problem. A. *glabra* and A. *hippocastanum* serve as suitable understocks.

Aesculus pavia Red Buckeye

SEED: Although described as having no dormancy, a one month cold stratification is recommended.

Other red flowered species (questionable if they are, in fact, distinct from A. *pavia*) include A. *discolor*, A. × *georgiana* (A. *pavia* × A. *sylvatica*) and A. *splendens*. All require the same treatment as A *pavia*.

GRAFTING: 'Atrosanguinea' and 'Humilis' are grafted on seedling A. *pavia* and, no doubt, could also be grafted on A. *hippocastanum*.

Aesculus sylvatica Painted Buckeye

SEED: Fall plant or 3 months cold stratification.

Aesculus turbinata Japanese Horsechestnut

SEED: Fall plant or 3 to 4 months cold stratification. This is a rather striking horsechestnut with large leaves and white flowers in June. The texture of the bold leaves is particularly interesting and the flowers appear after most *Aesculus* species (exlcuding A. *parviflora*) have passed.

Ailanthus altissima Tree of Heaven

SEED: Dioecious, fruits (samara) ripen in October but will remain on tree all winter. There may be a shallow dormancy and, if in doubt, provide cold stratification for 1 to 2 months. Seeds collected in November (Boston) and December (Urbana, IL) germinated without pretreatment. A limited amount of chilling hastens and unifies germination.

CUTTINGS: Root cuttings regenerate shoots readily. The shoots can be severed from the root systems and rooted like stem cuttings or the entire shoot-root potted and grown on.

GRAFTING: Can be field-budded or grafted in summer.

Akebia

The akebias are not well known in American gardens but make rather handsome covers for trellises, fences and pergolas. Leaves appear early in spring and hold late into fall (November). The polygamo-monoecious flowers appear with the leaves. Pistillate are chocolate-purple, staminate rose-purple, borne in the same pendent axillary raceme. Fruit is a 2-4″ long, bloomy, sausage-like pod of purple-violet color. Collect fruits in October and remove black seeds (by hand) and dry. Fruits seldom set under cultivation.

Akebia × pentaphylla
(A. *quinata* × A. *trifoliata*)

SEED: Seed requires a cold stratification although 30% germination was recorded for freshly sown seed. One month cold stratification is sufficient to induce good germination.

CUTTINGS: Summer softwoods will root. Late June and July are the best months. Treat with 1000 ppm IBA-solution. Rooting will take place in 2 to 3 weeks. Mid-July cuttings, 4000 ppm IBA-talc plus thiram, rooted 63%. No problem overwintering the rooted cuttings.

Akebia quinata Fiveleaf Akebia

SEED: See under A. × *pentaphylla*. It is important to note that additional cold stratification (2 or 3 months) did not improve germination over one month.

CUTTINGS: See under A. × *pentaphylla*. Root cuttings placed in warm greenhouse in mid-December developed many shoots by late January. Cuttings taken from the juvenile shoots, 4000 ppm IBA-talc plus thiram, rooted 67% in one month.

Akebia trifoliata Three-leaflet Akebia

SEED: Same procedure as for previous species. Germination will occur to some degree without cold stratification but it is erratic.

CUTTINGS: Same procedure as for previous two species.

Albizia julibrissin Silktree, Mimosa

Tree with bipinnately compound leaves with numerous small leaflets. Showy light to deep pink flowers in June-July are followed by 5 to 7″ long, 1″ wide pods. Seeds, when dry, have a hard seed coat.

SEED: Dormancy is controlled by impermeable seed coat. Seeds sown directly germinated 1%. Those treated for 30 minutes in H_2SO_4 germinated readily. A scarification time of 15 to 30 minutes is recommended. Seeds can be steeped in hot water and allowed to soak for 24 hours. This facilitates reasonable germination but is not as fail-safe as acid treatments. Seeds should germinate in 10 to 14 days. In one test, 30 minutes in sulfuric acid, 3 months stratification at 36°F, sand paper scarification, and 200°F hot water (initial temperature) soak for 24 hours produced 97, 7, 80 and 55% germination, respectively.

Stage of seed development makes a significant difference in ability to germinate without acid treatments. A fingernail test is one the authors like to use. If the seed coat can be broken with the nail then acid treatment is not necessary. Fall planting should suffice since alternate freezing and thawing as well as microbial activity will break down the impermeable seed coat.

CUTTINGS: Stem cuttings from adult trees will not root, however, root cuttings are extremely successful. Large root pieces produce the best shoots. Mid-February (Virginia), 4 to 6″ long, 1/2 to 1″ diameter, horizontally placed, 1″ deep, peat: perlite, moistened, shoots arose by mid-March. Root pieces were collected in mid-March (Boston) and placed in soil in the greenhouse. Shoots appeared in late May. These shoots are then severed from the root and rooted in the traditional manner. A hormone, 1000 to 3000 ppm IBA-talc or solution, is recommended. Interestingly, the shoots produced from the roots have bipinnately compound leaves which is a mature characteristic. Large root pieces continued to produce roots for 2 years. 'Union', a wilt resistant clone, and 'Rosea' a cold hardy form, are propagated by 3 to 4″ long dormant root cuttings inserted in vermiculite. Shoots root easily with or without IBA.

Alnus Alder

Wide ranging group of trees and shrubs generally native to moist or wet sites. Thirty-three of 35 species have definitely been shown to fix atmospheric nitrogen and thus are able to inhabit the most inhospitable and infertile soils. Flowers are monoecious with the male in narrow catkins; the female in a distinct egg-shaped strobile. Fruits should be collected when the scales of the strobile (green to brown) just start to part. Dry fruits inside and the winged seeds (nutlets) fall out on their own. Seeds can be stored in airtight containers at 35 to 40°F and remain viable for 1 to 2 years. Seed of most *Alnus* species requires 2 to 3 months cold stratification.

Alnus cordata Italian Alder

SEED: Three months cold stratification is considered the best treatment. Late fall planting is also recommended. It has been noted that some germination occurs without a cold treatment but results are variable.

CUTTINGS: Information is virtually non-existent for this species but cuttings (no time given), wound, 8000 ppm IBA-talc, rooted 25%. A wound and hormone were essential. Control (no treatment) cuttings did not root. Bulked hardwood cuttings from seedlings rooted 45%. Juvenility had a pronounced effect.

Alnus crispa American Green Alder

SEED: Three months cold stratification.

Alnus firma Japanese Green Alder

SEED: Three months cold stratification.

Alnus glutinosa European Alder

SEED: Best treatment is late fall planting or 3 months cold stratification.

CUTTINGS: Cuttings (no time given), wound, 8000 ppm IBA-talc, rooted 64%. Without treatment there was no rooting. In general, this species is produced from seed, however, two notable cutleaf selections, 'Imperialis' and 'Laciniata', can be rooted successfully. Mid-July cuttings of 'Imperialis' under mist rooted 34% in 7 weeks when treated with 5000 ppm 2, 4, 5-TP and 58% with 5000 ppm 2, 4, 5-TP plus Rootone. Late June 'Laciniata' cuttings rooted 34% in 7 weeks with a wound and 8000 ppm IBA-talc plus thiram, mist. Another report noted 'Imperialis' rooted easily from softwoods when treated with 3000 ppm IBA-talc.

GRAFTING: Many cultivars are known and the usual procedure is to graft them on seedling A. *glutinosa* using a side-veneer graft. Budding in July-August could probably be practiced but the authors could not find published reports.

TISSUE CULTURE: Lateral bud explants from juvenile stock plants using shoot tip culture led to multiple shoot development. *HortScience* 16:758-759 (1981), *HortScience* 16:453. Abst. 400 (1981), *HortScience* 17:533. Abst. 434 (1982).

Shoot tip and lateral bud explants using shoot tip culture yielded multiple shoots. Plant tissue culture 1982, Proc. 5th Int. Cong. Plant Tiss. Cell Culture, Japan, Jap. Assn. Plant Tissue Cult., Tokoyo, Fujiwara (ed), pp. 757-758 (1982).

Alnus hirsuta Manchurian Alder

SEED: Fall plant or 3 months cold stratification.

Alnus incana Gray Alder

SEED: Fall plant or 3 months cold stratification.

CUTTINGS: 65% rooting was obtained when cuttings were wounded and treated with 8000 ppm IBA-talc. Untreated cuttings rooted only 2%.

GRAFTING: The cultivars are grafted on potted seedling understock.

Alnus rhombifolia Sierra Alder

SEED: Fresh seed will germinate readily without pretreatment. If dry, fall plant or provide 3 months cold stratification.

Alnus rubra Red Alder

SEED: See A. *rhombifolia*.

TISSUE CULTURE: Shoot tip and lateral bud explants using shoot tip culture yielded multiple shoots. Plant tissue culture, 1982, Proc. 5th Int. Cong. Plant Tiss. Cell Culture, Japan, Jap. Assn. Plant Tissue Cult. Tokoyo, Fujiwara (ed), pp. 757-758 (1982).

Alnus rugosa Speckled Alder

SEED: See A. *rhombifolia*. The only report the authors found was a 6-month cold stratification after which seed germinated in 10 days.

TISSUE CULTURE: Shoot tip and lateral bud explants using shoot tip culture yielded multiple shoots. Plant tissue culture, 1982, Proc. 5th Int. Cong. Plant Tiss. Cell Culture, Japan, Jap. Assn. Plant Tissue Cult. Tokoyo, Fujiwara (ed), pp. 757-758 (1982).

Alnus serrulata Tag Alder

SEED: Fall plant or 3 months cold stratification.

CUTTINGS: Late June, 8000 ppm IBA-talc, vermiculite, mist, rooted 52%. Late May, 8000 ppm IBA, rooted 30%.

Amelanchier Serviceberry

Attractive, functional trees and shrubs for inclusion in almost any landscape. White (pink in a few forms) flowers in fleecy racemes, purplish 4 to 10-seeded berry-like pomes, excellent yellow to apricot and sometimes red fall color, and smooth gray bark are common characteristics. Fruits ripen in June-July and must be collected before birds remove them. The fruits are fleshy and the seeds must be removed by maceration. Seeds are the preferred method of propagation although in recent years cuttings and tissue culture have proven extremely successful.

Amelanchier alnifolia Saskatoon

SEED: Should not be allowed to dry out and should be stored in sealed container at 41°F. 3 months cold stratification or fall planting is necessary. The seed coat can become quite leathery and impervious. A short H_2SO_4 treatment (15 to 30 minutes) followed by 3 months cold stratification resulted in good germination.

If dried berries (entire fruit) are used the fruits may need a warm period (as much as one year) prior to cold stratification.

CUTTINGS: Softwood when new growth is several inches long, 3000 ppm IBA-talc and mist result in good rooting. Late June (Canada) rooted better than cuttings taken 2 weeks earlier or semi-hardwood cuttings taken later. Hormone had no effect. A minimum of 3 roots per cutting was necessary for survival.

A key to rooting this and subsequent species is timing. The cuttings must be soft, taken well before the end bud has set. For most parts of the U.S. late May into June would be the peak time.

Root cuttings would probably work on this species. Use root pieces 1/4″ wide and 2″ long.

GRAFTING: Applies to *Amelanchier* in general: Grafted on *Crataegus phaenopyrum* and *Sorbus*. Some propagators say successful. Others noted graft fails over time. Definitely not the best way to propagate serviceberries. With refinements in cutting and tissue culture propagation, there is little reason to graft other than to produce a large tree in a short time.

TISSUE CULTURE: Shoot tip explants produced rapid proliferation. *Can. J. Plant Sci.* 63:311-316 (1983).

Amelanchier arborea Downy Serviceberry

SEED: Fall plant or 3 to 4 months cold stratification. Interestingly, fresh seed (mid-June) when planted germinated sparingly.

CUTTINGS: Softwood as described under A. *alnifolia*. Root cuttings have been tried but failed to regenerate shoots. Early May (GA), peat:perlite, mist, rooted 100% with 1000 ppm IBA-solution in 5 weeks. June 3 (GA), same rooting environment as previous rooted 83% with 5000 ppm IBA-solution. All were transplanted, fertilized and produced new growth.

Amelanchier asiatica Asian Serviceberry

SEED: Fall plant or 3 to 4 months cold stratification.

CUTTINGS: Late May or mid-July cuttings, 3000 ppm IBA-talc or 50% Chloromone, sand, mist, failed to root.

Amelanchier canadensis Shadblow Serviceberry

SEED: Fall plant or 3 to 4 months cold stratification.

CUTTINGS: This is a suckering shrub and although root cuttings would appear to be a logical means of propagation the sole report indicated no success.

Amelanchier florida Pacific Serviceberry

SEED: Three months cold stratification resulted in good germination.

Amelanchier × grandiflora Apple Serviceberry

Northern Illinois grower uses June cuttings, 4000 ppm IBA-quick dip, sand, mist with good success. Cuttings are left in place after rooting and overwintered.

NOTE: Treat like A. *arborea* or A. *laevis*.

Amelanchier laevis Allegheny Serviceberry

SEED: Scarification followed by cold stratification (2 months +) resulted in good germination (61 to 74%).

CUTTINGS: Midsummer cuttings, 1% IBA, mist, root readily. Root cuttings did not work.

TISSUE CULTURE: Shoot tip explants using shoot tip culture yielded axillary shoot proliferation. Shoot tip explants using callus culture led to adventitious shoot regeneration [*HortScience* 16:406 Abst. 52 (1981), *HortScience* 16:453 Abst. 398 (1981)].

Amelanchier spicata Dwarf Serviceberry

SEED: Warm/cold stratification resulted in good germination.

CUTTINGS: Root pieces (early January) placed in sandy soil produced numerous shoots.

Amorpha Indigobush

Shrubby members of the legume family with bluish to purple-orange flowers and small kidney-shaped pods. Seeds are hard-coated, small, and can be stored in sealed containers for long periods. Seed of A. *fruticosa* has retained viability 3 to 5 years at room temperature. Apparently dormancy is caused by impermeable seed coat and a brief acid treatment (10 minutes) results in good germination.

Amorpha brachycarpa

SEED: Fresh seed will germinate reasonably well. If the seed coats become hard, a 10 to 15 minute acid treatment improves germination. If seeds are fall sown, no acid pretreatment is necessary.

CUTTINGS: Late July, 8000 ppm IBA-talc plus thiram, sand: perlite, mist, rooted 80%.

Amorpha canescens Leadplant Amorpha

SEED: 10 to 15-minute acid treatment results in good germination. Fresh seed germinates in reasonable percentages. A 2 to 8 week stratification period produced good germination.

CUTTINGS: Late July, 3000 ppm or 8000 ppm IBA-talc plus thiram, sand:perlite, mist, rooted 80 and 90%, respectively.

Amorpha fruticosa Indigobush Amorpha

SEED: Fall plant or 5 to 10 minutes acid treatment result in good germination.

CUTTINGS: All species treated here would probably root well from softwood cuttings although definitive studies have not proven this contention.

Untreated softwood cuttings of A. *f.* from 1, 2 and 5-year-old plants rooted 89, 88 and 89%, respectively. 5 to 8″ long, late-July cuttings treated with 240 to1440 ppm Ethrel, 1 peat:1 perlite, mist, rooted 80 to 100%. Control cuttings did not root in peat:perlite and rooted 33% in sand.

Amorpha nana Fragrant False Indigobush

SEED: Fall plant or 10 minute acid treatment.

NOTE: The small species, A. *brachycarpa*, A. *canescens*, and A. *nana* can be mounded or divided with a sharp spade if a few plants are needed.

Ampelopsis Porcelain-vine

Vines with inconspicuous flowers and rather stunning fruits (berry) that range from yellow to pale lilac and porcelain blue. Small seeds need to be cleaned from pulp. All species are easily rooted from softwood cuttings. Seed requires cold stratification with 3 months optimum for most species.

Ampelopsis aconitifolia Monks Hood Vine

SEED: Seed collected in September, cleaned and sown immediately germinated 30%. Those cold stratified germinated significantly better. One month cold improved germination to 60%; 7 months...76%.

Ampelopsis brevipedunculata Porcelain Ampelopsis

SEED: 3 months cold stratification is ideal. Apparently 'Elegans' will come true-to-type from seed and responds to the same treatment as the species. Senior author has raised numerous seedlings following the 3 month stratification period.

CUTTINGS: Untreated early summer cuttings, sand, mist, rooted 90% in 30 days. The senior author has rooted this species successfully from June to August. 'Elegans' — a pinkish white variegated leaf form can be easily rooted as described above. It is often difficult to find firm wood since growth is so rampant and stems are rather spindly.

Ampelopsis humulifolia Hops Ampelopsis

SEED: 3 months cold stratification.

Andrachne colchica

Rather unusual shrub with handsome rich green foliage. Grown only for its foliage as flowers and fruits are inconspicuous. Not well known but a good compact plant for rough and tumble areas.

CUTTINGS: Mid-July, either 4000 or 8000 ppm IBA-talc, rooted 100% in 8 weeks.

Andromeda glaucophylla Downy Andromeda

SEED: Will germinate without pretreatment. Best to sow on a sphagnum (milled) medium without covering seeds. Flats can be placed under mist or in a poly structure.

CUTTINGS: Late January, 2-sided wound, 4000 ppm IBA-talc, polytent, rooted 100% and had superb root systems.

Andromeda polifolia Bog-rosemary

SEED: Seed sown in June germinated sporadically. Same flats returned to cold then to warm showed heavy germination. Some cold may unify and hasten germination.

CUTTINGS: See A. *glaucophylla*. 'Nana' has been rooted successfully (100%) using early December cuttings (evaluated late February), 8000 ppm IBA-talc plus thiram, polytent. Late November cuttings rooted similarly with a similar treatment.

Anisostichus capreolata (*Bignonia capreolata*) Cross-vine

SEED: Requires no pretreatment.

CUTTINGS: Softwood root readily in June-July. Aerial roots are often present on the stems which is usually a good indication the species roots easily.

Aralia Walkingstick, Angelica-tree

Suckering shrubs or trees with immense, white, umbellose-panicles in July-August-September followed by purplish black drupes in September-October. Seeds (usually 2 per fruit) should be removed from the pulpy matrix, dried and stored at 40°F in sealed containers.

Aralia elata Japanese Angelica-tree

SEED: Can be fall sown after cleaning but spring germination may approach only 5%. The resultant plants approach 16" high and 2/5 to 3/5" diameter by fall. Seed may be doubly dormant and require a 3 month warm/3 month cold period, however, one study reported 70% germination after 3 months cold stratification.

CUTTINGS: Root cuttings offer the best method of vegetative propagation. The species suckers profusely which is an indication that root cuttings will work. Dig and store roots in fall, pieces 1/3 to 2/5" wide, 4 to 4 1/2" long are best, place vertically in pots in greenhouse or outplant 3 to 5" below ground vertically into warm soil in spring. A 6 to 8" plant is the result the first grown season.

The cultivars 'Aureo-variegata' and 'Variegata' can also be produced from root cuttings assuming they are on their own roots to start with. If the variegation is a periclinal chimera then root pieces will not work.

GRAFTING: The variegated forms must be produced by grafting but the percentage take is usually low. Patch buds are taken from scion wood collected in February. Patch is cut 1" above bud and 5" below, identical patch cut out of understock, wrapped, waxed, placed in sawdust at 60°F in greenhouse and kept moist. When bare root understock suckers (approx. 2 to 3 weeks), the plant is placed in a 6" pot. Shoot development from bud seldom exceeds 1 1/2 to 2 1/2" the first season. Winter protection is advisable the first season. In spring, plants are placed in pots and 24" long shoots develop by end of growing season.

Aralia spinosa Devil's-walkingstick

SEED: An embryo dormancy is present that is satisfied by 3 months cold stratification. Seeds sown without cold treatment germinated 1%; those subjected to 3 months cold, 55%.

CUTTINGS: See A. *elata*.

Arctostaphylos uva-ursi Bearberry

SEED: Fruit is a fleshy drupe that may persist on the plant through spring. Collect in fall and clean seeds. Seeds have impermeable coats and dormant embryos. Nursery practice involves 2 to 5 hours acid treatment then plant in summer with germination the following spring. Acid treatment for 3 to 6 hours followed by 2 to 4 months warm and 2 to 3 months cold stratification has proven effective.

CUTTINGS: Timing appears to be of importance and reports from Oregon indicate mid-September to mid-October and early March to early April are optimum for taking cuttings. Use current season's growth, 2 to 6" long, Dip 'N Grow (1:9) (1000 ppm IBA, 500 NAA), sand, 72°F bottom heat, mist, and rooting is 85% or greater. In New York (Long Island), early March cuttings, 1 part 8000 ppm IBA-talc: 1 part Benlate, sand, 65°F heat, 98% rooting. Late October cuttings, 8000 ppm IBA, sand, polytent, rooted 80% in 6 weeks. New York work indicates that side branches from the middle of the plant root 95%. The long shoots do not root as well.

TISSUE CULTURE: One centimeter segments from newest leaves of shoots were cultured on modified MS. Optimum growth occurred with 10 mg/l NAA plus 45% coconut milk [*Pharmacie* 37(7):509-511 (1982)]. Shoot tips have been successful on MS medium with transfer to Randolph and Cox medium for rooting. Also refer to: *Can. J. Plant Sci.* 63:311-316 (1983).

Ardisia japonica Japanese Ardisia

SEED: May germinate without any cold stratification.

CUTTINGS: February (Georgia), 1000 ppm IBA-solution, peat: perlite, mist, rooted in 4 weeks and transplanted readily. The species and cultivars are quite easy to root. Growth should be firm.

The plant is somewhat stoloniferous and can be easily divided. The Japanese have selected numerous cultivars over the years.

Aristolochia durior Dutchman's Pipe

SEED: 3 months cold stratification proved excellent.

CUTTINGS: July cuttings should be treated with a hormone (IBA-1000 to 3000 ppm). Plants can also be divided.

Aronia Chokeberry

Handsome suckering shrubs with white flowers in April-May followed by red or black berry-like pomes that persist into winter. Each fruit may contain up to five seeds. Percent sound seed is usually very high. Seeds should be removed from fruit before planting.

Aronia arbutifolia Red Chokeberry

SEED: Fall plant or 3 months cold stratification. The senior author collected fruits in January, cleaned seed, sowed, with tremendous germination occurring in 2 weeks. Apparently natural cold stratification was sufficient.

CUTTINGS: July softwoods root readily when treated with a hormone (4000 ppm IBA-solution), mist. Root cuttings in December-January also work well. 'Brilliantissima' — early June, 4000 ppm IBA-solution, peat:perlite, mist, rooted 80% with outstanding root systems and continued to grow (shoots) in the flats.

Aronia melanocarpa Black Chokeberry

SEED: Fall plant or 2 to 3 months cold stratification. Variety *elata* requires the same treatment.

CUTTINGS: No published American information of note but the authors suspect it should be similar to A. *arbutifolia*. Softwood, late May-early June, stuck in well ventilated frames rooted well (Russian). Hardwood cuttings, with piece of 2-year-old wood attached, rooted.

Aronia prunifolia Purple-fruited Chokeberry

SEED: Two months cold stratification. Seems to require less cold than A. *arbutifolia* to maximize germination.

Asimina triloba Pawpaw

SEED: Seeds (usually 2 to 3, large, dark-brown) should be removed from pulpy matrix. Delayed germination is caused by slowly permeable seedcoats and dormant embryos. Interestingly, seed from a southern source germinated 100% without cold treatment. Three seed lots, each given 2 months cold stratification, germinated 50, 62, 82%, respectively. Fall sowing of untreated seed did not improve germination. Fall planted seeds may germinate over a two-year period. Seedlings have a long taproot.

CUTTINGS: No report of stem cuttings being successfully rooted. The authors have observed colonies of pawpaw not unlike those of sumac (*Rhus*). It would seem that root cuttings might be successful.

GRAFTING: Professor Joseph McDaniel, University of Illinois, was a master grafter/budder and successfully chip-budded (late spring-summer) pawpaw cultivars on established trees. Seedling understock would also work.

Aucuba japonica Japanese Aucuba

SEED: Fruit is a bright red, one-seeded drupe. Senior author has planted whole fruits and cleaned seeds without any success. Seeds obviously require a cold treatment. A good starting point is the 3-month cold period.

CUTTINGS: Cuttings can be taken any time the new growth has hardened to the greenwood stage. The senior author has rooted cuttings in January, February, March with 90-100% success using 3000 ppm IBA-solution, peat:perlite, mist or polytent. Plants are easily transplanted. One report indicated that untreated January cuttings rooted 70%.

Baccharis halimifolia Eastern Baccharis
B. angustifolia False Willow, Narrowleaf Baccharis

Native shrubs with excellent salt tolerance. Hardy to Boston. Dioecious and the white cottony seeds ripen in fall on female plants. Will grow in pure sand or clay.

SEED: No pre-germination treatment is necessary. Can be sown when collected. Germination takes 7 to 15 days.

Berberis Barberry

Evergreen and deciduous shrubs with waxy, generally yellow flowers followed by red, reddish black or bluish berries. Fruits ripen in fall and may persist all winter. Ideally, seeds (often 2 per fruit) should be removed from pulp, dried and stored in sealed containers at 34 to 41°F. Seeds of B. *thunbergii* have kept well for 4 years. Percentage of sound seed is usually high, often 90% or above. Many plants with fleshy fruit or seed coats, if planted directly, do not germinate readily. Inhibitors are present in the fleshy coats. Apparently, barberry does not have this problem and whole berries may be sown in fall. Dried fruits should be soaked before fall planting. The best precaution, however, is the removal of the seed coat. Seedlings are subject to "damping off" and should be treated with appropriate fungicides. (D) = deciduous; (E) = evergreen

Berberis buxifolia (D) Boxleaf Barberry

SEED: Fall planting or 2 to 3 months cold stratification result in excellent germination.

CUTTINGS: In general, soft to greenwood cuttings of all deciduous types root well in June, July, August. In northern Ohio the last week in June and first two weeks in July prove best. Remove lower leaves and spines, 3000 to 5000 ppm IBA-talc or solution, well-drained medium, mist. As soon as rooting has taken place reduce mist. Barberries are sensitive to excess moisture. Barberries display an overwintering problem and, if possible, should not be potted until they have gone through a dormant period. The senior author has rooted many barberry species and has not noted the overwintering problem, however, it is discussed in the literature.

Many barberry cuttings are rooted in outdoor ground beds, plastic houses, and poly tunnels with mist. These can be readily left in place until spring and then transplanted. Rooting should approach 80 to 90% and higher for most species. Water management, as mentioned, is critical to success. Cuttings of most species root in 4 to 8 weeks.

Berberis candidula (E) Paleleaf Barberry

SEED: Some cold stratification (2 to 3 months) or fall planting is advised.

CUTTINGS: Easy to root about any time of year except when new growth is exceptionally soft. Senior author has used 1000 ppm IBA-solution, peat: perlite, mist, with 90% resultant success. The evergreen barberries can be taken September-October (November) and rooted in cold frames, polyhouses, etc., with 8000 ppm IBA-talc or solution. Bottom heat would probably be beneficial. They can be transplanted in spring.

Berberis × chenaultii (B. *gagnepainii* × B. *verruculosa*) (E) Chenault Barberry

SEED: 50% germination without cold treatment on seed sown in mid-January. Cold stratification of 2 to 3 months is recommended.

CUTTINGS: Mid-November, 8000 ppm IBA-talc, rooted 87% in 8 weeks.

Berberis circumserrata (D) Cutleaf Barberry

SEED: Fall plant or 2 to 3 months cold stratification.

CUTTINGS: See B. *buxifolia*.

Berberis darwinii (E) Darwin Barberry

SEED: Fall plant or 2 to 3 months cold stratification.

CUTTINGS: See B. *buxifolia* and B. *candidula*.

Berberis gagnepainii (E) Black Barberry

CUTTINGS: Early November, untreated or 3000 ppm IBA-talc, rooted 100 and 87%, respectively. Cuttings taken toward end of November (19th) did not root as well (Control — 12%, 3000 ppm IBA-talc — 42%). It is not known whether timing is this critical but the peak times for evergreen barberries according to an English reference are early October to early November.

Berberis gilgiana (D) Wildfire Barberry

SEED: One month of cold stratification resulted in 91% germination compared to 20% for untreated seed. Fall planting would also suffice.

CUTTINGS: Mid-November cuttings did not root regardless of treatment. This is a deciduous barberry and the timing was, no doubt, too late. June, July or early August would have improved rooting. This is a most handsome species resembling B. *koreana* in appearance.

Berberis × gladwynensis (E) 'William Penn' (B. *verruculosa* × B. *gagnepainii*)

SEED: The authors have not observed fruit on this species.

CUTTINGS: Mid-September, 8000 ppm IBA-talc plus thiram, rooted 100%. Late October, 3000 ppm IBA-talc, rooted 100% in 9 weeks.

Berberis julianae (E) Wintergreen Barberry

SEED: Fall plant or 3 months cold stratification. Untreated seed germinated 29%; 3 months cold stratification resulted in 100% germination.

CUTTINGS: Winter, 4 to 6″ long, 10,000 ppm IBA, 10-second dip, sand, 70°F bottom heat, greenhouse (72°F day/62°F night), 98% rooting in 6 to 7 weeks. B. *sargentiana* (E) and B. *verruculosa* (E) rooted better with 5000 ppm IBA-solution from early spring and late fall cuttings. In this study wounding was of no benefit. Summer cuttings of B. *julianae* did not respond to treatment. 'Nana' — Early December, 8000 ppm IBA-talc, rooted 80% in 12 weeks.

Berberis koreana (D) Korean Barberry

SEED: Fall plant or 2 to 3 months cold stratification. Seed without pretreatment germinated 48%; those subjected to 2 to 3 months cold germinated 80 to 88%.

CUTTINGS: Mid-August, 8000 ppm IBA-talc plus thiram, mist, excellent rooting. This species suckers and if a few plants are desired simple division with a sharp spade will suffice.

Berberis × mentorensis Mentor Barberry (B. *julianae* × B. *thunbergii*) (D)

CUTTINGS: Easy to root species. Early August, 8000 ppm IBA-talc, rooted 100% in 7 weeks.

Berberis × ottawensis (D) Ottawa Barberry

CUTTINGS: See B. *buxifolia*.

Berberis sargentiana (E) Sargent Barberry

CUTTINGS: November, untreated or treated with 3000 to 8000 ppm IBA-talc rooted 94%. Cuttings collected in early spring, late fall-winter, rooted a composite 91% in 46 days when treated with 5000 ppm IBA-solution with sand, 70°F bottom heat, mist.

Berberis × stenophylla (E) Rosemary Barberry

CUTTINGS: See B. *candidula*.

Berberis thunbergii (D) Japanese Barberry

SEED: Fall planting or 1 to 2 months cold stratification are necessary. Untreated seed germinated only 5%; 1 or 2 month cold-stratified seed averaged about 90%.

Seed propagation of var. *atropurpurea* is often practiced. The off-color (non-purple) types are often rogued. This is a great species to grow from seed to produce unusual variations. The senior author has observed seedlings from B. *t.* 'Kobold' that were yellow, green, purple, tall and dwarf. The maternal parent was growing next to other cultivars and obviously had been cross-pollinated by 'Aurea', var. *atropurpurea* and others.

CUTTINGS: This species and its cultivars are the bread and butter for many nurseries. More work has been conducted with this species than any other barberry. In general, rooting is not difficult. June, July, August are the best times and hormone levels from 1000 to 5000 ppm IBA-solution are best. Good water management is necessary to secure high percentages. Too much water can be disasterous. Ninety to 100% success has been reported with rooting occurring in 17 days (100% for B. *thunbergii*) and 28 days (90%) for B. *thunbergii* var. *atropurpurea*. 3000 ppm KIBA has been used with good success on 'Crimson Pygmy' and var. *atropurpurea*.

Berberis triacanthophora (E) Threespine Barberry

CUTTINGS: See B. *candidula*.

Berberis vernae (D) Verna Barberry

A beautiful, brilliant yellow-flowered barberry that the senior author has observed only once.

SEED: Fall plant or 3 months cold stratification.

Berberis verruculosa (E) Warty Barberry

CUTTINGS: See B. *candidula*. Early spring, late fall-winter, 5000 ppm IBA-solution, 10-second dip, sand, 70°F bottom heat, mist, compositely rooted 96% in 31 days. Other studies support these findings. Late July, 3000 ppm IBA-talc rooted 87%. Mid-November, 3000 or 8000 ppm IBA-talc rooted 88%. January, February, 8000 ppm IBA-talc, good rooting in peat.

This species is very closely allied to B. *candidula* and often confused with it.

Berberis vulgaris (D) Common Barberry

SEED: Fall plant or 1 to 2 months cold stratification. Forty days cold stratification resulted in 91% germination.

Betula Birch

Trees and shrubs of great beauty offering excellent fall color and handsome bark. Flowers are monoecious with male in hanging catkins; female in upright catkins. When pollinated and fruit develops, they generally become pendulous. Seeds are actually winged nutlets which are shed in fall or spring. Seed should be collected before the catkin shatters, dried and stored in sealed containers at 36 to 38°F. Seeds will remain viable for 1 to 2 years. Seeds will germinate without pretreatment. Heit reported that in the presence of light seeds germinate readily. If in dark, seeds need to be cold stratified for a month (±) before germination takes place. The crux is to plant seed shallow; never cover too deep. See B. *papyrifera*/B. *pendula* for the generalities of rooting birches from cuttings.

Betula albo-sinensis Chinese Paper Birch

SEED: Fall plant or 2 to 3 months cold stratification; light treatment should suffice.

CUTTINGS: A report indicated cuttings have been rooted but gave no specifics. See B. *papyrifera*/B. *pendula*.

Betula alleghaniensis Yellow Birch

SEED: No pretreatment, 1 or 2 months cold, light or fall plant.

CUTTINGS: English literature indicated that May through July cuttings could be rooted. Young plants (no age given) root readily. See B. *papyrifera*/B. *pendula*.

GRAFTING: *Betula pendula* serves as a suitable understock for this species.

TISSUE CULTURE: Axillary bud explants from young stem segments through shoot tip culture led to shoot proliferation with shoots rooting later. Plant cell cultures: results and perspectives. Elsevier, N. Holland, Sala et al. (eds), pp. 295-300 (1980).

Betula ermanii Erman Birch

One of the most handsome birches with a bloomy, tawny-cream colored bark. Easy to fall in love with this species.

SEED: Arnold Arboretum reported that seed collected in Japan germinated well without pretreatment.

Betula grossa Japanese Cherry Birch

A beautiful tree with glossy rich brown bark reminding of cherry bark. Bruised stems have odor of wintergreen. Fall color is a vibrant golden. Wonderful tree but not well known.

SEED: No pretreatment, 1 to 2 months cold, light or fall plant.

Betula jacquemontii Jacquemont Birch

Magnificent birch with excellent chalky white bark that develops on 2 to 3-year-old trees. Current reports indicate it is resistant to bronze birch borer.

SEED: No pretreatment, 1 to 2 months cold, light or fall plant.

CUTTINGS: English literature reported that rooting is successful if cuttings are taken mid-August and later.

Betula lenta Sweet Birch

SEED: Germinates after exposure to light at 68 to 86°F; however, a 1 month stratification period is required for germination when seed is maintained in dark [*The Plant Propagator* 26(4):7-9 (1980)].

Betula maximowicziana Monarch Birch

SEED: Untreated — no germination, 2 months cold — poor germination, under lights — good germination.

CUTTINGS: Early July, 1% IBA, rooted 50% with good root systems in 9 weeks. NAA failed to produce roots. Untreated cuttings rooted 30%.

Betula nigra River Birch

SEED: Matures in spring, shed at this time, and will germinate without any pretreatment.

CUTTINGS: Becoming more popular since the advent of 'Heritage'. June-July, 1000 ppm IBA-solution, peat: perlite, mist, good rooting. Cuttings are best left undisturbed and handled when growth ensues in spring. Can be rooted in fall before leaf drop. Overwinter cuttings in flats until following spring.

Terminal and sub-terminal (basal) cuttings, 6" long, lower leaves removed, wound, peat:sand, mist, 70°F bottom heat, greenhouse. Cuttings were taken from a 13-year-old tree in late July (Chicago). Ninety percent rooting of basal cuttings with 2% Benlate-talc without hormone after 6 weeks. Terminal cuttings rooted 60% with the same treatment. Basal cuttings broke bud and grew while most terminal cuttings did not. In all cases, terminal cuttings treated with auxin rooted better than basal cuttings treated with auxin [*The Plant Propagator* 23(2):5-7 (1977)].

River birch responds to extended photoperiod and cuttings that were rooted in June (Georgia), fertilized with 18-6-12 Osmocote and placed under interrupted night lighting (10 pm to 2 am) continued to grow. In fact, the growth was so fast and structurally weak that the plants assumed a weeping posture.

Betula papyrifera Paper Birch

SEED: Authors have read conflicting reports on birch seed germination. No pretreatment, cold stratification, light, or fall planting will result in good germination. In one controlled study, 3 months cold stratification resulted in good germination. No germination occurred without a cold treatment. To be on the safe side a month or greater chilling might be warranted.

CUTTINGS: Applies to birches in general (English reference). Timing is exceedingly important. Shoots should be still active with the base (of cuttings) just becoming firm. If the terminal bud is visible, results are usually poor. Six to 8" long nodal cuttings with a long shallow wound are best. 2000 ppm IBA-solution gives good results but may burn some of the softer cuttings. 8000 ppm IBA-talc resulted in 100% rootings of B. *papyrifera*. Peat: sand serves as a good medium. Roots are initiated in 12 days with results of 90 to 100%. If cuttings were taken 1 or 2 weeks later, rooting was nil. Early rooted cuttings can be potted; those rooted later should not be disturbed. Cuttings taken on August 18 and September 1 made 3 to 4 feet of growth the same season.

AFTER CARE: Seedlings and cuttings can be accelerated by placing under long days (16 hrs.), warm day (77°F)/night (65°F), good air movement and adequate moisture and nutrition. Possible to obtain white bark in 2 years rather than 3. This species is possibly more responsive to light treatment than B. *pendula*. Light treatment should be initiated in seedling stage. The first 6 weeks are the most important.

TISSUE CULTURE: Axillary bud explants from young stem segments through shoot tip culture led to shoot proliferation with shoots rooting later. Plant cell cultures: results and perspectives. Elsevier, N. Holland, Sala et al. (eds), pp. 295-300 (1980).

Betula pendula European White Birch

SEED: One large grower collects seed in fall, plants them in peat: perlite, in a greenhouse (65°F). Germination is rapid and with fertilizer (no mention of lights) the seedlings grow all winter. They are potted in 2¼" rose pots in May. This same grower noted that some seedlings of B. *p.* 'Purpurea' were reddish purple.

CUTTINGS: Late May and June, 8000 ppm IBA-talc, mist, root readily. Best rooting is obtained from young juvenile plants. When plants start to produce catkins rooting becomes more difficult. The following recipe has been tested on many birches and works quite well. 6 to 8" long, from current season's growth, shallow wound, last leaf at apex reaches full size and before last bud has fully developed, 8000 to 10,000 ppm IBA-talc, peat:perlite in peat pots. After rooting, cuttings are maintained in 32°F greenhouse. Again lights can be used to induce growth after rooting.

GRAFTING: Many cultivars are pot grafted in January-February. Budding (chip) is also practiced in July (England). See *The Plant Propagator* 25(3):8-9 (1979).

TISSUE CULTURE: B. *pendula* and B. *platyphylla* var. *japonica* have been successfully produced. Senior author observed thousands of B. *pendula* 'Dalecarlica' in Oregon being hardened off and made ready for field planting. See PIPPS 29:387 for details on B. *platyphylla* var. *szechuanica*. Cambial explants using callus culture produced shoots [*Silvae Genet*. 23:32-34 (1974)]. Shoot tip and nodal explants through shoot tip culture produced shoot proliferation and roots [PIPPS 29:387-393 (1979)]. Shoot bud explants using callus culture produced shoots [Z. *Pflanzenphysiol*. 103:341-346 (1981)].

Betula platyphylla var. ***japonica*** Japanese White Birch
Betula platyphylla var. ***szechuanica*** Szechwan White Birch

Two excellent birch selections, the first (Wisconsin source) being resistant to bronze birch borer. The Wisconsin clone has been named 'Whitespire'. Variety *japonica* is more graceful with a flat glossy green leaf compared to the stockier, leathery, waxy, deep green leaved var. *szechuanica*.

SEED: No pretreatment, 1 to 2 months cold, light, or fall plant. Three months cold stratification produced good germination.

CUTTINGS: Variety *japonica* is difficult and timing is critical as described under B. *papyrifera* (*q.v.*). Variety *szechuanica*, late June or late July, 5-6" long terminal, 4 to 5-year-old stock plants, basal 50% of foliage removed, wound, 5-second ethanol dips of IBA/NAA, peat:perlite, mist. June cuttings evaluated 1 month later, July-5 weeks later. 50% ethanol, 70% ethanol, 500, 1000, 2000, 4000 ppm IBA-50% ethanol, 8000 ppm IBA-70% ethanol produced 35, 10, 50, 60, 80, 80, 80% rooting, respectively, for the late June cuttings and 56, 10, 40, 40, 40, 42, 75% rooting, respectively, for July cuttings. NAA was inferior to IBA in rooting percentage and number. Best root numbers resulted from highest IBA treatments and earliest sampling date. The late June cuttings were still growing while the late July cuttings were more mature and leaf expansion had ceased [*The Plant Propagator* 31(1):9-10 (1985)].

TISSUE CULTURE: Stem tip and nodal explants from seedlings and mature plants using shoot tip culture produced multiple shoots. PIPPS 29:387-393 (1979); *Plant Sci. Lett*. 28:149-156 (1983).

Betula populifolia Gray Birch

SEED: No pretreatment, 1 to 2 months cold, light or fall plant.

CUTTINGS: July, 50 ppm IBA, 6 hour soak, rooted 30%.

Broussonetia papyrifera Paper Mulberry

SEED: No pretreatment is necessary. Seeds will germinate upon sowing.

CUTTINGS: July-August, short shoots with a heel, root readily. This is a suckering species and should root readily from cuttings collected from root suckers. Mid-April root cuttings produced shoots.

TISSUE CULTURE: Explants produced callus. No mention of shoot production. 3rd Symp. Pl. Tiss. Cult. Japan 31(1972).

Buddleia Butterfly-bush

Shrubs or a small tree (B. *alternifolia*) with beautiful fragrant flowers and ease of culture. *Buddleia davidii* and cultivars are best treated as herbaceous perennials and cut to ground in late winter. Flowering (July until frost) occurs on new growth of season. Fruit is a dehiscent capsule and should be collected when color is yellow-brown and dried inside. Seed requires no pretreatment. Softwood cuttings root readily.

Buddleia alternifolia Alternateleaf Butterfly-bush

SEED: No pretreatment necessary.

CUTTINGS: Mid-August, 8000 ppm IBA-talc, rooted 74% in 4 weeks. August (N.J.) greenwood with heel, light wound, 4 peat: 1 sand, mist, 0.8 to 1.0% IBA-talc works well. This is probably the most difficult of the butterfly-bushes to root and it is not overwhelmingly hard. Senior author has rooted this from July cuttings using 3000 ppm IBA-solution. Does not like excess moisture and once rooted should be weaned from the mist. Hardwood cuttings can also be rooted and var. *argentea* taken in late December, 3000 ppm IBA-talc, open bench in greenhouse with bottom heat rooted 75% in 10 weeks.

Buddleia davidii Orange-eye Butterfly-bush

SEED: Requires no pretreatment.

CUTTINGS: If all plants rooted as easily as this species and the cultivars, plant propagation would not be a challenge. June, July, August, 3000 ppm IBA-talc or solution, well-drained medium, mist, 100% rooting takes place in 2 to 3 weeks. Interestingly, cuttings continue to grow in mist bed and after transplanting. A 4" cutting may be 8" long when it is removed from the rooting medium.

TISSUE CULTURE: Meristem explants through meristem tip culture produced shoots and subsequent virus-free plants [*Ann. Phytopath*. 20:371-374 (1978)].

Buxus Boxwood

Evergreen shrubs (can become tree-like) of great beauty and functionality. The name is almost synonymous with hedges, parterres, and formal gardens. Flowers and fruits (capsule) are not showy. Cuttings root readily and are the principal means of propagation.

Buxus harlandii Harland Boxwood

CUTTINGS: June, 3000 ppm IBA, peat: perlite, mist, rooted 100%. Easy to root species.

Buxus microphylla Littleleaf Boxwood

SEED: Variety *koreana*. Cold stratification for 1 to 3 months is optimum although some seeds will germinate without pretreatment.

CUTTINGS: Senior author has excellent success anytime of year except when growth is extremely soft. 1000 ppm IBA-solution, peat: perlite, mist (summer), polytent (fall-winter). Rooting may take 2 months. One report noted optimum time is early August when cuttings are firm, 100% chloromone, 100% rooting. Cuttings (some) initiated roots after one month in bench. Varieties *japonica* and *koreana* and cultivars can be treated as above. Reports indicate bottom heat is beneficial with optimum rooting (70 to 73%) at 70 to 75°F.

Buxus sempervirens Common Boxwood

SEED: Requires no pretreatment (72% germination) although some cold (1 month or greater) unifies germination (82%).

CUTTINGS: Mid-July, no hormone, sand, root readily. Authors recommend a hormone in the range of 1000 to 3000 ppm IBA-solution; 100% rooting obtained on cuttings rooted in high humidity chamber. August through November is a good time to take cuttings. Late September (Virginia), 4 to 6″ long, mist, with media and hormones the variables. Best medium was sand followed by peat then perlite. 8000 ppm IBA-talc plus thiram gave best results. 2% IBA-talc and Rootone 10 caused burn. Hormodin 1, 2 and 3 proved reliable. In Oregon tests, 'Suffruticosa' rooted about the same (80%) regardless of hormone treatment (0, 2000 ppm IBA plus 1000 ppm NAA, 8000 ppm IBA-talc plus thiram, 2000 ppm IBA plus 1000 ppm NAA in 4% dimethylformamide).

Callicarpa Beautyberry

Shrubs with insignificant lavender-pink flowers and beautiful, rich metallic purple fruits in autumn. All flower on new growth of the season and can be cut to the ground in late winter. Flowering commences in June with fruit showing trace of color in late August-September. The fruit is a berry-like drupe with small seeds that will germinate without pretreatment. Softwood cuttings of all species root readily. White-fruited forms of all species listed root readily also.

Callicarpa americana American Beautyberry, French Mulberry

SEED: No pretreatment. Seeds might be removed from pulp but senior author has sown whole fruits in fall with excellent germination in spring.

CUTTINGS: Softwood cuttings anytime from June through September. 1000 ppm IBA-solution hastens and unifies rooting. Mist is necessary. 7 to 14 days for root initiation. Remove from mist and transplant into containers. Will continue to grow.

Callicarpa bodinieri Bodinier Beautyberry

SEED: No pretreatment. In one study, seed of var. *giraldii* sown immediately, germinated heavily.

CUTTINGS: Late June, 5 to 10 ppm IBA-24 hour soak, 100% rooting in 20 days.

Callicarpa dichotoma Purple Beautyberry

SEED: No pretreatment.

CUTTINGS: June-September, 1000 ppm IBA-solution, peat: perlite, mist, 100% rooting. January hardwoods rooted 100% in 8 weeks after treatment with IBA.

Callicarpa japonica Japanese Beautyberry

SEED: No pretreatment, although in a comparative test, germination was heavier after 2 months cold compared to unstratified seed.

CUTTINGS: October, untreated, sand, rooted 75%. IBA treatment hastened rooting. Terminal cuttings rooted better than basal cuttings. Senior author has rooted softwood cuttings with 1000 ppm IBA-dip with 100% success.

Calluna vulgaris Heather

SEED: No pretreatment; sow on milled peat moss, do not cover, place under mist or in high humidity environment, germination takes place in 3 to 5 weeks.

CUTTINGS: About as many ways to root heather as months in the year. Cuttings root readily in summer, fall and winter. Senior author had great success in August, 1000 ppm IBA-solution, peat:perlite, mist, with 100% rooting in 2 to 3 weeks. Untreated cuttings in September, November, January, rooted 90 to 100% in 6 to 8 weeks in 2 sand:1 peat. Hormone may be unnecessary but does hasten rooting. Early December, no hormone, rooted 100% in 72 days; with IBA 100% in 46 days. Winter, polytent, bottom heat, 92 to 100% in 21 days.

A large, Long Island, NY grower collects in mid-October to November, peat: perlite, mist, 90 to 100% rooting in 10 to 25 days depending on difficulty of cultivar in question. Same grower noted small cuttings (1 to 2″ long), lower leaves removed, washed in Captan or Benlate root well and produce same end product as larger cuttings. 200 cuttings can be put in 2′ by 2′ flat.

NOTE: There are 600+ cultivars varying in foliage, flower color, and flowering time. Seed grown material is fun to play with because the chance for new variations is enhanced and heaven knows the world needs another *Calluna* cultivar.

Calocedrus decurrens (*Libocedrus decurrens*) California Incensecedar

SEED: Collect cones when reddish brown, dry, extract winged seeds. Seed exhibits an embryo dormancy and two months cold stratification proves optimum. Senior author has tried commercial seed many times without success.

CUTTINGS: Senior author has been unable to put a single root on a cutting. A propagator mentioned that he was 100% unsuccessful in 50 or 60 experiments. One report noted that partially hardened August cuttings *may* be successful. Mid-

November (Boston), sand: perlite, 75°F bottom heat, polytent in 50 to 60°F greenhouse, evaluations in mid-April; numerous hormone combinations tested; best rooting from 2500 ppm NAA-5 second dip (92%) with excellent root systems; 2500 ppm IBA plus 2500 ppm NAA-5 second dip (66%); untreated cuttings rooted 8%. See Nicholson *The Plant Propagator* 30(1):5-6 (1986).

Calycanthus floridus Sweetshrub, Carolina Allspice

Several species of eastern U.S. origin including the above and *C. fertilis* and *C. mohrii*. Not a great deal of difference and probably best lumped under the above. The western sweetshrub, *C. occidentalis*, grows larger, flowers tend to be more wine red and fragrance is nil (senior author's opinion). Propagation comments apply to eastern species. Fruit is an achene that is enclosed in a receptacle. Fruits may be collected in October-November and achenes removed.

SEED: Interesting story here. Fall plant or 2 to 3 months cold stratification of dried seed. Collect urn-shaped receptacle when changing from green to brown (August, Georgia). Seed coats can be easily broken with fingernail; seeds germinated 90% in 3 weeks after planting (no pretreatment). Seed collected later from withered receptacles had bullet-hard seed coats and out of 75 seeds planted, one germinated (1.3%). In numerous studies, 3 months cold stratification produced excellent germination.

CUTTINGS: July, 8000 ppm IBA-talc, sand, rooted 93% in 6 to 7 weeks. Another report indicated 30% success at this same time. Senior author had miserable success using 1000 or 3000 ppm IBA-solution (50% alcohol solvent). In general, alcohol causes leaves to drop ahead of talc-treated cuttings. Success averages 30 to 50%. 'Athens' (a yellow flowered form) falls in the same category. Arnold Arboretum had success (*C. fertilis*) with mid-July, light wound, 1 or 2% IBA-talc, 100% rooting. All rooted cuttings were overwintered, broke bud, assumed normal growth and survived. In a mid-August study, senior author tested 1% KIBA quick dip, 1% IBA in alcohol (quick dip) and 8000 ppm IBA-talc. Rooting averaged 93, 82, and 74%, respectively. The wood was firm which may have reduced alcohol burn. Root systems were profuse in all treatments and all rooted cuttings were transplanted to a bark medium and successfully overwintered.

Camellia Camellia

Small evergreen shrubs/trees with showy flowers from fall to early spring. Fruit is a woody capsule that houses 1 to 3 brown seeds.

Camellia japonica Japanese Camellia

SEED: Collect in fall before seed coat hardens and germination will occur. If seed is dry, soaking in warm (180°F) water for 24 hours is recommended with subsequent planting. One nursery pre-chills the seed until root radicle emerges and then plants, discarding seeds that produce no radicle. 1000 ppm GA-24 hour soak, embryos excised and placed on nutrient agar germinated 100% in 6 days.

CUTTINGS: Many recipes for success which at times may be elusive. Senior author collected cuttings in November, treated them with 1.0% IBA-solution, peat: perlite, mist, with no rooting evident 5 months later. Cuttings did not look bad, just did not root. Wood should not be too hard (greenwood, July), 4000 to 10,000 ppm IBA-solution, peat: perlite, mist, reasonable rooting. Rooting may take 8 to 12 weeks.

American Camellia Society reported May to September (and November), current season's growth just below 5th node, 3000 to 5000 ppm IBA, wound, sand: peat or peat: perlite, mist or polytent. Patience may be the key with camellias.

Early November (Alabama), 25 different cultivars, bark shavings, sand and heat in humid poly-house, rooted 90% by spring. November (Boston), wound, 8000 ppm IBA-talc plus thiram, peat: perlite, polytent, rooted 100%.

One grower in a pooled sample of cuttings collected from 128 cultivars achieved 13 to 73% rooting depending on treatment. Color of wood appears to be a good index for taking cuttings. A change from green to brown is supposedly the best time. Single node or multiple node cuttings are best. The younger the stock plant the better the rooting response. 3000 to 8000 ppm IBA-talc produced best results. Hormone might delay bud break on cuttings, especially single node.

GRAFTING: Pot grafting (whip) in January-February on *C. japonica*, *C. saluenensis* and *C. reticulata* seedlings. Wedge grafts can be used. Scions were placed on understock (*C. hiemalis* 'Kanjiro') and rooted in sand: peat with the union 10 to 12″ below surface. Plants are ready for sale after 3 years compared to 5 years [*Intern. Camellia* J. 8:50-51 (1976)].

TISSUE CULTURE: Successful. See *Intern. Camellia* J. 12:31-34 (1980). Plantlets were produced from callus derived from excised cotyledons on Knop's low salt medium with 0.5 to 5.0 mg/l BA. Shoot tips, 4 to 5 month seedlings, MS medium with 1.0 mg/l BA; shoots were rooted using 0.5 to 1.0 mg/l IBA dip and transferred to basal medium with no growth regulators [*HortScience* 19(2):225-226 (1984)].

Camellia sasanqua Sasanqua Camellia

SEED: See *C. japonica*

CUTTINGS: Mid-January, 3000 ppm IBA, sand:peat, 100% rooting. This species should be handled as described for *C. japonica*. The senior author has had variable results with various cultivars. Largely a trial and error approach for each cultivar.

TISSUE CULTURE: Fresh explants from juvenile material initiated shoots best when cultured on half-strength MS medium with 0.1 or 0.5 mg/l BA. GA_3 at 5.0 mg/l added to the medium with the BA prevented dwarfing of the subsequent shoots [*HortScience* 20(3):592 (Abst. 505)].

Camellia sinensis Tea

A rather attractive evergreen shrub with handsome lustrous foliage and white flowers in September-October. Definitely hardier than either of the two previous species.

CUTTINGS: July-August, 1000 ppm IBA-solution, peat: perlite, mist, root readily. Cuttings soaked in 50, 100, and 250 ppm IBA-water for 20 hours rooted 88, 92 and 100% in 4 months. Quick dips of 5000, 7500 and 10,000 ppm IBA-alcohol produced 56, 76 and 84% rooting, respectively.

Campsis radicans Common Trumpetcreeper

SEED: No pretreatment is necesssary although seed germination is hastened and unified after 2 months cold stratification. Fruit is a large capsule which ripens in October and is full of flattened seeds with transparent wings.

CUTTINGS: Easily rooted from June-September (October) softwood. Hormone is not necessary. Well-drained medium and mist are important. Root cuttings work quite well for this species. Fall and spring-dug root pieces work. Layering is also successful. *Campsis* forms aerial roots which is a good indication of ease of rooting. Several selected cultivars are superior to species and are vegetatively propagated ('Flava' and 'Crimson Trumpet').

Campsis × *tagliabuana* 'Mme. Galen'

CUTTINGS: A hybrid with large, wide-flaring, trumpet-shaped orange flowers. It must be vegetatively propagated to maintain floral chracteristics. Early summer cuttings rooted 68% in sand in 104 days. In general, it can be handled like C. *radicans*. The use of 1000 to 3000 ppm IBA-talc or solution might prove beneficial.

Caragana Peashrub

Shrubs or small trees of the legume family with yellow flowers and short pencil-thick pods that ripen in fall. Plants are exceedingly tough and survive the coldest and harshest conditions. Fruits should be collected before pods split, dried inside and seed extracted. Dormancy is caused by hard (impermeable) seed coats.

Caragana arborescens Siberian Peashrub

SEED: Untreated seed will germinate but germination is erratic. Cold or hot (180°F) water soaks for 24 hours have resulted in 87 to 100% germination in 5 days. Unsoaked seeds took 15 days. A light acid scarification (15 minutes), 2 to 4 weeks cold stratification or fall planting would also suffice.

CUTTINGS: Late July, untreated, rooted 80% in sand. IBA or NAA caused injury or retarded shoot growth. Several reports appear in literature alluding to delayed bud break caused by high hormone treatment. Appears a trade-off is necessary for rooting quality can be improved with hormone treatment. May, June, July are best months to take cuttings. 'Nana', late May, 3000 ppm IBA-talc, rooted 85% in 6 to 7 weeks. Cuttings of 'Nana', mid-June, vermiculite, rooted an average of 86%. 'Lorbergii' (fine foliage form), 3000 or 8000 ppm IBA-talc, rooted 83 and 92%, respectively, in 3 to 4 weeks. 'Pendula' rooted 75 and 50% with same treatment as 'Lorbergii'.

NOTE: C. *arborescens* responds to supplemental light and growth is increased over non-lighted controls.

GRAFTING: The cultivars, especially 'Pendula', 'Lorbergii' and 'Walker', are top worked at 4 to 6′ height on seedling C. *arborescens*. One nursery does this in spring when frost is out of ground. These cultivars can also be rooted from cuttings.

Other *Caragana* species

Many other species exist but are largely confined to botanical gardens. The following species should be handled as described for C. *arborescens*.

C. *aurantiaca* — Dwarf Peashrub
C. *frutex* — Russian Peashrub
C. *maximowicziana* — Maximowicz Peashrub
C. *microphylla* — Littleleaf Peashrub
C. *pygmaea* — Pygmy Peashrub
C. *sinica* — Chinese Peashrub

Carpinus Hornbeam

Beautiful landscape trees with handsome foliage and bark. Flowers are monoecious with male in slender birch-like catkins and female in catkins but each flower surrounded by a 3-lobed bract. Fruits should be collected in September-October as bract color turns greenish yellow and wings are still pliable.

Carpinus betulus European Hornbeam

SEED: Variable results have been reported with different treatments. Seed should be collected green (probably September) and sown immediately or cold stratified for 3 to 4 months. If seed dries, the wall becomes extremely hard and a double dormancy results. 1 to 2 months warm/3 to 4 months cold stratification may suffice but there is no guarantee. One report (U.S.) noted that the problem is not so much getting seed to germinate as finding solid seed. Work in California with eastern U.S. seed (dry) resulted in only 20% germination after cold stratification. Belgium trials reported 70% germination for seed sown in spring following 1 month warm/4 months cold stratification. In viability tests, seeds soaked for 6 hours in water sunk (viable) or floated (hollow).

CUTTINGS: The species is usually not rooted but the many cultivars are of interest since grafting is labor intensive and results unpredictable. The best recipe to date includes the following details:

1. Select healthy, vigorous stock plants. Cuttings should be six to eight inches long and wounded.
2. Cuttings should be taken about the time the last leaf reaches mature size and the last bud has not fully developed. This would probably coincide with the month of July.
3. Many rooting mediums were tested; however, the best was a mixture of perlite and peat moss.
4. Hormone concentrations must be high. Hormodin No. 2 and No. 3 are not sufficiently high and a concentration of 2% IBA (20,000 ppm) was required to successfully root *Carpinus betulus* 'Fastigiata'.
5. After rooting, the cuttings require a dormancy period. Placing them at a temperature of 32° during the winter months satisfied the dormancy requirements, and when budbreak ensued in March or April, the rooted cuttings were transplanted to containers.

Cuttings of other cultivars should respond similarly, however, another report noted that 1, 1.6 and 3% IBA-talc resulted in excellent rooting of 'Fastigiata' and poor to no rooting for 'Globosa' and 'Compacta'. Root cuttings have not worked. Interesting work with etiolation of stock plants and localized banding which produced 43% rooting compared to 15% for non-etiolated plants. See PIPPS 34:543-550 (1984).

GRAFTING: January-February pot grafting using C. *betulus* or C. *caroliniana* understocks. A side veneer graft is used.

Carpinus caroliniana American Hornbeam

SEED: In general easier to germinate than C. *betulus*. Green seed, Vermont (early September), planted immediately germinated 24% the following spring; late September seeds germinated less than 1%. 15 and 18 weeks cold stratification increased germination to 43 and 58%, respectively. If seed becomes dry, 2 months warm/2 months cold stratification is recommended. See *HortScience* 14(5):621-622 (1979).

CUTTINGS: No report on species, however, June (Boston)

cuttings of 'Pyramidalis', 1 and 1.6% IBA, rooted well. Root cuttings did not work.

Carpinus cordata Heartleaf Hornbeam

SEED: 3 to 4 months cold or plant as soon as collected.

CUTTINGS: Variety *chinensis*, June, 1.6 and 3.0% IBA-talc, excellent rooting.

Carpinus japonica Japanese Hornbeam

SEED: Collected mid-November, 4 months cold stratification, sown in mid-March with resultant uniform germination.

CUTTINGS: Follow procedure under C. *betulus*. Late July (Boston), 3000 ppm IBA-talc plus thiram, sand: perlite, mist, rooted.

Carpinus laxiflora Loose-flower Hornbeam

SEED: 4 to 5 months cold stratification resulted in good germination. Seed was put into stratification in mid-September.

CUTTINGS: Follow procedure under C. *betulus*.

Carpinus orientalis Oriental Hornbeam

SEED: 2, 3 or 4 months cold stratification worked reasonably well. Seed sown without pretreatment did not germinate. 2 months warm/3 months cold stratification produced good germination.

Carpinus turczaninovii

SEED: 2 months cold stratification produced 60% germination.

CUTTINGS: June, 1, 1.6, 3% IBA-talc, poor to no rooting.

Carya Hickory, Pecan

Monoecious trees with edible fruits in some species, particularly pecan. Fruit is a nut with a dehiscent involucre-husk that splits along defined suture lines. Fruits should be collected in fall and kept moist in refrigerated storage. Seeds are high in oils and can dry easily. Nuts (seeds) of most species can germinate without pretreatment, although 1 to 3 months cold stratification is recommended.

Carya cordiformis Bitternut Hickory

SEED: 3 months cold stratification produced high germination.

Carya glabra Pignut Hickory

SEED: No pretreatment resulted in 79% germination. 3 months cold stratification resulted in uniform germination (91%).

Carya illinoensis Pecan

SEED: May start growing in hulls. Store seeds at 32°F immediately after harvest until planting. If seeds become dry, soak. 1 to 3 months cold stratification is recommended. Seed is described as possessing no embryo dormancy and germinates at any time under favorable conditions, but germination is erratic. Cold stratification hastens and unifies germination within 20 days. Two to 3 months at 32 to 35°F promoted earliness and uniformity of germination. Nuts soaked in H_2O for 2 days, stratified in moist sand, germinated in 4 weeks. Numerous germination studies indicate that pecans germinate readily at warm temperatures. Nuts incubated at 86°F germinated 96% in 15 days, while germination at 68°F was poor. Stratification at 41°F for 3 months resulted in 86% germination after 20 days at 68°F.

CUTTINGS: Success has been minimal but several published reports indicated that cuttings could be rooted. Unfortunately, they did not overwinter. One author speculated that keeping the leaves on the cuttings throughout the rooting process and into fall was essential. This harkens back to adequate carbohydrates to sustain rooted cuttings through an overwintering period.

'Posey', hardwood from 2 to 4-year-old plants, early April, callused in sphagnum at 68 to 75°F bottom heat, large cuttings, rooted 63%; smaller rooted not as well; untreated or small did not root at all. Another study with 8-year-old 'Stuart' pecan softwood, 6" long, 3/5" wide, vermiculite in 2¼" peat pots, mist, rooted best with 1% IBA-talc followed by 1000 ppm IBA in 0.5% DMSO-15 minute soak. Percent rooting was not given but mean number of roots was 23.2 and 20.7, respectively, for above treatments. Rooting was first observed 6-7 weeks after sticking.

Juvenile wood produced more roots and rooted cuttings than mature wood when treated with a 0.5, 1 or 2% IBA-solution in February, 1 or 2% IBA in June, and 0.5% or 1.0% IBA in August. See *HortScience* 15(5):594-595 (1980).

Root pieces, 4, 6, and 8" long with an average basal diameter of 2/5 to 3/5" were taken from 2-year-old 'Barton' seedlings in August (S. Africa). Cuttings were planted in a nursery under 50% shade. Average shoot production was 60, 80 and 100% for the three lengths, respectively, and shoot length averaged 6, 8 and 12", respectively. This implies that food reserves in the root pieces are essential for increased shoot production.

Kansas work has shown that hardwood and softwood cuttings could be rooted but percentages were low [*The Plant Propagator* 24(2):6-8 (1978)].

GRAFTING: According to many growers pecan is not the easiest plant to bud. A patch bud is recommended and has produced 90% takes for good propagators. This is accomplished in August in the field. The principal understock is C. *illinoensis* although selected cultivars are used as seed sources since they produce more fibrous root systems and better grafting stock in a shorter time frame.

TISSUE CULTURE: Lateral bud stem section explants through shoot tip culture led to shoot development [*HortScience* 17:487. Abst. 109 (1982)].

Carya laciniosa Shellbark Hickory

SEED: 3 months cold stratification has proven optimum.

Carya ovalis Red Hickory

SEED: 3 months cold stratification.

Carya ovata Shagbark Hickory

SEED: 3 to 4 months cold stratification.

Carya tomentosa Mockernut Hickory

SEED: 3 to 5 months cold stratification.

Caryopteris × *clandonensis* Bluemist Shrub

SEED: Germinate without pretreatment. Collect in late summer-fall, dry inside, store in sealed containers and plant to coincide with lengthening days of late winter and early spring.

CUTTINGS: May-June until flowering, actually as soon as new growth is substantial. Senior author uses 1000 ppm IBA-solution, peat: perlite, mist, 100% rooting. Cuttings should be weaned from mist immediately after rooting (7 to 14 days). Cuttings will continue to grow. Can be rooted into October and thereafter. Cuttings taken later in season have flowers and fruits, and at times it is difficult to find decent leafy cutting wood. Also these rooted cuttings may prove difficult to overwinter. Many named selections and all are easily rooted from softwood cuttings.

Caryopteris incana Common Bluebeard

CUTTINGS: See *C. × clandonensis*. One report noted hormone did not improve rooting; another said it was beneficial. 1000 ppm IBA-talc or solution are adequate.

Castanea dentata American Chestnut

SEED: 3 months cold stratification.

TISSUE CULTURE: Cambial tissue explant through callus culture have produced bud-like structures. Propagation of higher plants through tissue culture. Tech. Info. Center, U.S. Dept. Energy, Springfield, VA. Hughes et al. (eds) p. 259 (1978). Etiolated epicotyls through callus culture have produced shoot initials. Proc. 26th Northeastern. For. Tree Improv. Conf. (1979).

Castanea mollissima Chinese Chestnut

SEED: May germinate without pretreatment although 2 to 3 months cold stratification is optimum. Seed deteriorates rapidly and should not be allowed to dry. Store in sealed containers or poly bags under refrigerated conditions. Seed will survive up to 12 months. In general, as soon as nuts are collected in fall they can be sown. Weevils may infect nuts and will destroy embryo. Hot water (120°F) for 30 minutes will destroy them.

CUTTINGS: Most attempts have failed, however, cuttings from young plants forced in the greenhouse in early April, 8000 ppm IBA-talc, rooted 45% in 3 weeks. No mention made of survival. The use of juvenile material is important. 7000 ppm IBA in 95% ethanol promoted rooting.

GRAFTING: Difficult to graft although pot grafting in January-February and dormant grafting in the field in early spring are practiced. Nurse-seed grafting has also been successful in the 30 to 50% range. Bark grafting and inverted T-budding have given good results.

TISSUE CULTURE: Axillary shoot proliferation was achieved on Woody Plant Medium containing BA. Rooting of microshoots was stimulated by a 1-second basal dip in 1000 or 3000 ppm IBA |*HortScience* 20(3):593 (Abst. 508)|.

Castanea sativa Sweet Chestnut

CUTTINGS: Mature cuttings are difficult to root but shoots from 3-month-old greenhouse plants treated with 1 part NAA: 10 parts IBA-talc mixture, mist, rooted 80% in 4 weeks.

GRAFTING: Numerous approaches that work. In one study-cleft, tongue grafting and chip, patch and shield budding were compared. Tongue grafting and chip budding produced 85 to 90% take.

TISSUE CULTURE: Considerable work has been accomplished with explants from seedlings. Lateral buds from 3 to 4-month-old seedlings developed axillary shoots on 1 mg/l BA. Subcultures proliferated best on 0.1 mg/l BA |*J. Hort. Sci.* 55(1):83-84|. Cotyledonary tissue has also been successfully cultured |*Scientia Hortic.* 8(3):243-247 (1978)|. Cotyledon explants have produced callus with no mention of shoot production |*Experientia* 31:1163-1164 (1975)|. Embryonic shoot axis explants through shoot tip culture produced axillary shoots which rooted |*Physiol. Plant.* 50:237-130 (1980)|. Seed explants produced multiple shoots. Plant tissue culture. |Proc. 5th Int. Cong. Plant Tissue Cell Culture, Japan, Japan Assn. Plant Tissue Cult., Tokoyo, Fujiwara (ed) pp. 141-142 (1982)|. Axillary bud explants of juvenile shoots through shoot tip culture led to axillary shoot proliferation |*Scientia Hort.* 18:343-351 (1983); *HortScience* 17:888-889 (1982)|.

Catalpa Catalpa

Large rather coarse trees with immense leaves and stiff branches. White to lavender flowers in large panicles during June-July are followed by elongated narrow-cylindric capsules full of fringed seeds. Fruits can be collected in October and dried so capsule splits; some seeds will fall out naturally but best extraction is accomplished by hand. Store seed in refrigerated storage and sow in spring.

Catalpa bignonioides Southern Catalpa

SEED: No pretreatment although dry seed might benefit from brief cold stratification.

CUTTINGS: Late-December, hardwood, 8000 ppm IBA-talc plus thiram, open bench, rooted 40% with terminal bud present, 100% without, by early April. No information was given on age of these trees. Summer softwoods will also root but no specifics were given. 'Nana', mid-July, 3000 ppm IBA-talc plus thiram, rooted 100% with good root systems in 3 weeks.

GRAFTING: Several cultivars: 'Aurea', 'Nana', etc., exist and are usually pot grafted in January-February and placed in a closed case. Side or splice grafting on seedling *C. speciosa* is also practiced. To produce 'Nana' on a 4 to 6′ standard, budding can be accomplished in August.

Catalpa bungei Bunge Catalpa

SEED: No pretreatment.

Catalpa fargesii Farges Catalpa

SEED: No pretreatment.

Catalpa ovata Chinese Catalpa

SEED: No pretreatment.

Catalpa speciosa Western or Common Catalpa

SEED: No pretreatment.

CUTTINGS: Late-December, hardwood, open bench, 8000 ppm IBA-talc plus thiram, rooted 50 to 75%.

Ceanothus (non-cold hardy types)

Ceanothus americanus New Jersey Tea

Ceanothus ovatus Inland Ceanothus

SEED: Capsules should not be collected until ripe. Since they dehisce naturally, timing is critical. One report noted that fruits collected before they ripened naturally produced very few viable seed. Most *Ceanothus* species have either seed coat and/or embryo dormancy. A hot water soak and/or 2 to 3 months cold stratification have induced good germination. C. *americanus* responded well to 3 months cold stratification. Seedlings of all *Ceanothus* species are very susceptible to damping off and a preventative fungicide program should be used. See PIPPS 12:214 for West Coast species.

CUTTINGS: Softwood cuttings root readily (100%). A hormone will hasten and improve rooting. Spring to fall is appropriate. Mist is necessary. Late July cuttings of C. *ovatus*, 8000 ppm IBA-talc, sand, rooted 100% but failed to overwinter. See PIPPS 12:226 for specific West Coast species.

Cedrela sinensis Chinese Cedrela or Toon

SEED: No pretreatment, seed sown directly germinated 71%. Three months cold stratification produced excellent germination. Reports with untreated seed indicated poor to reasonably good germination. A one-month cold period would be a wise safeguard.

CUTTINGS: This species has a propensity to sucker from roots and is an ideal candidate for root cuttings. Root cuttings, October to February, will work. One grower produces the tree commercially by cleaning root pieces in sodium hypochlorite, cutting into pieces ½" to 1" long, dipping in Captan, placing in medium in an upright position (maintain polarity) so top of root protrudes 1/8" above medium surface, placing on bench in greenhouse at 70°F, after 2 or 3 weeks adventitious buds develop. Plants are then potted. Terminal stem cuttings taken from shoots root in 4 weeks when treated with 1000 ppm IBA-talc, sand, mist, bottom heat.

Cedrus Cedar

Splendid evergreen conifers with dark green to rich blue needles...specimen trees of the first order. Pollen is shed in late summer/fall and cones take two years to mature. Cones should be collected before they turn brown, allowed to dry until the scales loosen and the seeds can be removed. Seeds are oily and do not store well although when moisture content is reduced to 10%, viability is retained for 3 to 6 years when stored in sealed containers at 30 to 38°F.

Cedrus atlantica Atlas Cedar

SEED: The cedars have essentially no cold requirement and will germinate immediately upon sowing. However, variable degrees of dormancy may be observed within a single lot of seeds. Controlled studies have demonstrated that seeds sown without treatment germinated erratically over several months; seed provided one month cold stratification germinated in 2 weeks and seed given 2 months cold germinated in 4 to 7 days. A short 2 week cold period has been recommended to hasten and unify germination.

CUTTINGS: Difficult to root with any kind of consistent high percentage. Late February, 8000 ppm IBA-talc, rooted 20% but had a single long root. Mid-December, 6" long, 3 to 4% IBA-solution, perlite, will work. Late summer or fall, hormone, bottom heat, polytent, some rooting may occur.

GRAFTING: Cultivars of all three species are pot grafted in December, January and February on seedling C. *deodara* using side veneer grafts. C. *deodara* has a more compact and fibrous root system.

Cedrus deodara Deodar Cedar

SEED: As described under C. *atlantica*. Damping-off can be problematic and an appropriate fungicide should be utilized. Usually *Cedrus* seeds are spring sown. This necessitates storing seed, providing a short chilling period and sowing in March-April. Stratified seed (40°F) in moist sand for 30 days germinated 45%; 16% after 15 days stratification; and 11% without stratification.

CUTTINGS: This species can be rooted successfully from October to December using last season's growth, stripping basal needles, wound, well-drained medium, bottom heat, polytent. Rooting will range from 60 to 90%. Most of the work indicated a low level hormone did not improve rooting over untreated cuttings. 1.0% IBA-solution has given good results. There is an adage applied to this species that cold tops and warm bottoms root well. Cold frames with bottom heat have been used to successfully root this species. Winter rooted cuttings can be expected to make good growth during spring and summer. 'Kashmir', October to mid-November (Oregon), 2000 ppm IBA plus 1000 ppm NAA-solution, good rooting. 'Shalimar', mid-November, sand: perlite, 75°F bottom heat, polytent in 50 to 60°F greenhouse, evaluations were made in mid-April, numerous hormone combinations were tested; best rooting obtained with 5000 ppm IBA-5 second dip (67%) and 10,000 ppm IBA-5 second dip (50%). NAA appeared to be inhibitory and untreated cuttings did not root [See *The Plant Propagator* 30(1):5-6 (1984)].

GRAFTING: This is the universal understock for all cedars. In colder climates, the hardy cultivars, 'Kashmir' and 'Shalimar', might be rooted and used as understock. The species is considered cold hardy to Zone 7 (USDA) while 'Shalimar' should be hardy in 5/6.

TISSUE CULTURE: MS medium plus 1 to 7 mg/l phloridzin produced callus from young stem segments [*Plant Physiol. Comm.* 4:37-39 (1983)].

Cedrus libani Cedar of Lebanon

SEED: As described under C. *atlantica*. The treatments described, thereunder, were performed with this species. This species is particularly susceptible to damping-off diseases.

CUTTINGS: Difficult. November cuttings rooted 30% in sand: peat without treatment.

GRAFTING: The cultivars are pot grafted in January-February on C. *deodara*.

Celastrus x *loeseneri* Loesener Bittersweet

CUTTINGS: Mid-December root pieces, placed horizontally in medium, produced shoots on all pieces.

Celastrus orbiculatus Oriental Bittersweet

SEED: 3 months cold stratification resulted in 100% germination in one test.

CUTTINGS: Early August, 8000 ppm IBA-talc plus thiram, mist, rooted 100% in 3 weeks. Root cuttings will work. Hardwood cuttings can also be rooted.

Celastrus scandens American Bittersweet

SEED: Fruit is a dehiscent capsule. Seeds are covered by a bright orangish to reddish aril. The aril should be removed since there is evidence it may have an inhibiting effect on germination. Three months cold stratification has resulted in good germination. Nursery practice would involve fall planting.

CUTTINGS: Early August, 8000 ppm IBA-talc, rooted 83%. July softwoods of all species treated with IBA root readily. Cuttings will root without treatment but it is hastened or improved by treatment.

Celtis Hackberry, Sugarberry

Trees of tough, durable constitution with inconspicuous polygamo-monoecious flowers and a fleshy orange-red to dark purple, ⅓″ diameter drupe. Fruits ripen in October and can be collected then or later, cleaned, and stored in sealed containers in refrigerated storage. Seeds have been stored for 5½ years without loss of viability. Removing the pulp is not essential but has been reported to aid germination of all species. Seed of most species benefits from 3 months cold stratification.

Celtis jessoensis Jesso Hackberry

Possibly the best ornamental hackberry. Habit is not unlike American elm and the bark is smooth like beech. Additionally, "witches broom" does not seem to be a problem.

SEED: 3 months cold resulted in good germination. Fall planting would also suffice.

CUTTINGS: Early August, 3000 ppm IBA-talc, rooted 38%. No mention is given relative to survival but authors suspect overwintering might be a problem.

Celtis laevigata Sugar Hackberry

SEED: Germination may occur without stratification but 2 to 3 months cold is recommended. Fall plant.

CUTTINGS: There is the possibility of rooting cuttings but no published results were found.

GRAFTING: Chip budding in summer has been successful. *C. occidentalis* or *C. laevigata* seedlings serve as suitable understocks.

Celtis occidentalis Common Hackberry

SEED: 2 to 3 months cold stratification or fall plant.

CUTTINGS: Tremendous interest in rooting this species from cuttings. Suspect summer softwoods taken just as end bud forms, 5000 to 10,000 ppm IBA-solution, peat: perlite, mist, would work. Handle rooted cuttings like *Acer palmatum*.

Root cuttings, 4-year-old seedlings, fall, 5-6″ long, 1/5 to 2/5″ diameter, 3 peat: 1 sand: 1 soil mixture, greenhouse at 65°F air temperature, shoots appeared after 10 weeks. 30% of the shoots rooted. Root cuttings might be a possibility for this species.

GRAFTING: Not the easiest plant to bud and chip budding in summer has been used on the cultivars.

Cephalanthus occidentalis Buttonbush

SEED: Fruits should be collected when reddish brown and before they disintegrate. Dry, separate seeds, and either store in sealed containers or sow immediately. No pretreatment is required and seeds will germinate in 10 to 14 days.

CUTTINGS: Late July to early August, sand: peat, rooted 100% in 4 weeks without hormone treatment. Softwood should root without difficulty. Hardwood cuttings will also root. A hormone will unify rooting but is not absolutely necessary.

Cephalotaxus Plum-yew

Unusual dioecious, yew-like evergreens with longer dark green needles and larger (3/4 to 1″ long) plum-shaped fruits that take 2 years to mature. Heat and shade tolerant, and better adapted to southern climates. Fruit is actually a naked seed with a fleshy aril and should be collected in late September-October, cleaned and fall planted.

Cephalotaxus harringtonia Harrington Plum-yew

SEED: Fall plant or 3 months cold stratification.

CUTTINGS: Not the easiest plant to root. Senior author has had excellent or virtually no success. Collect in fall, 5000 to 10,000 ppm IBA-talc or solution, rooting will occur in 2 to 3 months. The procedure is similar to that used for *Taxus*. June softwoods have also been rooted but no percentage was given. 'Duke Gardens', a spreading sport of *C. harringtonia*, was collected in mid-March, 8000 ppm IBA-talc, 2.0% IBA-talc, or no treatment, examined 2 years later and percentages were 63, 70, 73%, respectively. There was no benefit to using a hormone according to author. However, this was a juvenile branch sport from the base of a mature plant. Interestingly, cuttings taken from the parent plant, 8000 ppm IBA-talc, rooted 40% and root quality was inferior to that of all the juvenile rooted cuttings.

TISSUE CULTURE: Callus cultures were maintained but no shoot production was reported [*Planta Medica* 40(3):237-244 (1980)].

Ceratostigma plumbaginoides Blue Ceratostigma

CUTTINGS: Softwoods root readily in June to August. Division is a suitable means of producing small numbers. Root cuttings will also work. Root cuttings taken in March-April produced 70 to 96% shoots.

Cercidiphyllum japonicum Katsuratree

SEED: Seeds are small and winged. Capsules should be collected in late October or November before they start to split. The fruits can be dried inside, seeds shaken loose and stored in a glass or plastic vessel at 40 to 45°F. Seed remains viable for at least three years. Seed requires no pretreatment and germinates within 7 to 14 days. If seeds are started in March, a 3 to 5′ high plant can be produced in a single growing season. Katsura seed will germinate at temperatures of 41°F to 99°F with an optimum from 68 to 77°F.

CUTTINGS: Mid-July (Canada), firm growth, 2% IBA, rooted. Late June (Canada), 6″ soft cuttings, lower leaves removed, no hormone, cold frame, syringed twice weekly, rooted 76% in 9 weeks. A report from Denmark indicates that softwood cuttings, June, 500 ppm IBA, can be rooted. Cuttings should be from branch tips and include the first two pairs of fully developed leaves. Rooting takes about 10 weeks. Senior author has rooted cuttings from seedlings but never from mature plants.

GRAFTING: The weeping cultivars can be chip budded, wedge or cleft grafted on seedling understock during the summer months. Bench grafting can also be practiced if potted seedling material is available.

Cercis Redbud

Small trees with bright lavender-pink flowers in early spring followed by abundant 2 to 3" brown pods, each containing several hard-coated brown seeds. Collect in September-October, dry, clean, remove weevil infested seeds (often many), and store in sealed containers under refrigeration. Redbud seed should approach 85 to 90% soundness.

Cercis canadensis Eastern Redbud

SEED: In general, seed possesses an impermeable coat and embryo dormancy. Many nurserymen plant in fall and achieve good germination. A 30-minute sulfuric acid soak followed by 3 months cold stratification produces excellent germination [*The Plant Propagator* 25(2):4-6, 1979]. Hot water (190°F) soaks have also been used in place of acid. A cold stratification period must follow.

Variety *alba*: Since most redbuds are self-incompatible, any seed collected from the white form will probably be the result of cross pollination by the species. The normal lavender-pink flower color will result in the offspring. Different clones of the white flowered form should be established in an isolated seed orchard. Seedlings will come true-to-type and the absence of anthocyanin pigmentation in the young leaves is a good indication of potential white-flowering seedlings [*The Plant Propagator* 17(2):21-22 (1971)].

CUTTINGS: Although the authors read reports that state this species can be successfully rooted, we have serious reservations. Senior author has never rooted a single cutting. Untreated June-July cuttings, 72°F bottom heat, mist, rooted 75 to 90% in 4 weeks. No mention was made relative to age of stock plants. Cuttings would be a tremendous advantage with this species since many cultivars exist and budding is not particularly reliable. One report noted softwoods, 8000 ppm IBA-talc, mist, rooted 90% at 75°F (air temperature) and 0% at 65°F (air temperature). Photoperiod responsive and extended photoperiod results in significant growth increase over unlighted controls. No mention was ever made of overwintering aspects. Authors have never seen a redbud on its own roots. Oklahoma nursery has propagated cultivars by cuttings but noted root systems were not as good as trees produced by budding.

GRAFTING: T-budding is traditionally practiced in late July to early August but results can vary from year to year. Pot grafting in winter using a side graft is another possibility. One grower uses the following method to achieve 90% success: 18 to 24", 2-year seedlings are lined out in spring, side dressed with N, cut to ground, one shoot from the regrowth is retained, a T-bud is used, 3 weeks later understock is cut back to bud (eye), 12 to 15" regrowth (from bud) occurs same season. If not cut back buds will be dead next spring. Success is predicated on plump budwood with large buds.

T-bud, July-August (Oklahoma); second bud (rebudding) — 2 to 3 weeks later with bud placed directly above the other; 70 to 95% take; understock cut back in March, suckering from understock is common and they are removed. Order of budding ease (highest to lowest): 'Flame', 'Rubye Atkinson', 'Alba', 'Pink Charm', C. *racemosa*, C. *siliquastrum* 'Alba', C. *reniformis* 'Oklahoma', C. *chinensis*, C. *reniformis* 'Alba' and C. *canadensis* 'Forest Pansy'.

TISSUE CULTURE: Nodal explants had best proliferation on Woody Plant Medium containing 2.0 mg/l BA [*HortScience* 20(3):592 (Abst. 503)].

Cercis chinensis Chinese Redbud

SEED: Senior author has fall planted with excellent germination in spring. Acid, 15 to 30 minutes, followed by 2 months cold stratification.

CUTTINGS: See C. *canadensis*. June-July softwoods rooted 75 to 90% without hormone treatment. Timing is probably critical with this species.

Cercis occidentalis Western Redbud

SEED: Same as C. *canadensis*.

Cercis reniformis 'Oklahoma'

A particularly attractive selection with thick-textured, rich glossy leaves.

GRAFTING: T-budding in July-August on C. *canadensis* seedlings has proven most successful.

Cercis siliquastrum Judas-tree

SEED: Fall sow and will germinate in spring. An English reference mentioned that this species does not exhibit embryo dormancy which would only necessitate breaking seed coat with acid before germination would occur. Variety *alba* will come true from seed if seed is taken from an isolated planting. One report noted that seed given a hot water soak produced only one seedling. Same treatment plus 3 months cold stratification produced excellent germination in 22 days.

TISSUE CULTURE: Axillary buds were proliferated with 2 mg/l BA. Sixty percent rooting was obtained in liquid culture with 0.3 mg/l IBA. The resulting plantlets survived in the greenhouse. See *Informatore Agrario* 40(11):103-15 (1984).

Chaenomeles japonica Japanese Floweringquince

SEED: Fall plant or 2 to 3 months cold stratification. Fruit is a pome and should be collected in fall, allowed to soften, seeds extracted from pulp (mash), dried and refrigerated in sealed containers.

CUTTINGS: Easily rooted from June-July softwoods, 1000 to 5000 ppm IBA-solution, peat: perlite, mist. A hormone treatment and bottom heat are beneficial. Cuttings have been rooted in September. Root cuttings, late fall, ¼" diameter, 2 to 4" long, refrigerated until spring, lined out horizontally in nursery row, produced plants. Division and layering may also be used if a few plants are desired.

TISSUE CULTURE: Shoot tip explants led to axillary shoot proliferation [*HortScience* 15:432, Abst. 455 (1980)]. Shoot tip explants through shoot tip culture led to shoot proliferation with subsequent rooting [*HortScience* 17:190-191 (1982)].

Chaenomeles speciosa Common Floweringquince

Chaenomeles × *superba*

SEED: In one study seeds sown without pretreatment did not germinate. Two months cold stratification resulted in 63% germination. Fall plant.

CUTTINGS: June-July, 1000 to 5000 ppm IBA-solution, well-drained medium, mist, root readily. Root cuttings, ¼″ by 3″ long, pot vertically.

Chamaecyparis Falsecypress

Impressive, elegant evergreens found in western and eastern North America and Asia. The species are monoecious with small 2/5 to 1/2″ diameter cones composed of 6 to 8 (12) scales. The seeds (1 to 5 per scale) are quite small and winged. In general, seed quality is low and seeds require 2 to 3 months cold stratification. Most species and cultivars are rooted from cuttings and a few difficult-to-root types are grafted.

Chamaecyparis lawsoniana Lawson's Falsecypress

SEED: Cones mature in one season and should be collected before dehiscing naturally. Cones can be dried inside and seeds shaken free (usually 20 to 30 seeds/cone). English reports indicate seeds will germinate without stratification; however, rate and uniformity of germination are enhanced by cold moist stratification. Plants grown from seed tend to be extremely variable and unless rootstock or hedging material are desired, vegetative propagation is logical. Seedlings suffer from damping-off organisms and preventative measures should be practiced.

CUTTINGS: Cultivars are traditionally rooted from cuttings collected in September through April. The cuttings will vary in size (from 1 to 4 to 6″) due to the numerous cultivars. Basal foliage should be removed, IBA in the range of 3000 to 8000 ppm, well-drained medium (peat: perlite), mist or polytent with bottom heat (during winter months). Rooting takes 2 to 3 months after which cuttings can be easily lifted and transplanted. July (Spain), 8000 ppm IBA-dip, peat: perlite, produced the best rooting compared to other times of the year. In Holland tests, cuttings rooted best in October when sampled from August to March using outdoor cold frames.

GRAFTING: For difficult-to-root cultivars, side grafting on seedlings of *C. lawsoniana* is practiced. Winter grafting is the traditional method but summer (August) grafting is practiced in Europe. Understocks should be about pencil thickness. In the south, *C. lawsoniana* does not perform well. There is some indication that the problem is root related for plants grafted on *Thuja* rootstock perform well. 57 to 98% success was obtained with winter grafts of 'Patula', 'Silver Queen', 'Stewartii' and 'Triumph van Boskoop'. In all cases winter grafting was superior to summer grafting.

Chamaecyparis obtusa Hinoki Falsecypress

SEED: As described for *C. lawsoniana*. One of the most fascinating examples of seed variability occurred at the Arnold Arboretum where Mr. Al Fordham germinated seeds of *C. obtusa* 'Nana Gracilis'. The resultant seedlings after 21 years ranged in size from 3″ to 10 and 12 feet. Each plant was sufficiently different that a cultivar name could have been attached. Thankfully, Mr. Fordham did not do this.

CUTTINGS: This species is considered more difficult to root than *C. lawsoniana* but senior author has had excellent success with many of the cultivars. The time frame is similar to that described under *C. lawsoniana*, except cuttings taken in late winter or early spring have been reported to root better than those taken in fall and mid-winter. In Georgia studies, 80 to 100% rooting with *C. obtusa* 'Gracilis' was obtained with March cuttings, 1% IBA-quick dip, peat: perlite or bark, 70°F bottom heat, in outdoor poly-covered ground beds. No burn was observed on the cuttings even with the high rate of IBA.

Cuttings of *C. obtusa* and 'Nana', 'Compacta', 'Lycopodioides', 'Filicoides', 'Gracilis' and 'Magnifica' were taken 11 times between late September and late January. Untreated cuttings averaged 41% rooting. Ninety-six percent rooting occurred on cuttings treated with 8000 ppm IBA-talc. Cutting wood was collected from current season's growth although 2 and 3-year-old wood also rooted. Untreated cuttings rooted equally well from September to January.

GRAFTING: As described for *C. lawsoniana*. Some American nurseries top work the dwarf forms of *C. obtusa* on older plants of *C. obtusa* thus producing novelty plants. Summer grafting would be the best way to do this. Some dwarf forms of *C. obtusa* are so slow growing and when established on seedling understock will grow faster. 'Pygmaea Aurescens' was 4″ high and 8″ wide after 5 years from a rooted cutting. Scions similar in size to the cuttings were 6″ by 8″ six months after grafting. The same type of response was noted for 'Nana Gracilis', a notoriously slow-growing form.

Chamaecyparis nootkatensis Nootka Falsecypress

SEED: More difficult than the other species and stratification for up to one year may be necessary. One month warm stratification at fluctuating 68° to 86°F temperatures followed by one month stratification at 40°F produced good results. Seeds that were sown in the spring did not germinate the following spring indicating a double dormancy mechanism.

CUTTINGS: A difficult species to root and higher hormone (8000 ppm IBA and greater) along with late winter-early spring collection times are recommended. Combinations of IBA and NAA might also be tried. In Ireland, 75% rooting of 'Pendula' was obtained from February cuttings under mist after 14 weeks. Hormone was not mentioned.

GRAFTING: 'Pendula' is a beautiful plant and in the south could possibly be grafted on *Thuja orientalis* or perhaps even × *Cupressocyparis leylandii*. *Thuja orientalis* is frequently used as an understock in Europe for grafting cultivars of *C. nootkatensis*.

Chamaecyparis pisifera Japanese Falsecypress

SEED: See *C. lawsoniana*.

CUTTINGS: This species and the many cultivars are relatively easy to root. Follow directions under *C. lawsoniana*. Senior author has rooted 'Boulevard' (a juvenile form), from July through April. IBA from 1000 to 10,000 ppm quick dip has been succcessful with a wide range of cultivars. Possibly 3000 to 5000 ppm would provide the maximum broad spectrum rooting. In Spanish tests, July was the optimum month for rooting with 8000 ppm IBA dip and peat: perlite medium.

GRAFTING: Since most cultivars root readily grafting is not practiced to any significant degree. 'Boulevard' is easily rooted and has been used as a rootstock for grafting various falsecypress. In fact, cuttings inserted in October-November are ready for veneer or side-grafting by January or February. In the south, older plants of 'Boulevard' and 'Plumosa' are evident, indicating this species might be more heat tolerant and root rot resistant than *C. lawsoniana* or *C. nootkatensis*, thus making it a good understock for these species and their cultivars.

Chamaecyparis thyoides Atlantic White Cedar

SEED: Not a particularly important ornamental species although a few cultivars exist. Like *C. nootkatensis*, the best stratification treatments have not been developed. Fall sown seed germinated about 50% the first spring with the rest taking place the second year. A warm/cold stratification treatment is recommended.

CUTTINGS: Follow procedures described for *C. lawsoniana*. A higher IBA level (1.0%) is perhaps beneficial on this species. Two reports noted that mid-November, 125 ppm IBA/24-hour soak, sand: peat, rooted 96% in 6 months; mid-December, 8000 ppm IBA-talc, sand: peat, rooted 70% compared to 14% untreated.

Chamaedaphne calyculata Leatherleaf

SEED: No pretreatment is required. See procedure under *Calluna*.

CUTTINGS: Softwood, late July, 4000 ppm IBA-solution, sand, perlite, mist, rooted 100% with profuse root systems.

Chimonanthus praecox Fragrant Wintersweet

SEED: Flowering occurs in January-March with fruits ripening in May-June. The seeds (actually achenes) are enclosed in small cylindrical to urn-shaped receptacles. If achenes are sown in the green to brown color state germination is 90% or above. If the fruit wall hardens (July and later) germination may approach 5%. Seed at this stage requires a 3 month warm/3 month cold period. Seed-grown plants are often slow to reach flowering size.

CUTTINGS: Considered difficult. Senior author collected cuttings in late July (Georgia), 3000 ppm IBA-solution, peat: perlite, mist, wood was firm and growth had stopped, 70% rooting resulted. Cuttings were collected from a mature 15′ high shrub. Late August, 3000 ppm IBA-talc plus thiram, sand: perlite, rooted 66%. Best to collect cuttings from young plants (juvenile).

TISSUE CULTURE: Explants produced shoots which eventually rooted [*Plant Physiol.* 63(5), Abst. 756 (1979)].

Chionanthus Fringetree

Graceful polygamo-dioecious shrubs or trees with lacy white, fragrant flowers and bluish purple grape-like drupes. Fruits should be collected in August (South) and September-October (North) before birds clean the plants. The fleshy pulp can be removed by maceration, seed dried and refrigerated in sealed containers for long periods. Seed shows high degree of soundness.

Chionanthus retusus Chinese Fringetree

SEED: Doubly dormant and warm/cold stratification is required. The best method is to place in warm and watch for general radical emergence. One to 3 months is the usual time frame. Place in cold for 2 to 3 months and then sow. The dormancy in *C. retusus* is not as deep seated as in *C. virginicus*. Seeds can be fall sown with germination taking place the second spring. Reports noted that if seed is picked early, and planted immediately, some (reasonable) germination may occur the following spring.

CUTTINGS: Rooting is easier than with *C. virginicus* but still no "piece of cake". June to mid-July as growth hardens, 1.0% IBA-solution, peat: perlite, mist, rooted about 50%. Rooted cuttings can be transplanted and will overwinter quite well. New growth often occurs on early-rooted cuttings thus assuring overwintering success. Senior author has taken August cuttings without much success. Although wood appeared similar in degree of firmness to June to mid-July cuttings, physiologically it was different. A report from Tennessee noted cuttings should be taken just as they start to harden, wound, 8000 ppm IBA-talc, early July (85%) and mid-July (95%) rooting. Same report noted that cuttings from old trees rooted poorly but cuttings from these rooted cuttings rooted readily. The importance of keeping the stock plants juvenile cannot be overemphasized. Alabama work with juvenile shoots from the base of 25 to 30-year-old stock plants indicated the following. Terminal leaf should have reached full size and thickness, 5 to 6 leaves per cutting, wound, sand: peat: perlite, Captan drench, mist. Rooting percentages increased with maturity date (mid-June), cuttings from basal portion of shoots, treated with IBA (0.8% or 2.0% in talc) rooted 87 to 97%.

Chionanthus virginicus White Fringetree, Grancy Graybeard

SEED: Fall plant with germination occurring the second spring. 3 months warm/3 months cold stratification is ideal. Dormancy is deep seated and appears to involve hard bony endocarp, inhibitors in the endosperm and dormancy in the shoot portion of the embryo. Embryos extracted from fresh seed, incubated with gibberellic acid, greened up and started to grow. This, however, is a painful way to grow fringetree.

CUTTINGS: Senior author has tried numerous combinations and permutations using time, hormone and age of stock plant with no success. Not a single root. Alcohol quick dips result in rapid defoliation and talc preparations appear necessary. The K-salt of IBA would be worth testing. Wood taken from 3-year-old seedlings failed to root. Either juvenility is lost at an early age or the plant is almost impossible to root. One propagator said he rooted a "few" but never went beyond that.

GRAFTING: Has been grafted on bare root *Fraxinus ornus* using a side graft.

Cladrastis lutea American Yellowwood

SEED: Two schools of thought here. One says the dormancy is caused by hard seed coat only; the other by that and embryo. Seeds steeped in hot water (120°F) for 24 to 36 hours until swollen, dried so they do not stick together, planted in nursery rows germinated readily in spring. Acid treatment will satisfy seed coat dormancy. Based on Frett and Dirr's work [*The Plant Propagator* 25(2):4-6, 1979], embryo dormancy does not appear to exist. Acid treatment for 0, 30, 60, and 120 minutes produced 5, 41, 92, and 96% germination, respectively.

CUTTINGS: December root cuttings have worked.

GRAFTING: 'Rosea', a pink-flowered form, can be budded on seedling understock.

Clematis Clematis

Large group of vines and herbaceous perennials offering magnificent flowers, grace, and dignity. Seeds (achenes with hairy style) ripen in summer and fall.

SEED: Rather than attempting to quantify each species and cultivar, a general discussion is offered. The fruits (achenes) ripen in summer through fall and can be collected immediately and sown. Seeds have dormant embryos and require stratification.

Two to 6 months stratification at 40°F is recommended. Some species, C. *vitalba* and C. *viticella*, need warm/cold stratification. Senior author germinated C. *paniculata* seeds by collecting in fall and sowing them in flats in a greenhouse. Germination took four months. The seeds were not exposed to any cold treatment. Seeds may require a period of after-ripening in order for embryo development to take place. Water soluble inhibitors may also be present in the pericarp wall which are leached out with time. Researchers have shown more rapid germination of C. *microphylla* following pericarp removal or a cycle of wetting and drying. Embryos extracted from C. *orientalis* completed development in 2 months on agar medium at 41 to 46°F. They were then transferred to 65°F temperature where normal growth occurred. This is again an example of an immature embryo which will not grow into a complete plant until it has matured. Senior author collected *Clematis paniculata* on Nov. 15, 1985, stylar necks removed, sown immediately, placed under mist in a warm greenhouse (68 to 86°F), after 2 months seeds germinated evenly and in high percentages. C. *orientalis* and C. *tangutica* have germinated 3 to 4 weeks after sowing.

CUTTINGS: Most of the large-flowered hybrids are grown from single node cuttings collected when the tight flower buds are developing. Cut just above a node and shorten internode to 1 to 2″ in length. 8000 ppm IBA-talc is recommended. Medium should be well drained. 60 to 70% rooting occurs in 4 to 5 weeks. Two node cuttings are also used to insure that buds will break and grow after rootings. Semi-hardwood cuttings can also be used later in the season for the easier-to-root types.

GRAFTING: In Europe, grafting is practiced for some large flowered, difficult-to-root cultivars. *Clematis vitalba* is a suitable rootstock. The grafting is done in February.

Clerodendron trichotomum Glory-bower

SEED: 3 months cold produced good germination. Variety *fargesii* also germinated with similar treatment.

CUTTINGS: Softwood cuttings are extremely easy to root. 1000 ppm IBA-quick dip, peat: perlite, mist, with 90 to 100% rooting in 2 to 3 weeks. Root cuttings in December offer an alternative method.

Clethra Clethra

Shrubs or small trees with handsome foliage, white fragrant summer flowers followed by dehiscent capsular fruits. Collect in fall, dry, seed can be shaken from capsules and provided refrigerated storage.

Clethra acuminata Cinnamon Clethra

SEED: No pretreatment is required. Seed is quite small and should be sown in flats under mist much in same fashion as rhododendron seed.

CUTTINGS: June-July, 1000 ppm IBA-solution, peat: perlite, mist, root in high percentages and transplant readily. December-January root cuttings produce shoots in 10 to 12 weeks. This species suckers prolifically and is a good candidate for root cutting propagation.

Clethra alnifolia Summersweet Clethra

SEED: No pretreatment required. See C. *acuminata* for details.

CUTTINGS: Exceedingly easy to root. Senior author has rooted the species, var. *rosea* and 'Pink Spires' using summer softwoods, 1000 ppm IBA-solution, sand or peat: perlite, mist. Cuttings root 90 to 100% in 2 weeks and continue to grow in the flat and cuttings can be taken in 2 weeks (after rooting) from the rooted cuttings. Root systems are unbelievably profuse. Authors suspect hardwood cuttings would also work. Root cuttings in December-January work quite well.

Clethra barbinervis Japanese Clethra

A beautiful species with smooth, polished gray to rich brown bark that may display an exfoliating character. Many splendid forms and clonal selections should be made.

SEED: No pretreatment required. Good germination results if fresh seed is sown immediately.

CUTTINGS: June, 3000 ppm IBA-talc, root readily. This species is slightly more difficult than C. *alnifolia* but still relatively easy. The hormone level might be elevated (5000 to 7500 ppm IBA).

Cleyera japonica (*Ternstroemia gymnanthera*) Japanese Cleyera

SEED: The fruits are collected in late summer or early fall, cleaned and sown. No stratification is necessary. A large Georgia commercial nursery produces thousands of plants with this procedure.

CUTTINGS: Late June, early October and late February (Alabama), 4″ terminal cuttings, leaves removed from basal one-third, approx. 1″ wound, peat: perlite, mist, 70°F bottom heat, untreated or various IBA and NAA talc preparations plus 5% Benlate. Untreated cuttings rooted 67% in June and 13% in February. 2% IBA produced 70% rooting in October; 70% in Feburary.

In another study (Alabama), cuttings were sampled at 2-week intervals from June through November. All cuttings received 3000 ppm IBA plus 5% Benlate-talc preparation. Ninety to 99% rooting occurred in late September through early November.

Colutea arborescens
Colutea × media
Colutea orientalis
The Bladder-sennas

SEED: Fruit is an inflated pod that ripens in August-October. Pods are collected when straw brown, dried, and small seeds extracted. Dormancy results from a hard seed coat. One hour sulfuric acid treatment has resulted in good germination. Steeping seeds in hot water (190°F) for 24 hours will also work. Fall planting is another alternative.

CUTTINGS: Softwood cuttings are easy to root. Half-ripened, early November (England), rooted 29% without treatment; failed to respond to NAA; rooted 73% after treatment with 100 ppm IBA-solution/18 hour soak. Summer softwoods should be treated with about 1000 to 3000 ppm IBA-solution or talc. Root systems are stringy and container growing is the preferred production method.

Comptonia peregrina Sweetfern

SEED: Difficult. Scarified seed that was soaked for 24 hours in 5000 ppm gibberellic acid (GA_3) germinated 80%. Non-scarified seed treated with GA_3 germinated 20%. An inhibitor in the seed coat, possibly abscisic acid, is believed responsible for dormancy [See *Bot. Gaz.* 137(3):262-268 (1976)].

CUTTINGS: Stem cuttings from mature plants root poorly. Root cuttings work well and the following recipe (PIPPS 24:364) is foolproof. Roots dug in late winter-early spring, size 1/16″ wide and 4″ long or 3/8 to 1/2″ wide and 2″ long, fine sand: peat medium, pieces placed horizontally 1/2″ deep, develop shoots and additional roots. These can be potted. Young shoots can be easily rooted using 3000 ppm IBA-talc, mist. Cuttings 3″ long or less work best.

TISSUE CULTURE: Roots have been cultured but shoots did not develop [*Amer. J. Bot.* 64(4):476-482 (1977)].

Cornus Dogwood

A large genus of trees and shrubs with beautiful flower, fruit, fall color, bark, and habit. *Cornus florida* and *C. kousa* are among the most popular flowering trees. A single nurseryman may bud as many as 200,000 *C. florida* in a season. The fruit is a drupe and usually ripens in fall although some species (*C. alba, C. racemosa, C. sericea, C. amomum, C. mas* and others) ripen fruits in summer which must be collected before birds remove them. The fleshy exterior should be removed, the seeds dried, and placed in refrigerated storage in sealed containers. Seeds may be stored for 2 to 4 years. In general, seeds require a variable cold period or combination of warm and cold to facilitate good germination.

Cornus alba Tatarian Dogwood

SEED: Fall plant or 2 to 3 months cold stratification. Most dogwood seed should be sown after cleaning. For this species, the fruit ripens in July-August.

CUTTINGS: Almost disgustingly easy to root from summer softwoods. 1000 to 3000 ppm IBA-solution or talc hasten rooting but are not necessary. Hardwood cuttings also root readily. Long branches are often gathered in winter, cut 6 to 12″ long with a saw, stuck directly into nursery rows in late winter-early spring. Four to 6″ long, late June to early July (Canada), 3000 ppm IBA-talc, sand: peat: perlite, mist, 72°F bottom heat — 'Argenteo-marginata' rooted 98%, 'Gouchaultii' rooted 97% in 3 weeks. Hardwood cuttings of *C. alba* collected from November to April and stored in damp peat moss (except April) until lined out in April rooted 62, 64, 76, 78, 66, 68%, respectively. All cultivars can be rooted easily using the procedures described.

Cornus alternifolia Pagoda Dogwood

SEED: 2 to 5 months warm/2 to 3 months cold stratification. Fruits ripen in late August-September and should be cleaned and planted immediately. Possibility reasonable germination might occur in spring since the early planting would provide warm stratification.

CUTTINGS: July-August, 8000 ppm IBA-talc, roots readily. Rooted cuttings should be taken through a dormant period and then transplanted. Senior author has had variable success. Wood should be reasonably firm and end bud should be well-formed.

Cornus amomum Silky Dogwood

SEED: Fall plant or 3 to 4 months cold stratification. Fruit ripens in August.

Cornus canadensis Bunchberry

SEED: One hour sulfuric acid treatment followed by 2 to 3 months cold stratification resulted in 70% germination. A 3 month warm/3 month cold stratification period will also work.

CUTTINGS: Dividing pieces of sod is the best approach. This is a ground cover species. English reference noted it can be propagated by forced softwood cuttings in the greenhouse.

Cornus controversa Giant Dogwood

SEED: Sow as soon as collected and cleaned in late summer or provide 5 months warm/3 months cold stratification. Much like *C. alternifolia* in germination requirements. Work in Korea reported that 3 to 4 hours sulfuric acid scarification followed by 3 months stratification at 37°F facilitated germination.

CUTTINGS: Late August, 8000 ppm IBA-solution, sand: perlite, mist, 100% rooting with unbelievable root systems. Rooted cuttings should not be disturbed but overwintered and transplanted when growth ensues in spring. In youth the growth is extremely vigorous. The cutting wood described above was collected from a mature specimen. *C. c.* 'Variegata' has proven more difficult to root.

Cornus florida Flowering Dogwood

SEED: Collect when pressure on the fruit causes the seed to pop through. Birds can be a problem. Clean seed, dry, store in refrigerated storage. No appreciable loss of viability for 3 years, possibly more. Fall plant or provide 3 to 4 months cold stratification. Germinating dogwood seed is no problem. Senior author germinated numerous seeds with 3 months cold stratification. One report noted that one hour scarification in sulfuric acid followed by 3 months cold stratification produced good germination.

CUTTINGS: There are as many recipes as propagators. Several papers appear in PIPPS (30:405 and 28:360) specifically addressing *C. f.* and *C. f.* var. *rubra* cutting propagation. The following is a synthesis of this information and numerous other references. June-July softwood terminal cuttings, 5-6″ long, terminal pair of leaves and possibly up to four leaves, leaf surface is reduced in some cases, not in others, wound, IBA from 0.3% to 2.0% talc or solution produced excellent results. Two percent IBA was recommended by several growers. Well drained medium, mist, 90 to 100% rooting in 5 to 8 weeks. Do not disturb cuttings, allow to go through dormant period and transplant after growth ensues in spring. 90% mortality was reported on cuttings transplanted immediately after rooting. If rooted cuttings produce a growth flush after rooting chances of suvival are much enhanced. Extended photoperiods of 18 hrs or 9 hrs during the rooting period resulted in twice and 1½ times the roots of cuttings produced under natural day length. Flowering dogwood requires 1000 hours (northern seed source/cultivars) cold between 32-45°F before bud dormancy is satisfied.

In general var. *rubra* is more difficult to root than the white-flowered forms. Differences also occur in degree of rootability among cultivars. In a media study, 83% rooting (best) occurred in 1 peat: 1 perlite compared to 18% in 2 bark: 1 sand. Peat: sand proved to be a reasonably good medium (68% rooting).

Tip (terminal) cuttings are very important since their rooting percentages are highest and they produce a straight-trunked tree. Sub-terminal cuttings are harder to root (85%) and tend to produce a crook/bend much like a budded tree. At the end of two years, a tree produced from a rooted cutting should approximate 2½ to 3′ high.

Hardwood cuttings have been rooted but percentages (30% average) have not been good and some cultivars did not respond at all. Midwinter (February), wound, 0.8% to 2.0% IBA-talc or solution, ground beds or perlite (greenhouse), root by late May and grow 4 to 6″. They can be overwintered and lined out in spring or transplanted. One nursery considered hardwoods as a viable alternative to budding and summer rooting but after 3 years testing scrapped the idea and went exclusively to softwood production.

In Alabama 5000 ppm IBA-10 second dip, full strength and 1:5 Jiffy Grow (10,000 ppm IBA plus 5000 ppm NAA) produced 100% rooting of the species and var. *rubra*. Senior author collected June cuttings (Illinois), terminal, 1.0% IBA-solution, sand, mist, with 56% rooting in 8 weeks. Same clone rooted 93% in 10 weeks in peat: perlite with same procedure. June 25 (Georgia), peat: perlite, mist, rooting averaged 24% (50% ethanol), 64% (0.25% IBA in 50% ethanol), 80% (0.5% IBA in 50% ethanol) and 60% (1.0% IBA in 50% ethanol). Best root systems occurred at 0.25 and 0.5% levels. A 30-second dip of 5000 ppm K-IBA has produced good rooting of var. *rubra*.

The key to rooting dogwood is collecting early in season and either producing a growth flush or allowing to overwinter without disturbance until growth ensues. These two prerequisites have been mentioned numerous times in the PIPPS by researchers and nurserymen.

GRAFTING: Numerous cultivars are T or shield budded in summer on seedling understock. Seeds are fall sown or stratified and spring sown, seedlings are thinned, spring fertilized, weeded, irrigated if necessary and ready for budding in summer. Seedlings can also be lined out in spring or even the previous fall. The result is a two-year, possibly stockier seedling. A shield bud is placed in a T-cut in late July, August into September (Tennessee), understock can be root-pruned in December, tops of understock are cut off before early April, and the growth of bud (depends on cultivar) should average 1 to 3′ by end of season.

Pot grafting (whip grafts) can be accomplished in winter on seedling understock.

Cornus kousa Kousa Dogwood

SEED: Exceedingly easy to germinate. Collect large raspberry-like fruits in September-October, remove shiny brown seeds from pulp, fall plant or provide 3 months cold stratification. Seeds are generally sound and germination is high. Senior author has an 8-10′, 6-year-old tree grown from seed that has not set one flower bud. Tremendous variation in seed-produced material. Wise to use a proven clone (i.e. var. *chinensis*, 'Summer Stars', etc.) for reliable flowering response.

CUTTINGS: Should be handled exactly as described under C. *florida*. Rooting percentages are generally not as high. Mid-June-July (Boston), 8000 ppm IBA-talc plus thiram or 2500 ppm IBA plus 2500 ppm NAA-5 second dip have proven successful. Senior author has achieved 50% rooting in 12 weeks with June (Illinois) softwood terminal cuttings, 1.0% IBA-solution, peat: perlite, mist. Other investigators reported 55 to 100% success with softwood cuttings. Overwintering should be handled as described for C. *florida*. A classic paper is PIPPS 34:598-603 (1984). A grower noted that C. *kousa* on its own roots was not as hardy as seedling grown material. Plants 4 to 5′ from cuttings were killed in a severe winter.

GRAFTING: Can be treated like C. *florida*. Several weeping cultivars have been introduced and are top worked on C. *florida* understock. Ideally, C. *kousa* understock would be best but 25- to 30-year-old C. *kousa* grafted on C. *florida* are healthy and vigorous. C. *kousa* has also been used as an understock for C. *florida* cultivars.

Cornus macrophylla Bigleaf Dogwood

SEED: White flowers appear in July-August (Boston) and are followed by purplish fruits. Collect in fall, clean and provide warm/cold stratification. 3 months/3 months resulted in 54% germination; 5/3-62% germination.

CUTTINGS: Amazingly easy from June-July (Boston) softwoods, terminal, 4000 ppm IBA-solution, sand: perlite, mist, rooted 100% in 3 to 4 weeks. Handsome tree that might be used more if better known.

Cornus mas Corneliancherry Dogwood

SEED: The cherry-like drupes ripen in June-July. Collect, clean, and sow immediately in outdoor beds and mulch. Seed is doubly dormant and requires a 4 to 5 month warm/3 month cold period. One grower noted that fall planted seed does not germinate until the second spring. European work indicates 2 to 3 years of warm/cold cycles before germination is completed.

CUTTINGS: Not the easiest thing to root. Senior author has had minimal success with the species, although softwoods treated with IBA have been reported to root in high percentages. Mid-July (Boston), 8000 ppm IBA-talc plus thiram, peat: perlite, mist, rooted 44%. A procedure similar to that described for C. *florida* might be effective. C. *mas* 'Flava' early August (Boston), terminal, 4 to 6″ long, 4000 ppm IBA-solution, sand: perlite, mist, rooted 100% in 8 weeks. Early June (Boston), 3000 ppm IBA-talc plus thiram, sand: perlite, mist, rooted 13%. Hardwood cuttings, mid-February, 8000 ppm IBA, perlite, rooted 40%. Work in Europe indicates that high IBA levels are necessary for successful rooting. Cuttings from seedlings taken in first half of August rooted 80% when treated with 1.0% IBA in 8 weeks. Untreated cuttings did not root. Plants selected for fruit production rooted 40 to 60% with 1% IBA, indicating that ease of rooting is lost with maturity of the plant.

GRAFTING: The cultivars 'Aurea' and 'Variegata' could be pot grafted on seedling understock in winter although no mention of this was found in literature. Budding is also a possibility in summer if suitable understocks are available.

Cornus officinalis Japanese Cornel Dogwood

Very similar to C. *mas* in attributes. Superior forms have larger flowers and beautiful, exfoliating bark. These types need to be propagated vegetatively. Cuttings and seed should be handled as described under C. *mas*.

Cornus racemosa Gray Dogwood

SEED: White drupes ripen in August-September, collect, clean and plant seed immediately or provide 2 to 5 months warm/3 to 4 months cold.

CUTTINGS: July, untreated, rooted 8%; 1000 ppm NAA-talc, rooted 100% in sand in 5 weeks. 80 to 100 ppm IBA-4 hour soak, 66% rooting in 6 weeks. May be some credence to fact that NAA is more effective than IBA for this species and perhaps other dogwoods. July (Boston), 8000 ppm IBA, outdoor beds with bottom heat, rooted 100%. Could possibly be propagated by root cuttings since this plant suckers profusely.

Cornus sanguinea Bloodtwig Dogwood

SEED: Fruits ripen in August-September, collect, clean, plant immediately or provide 3 to 5 months warm/3 months cold stratification.

CUTTINGS: Late June, 44% rooting without treatment; 68% in 3 weeks, 30 ppm IBA-12 hour soak. The older literature emphasized the soak method. A 3000 ppm IBA-solution (quick dip) or talc would accomplish the same or superior end result.

Cornus sericea (C. *stolonifera*) Red-osier Dogwood

SEED: Fall plant or 2 to 3 months cold stratification. Some literature indicated seed is difficult and may require a warm/cold period.

CUTTINGS: Piece of cake! June-July into early fall, 1000 ppm IBA-solution, peat: perlite, mist, 90 to 100% rooting in 4 to 5 weeks. Like C. *alba* (*q.v.*) an exceedingly easy species to root. The cultivars are also easy. In one study, 'Flaviramea', June-July, 8000 ppm IBA-talc, rooted 98 and 96% in separate years. Hardwood cuttings also root easily. See methodology under C. *alba*.

Cornus walteri Walter Dogwood

Interesting large (40′) tree much like C. *macrophylla* in flower and fruit.

SEED: Difficult. 4 months warm/4 months cold stratification.

CUTTINGS: More difficult to root and procedure for C. *florida* should be followed.

Corylopsis Winterhazel

Beautiful, elegant shrubs, especially in flower when the soft to deep yellow flowers flutter in the early spring breezes. Fruit is a two-valved dehiscent capsule that ripens in late summer. Fruits must be collected before capsule dehisces. A yellow-green to brown transition is fair warning to collect. Bring inside and put in a cardboard box with a lid. Seeds will be expelled from capsules and can be collected and placed in sealed containers in refrigerated storage. See PIPPS 19:204.

Corylopsis glabrescens Fragrant Winterhazel
C. pauciflora Buttercup Winterhazel
C. sinensis Chinese Winterhazel
C. spicata Spike Winterhazel
C. willmottiae Willmott Winterhazel

SEED: Difficult. Finding good seed is difficult. A 5 month warm/3 month cold stratification period will work. Freshly collected seed might be planted immediately with some germination occurring the following spring. If seed dries, germination will take 2 years if seed is fall sown.

CUTTINGS: Treated as a group since propagation is similar and literature on individual species is not abundant. In general, softwood to greenwood in June-July-August are easy to root. Cuttings should still be growing and the end bud should not have formed. Senior author has had 90 to 100% success with 1000 ppm IBA-solution, peat: perlite, mist in 4 to 6 weeks. Root systems are profuse under the above conditions but cuttings resist moving. They should be induced to produce additional growth or taken through an overwintering period and transplanted after growth ensues. Even these procedures do not guarantee absolute success (Again, senior author's experience).

An English reference emphasized young, vigorous cutting wood and recommended stock plants be maintained to produce such wood. If cuttings were taken in June, 3000 ppm IBA-talc proved sufficient; however, July-August cuttings required 5000 ppm IBA-solution to produce good rooting. Cuttings are also rooted in pots to avoid root disturbance. Success ranged from 70 to 90% with rooting occurring in 6 to 10 weeks.

Corylus Filbert, Hazel

Shrubs or trees with monoecious flowers in early spring, the male in slender, cylindric catkins, female with the stigma/style protruding from the bud and almost invisible except to the initiated. The fruit is a nut surrounded by an involucral bract (husk) which fully ripens in October. A pitched battle usually ensues between the propagator and squirrels since the latter seem to relish and ravage the fruits. Collect as husks begin to turn brown in September. Dry fruits inside until husks open so nuts can be removed. Store seeds immediately after cleaning in sealed containers under refrigeration. See PIPPS 20:357.

Corylus americana American Hazelnut

SEED: Fall planting or 2 to 6 months cold stratification are best. Two seed lots planted in late fall germinated 64 and 48%, respectively, in spring. Three months cold stratification has proven effective.

CUTTINGS: No literature was unearthed relative to this species' requirements. Layering and stooling would work. Use procedures described for cuttings under C. *avellana*.

Corylus avellana European Filbert

SEED: Fall plant or 2 to 6 months cold stratification. 3 months would prove the best estimate. Russian work noted that freshly harvested seed that were warm stratified for 3 weeks followed by 3 weeks at 40°F germinated best.

CUTTINGS: An important commercial species because of nut-producing capabilities. Timing is critical with mid-June to mid-July (Oregon) being optimum; cuttings should be in active growth (no end bud set), IBA essential in range of 5000 to 10,000 ppm -solution or talc, well drained medium, even mist, preferably fog, root 60 to 65%. Too much water results in decline. Cuttings should be kept in active growth after rooting and handled as described for *Cornus florida*. Successful handling after rooting is the most important part of the process. Etiolation of stock plants followed by localized etiolation of actively growing shoots has been used to ensure stable and high rooting percentages. The best time for taking cuttings was during intensive shoot growth.

In Danish tests, early June cuttings of 3 cultivars from new shoot growth, 2″ long, 1 leaf retained, untreated, 1000 or 2000 ppm IBA-dip, 2 peat: 1 perlite, mist, greenhouse, 72°F bottom heat, rooted consistently better with IBA treatment. Root numbers averaged 4, 14, and 7 for untreated, 1000 and 2000 ppm IBA, respectively, after 38 days.

C. *a.* 'Aurea' and 'Contorta' mid-July (Canada), heavy tip and shoot cuttings, firm but still growing, 8000 ppm IBA-talc, sand, mist, rooted over 50%. Cuttings were pottted or transplatned to ground beds with good survival.

Hardwood cuttings mid-February (England) from stooled plants (juvenile) in hedgerows, 6″ long, 1% IBA-solution, 70°F heat in rooting bin, rooted in 4 to 5 weeks and were then bedded into a frame. C. *avellana* 'Aurea', 'Contorta', 'Heterophylla', C. *chinensis*, C. *colurna*, and C. *axina* have been rooted successfully. No mention was made of percent rooting or survival although gleanings from the literature indicated it was low (approx. 30% or less) and subsequent survival even worse.

Stooling is practiced with success and might be feasible for some ornamental and fruit cultivars. Stooled shoots of pencil thickness were ringed at base and smeared with 1000 or 2500 ppm IBA paste and covered with soil (May). The percentage of shoots rooted (December) was increased from 32 to 47% (untreated) to 80 to 100% (IBA at either concentration). Root numbers and lengths were also increased by IBA treatment.

GRAFTING: The commercial nut-producing cultivars are grafted on C. *avellana* seedlings in winter using a 3 to 4 bud scion and a whip graft. This can be accomplished on potted or bare root understock. Finished grafts are then placed in a grafting case at 70°F until union calluses properly.

'Barcelona' grafted on C. *colurna* (minimal to no suckering) declined in nut production over a 20-year period compared to plants grafted on C. *avellana*. One serious problem with C. *avellana* as an understock is the propensity to sucker and over-grow the scions. This occurs frequently with 'Contorta'. Might be advisable to put 'Contorta', 'Aurea', etc., on C. *colurna*.

Scions of 4 cultivars were collected in January and stored until March when they were tongue-grafted or cleft-grafted on 1-year-old seedling rootstocks. Success ranged from 80 to 100% with ultimate survival from 60 to 95%. Method made little difference. In a budding test, chip, patch and shield (August) yielded 75, 65 and 45% take, respectively. Budded plants were more vigorous than grafted plants.

TISSUE CULTURE: PIPPS 33:132-137 (1983), outstanding paper. Best multiplication on Anderson's inorganics plus 1 mg/l 2ip and 2 mg/l BA. Immature embryo explants through callus culture produced meristemoids and embryogenesis [Z. *Pflanzenphysiol.* 77:33-41 (1975)].

Corylus colurna Turkish Filbert

SEED: Fall plant or 2 to 6 months cold stratification. Senior author germinated a few seedlings using 3 to 4 months cold stratification but percentage was not high. One nurseryman had a 20-year-old seedling orchard that had not produced a nut.

CUTTINGS: Only reference was to stooled hedgerow plants (juvenile) rooted as hardwood cuttings (See C. *avellana*). No rooting percentages or survival rates were listed. One reference reported 7% rooting on soft growth without any additional specifics.

GRAFTING: Has been grafted successfully on C. *avellana* but the suckering problem that ensues makes this approach foolish.

Corylus cornuta Beaked Filbert

A small shrubby species used for naturalizing that should be propagated by seeds sown in fall or provided 3 months cold stratification.

Corylus maxima var. *purpurea* Purple Giant Filbert

SEED: Fall plant or 2 to 6 months cold stratification.

CUTTINGS: Procedure for softwood (Canada) described under C. *avellana* applies to this species. Late July (Boston), 1% IBA, mist, rooted 100%. Interestingly, July cuttings rooted 52% when treated with 4000 ppm IBA-talc and 0% without hormone. Other studies indicated variable success with this plant.

For all the *Corylus* species the use of hormone is essential. Highest rates the authors discovered in the literature were 1 and 2%. Might be worthwhile to experiment with NAA or IBA/NAA combinations and several reports indicate NAA will improve rooting of C. *avellana*.

Cotinus Smoketree

Dioecious shrub or small tree with large furry panicles giving the appearance of smoke because of hairy nature of inflorescence. Yellowish green rather insignificant flowers appear in June-July with the small, 1/16″ long kidney-shaped drupe ripening in fall. One reference noted that seed can be stored for several years in sealed containers at room temperature. However, it is almost always judicious to provide refrigerated storage.

Cotinus coggygria Common Smoketree

SEED: Several recipes will work. Seed treated with 30 to 60 minutes acid, 3 months cold, germinated 86%. Seed collected green, late August-September, planted, germinated in high percentages the following spring. Seed collected from purple-leaf forms produces an interesting mixture of green- and purple-leaf seedlings. With proper water and fertilizer they make 2 to 3′ of growth in one season.

CUTTINGS: The species is not used to any degree in contemporary landscaping and has been superseded by the purple-leaf cultivars. Rooting cuttings is not difficult but overwintering successfully is another story. Cuttings have been direct stuck in pots, in flats, greenhouse benches and in outdoor ground beds with 80 to 100% success. Key to success includes using softwood, actively-growing cuttings in June (preferably), 4 to 6″ long, 1000 to 3000 ppm IBA-solution, well drained medium, mist, rooting takes place in 4 to 8 weeks. Early June, 'Royal Purple' (Iowa) rooted 86%, while late July (Iowa) rooted 33% using 1425 ppm IBA plus 1425 ppm NAA and 50 ppm B (H_3BO_3). Cuttings must be overwintered without disturbance and transplanted in spring. Transplanting immediately after rooting resulted in 70% and great losses. One grower overwinters rooted cuttings in 33 to 38°F storage. Hormone is definitely needed to produce vigorous root systems. In one study, 8000 to 10,000 ppm IBA-talc proved optimum. 2% IBA-talc burned the cuttings. Untreated cuttings rooted similarly to 8000 and 10,000 ppm IBA cuttings (80-90%) but had only one or two stringy roots. Hardwood cuttings of 'Royal Purple', early December, treated with Jiffy Grow, rooted 100%. Successful handling after rooting was problematic. Other instances of hardwood cuttings were reported but success was variable.

GRAFTING: Grafting has been reported on the better purple-leaf types but it seems unnecessary in view of cutting propagation successes.

Cotinus obovatus (*C. americanus*)
American Smoketree

SEED: Treat as described for *C. coggygria*. In one test, acid treatment ranged from 20 to 40 minutes with 2 months cold stratification resulting in 32% germination in 11 days.

CUTTINGS: A procedure similar to *C. coggygria* should be followed but rooting and overwintering may be more difficult. Senior author has observed excellent rooting in one Tennessee nursery where softwoods (again timing important), 1.0% IBA-solution, sandy soil, outdoor beds, mist, overwintered in place were used. Growth must be soft and there is a critical period for good success.

Cotoneaster Cotoneaster

Deciduous or evergreen shrubs or small trees with white or pink flowers and red or black fruits. The berry-like pome contains 1 to 5 seeds. The fruits should be collected when ripe (usually October), and seeds extracted from flesh. Seeds can be stored dry in sealed containers in a cool place. Some seeds have a hard impermeable seed coat and dormant embryo. Acid scarification and cold stratification are required to facilitate good germination. A 5 to 6 month warm/3 month cold stratification period will also work. Most cotoneasters root reasonably well from cuttings. The senior author finds the evergreen/semi-evergreen types easier than the deciduous. 1000 to 3000 ppm IBA-solution, sand or peat: perlite, mist, June-July-August cuttings root readily and are easily transplanted and overwintered. Specific recommendations follow. (D) — deciduous; (E) — evergreen.

Cotoneaster acutifolius (D)
Peking Cotoneaster

SEED: Acid, 3 months cold treatment or sow seed as soon as cleaned in fall.

CUTTINGS: July cuttings rooted well. Follow directions under *Cotoneaster*.

Cotoneaster adpressus (D)
Creeping Cotoneaster

SEED: 2 hours acid and fall plant or 2 hours acid and 3 months cold stratification. This treatment also worked for var. *praecox*.

CUTTINGS: June, July, August, 1000 to 3000 ppm IBA-solution, peat: perlite or sand, mist, root 90+%. Late July (Boston), 4000 ppm IBA—talc plus thiram, sand: perlite, mist, rooted 100% in 8 weeks.

Cotoneaster apiculatus (D)
Cranberry Cotoneaster

SEED: Acid treatment for 1 to 2 hours plus 2 to 3 months cold stratification. Acid treatment and fall planting accomplish same end.

CUTTINGS: Easily rooted as described under the generic heading. This is one of the most popular and cold hardy species for midwestern and eastern gardens. When properly embellished with rich red, cranberry-like fruits, it has few rivals.

Cotoneaster congestus (E)
Pyrenees Cotoneaster

A handsome small-leaf species with white flowers and red fruits. Authors have observed it in many European gardens. Cuttings and seed should be treated like *C. apiculatus*.

Cotoneaster conspicuus (E)
Wintergreen Cotoneaster

Another small leaf species of variable habit. See *Cotoneaster* for propagation details. 4000 ppm IBA promoted good rooting.

Cotoneaster dammeri (E)
Bearberry Cotoneaster

SEED: 3 months cold stratification produced good germination. Fall planting will accomplish the same end.

CUTTINGS: Possibly the easiest cotoneaster to root. Senior author (Illinois experiences) has collected cuttings from June to September, 1000 ppm IBA-solution, peat: perlite, mist, 100% rooting in 3 weeks. Cuttings collected in Atlanta during February (leaves purplish) rooted about 80% with same treatment. Some growers (Georgia) root directly in one gallon containers in May and have an 18 to 24″ plant for October sales. NAA solution produced excellent rooting of *C. dammeri* and *C. d.* 'Lowfast'. Interestingly, *C. dammeri* rooted only 83% in a high humidity chamber. A New York grower takes November cuttings and has a full 18 to 24″ plant in a 2 gallon container at the end of the first growing season. High IBA (2.33% in talc) killed this species.

Interesting results have been obtained with the use of 18-6-12 or 14-14-14 Osmocote incorporated into the rooting medium at 4 oz./bushel of medium. 'Skogholm' (August cuttings) were similar after 3 weeks but markedly improved in Osmoccote amended medium after 5 and 7 weeks. Rooting percentages were similar but greater root systems and new breaks occurred in the Osmocote rooting medium. After transplanting, the Osmocote treated cuttings grew 25% larger in the same time period.

TISSUE CUTLURE: Shoot tip explants through shoot tip culture led to shoot proliferation [*HortScience* 17:190-191 (1982)].

Cotoneaster divaricatus (D)
Spreading Cotoneaster

SEED: 2 hours acid/3 months cold or acid and fall planting.

CUTTINGS: Senior author's experience: early June (Illinois), 1000 ppm IBA-solution, sand, mist, rooted 90% in 3 months. Another worker reported 100% rooting in 6 weeks with early July untreated cuttings. This species seems to take longer than the creeping/low-growing types and does not produce as fibrous a root system. However, it is easily transplanted and grown. August, greenwood with heel, peat: sand, heavy wound, 1 to 2% IBA-talc was successful.

Cotoneaster glaucophyllus (E) Brightbead Cotoneaster

CUTTINGS: A pubescent-leaved species with good red fruits. Water management might be important in the rooting process. Treat as described under *Cotoneaster*.

Cotoneaster horizontalis (D)
Rockspray Cotoneaster

SEED: 1½ hours acid/3 to 4 months cold stratification or acid followed by fall planting. This is not the easiest to grow from seed and 67% germination was achieved when seeds were given 30 minutes sulfuric acid treatment followed by 11 months cold stratification; 59% germination resulted from 2 months warm/9 months cold stratification.

CUTTINGS: Late July (Rhode Island), 3000 ppm IBA-talc, sand, mist, excellent rooting. There is some indication the cutting should approach the greenwood (reasonably firm) rather than exceedingly soft or firm stage. Generally treat as described under *Cotoneaster*.

Cotoneaster × hybridus 'Pendula' (E)

Often grafted on a standard (C. *phaenopyrum* or suitable *Cotoneaster*) to produce a small evergreen weeping tree. Easily rooted from cuttings as described under C. *dammeri*.

Cotoneaster lacteus (C. *parneyi*) (E)

Large leaf evergreen species used sparingly in southern states. Seems to resist fireblight better than similar species. Easily rooted as decribed under C. *dammeri*.

Cotoneaster lucidus (D) Hedge Cotoneaster

SEED: 5 to 20 minutes acid followed by 1 to 3 months cold stratification or acid and fall planting.

CUTTINGS: Comparable to C. *acutifolius* in rooting response. Not as easy as the low-growing species. Early July cuttings rooted well but no specifics were given.

Cotoneaster microphyllus (E)
Littleleaf Cotoneaster

SEED: Interestingly, imported seed of the species and var. *thymifolia* behaved in the words of Mr. Alfred Fordham, the great propagator at the Arnold Arboretum, "like a handful of grass and germinated without pretreatment". The seed was from a foreign source and pretreatment (if any) was not known.

CUTTINGS: See C. *dammeri*. This species responded well to NAA treatment. Tends to develop irregular upright shoots which, if used for cutting material, results in a more upright form than true C. *microphyllus*. Cuttings should be taken from normal spreading shoots.

Cotoneaster multiflorus (D)
Many-flowered Cotoneaster
C. m. var. ***calocarpus***

SEED: See C. *apiculatus*.

CUTTINGS: Not the easiest to root. Senior author, June, 1.0% IBA-solution, sand, mist, 1% rooting. An Illinois nurseryman reported June softwoods root easily. Literature noted this species roots with some difficulty. October (Boston), 8000 ppm IBA-talc, rooted 52%.

Cotoneaster racemiflorus var. ***soongoricus*** (D)
Sungari Redbead Cotoneaster

SEED: Acid (one report said 5 hours), 1 to 2 hours best estimate followed by 3 months cold stratification.

CUTTINGS: Similar to C. *multiflorus*. Again not the easiest cotoneaster to root. Late July (Fort Collins, Colorado), sand or peat: perlite, mist, 70°F bottom heat, hormones were Ethrel and IBA in various solution concentrations. Untreated cuttings did not root. In sand, 50% rooting at 960 ppm Ethrel and 80% at 1440 and 1920 ppm. 65% rooting at 480 ppm Ethrel plus 5000 ppm IBA. Essentially no cuttings rooted in peat: perlite.

Cotoneaster salicifolius (E)
Willowleaf Cotoneaster

SEED: See C. *apiculatus*.

CUTTINGS: Wonderfully easy to root much in the mold of C. *dammeri* (*q. v.*) and it can even be rooted in late fall or late winter as long as the leaves have not been damaged severely. The low growing selections 'Gnom', 'Repens' ('Repandens'), 'Autumn Fire', 'Scarlet Leader', etc., can be rooted in early spring and by fall will fill a one gallon container.

Crataegus Hawthorn

Small trees usually armed with prominent spines. The white flowers (rarely pink) are followed by red fruits that ripen in September through October. *Crataegus mollis* is the first to ripen and abscise, while the bright red fruits of C. *phaenopyrum* hold all winter. The fruit is a drupe that contains a number of hard-coated seeds. The fruits should be macerated in water and the seeds removed, cleaned, dried and stored. The seeds (nutlets) may be stored for 2 to 3 years in sealed containers under refrigeration. The seeds have an embryo dormancy and some require an acid treatment to break down the impermeable coat (endocarp). The coat may be 3/16" thick in some species and require as much as 7-8 hours acid treatment. Cuttings are rarely successful and budding is used to produce the cultivars. Seedlings should be transplanted after one year for with age some species become difficult to transplant. See PIPPS 32:205 (1982) and *Contrib. Boyce Thompson Inst.* 9(5):409 (1938).

Crataegus arnoldiana Arnold Hawthorn

SEED: A 4½ hour acid treatment followed by 6 months cold stratification resulted in 35% germination.

GRAFTING: Work at Morden Experiment Station indicated this was a good, cold-hardy understock with a root system that permitted excellent transplanting.

Crataegus crusgalli Cockspur Hawthorn

SEED: A 2 to 3 hour acid treatment (up to 4 hours in British work) followed by a variable warm and 3 to 4 month cold period will work. This species has one of the thickest seed coats (endocarp walls) and may require longer acid treatments. The warm period after the acid ensures that the coat is broken. This approach produced 73% germination in one study. If seed is fall sown without acid treatment, germination will be sparse in spring and occur in the second and third years.

GRAFTING: Var. *inermis* and 'Crusader' (thornless forms) are budded on seedling understock in August-September. C. *phaenopyrum* serves as a suitable understock for top working or budding. Budded stock should have 3 to 4' of top growth after the first growing season.

Crataegus laevigata English Hawthorn

SEED: Acid treatment followed by cold stratification is necessary. Acid treatment and fall planting would work. This species does not require as much scarification as C. *crusgalli*; perhaps one hour.

GRAFTING: Numerous cultivars exist ('Paul's Scarlet', 'Crimson Cloud'). They can be bench grafted in January on C. *laevigata* understock. Interestingly, C. *laevigata* is the best (perhaps only) rootstock for the cultivars. A West Coast grower noted poor development when budded on C. *phaenopyrum*. North American species or cultivars are best grafted or budded onto North American species while the reciprocal is true of European species.

Crataegus mollis Downy Hawthorn

SEED: If relatively fresh, seed can be fall planted or given a 3 month cold period. Usually, the seed coat wall is sufficiently thin that no acid treatment is required.

Crataegus monogyna Singleseed Hawthorn

SEED: Acid treatment (30 minutes to 2 hours) followed by cold stratification or fall planting.

GRAFTING: The cultivars ('Stricta', 'Semperflorens') are budded on seedling understock of the species. C. *laevigata* should also work.

Crataegus phaenopyrum (C. *cordata*) Washington Hawthorn

SEED: Fall plant or 3 months cold stratification. No acid treatment is necessary as long as seed is reasonably fresh.

GRAFTING: Virtually the universal rootstock for most hawthorns. It is easy to germinate, has a good root system, transplants easily and bark peels over a long period. Budding procedure includes lining out seedlings in spring, bud in August, early to mid-September, remove top of understock in late winter. Bud should grow 3 to 4′, perhaps longer by end of first season.

Crataegus prunifolia Plumleaf Hawthorn

SEED: Acid 1 to 2 hours (British work indicates up to 4 hours) followed by fall planting or 3 to 4 months cold stratification.

Crataegus punctata Dotted Hawthorn

SEED: Acid (1 to 2 hours) followed by fall planting or 4 months cold stratification. 4 months warm/4½ months cold stratification resulted in 60% germination.

GRAFTING: 'Ohio Pioneer' is a thornless (essentially) form that must be budded to retain desirable traits. C. *punctata* is considered somewhat difficult to transplant so C. *phaenopyrum* would be the rootstock of choice.

Crataegus succulenta Fleshy Hawthorn

SEED: 30 minutes acid followed by 3 months cold stratification resulted in 35 to 40% germination.

Croton alabamensis

CUTTINGS: Late November (Alabama), taken from plants in a native stand near Tuscaloosa, 3000 ppm IBA-talc, sand, 70°F bottom heat, 70°F night air temperature, mist, with 70% rooting in 4 weeks.

Cryptomeria japonica Japanese Cryptomeria

SEED: Collect cones when they change from grayish brown to reddish brown. Spread cones in trays or on paper in a dry place. Shake and seeds fall out. Dry seed should be stored in plastic bags under refrigeration. Seed should be soaked in cold water (32°F) for 12 hours, placed in plastic bags and provided 2 to 3 months cold stratification. Bags should be left open for adequate aeration. Three months warm/3 months cold stratification results in good germination. A germination rate of 30% is considered normal.

CUTTINGS: Based on literature and senior author's experience any time from August to March is probably acceptable. Not all cultivars root with equal facility. 'Lobbii' and 'Lobbii Compacta' are difficult although one grower achieved 40 to 50% success with late July cuttings of 'Lobbii Compacta'. Most compact cultivars taken from November to February, 8000 ppm IBA-talc or solution, sand or peat: perlite, preferably polytent (mist will work), bottom heat, root in 8 to 10 weeks. Rooting should approach 70 to 90% depending on cultivar. 'Bandai-Sugi', 'Compressa', 'Compacta', 'Cristata', 'Elegans', 'Nana', 'Spiralis', 'Falcata' and 'Vilmoriniana' root in high percentages following the above recommendations. Senior author believes polytent is best type of rooting structure for this species and the cultivars. In Japan, rooted cuttings (grafts also) are used in reforestation. September, wound, 25 to 100 mg/l NAA soak is recommended for dwarf types (Holland). Late October (Oregon), 4-6″ long, slant cut, soaked in Captan, Terrachlor, Streptomycin, Jiffy-Grow 1:9-5 second dip, sand, 70°F bottom heat, greenhouse, water once a day, cuttings root in 3 months.

GRAFTING: Difficult-to-root cultivars are probably best side-grafted on seedling C. *japonica* in winter. One report mentioned use of a veneer graft from winter to August.

TISSUE CULTURE: Plantlets have been successfully regenerated using cotyledons as the explant sources. Hypocotyl explants led to direct adventitious bud formation and callus. *Bot Mag.* 87:73-77 (1974), *Mokuzai Gakkaishi* 21:457-460 (1975), *Physiol. Plant.* 45:127-131 (1979).

Cunninghamia lanceolata China-fir

SEED: Cones should be collected when turning brown, dried inside and seed collected. Seed probably benefits from a short (one month) cold stratification but no published information was found.

CUTTINGS: November and later, 8000 ppm IBA-talc plus thiram, sand or peat: perlite, polytent, reasonable rooting results. Cuttings collected from lateral branches grow horizontally; from vertical shoots...upright. Published reports indicated 50 to 70% rooting with the above approach. November 'Glauca' cuttings rooted 100%. The younger the plant, probably the better the chance for success.

× *Cupressocyparis leylandii* Leyland Cypress

An intergeneric hybrid between *Cupressus macrocarpa* and *Chamaecyparis nootkatensis*. Many cultivars are available and there are distinct differences in rootability among the clones.

CUTTINGS: This fast growing, rather handsome evergreen has caught the imagination of the nursery trade and is gradually finding its way into garden centers and landscapes. Cuttings are easy to root if a few basic rules are followed. Some investigators believe the plant can be rooted any time of year; others report certain specific times for successful rooting. Large cuttings, 4 to 6" to 6 to 8" (some advocate 8 to 12"), with brown wood at the base, strip bottom needles, wound (taking needles off accomplishes this), treat with IBA-talc or solution in range of 3000 to 8000 ppm, well drained medium (peat:perlite), bottom heat (70 to 75°F), polytent or mist depending on the time of year, rooting generally takes 10 to 12 weeks. Senior author takes cuttings in late February-early March (Georgia) with 90% plus success [*HortScience* 18:204-205 (1983)]. A polytent at this time of year works exceedingly well. 1500 of 1600 cuttings rooted by the end of May using the above procedures and the polytent. Three things are exceedingly important: (1) vigorous cuttings from young stock plants, (2) wounding, and (3) hormone. Leyland is slow to root and a well drained medium coupled with bottom heat will speed up process by 2 to 4 weeks. Outdoor ground beds (*q. v.*, Chapter 2) with bottom heat can also be used. Cuttings have stayed in the rooting bench 6 months without rooting and still looked good.

A North Carolina researcher reported good rooting with March, July, August, October, November and January cuttings. An English researcher reported cuttings rooted best in February followed by May, August and November. Irish researchers noted that March-October cuttings are slower to root and root in lower percentages. Several growers commented on the need for a good frost or cold spell (October-November) before taking cuttings. In an English study, cuttings from 5-year-old trees rooted 94%, 20-year-old trees — 34%, and a 50-year-old tree 5%. A Georgia nurseryman has had good success with summer rooting, cuttings from container plants, and 1% IBA.

'Silver Dust' is exceedingly easy to root and roots faster than 'Haggerston Grey'. In a controlled study with all factors constant except the cultivar, 'Stapehill' rooted 84%, 'Haggerston Grey' — 71%, 'Leighton Green' — 57%, and 'Naylor's Blue' — 51%. 'Green Spire' is considered the "toughest" to root.

Cuttings can be transplanted easily after rooting. Senior author uses a 4" pot, pine bark medium, with Osmocote 18-6-12. Leyland responds fantastically to good nutrition and a February-March cutting will be salable in a one gallon container by October.

Cupressus Cypress

The true cypresses are not utilized to any degree in the eastern/southern United States but are common fixtures in parts of California and the southwest. A canker (*Coryneum cardinale*) is problematic in the southeast. *Cupressus arizonica*, Arizona cypress; C. *lusitanica*, Mexican cypress; C. *macrocarpa*, Monterey cypress; and C. *sempervirens*, Italian cypress, are the most common. Cypresses are monoecious with female cones ranging from ½ to over 1 inch in diameter. Seeds are shed the second season after pollination. Collect cones and dry at room temperature until scales split and seed can be shaken loose. Each cone averages 12 to 150 seeds. Seed viability can be maintained for 10 to 20 years at temperatures of 34 to 41°F. Seed requires 30 days stratification at 34°F. Seeds should be treated with Captan. Newly germinated seedlings are particularly susceptible to damping-off fungi. Cypresses can be outplanted as 1 or 2-year-old seedlings.

Cupressus arizonica Arizona Cypress

CUTTINGS: Not the easiest plant to root and cuttings should be taken from juvenile plants. Senior author has never rooted a single cutting from a large 70-year-old campus tree. Cuttings from younger trees (5-year-old), March, peat: perlite, 5000 ppm IBA-dip, did not root.

In French work cuttings from 50 clones were treated with 2000 to 5000 ppm IBA, peat: sand: perlite, and only 15% of the clones showed good rooting ability. November to December were the best months. Cuttings from lower ⅓ of trees (again more juvenile) rooted best.

GRAFTING: Var. *conica* was grafted in mid-January on potted rootstocks of C. *macrocarpa*, C. *arizonica* or *Thuja orientalis*. Best survival and growth was obtained on C. *macrocarpa*. Good survival but less vigorous growth with C. *arizonica*. *Thuja orientalis* proved unsatisfactory.

TISSUE CULTURE: Pollen explants produced callus [*Compt. Rend. Acad. Sci.* Paris 279D:651-654 (1974)].

Cupressus macrocarpa Monterey Cypress

CUTTINGS: The gold forms are often produced by cuttings. In general, rooting is low. Irish work, February, warm bench and plastic, mist, 8000 ppm IBA-talc, 25% rooting in 10 weeks. Good results were reported when cuttings were taken from young container-grown plants. In fact, 100% rooting was achieved when cuttings were taken from young plants of 'Lambertiana Aurea' in December, July or September.

Cupressus sempervirens Italian Cypress

CUTTINGS: April, untreated, rooted 100%; May — 80% untreated, 100% with IBA; February, March, June and July, untreated, rooted 60, 50, 45 and 65%, respectively; IBA increased rooting to 80, 70, 50 and 80%, respectively.

Cyrilla racemiflora Swamp Cyrilla, Leatherwood

SEED: Flowers appear in July-August and are followed by dehiscent capsules in September-October. Collect in yellow-brown transition, dry, allow seed to fall out, store in sealed containers under refrigeration. Seed requires no pretreatment.

CUTTINGS: August, 1.0% IBA, rooted 100%. Senior author rooted July-August cuttings from a 4-year-old plant with 100% success using 3000 ppm IBA-solution, peat: perlite, mist. The cuttings transplant readily and grow off quickly. Root cuttings also work when taken in winter.

Cytisus Broom

"Broomy" branched shrubs with small leaves (stems actually chief photosynthesizing organ in some species), angled green stems and white to yellow (pink, garnet) flowers. Fruit is a small pod that ripens in late summer-early fall. Collect fruits, dry inside, extract seed, store in sealed containers under refrigeration. Viability after storage for 81 years has been reported.

SEED: No embryo dormancy but seeds have an impermeable seed coat that can be softened by hot water soaks or acid treatment. 30 minutes acid has been recommended prior to sowing. Senior author obtained good germination of *Cytisus purpureus* after 15 minutes acid treatment.

CUTTINGS: There are many species and numerous cultivars. A general outline for C. *beanii*, C. *decumbens*, C. *hirsutus*, C. × *praecox*, and C. *scoparius* is presented. July-August cuttings, 8000 ppm IBA-solution or talc, well drained medium (root systems are often brittle), mist, root in the 70 to 90% range. Untreated cuttings of C. *scoparius* 'Dragonfly' rooted 35%; 70% with 8000 ppm IBA-talc. Cuttings can be rooted in greenhouse or outside beds. Rooting takes 4 to 5 weeks. Plants should be grown in containers since field grown material does not transplant easily.

All types of recipes for *Cytisus* have been recited including 2 to 4" cuttings, slender side shoots, heel cuttings, etc. In one study 2 to 4" long to 40" long November cuttings of C. × *praecox*, wound, hormone (no mention of rate), in containers on open bench, all rooted in 6 weeks. The bigger the cutting the better the root system. December-January hardwood cuttings in an open bench with bottom heat also work.

C. *beanii*, August and December, 8000 ppm IBA-talc and 3000 ppm IBA-talc, rooted 100%. C. *decumbens*, August, 3000 ppm IBA-talc, rooted 100%. June, 3000 ppm IBA-talc, sand, bottom heat, rooted 96%. C. × *praecox*, December, open greenhouse (cool) bench, bottom heat, sand, rooted 80% but the roots were brittle due to use of sand. Again, this points to the need for a light medium like peat: perlite. C. *scoparius*, October heel cuttings, 3000 ppm IBA-talc and 8000 ppm IBA-talc rooted 62 and 66%, respectively. Mid stem cuttings, October, 8000 ppm IBA-talc, 88% rooting. Most of the prostrate forms root easily with good root systems, however, upright types like C. × *praecox* produce 1 or 2 stringy, brittle roots that make transplanting difficult. A ½" wound induces a heavier root system.

Stock plants should be pruned and fertilized in late winter to induce strong vigorous growth for summer or winter cuttings. The juvenility factor although sometimes subtle, can make a large difference in success with a relatively easy-to-root plant like *Cytisus*.

Danae racemosa Alexandrian-laurel

SEED: Reports indicate difficulty with germination. Seed is slow to germinate and seedlings take a long time to grow off. Possibility the embryo is immature or seeds contain inhibitors.

CUTTINGS: An attractive evergreen shrub that is impossible from "stem" cuttings but can be produced by division.

Daphne Daphne

Large group of deciduous and evergreen shrubs, some species with showy, fragrant flowers that bring romance to the garden-making process and frustration to the plant propagator. The fruit is a fleshy drupe of various colors. Seeds are rarely used and cuttings give variable results. The daphnes are rather unpredictable from year to year. Over two years, the senior author rooted D. *odora*, late July (Georgia), 3000 ppm IBA-solution, peat: perlite, mist, 100% in 8 weeks. The wood was reasonably firm, green to slight brown at base of 3" cuttings and the end bud had set. The third year, rooting was about 30% using same procedures and same stock plants. Unfortunately, D. *odora* is known to carry at least 10 viruses, cucumber mosaic being most common. Virus-infected plants tend to drop leaves in the propagation process. Poor rooting of many species has been associated with degree of virus infection. Senior author lost D. *odora*, 'Alba', 'Aureo-marginata', and D. × *burkwoodii* 'Carol Mackie' to virus. Daphne tends to deteriorate under mist and rooted cuttings should be removed and transplanted as soon as possible. Rooting often takes 8 to 12 weeks.

A universal recommendation for success is clean stock maintained with good fertility and spray practices. One grower noted aphids must be kept under control to prevent spread of viruses. A good general reference on *Daphne* propagation is PIPPS 29:248. New York work indicated D. *cneorum* can be rooted well from softwood cuttings under double shade without mist. A well drained medium is important and one grower uses perlite with a touch of peat. A wound is also beneficial.

GRAFTING: D. *laureola* or D. *mezereum* are used for rootstocks.

Deciduous (D), Evergreen (E)

Daphne × burkwoodii (D to semi-evergreen) Burkwood Daphne

CUTTINGS: 'Somerset' is considered one of the easiest daphnes to root. June through early August, tip cuttings, 3" long or less, 2 sand: 1 peat, shaded polytent or mist, root in 6 to 7 weeks. Poor rooting of 'Somerset' was associated with virus infection. At least three viruses have been identified infecting this species. August, 3000 ppm IBA-talc, rooted 50%. July, 8000 ppm IBA-talc plus thiram, rooted 33% under poly, 67% under mist. Another report noted that D. × *b*. 'Somerset' rooted but from second flush shoots collected one month after flowering. 'Somerset', September, greenwood, 1 peat: 4 sand, heavy wound, 2% IBA-talc. Hormone may not be necessary according to several growers.

TISSUE CULTURE: Multiplication was achieved on M & S salts, kinetin & NAA at 0.3 mg/l and 0.1 mg/l, respectively [*Acta Hort.* 78:381-388 (1977); PIPPS 26:330-333 (1976)].

Daphne cneorum (E) Rose Daphne

CUTTINGS: One of the more difficult species to root. Green wood, firm cuttings in late June-July (England), 2-2½" long, 2 sand: 1 peat, cold frame, rooted in 6 to 8 weeks. One grower forced growth in greenhouse and used the soft cuttings with no success; firm cuttings taken from neglected plants rooted reasonably well. June-July (England), 8000 ppm IBA-talc, rooted 55 to 100% in 7 weeks. December, untreated, did not root; 56% with 100 ppm IAA-16 hour soak. July, 74% without treatment, 93% with 1000 ppm IBA-talc. Late August, 8000 ppm IBA-solution, failed to root. Obviously the results are all over the map and no clear cut recipe can be pieced together. Based on some of the above results, a hormone treatment appears necessary.

Daphne genkwa (D) Lilac Daphne

CUTTINGS: Root cuttings, December (England) (from plants in pots that were preconditioned by keeping in cold pit), ½" long, placed horizontally in sandy compost, produced sporadic results. This is practiced for D. *genkwa*, D. *mezereum* 'Grandiflora', and D. *mezereum* 'Plena'.

Daphne giraldii (D) Giraldi Daphne

SEED: One report noted fresh seed sown in fall germinated the following spring. Three months cold stratification produced heavy germination.

CUTTINGS: August, 8000 ppm IBA-talc plus thiram, rooted 17%.

Daphne mezereum (D) February Daphne

SEED: Bright red drupe which ripens in June-July. Birds relish the fruits so collect as soon as possible close to ripening, clean and sow. Germination will occur over a two year period. Seed sown in mid-June and evaluated the following spring showed the following germination rates. Red fruits — 17%, green — 42%, red soaked in water for one hour — 29%, red with outer coat removed — 79%. In another test red fruit soaked in water or gibberellic acid germinated 69%. Apparently, there is a water soluble inhibitor in fruit covering that is removed by soaking. It is always wise to clean the pulpy matrix off fruits. Seeds of var. *album* when sown in July (fresh) germinated 58%. Fruits or seeds that have been allowed to dry take 2 years to germinate if fall planted.

CUTTINGS: Difficult to root from stem cuttings but root cuttings have been used (see D. *genkwa*).

Dapne odora (E) Fragrant, Winter Daphne

SEED: Interestingly, this species sets virtually no fruit at least in the United States.

CUTTINGS: Perhaps the most fragrant of the cultivated daphnes and certainly one of the most popular. Unfortunately, it can be tempermental under cultivation and propagation. See description under *Daphne* for propagation specifics. In England, rooted as described under D. *cneorum* except 3″ cuttings are taken late July to late August. D. *collina*, D. *retusa*, D. *tangutica* follow a similar procedure. Cuttings collected from greenhouse stock plants rooted 100% compared to less than 80% from plants grown outside.

TISSUE CULTURE: Shoot tip and nodal explants through shoot tip culture led to axillary shoot proliferation [PIPPS 26:330-333 (1976); *Acta Hort*. 78:381-388 (1977)].

Daphne retusa (E)

SEED: 100% germination was obtained after 1 month warm (79°F)/2 months cold (36°F), followed by germination temperature of 51°F.

Davidia involucrata Dove-tree

SEED: A rather large, ovoid (rugby-ball shape), 1½″ long drupe that ripens in October. Fruit should be collected and cleaned if possible. The fruit has an outer coat that can be softened by fermenting in a plastic bag. The endocarp is ridged and extremely hard. Dormancy is caused by the hard wall and epicotyl. Five months warm followed by 3 months cold stratification proved excellent. Seeds should be placed in warm stratification until radicles emerge. At this time, place in cold for 3 months. If fruits are fall planted, germination occurs the second spring.

CUTTINGS: Certainly one of most famous trees because of its historical ties to Wilson, David and other plant explorers. Cuttings can be rooted but juvenility plays a part. Softwood from a 25-year-old tree (Canada) did not root. Mid-July (Canada), from 6′ seedling tree, 6″ long, leaves reduced, 1 part 8000 ppm IBA-talc: 1 part Captan 50W, 4 sand: 1 peat, 70°F bottom heat, mist, rooted 36%. Wounding improved the root quality. Mid-January, hardwood, rooted 20% without treatment, no response to IBA. Leaf-bud cuttings, September, 3000 ppm IBA-talc, sand, shaded, syringed, rooted 85% in 5 weeks. Senior author rooted several cuttings from an old tree at Arnold Arboretum but was unable to keep them alive. Might be wise to direct root in containers since overwintering is a problem. Avoid root disturbance after rooting. A two-year-old plant (from a rooted cutting) should be 2′ in a gallon container. Seven years may transpire before a rooted cutting will flower profusely. In Europe, 50% rooting is obtained on 4-leaf (fully developed) cuttings of the current season's growth, wound, 8000 ppm IBA-talc, 2 peat: 1 sand, mist. Rooted cuttings are potted and overwintered in an unheated plastic house.

It has been our observation that all dove trees do not flower with equal facility and vegetative propagation of superior forms is warranted.

GRAFTING : Some nurserymen have grafted dove-tree on *Nyssa sylvatica* seedlings but the grafts decline with time.

Deutzia Deutzia

Shrubs with a bulldog tenacity that are given second class garden citizenship by many gardeners and nurserymen. All need to be pruned after flowering every 2 or 3 years to keep them neat. White or rose-pink flowers are followed by capsular fruits. To authors' knowledge, no propagators grow deutzias from seed. Seed requires no pretreatment. Cuttings are the preferred and practical means of increase. June-July when the stems are green but slightly firm, a light hormone (1000 to 3000 ppm IBA-talc or solution), well drained medium, mist, rooting takes place in 3 to 4 weeks. Rooted cuttings can be easily transplanted and with fertilization will continue to grow.

Hardwood cuttings can be successfully rooted. One large producer maintains stock blocks, fertilizes, produces long growth, takes cuttings and uses a ban saw to cut them into 8″ lengths (leaves are not removed) and plants them in fall or spring with the top one inch above ground. No hormone is used.

Deutzia crenata var. *nakiana*

CUTTINGS: A dainty form with white flowers and purple fall color. Grows 18″. Easily rooted from summer softwood.

Deutzia gracilis Slender Deutzia

CUTTINGS: Nice compact form and certainly one of the best for northern and southern gardens. Easily rooted as described above. Layers have been used but this approach is largely outdated.

Deutzia × *lemoinei* Lemoine Deutzia

CUTTINGS: Larger than D. *gracilis* and quite floriferous. See *Deutzia* for cutting directions.

TISSUE CULTURE: Multiplication was achieved using actively growing shoot tips (explant) from greenhouse grown plants. M & S medium plus 0.1 mg/l BA and 0.1 mg/l IAA. Largest number of shoots were produced with 5 or 10 mg/l BA using same basal medium. Shoots rooted easily and were established in greenhouse [*J. Hort. Sci*. 59(4):545-548 (1984)].

Deutzia × *rosea* (D. *gracilis* × D. *purpurascens*)

CUTTINGS: A rather dainty shrub with handsome pink, almost lavender-pink flowers. Easily rooted as described under *Deutzia*.

Deutzia scabra Fuzzy Deutzia

CUTTINGS: Larger-growing than previous two species. Propagated the same way. *Deutzia* × *magnifica* and 'Pride of Rochester' are handled the same way.

Diervilla lonicera Dwarf Bush-honeysuckle
Diervilla rivularis Georgia Bush-honeysuckle
Diervilla sessilifolia Southern Bush-honeysuckle

SEED: Collect capsules in fall; seed has no dormancy and can be sown directly.

CUTTINGS: Certainly not the most popular plant in the nursery and landscape trades but worthwhile for naturalizing and large mass plantings. Softwood, June-July, 1000 ppm IBA-solution, peat: perlite, mist, root readily in 2 to 3 weeks. Rooted cuttings can be transplanted immediately and should continue to grow. A light water soluble fertilizer application is recommended. In controlled studies, June cuttings of D. *s.* rooted 85% without treatment, 98% with 1000 ppm NAA-talc, and 93% with 1000 ppm 1-naphthyleneacetamide. September, 1000 ppm IAA-talc, rooted 95%. Untreated cuttings root in lower percentages than auxin-treated cuttings.

Diospyros kaki Japanese Persimmon

SEED: No strong dormancy. In one study, seeds of 18 cultivars were collected from ripe fruits and sown immediately. Germination ranged from 20 to 77%. Other studies have shown that seeds harvested from ripe fruit and sown immediately germinate best.

GRAFTING: A shield bud or modified patch bud as described under *Carya illinoensis* or chip bud in August-September are successful. Seedling D. *virginiana* is the preferred understock. Winter pot grafting was also used with excellent success using 2-year-old seedlings of D. *kaki*. Side-veneer or tongue grafts are used.

TISSUE CULTURE: Shoot plumules proliferated on modified M & S medium plus 1 mg/l NAA and 0.1 to 1 mg/l Kinetin. *Phytomorphology* 26(3):273-275 (1976). See also PIPPS 34:118-124 (1984).

Diospyros virginiana Common Persimmon

SEED: Fruit is a 1 to 1½" diameter berry containing 3 to 8 flat brown seeds. Fruits ripen from September to November. Collect fruits and allow to soften in plastic bags, strain pulp, remove woody calyx and save seeds. Allow seed to dry and store in sealed containers under refrigeration. This procedure applies to D. *kaki*. Seed approximates 90% plus soundness. Seed should be fall planted or given 2 to 3 months cold stratification.

CUTTINGS: Root cuttings have been reported successful. This species tends to sucker naturally which is a good indication that it is amenable to propagation by root pieces.

GRAFTING: Chip budding on seedlings of the species in late July, August, early September.

Diospyros texana Texas Persimmon

SEED: Fresh seed should be sown as soon as extracted from fruit. Seed germinated 33%; all other treatments reduced germination [*The Plant Propagator* 29(4):14-15 (1983)].

Dipelta floribunda

SEED: Direct sow. No pretreatment is required. Fruits look like those of elms.

CUTTINGS: Softwood, June-July, 1000 to 5000 ppm IBA-talc or solution, peat: perlite, mist, root readily.

Dirca palustris Atlantic Leatherwood

SEED: A greenish with yellow tinge drupe that ripens in June-July and is either consumed by birds or falls soon after ripening. Fresh, cleaned seed sown immediately outside germinated about 54% in spring. All other treatments produced lower results. One report mentioned that 3 months cold stratification resulted in good germination. Interestingly, seeds that were not cleaned did not germinate as well as those cleaned and planted outside [*Arnoldia* 44:20-23]. Senior author has observed numerous seedlings germinating under a plant in the University of Georgia Botanical Garden.

CUTTINGS: A first rate native shrub for shade or partially shaded area. One of the first shrubs to flower (yellow) and leaf out in spring. Cuttings have been virtually impossible to root. Authors do not know of anyone who has succeeded.

Disanthus cercidifolius

SEED: Flowers are purple and appear in October. The 2-valved, capsular fruit contains several glossy black seeds in each cell. No published information on germination requirements is available but a warm/cold period is no doubt necessary. A good starting point would be 5 months warm/3 months cold. If this does not work return seeds to warm then cold for additional time. Seed has a thick leathery coat and if acid scarification is used follow procedure described for *Hamamelis*.

CUTTINGS: The Morris Arboretum has successfully rooted the plant and induced growth after rooting. Senior author used mid-June cuttings, 1% IBA-solution, peat: perlite, mist, with 100% rooting in 5 weeks and extremely strong root systems. Rooted cuttings were transplanted (a mistake since most witch-hazel members do not like to be disturbed after rooting) into 4" plastic pots in pinebark. Half were fertilized with Osmocote (18-6-12) at the recommended rate. Those fertilized produced new growth in 3 weeks and were 10 to 14" high by fall. Interestingly, the unfertilized plants did not produce new growth. This whole episode came about because we ran out of Osmocote while fertilizing. The non-fertilized plants were fertilized eventually and made new growth but never caught up with the first batch. This is considered one of the most difficult members of the Hamamelidaceae to successfully overwinter after rooting. However, it has not proved to be a problem when new growth is produced and the plants are hardened naturally and overwintered in a polyhouse.

Distylium racemosum

CUTTINGS: Fall (England), wound, 0.5 to 1.0% IBA-talc, sand, 68°F bottom heat, mist, root in 8 weeks.

Elaeagnus Elaeagnus

Shrubs or trees with fragrant flowers and scaly red fruits. The shrubby species are either deciduous or evergreen and flower in spring or fall. The fruits of most species ripen in fall and should be collected, cleaned (maceration) and the hard-coated seed stored under refrigeration. Seed remains viable for 2 to 3 years. See PIPPS 32:181 for cutting details. Deciduous (D) or evergreen (E).

Elaeagnus angustifolia (D) Russian-olive

SEED: Fall plant or 2 to 3 months cold stratification. Seeds soaked for 6 to 18 hours in 300 to 600 ppm Ethrel (no prior cold stratification) germinated 75 to 100%. Seed coats may be exceptionally hard and require 30 to 60 minutes acid treatment prior to fall planting or cold stratification. Two to 3 months stratification at 41°F promotes best germination. If the fruit wall and endocarp were removed, 50 to 60% germination occurred without cold stratification.

CUTTINGS: Some success may be obtained through this means but most are produced from seed. August, 3000 ppm IBA-talc, rooted 28%; mid-October; 40 ppm IBA-soak for 2 hours, rooting percentage was poor.

Elaeagnus commutata (D) Silverberry

SEED: Fall plant or 3 months cold stratification.

Elaeagnus × ebbingei (E) (E. *macrophylla* × E. *pungens*)

CUTTINGS: October to December, terminal, 5″ long, wound, 4% IBA-talc diluted with equal quantity of Captan, 5 sand: 1 peat, 65°F bottom heat, poly-type covering, 90% rooting takes place in 3 to 4 months. After rooting, cuttings should be left in place until spring when they are transplanted. 12 to 18″ additional growth is made during first growing season. July-August cuttings are also successful.

Elaeagnus macrophylla (E)

CUTTINGS: See E. × *ebbingei*

Elaeagnus multiflora (D) Cherry Elaeagnus

SEED: Fall plant or 1 to 2 months cold stratification.

CUTTINGS: June (Boston), rooted 60 and 40% when treated with 3000 and 8000 ppm IBA-talc, respectively. September, 3000 ppm IBA-talc, rooted 95% and November cuttings did not root.

Elaeagnus pungens (E) Thorny Elaeagnus

SEED: Absolute requirements not known but since it ripens in spring direct sowing would be worth a chance. If germination does not occur allow to remain until next spring.

CUTTINGS: See E. × *ebbingei*. A fall flowering species, the fruits ripening in April (Georgia). Extremely popular in southern states. February-March, 6″ nodal cuttings from tips of past season's growth, wound, 8000 ppm IBA-talc, 2 peat: 1 sand, bottom heat, polytent or mist, 90 to 100% rooting. Untreated January cuttings rooted 30%, 75% when treated with 40 ppm IBA-6 hour soak. Untreated, October, sand, rooted 72%; 100% with 30 ppm IBA-soak for 4 hours. Most literature indicated IBA treatment improved rooting significantly. Apparently some cold is helpful in preconditioing the cuttings for the rooting process. In southern Alabama, a large nursery successfully roots the species and cultivars using 2% IBA-talc. In Europe, 6″ long cuttings from current season's growth, wound, 8000 ppm IBA-talc, peat: sand, mist, root 100% in 8 weeks. This same process is also practiced on late winter (February) cuttings with good success. These cuttings can be containerized and will make salable one gallons in one season.

Numerous cultivars exist and ease of rooting might differ among clones, however, literature indicated most are easy to root.

Elaeagnus umbellata (D) Autumn-olive

SEED: Fall plant or 2 to 3 months cold stratification. 'Cardinal' a USDA-SCS introduction is a seed-produced cultivar and described as superior to the normal type.

CUTTINGS: Treat like E. *multiflora*. This is a weedy species and there is no real reason to propagate it from cuttings since it is easily grown from seed.

Elliottia racemosa Georgia-plume, Elliottia

SEED: Fruit is a 4 to 5-lobed dehiscent capsule that contains dish-shaped winged seeds. Only a small percentage of the seeds are sound. Fresh seed should be provided 3 months cold stratification which produces a uniform germination. Dry or old seed may acquire secondary dormancy and behavior is unpredictable.

CUTTINGS: Difficult (impossible ?) from mature plants. The best description of the plant as well as its propagation appears in PIPPS 31:436. Root cuttings are the preferred method — 4″ long and 3/8″ diameter from dormant plants (Boston), placed ½″ deep in sandy soil in flats in a greenhouse in late March, by mid-May multiple shoots appeared. Shoots were harvested in early July and treated with 3000 ppm IBA-talc plus thiram or 8000 ppm IBA-talc plus thiram and rooted 100% very quickly. Root pieces produced shoots for 3 years. When dormant in fall, roots were transferred to cold storage, brought out in spring and the process started again. The rooted shoots present no survival problems.

Elsholtzia stauntonii Staunton Elsholtzia

SEED: No pretreatment is required.

CUTTINGS: June-July, terminal, 1000 ppm IBA-solution, peat: perlite, mist, roots readily. Cuttings will root without hormone treatment. Remove from mist as soon as rooted since excess moisture can be detrimental. In Russia, this plant is grown for its essential oils. Rooting tests have given only 38% with 100 ppm IAA, peat: sand, mist, poly-tunnel.

Enkianthus Enkianthus

Elegant shrubs with refined foliage and lovely white to pinkish (almost red in one cultivar) flowers. The fall color can be striking, yellow to flaming red. Five-valved dehiscent capsules should be collected in the yellow-brown stage, dried inside and the small seeds stored in sealed containers.

Enkianthus campanulatus Redvein Enkianthus

SEED: No pretreatment. Senior author has grown numerous plants from seeds. Sow seed on milled sphagnum in flats, do not cover, place under mist or where high humidity can be maintained, germination occurs in 2 to 3 weeks. Seedlings can be transplanted in 6-8 weeks. Seeds are very small like azalea and rhododendrons so care must be excercised not to plant too densely. Seed should be sown to coincide with the increasing day length of late winter-early spring. Early March would be ideal and the plants should average ±12″ by fall.

CUTTINGS: Rooting is no problem but keeping them alive after the fact can be. Should be done in June-July. June, untreated, sand:peat, rooted 83% in 6 weeks; 100% with 50 ppm IBA-soak. Variety *albiflorus* rooted 88% from May cuttings treated with 8000 ppm IBA-talc. Senior author has rooted July softwoods, 1000 ppm IBA-solution, peat:perlite, mist with 100% success. After transplanting, all died over winter. One report stated to leave cuttings alone and harden off. Overwinter and when growth ensues in spring then transplant. 88% survival was achieved with this approach.

Interestingly, cuttings produce enormous masses of roots and one would think that overwintering would present no problem after transplanting. Possibly, root growth continues after transplanting as the roots fill their new container. This process "robs" the shoots of most photosynthetic materials which are shunted to growing roots. Thus, the above-ground portions are less able to withstand the overwintering process and do not push out new growth in spring. Often if buds do break they occur close to the base of the stem which logically contains higher reserves.

Enkianthus cernuus

SEED: See E. *campanulatus*.

CUTTINGS: Root as described under E. *campanulatus*. Overwintering presents the same problem. Var. *rubens* responds in a similar way.

Enkianthus deflexus Bent Enkianthus

No information was found but directions for E. *campanulatus* should suffice for this species.

Enkianthus perulatus White Enkianthus

SEED: In Arnold Arboretum trials, germination has been extremely low.

CUTTINGS: July, 3 to 4" long terminal shoots, 8 hour water soak (control), rooted 100%; 90 ppm IBA-8 hour soak resulted in 100% rooting in 35 days. IBA treatment hastened rooting and produced superior root systems. In another study, late May (Amherst, MA), rooted 80% without treatment. August (Boston), 8000 ppm IBA-talc, rooted 43%. Overwinter as described for E. *campanulatus*. Mid-June (Boston), wound, sand: perlite, mist, both terminal and lateral (with a heel), overwintered in flat and evaluated in early March. Lateral shoots rooted best with 5000 ppm IBA-dip (100%), 1.0% IBA-dip (93%), 8000 ppm IBA-talc plus thiram (100%), untreated (93%). Interestingly, 5000 ppm NAA-treated cuttings rooted only 6%.

Epigaea repens Trailing Arbutus

SEED: Dehiscent capsule that should be collected before seeds abscise. Seeds require no pretreatment and should be handled as described for *Enkianthus campanulatus*. Germination should occur in 3 to 5 weeks. Amazingly, a pound of cleaned seed contains 10,300,000 seeds.

CUTTINGS: Considered difficult but authors are not so sure if this is valid. Division of sod is often used but stem cuttings can be rooted with good success. August, untreated, sand: peat, rooted 94% in 5 weeks. In general, August-September, cuttings with current and part of last season's growth, untreated, rooted in high percentages in sand: peat, peat or native soil. August-September cuttings root better than those taken in spring. October, 50 ppm NAA - 24 hour soak, hastened and improved rooting over control.

Erica Heath

SEED: Treat exactly as described for C. *vulgaris*.

CUTTINGS: A large group of evergreen shrubs with needle-like foliage and attractive white to reddish flowers. It seems there is a heath in flower any time during the year. Their propagation is virtually identical with *Calluna* (*q.v.*). July-August, October through March are best times. Polytent can be used to great advantage with this group. Senior author rooted late summer cuttings of *Erica darleyensis* with 1000 ppm IBA-solution, peat: perlite, mist, 100% rooting in 5 to 6 weeks. Root systems were excellent. Much of the literature indicates good rooting can be obtained without hormones. Again, a low level is advisable. *Erica carnea*, E. *ciliaris*, E. *cinerea*, E. *tetralix*, and E. *vagans* are rooted easily. An English report noted that a 1 peat: 1 perlite medium was superior to 1 peat: 1 sand for rooting heaths.

Eriobotrya japonica Loquat

SEED: Requires no cold treatment.

CUTTINGS: No specific directions were found but one reference mentioned success with June-July cuttings. The leaves are large (8") and leathery with a pubescent lower surface. No doubt a hormone would be helpful. A well drained medium and mist are necessary. Since this is a broadleaf evergreen, cuttings might benefit from cold and October-November would be a good time to attempt.

GRAFTING: Cultivars are side grafted on seedlings in December-January. Budding is also practiced.

TISSUE CULTURE: MS medium with 0.5 mg/l BA, 0.1 mg/l NAA and 0.1 to 0.2 mg/l GA promoted growth of seedling shoot tips. Best shoot multiplication in MS with 0.25 mg/l BA plus 0.5 mg/l IAA. Rooting was accomplished in ½ strength MS. *Acta Hortic. Sivica* 10(2):79-86 (1983).

Eucommia ulmoides Hardy Rubber Tree

SEED: A notched, capsule-like fruit much like those of elms. Collect in October-November, fall sow or provide 3 months cold stratification. Senior author has fall planted with good germination the following spring. One report indicated 40% germination after 2 months cold stratification.

CUTTINGS: Interestingly, one reference reported that cuttings from current year's growth in summer, rooted in sand in a few weeks. This is a bit difficult to believe and an interesting controlled study at the University of Illinois indicates why. First, there is very limited time that cuttings can be successfully harvested. Late May-early June (Urbana, IL) is optimum. Early June, 50-year-old tree, wound or no wound, 1000, 3000, 8000 ppm IBA-talc or chloromone (*q.v.*), sand, mist, rooted 3 to 4% in all treatments except chloromone which rooted 57% in 14 weeks. The work was repeated with different media and only 8000 ppm IBA-talc and chloromone. Late May cuttings rooted 50% with chloromone treatment and calcined clay: peat mix. The leaves remained dark green and the root systems were heavier in the calcined clay: peat medium. Chloromone is a plant extract containing 1000 ppm 1-Naphthyl-acetamine. IBA was not at all effective. In a North Carolina study, 30-second dip in chloromone resulted in 85% rooting. All other treatments less than 5% [*Amer. Nurseryman* 156(11):85 (1982)].

Euonymus Euonymus

A large genus of deciduous and evergreen trees, shrubs and ground covers with worldwide landscape popularity. Flowers are basically inconspicuous but the capsular fruits with their bright orange or red seeds are beautiful. Collect fruits just as the capsules start to split and dry at room temperature until seeds can be easily removed. The fleshy seed coat (aril) is rather difficult to remove without damaging the seed (endosperm tissue). Usually, the aril is only partially removed by rubbing the seed through a coarse screen. Seed can be dried and stored in sealed containers under refrigeration. Many species are easily rooted from cuttings and specifics are provided below.

Euonymus alatus Winged Euonymus

SEED: Fall plant or 3 months cold stratification. In Korean tests, one month at 32°F facilitated good germination.

CUTTINGS: Thirty-nine references were checked for the species and 'Compactus' and two had specific recommendations. In general, the species and 'Compactus' are easy to root virtually anytime the plant is in leaf. June-July-August, 4 to 6" long, leave top two nodes (4 leaves), 1000 to 3000 ppm IBA—solution, peat: perlite, mist, 90% plus rooting in 8 weeks. Cuttings can be transplanted after rooting or left in the flat. The bud dormancy is quite deep and in a controlled experiment 90 to 100 days cold (40°F in cooler) was sufficient to induce bud growth. Any time period less than this resulted in terminal bud break or no growth. Several reports indicated variable success with this species from 50 to 75%. Personal experience indicates one has to work hard to achieve less than 90%. Senior author has had no success with hardwood cuttings although it can be done.

GRAFTING: One nurseryman successfully grafted E. *alatus* on 4' and 6' standards of E. *europaeus*.

Euonymus americanus Strawberry-bush

SEED: Apparently doubly dormant and 3 months warm/3 months cold is recommended. A 4 month plus cold stratification induced 15% germination in 14 days. Obviously some factor is missing.

CUTTINGS: This small, native, green-stemmed shrub appears like it would be easy to root. Tends to sucker in nature and either division or root cuttings would suffice. One report noted September, untreated, rooted 76%; 20 or 60 ppm IBA — 24 hour soak rooted 98% in 4 to 5 weeks.

Euonymus atropurpureus Eastern Wahoo

SEED: 2 months warm/3 months cold resulted in 40% germination.

CUTTINGS: Possibly like E. *europaeus* although no published literature was found.

Euonymus bungeanus Winterberry Euonymus

SEED: Fresh seed was cleaned (arils removed) and sown immediately with 90% germination. If seed dries fall planting or 3 months cold should suffice.

CUTTINGS: September, 3000 ppm IBA-talc, rooted 87%.

Euonymus europaeus European Euonymus, Spindle Tree

SEED: Fall plant or 2 to 4 months cold stratification.

CUTTINGS: Supposedly easy from July-August cuttings. November cuttings rooted 100%. Untreated August cuttings did not root; 80% rooting in 6 weeks with a 100 ppm IBA-21 hour soak. A hormone, particularly IBA, hastens and improves rooting. Several cultivars are known and should be treated like species. Ideally, 3000 to 5000 ppm IBA-solution should produce good rooting.

GRAFTING: Seedlings of the species are used as understock for the cultivars and to produce novelty plants. E. *fortunei* types like 'Kewensis', 'Emerald Gaiety', 'Vegetus', etc., are top-worked in summer or winter.

Euonymus fortunei Wintercreeper Euonymus

SEED: The "species" is a juvenile form and does not fruit until developing the mature phase. This occurs when the plant is allowed to grow on a structure, wall or tree. Three months cold stratification is recommended but one study used fresh seed of 'Vegetus' and reported 83% germination. Three months cold stratification produced 90% germination.

CUTTINGS: June, July-August (almost any time of year), 1000 to 3000 ppm IBA-solution or talc, peat: perlite, mist, root 90 to 100% in 3 to 6 weeks. A southern (Alabama) nurseryman propagates 'Coloratus' year-round. In general, the numerous cultivars are exceedingly easy to root although differences may occur among cultivars. Aerial roots are often evident along the stems which is an indication of ease of rootability. A nominal hormone treatment (IBA) definitely improves rooting response, however, excessive levels can result in stunting. In one study, untreated cuttings had good roots, no shoot/root stunting (rooting % not given); 6000 ppm IBA-talc induced severe root and shoot stunting. No mention was made of the cultivar. In Oregon work, 'Gracilis' rooted 90% to 100% when 8000 ppm IBA-talc or liquid preparation (2000 ppm IBA plus 1000 ppm NAA) were used compared to 63% rooting of untreated cuttings. Root quality was improved 2 to 3 fold with hormone treatment.

E. *fortunei* types can be transplanted immediately after rooting and if fertilized will continue to grow. Many growers root in small cells or containers.

One note of caution concerns the collection of cutting wood. Supposedly, cuttings from horizontal branches become sprawly; those from upright branches are more vertical in habit.

TISSUE CULTURE: Actively growing shoot tips from 2 to 3-year-old greenhouse plants of 'Emerald Gaiety' and 'Niagara' were established and multiplied on MS medium with 0.1 mg/l BA for shoot initiation and 2.0 mg/l for multiplication. Interestingly, 'Emerald Gaiety', a creamy variegated form, produced only 2 albino plants out of 1,146 shoots [*The Plant Propagator* 31(2):9-11 (1985)].

Euonymus hamiltonianus var. *sieboldianus* Yeddo Euonymus

SEED: Fall plant or 3 months cold stratification.

CUTTINGS: Root cuttings are effective. July-August greenwood cuttings have been rooted.

Euonymus japonicus Japanese Euonymus

SEED: Fall plant or 3 months cold stratification. In Korean tests, 8 weeks at 0°F promoted germination.

CUTTINGS: This species, like E. *fortunei*, has a wealth of cultivars which can be propagated from June, July, August cuttings as well as hardwood later in the season. A hormone in the range of 1000 to 3000 ppm IBA is beneficial. Mist early; mist or polytent later are necessary. All published work indicated an IBA treatment improved rooting response. In one study, November, untreated, rooted 10%; 100% with 50 ppm IBA-18 hour soak. This is powerful proof for the benefit of a hormone.

Euonymus kiautschovicus Spreading Euonymus

SEED: Fall plant or 3 months cold stratification.

CUTTINGS: See E. *fortunei* for details. In northern gardens the species is largely deciduous. Remains evergreen where temperatures do not drop below 0°F for extended periods. In Georgia, it has remained evergreen at −3°F. June, July, August are best months to root. 'Manhattan' is easier to root than species. Interestingly, untreated July and October cuttings rooted 16 and 40%, respectively; IBA treatment improved rooting to 100 and 91%, respectively, in 2 to 4 weeks.

Euonymus nanus var. *turkestanicus* Dwarf Euonymus

SEED: Untreated seed germinated 71%. Brief cold stratification would possibly prove beneficial.

CUTTINGS: June-July, 1000 ppm IBA-solution, peat: perlite, mist, root like a weed. Not too much going for the plant except as a novelty. Fall color can be quite nice (red).

Euonymus oxyphyllus

SEED: No pretreatment is necessary although a short cold period unifies and hastens germination.

CUTTINGS: Rather handsome species with beautiful rich red fruits and handsome fall color. Softwood (June-July) cuttings can be rooted as described under E. *europaeus*. 50% rooting has been reported.

Euonymus sachalinensis Sakhalin Euonymus

SEED: Another species with brilliant red, 5-valved, capsular fruits which open to expose bright orange seeds. Collect seeds (do not clean) and fall plant (3 months cold will work). Germination takes place early in spring.

Eurya japonica Japanese Eurya

CUTTINGS: A species that is not well known in this country but has been grown for a great number of years at Callaway Gardens, Pine Mountain, GA. Senior author collected August cuttings, 1000 ppm IBA-solution, peat: perlite, mist, 100% rooting in 4 to 5 weeks. 'Winter Wine' is a selection with burgundy winter foliage color. Untreated January cuttings of the species did not root; 150 ppm IAA-18 hour soak resulted in 100% rooting in 6 to 7 weeks.

Evodia daniellii Korean Evodia
Evodia hupehensis Hupeh Evodia

SEED: Interesting trees with creamy-white flowers in late June-July followed by reddish, 4 to 5-valved dehiscent capsules in September-October. Collect fruits and dry inside, BB-like, lustrous, brownish black seeds can be extracted easily. Seeds require no pretreatment and germinate like beans. Senior author once received a parcel of seeds from the Morris Arboretum and virtually every one germinated. Seedlings make exceptionally rapid growth after transplanting. Evodias are fine trees but are not really popular in the nursery trade.

Exochorda Pearlbush

Shrubby species of carefree status with "pearl-like" buds and showy 5-petaled white flowers in April-May. The 5-valved dehiscent capsular fruits should be collected in the yellow-brown stage, dried and the seed extracted and stored or placed in stratification. The species leafs out early and good cutting material should be available in May (South).

Exochorda giraldii Redbud Pearlbush

SEED: Embryo dormancy exists and 1 to 2 months cold stratification is recommended. Fresh seed germinated 39%; cold treated 90%. If seed is received dry, cold stratification is recommended.

CUTTINGS: Variety *wilsonii* (large flowers on a 12′ shrub) is the preferred landscape plant. July, 8000 ppm IBA-talc, rooted 67% in sand; 94% in peat: perlite. June-July (August) is possibly the best time for rooting. A hormone is important for untreated July cuttings rooted 40%.

Exochorda korolkowii Turkestan Pearlbush

SEED: 64% germination on freshly sown seed. Seed that had been cold stratified germinated 40%.

CUTTINGS: Perhaps more difficult than above species but same procedure should yield good results.

Exochorda × *macrantha* (E. *racemosa* × E. *korolkowii*)

SEED: 3 months cold stratification produced 96% germination. Cold period does not have to be this long but some cold is necessary for good germination.

CUTTINGS: See E. *giraldii*. 'The Bride' is a selected clone that grows only 3 to 4′ high. It is successfully propagated from cuttings.

Exochorda racemosa Common Pearlbush

SEED: A brief cold stratification proves beneficial.

CUTTINGS: See E. *giraldii*. Late April (England), 8000 ppm IBA-talc, mist, rooted 80 to 90% in 6 weeks. Cuttings can be potted after rooting without any signifcant loss.

Fagus grandifolia American Beech

SEED: Fall plant as soon as seed is available. Three months cold stratification can be used.

CUTTINGS: Although considered difficult to root, two reports have indicated success from summer (late July) cuttings, perlite, mist. No mention of hormone, age of stock plant, percentage, or overwintering success. If cuttings were taken from young trees (juvenile) there may be some validity. No cultivars have been named, unlike European beech, and vegetative propagation is not as important. American beech produces numerous shoots (suckers) from roots which indicates root cuttings might work.

GRAFTING: Apparently, F. *grandifolia* is not a suitable understock for F. *sylvatica* cultivars. One nurseryman has grafted F. *sylvatica* 'Pendula' on F. *grandifolia*.

Fagus sylvatica European Beech

SEED: Fresh seed should be fall sown and not allowed to dry out. Apparently, seed crops are variable from year to year, and a good crop cannot be expected every year. Some authorities cite 1 in 7 as an average for a good crop. The seed (nut) is enclosed in a prickly involucre. The fruit can be picked, dried and the seed shaken or screened out. Senior author conducted cut tests on seeds from midwestern beech trees on the Illinois campus and never found a solid seed. Additionally, he never found a stray European beech seedling. Seed with a 9% moisture content has been stored in refrigerated sealed containers for 3½ years. Although seed will germinate without cold stratification a 3-month period is recommended. Fresh seed germinated 14%, 67% with 3 months cold. Fresh and stored seed germinated 100% after exposure to 5 months cold stratification. The key is to have sound seed. Sound seeds are plump with shiny brown coats. All the pregermination treatments in the world do no good if the seed is hollow.

CUTTINGS: Generally considered difficult to impossible to root. July cuttings when last leaf on twig was developing, untreated, did not root; 200 ppm IAA-24 hour soak, sand: peat, rooted 50% in 5-6 weeks. This also applied to 'Pendula'. This report was from Czechoslovakian or Russian literature.

Interesting work in PIPPS 34:543-550 (1984) with etiolation of a beech hedge showed that cuttings rooted 69% with shade and localized banding, 42% with shade and no banding and 5% without either. Cuttings were treated with 3000 ppm IBA-talc, mist and rooting assessed after 5 weeks.

Layering occurs commonly with old beeches when the branches touch the ground and gradually become covered with organic matter. The central trunks of some weeping beeches have died and the outer branches layered creating an open coliseum effect in the interior with a stockade-like effect of layered (rooted) branches on the outside.

GRAFTING: Usually pot grafted in January-February on seedling understock of F. *sylvatica*. There are many recipes and success is variable ranging from 25 to 30% up. Understocks are potted the year before in 3 to 4" pots, maintained in shade, cold frame, pit or under lath. May be brought in during November-December, white roots should show at soil: pot interface before grafting commences. Scion wood may be one-year or two-year wood. Several propagators indicated the latter was the best. Understock should average 1/4" to 5/16" diameter. Cleft, side veneer, modified side veneer, and whip and tongue grafts are used. Grafts are tied (rubber), waxed, plunged in peat (union not covered) in a grafting case, watched and ventilated as necessary; grafts heal in 6 to 8 weeks. Small grafts can be gradually hardened and moved outside under shade in spring. Some growers go to field with the new grafts; others keep them 3 to 4 years in containers under light shade.

One nurseryman grafted outside in spring by cutting 5- to 6-year-old trees back to 12", grafting, waxing and obtained 3 to 4' River's beech in one season.

Chip budding has not proven successful based on English research work.

× *Fatshedera lizei*

SEED: Authors do not know of fruiting specimens.

CUTTINGS: Broadleaf evergreen intergeneric hybrid between *Fatsia japonica* 'Moseri' and *Hedera helix* 'Hibernica'. Not exceptionally cold hardy but has withstood 10°F in senior author's garden. Cuttings root readily in summer or fall. Use mist in summer and polytent late in fall. 1000 ppm IBA-talc or solution and well drained medium like peat: perlite are beneficial. Bottom heat may be helpful but is not necessary. The plant does not produce a great number of lateral branches and cutting wood is often at a premium. Leaf bud cuttings have been used quite successfully. Once rooted they take longer to produce a salable plant. Rooted cuttings can be transplanted immediately after rooting and if fertilized will continue to grow. This species is used as a house plant in the north and grown outside in many semi-tropical to tropical areas. It displays excellent shade tolerance and develops root-like holdfasts with which it cements itself to structures. The variegated form should be treated like the species.

GRAFTING: Has been used as an understock for *Hedera helix* cultivars.

Fatsia japonica Japanese Fatsia

SEED: No pretreatment necessary and best germination occurs at 80°F. Seeds [collected in late February (Alabama)] extracted from green (unripened) and blue-black (ripened) fruits, sown immediately germinated 58%. When green and blue-black fruits were sown whole, germination was 19 and 25%, respectively. This indicated the presence of an inhibitor in the fruit wall.

CUTTINGS: Apparently most plants are seed grown. The only pertinent recommendation noted to take cuttings when the wood was firming (July-August) and use bottom heat. Low hormone and mist, as well as possible reduction in leaf area (leaves are immense), would be advisable. The variegated form has to be produced vegetatively.

Feijoa sellowiana Pineapple Guava

SEEDS: Apparently no dormancy exists and seeds when removed from the pulp can be sown directly and will germinate.

CUTTINGS: November, 3 nodes long, top two leaves remaining, wound, 3000 ppm IBA-solution, peat: sawdust: pumice: sand, in beds with 68°F bottom heat. Cuttings were taken from 15 different clones that averaged about 10 years of age. When evaluated about 7 months later rooting ranged from 4 to 76%. In general, rooting is strongly influenced by the parent plant. The researcher concluded that the rootability of the clone is the most important factor affecting successful rooting. December hardwood rooted 31% when treated with IBA, 0% without.

The published literature is not very encouraging or specific relative to success. January cuttings, 50 ppm IAA-48 hour soak rooted 20% more than the control but no absolute numbers were available. July-August cuttings can be rooted. This plant is frequently injured in Zone 8 (+10 to +20°F) and acts almost like a herbaceous perennial. The soft shoots that develop might be amenable to treatment as softwood cuttings under mist. A report from Florida indicated that cuttings are the preferred method of propagation. 3 to 4" long, 1/5" diameter, 2-leaf cuttings taken in August rooted 75% (no mention of hormone). Cuttings from lower and middle parts of stock plants rooted 87 and 63%, respectively. Those from upper part rooted only 9%.

Ficus carica Common Fig

CUTTINGS: In several tests, 1000 ppm IBA-5 second dip proved optimum. February, terminal bud removed, 50 ppm lanolin paste (control) rooted 50% with an average of 4 roots; 770 ppm IAA-lanolin paste rooted 100% with 23 roots per cutting in 28 days. March, untreated, rooted 80% with an average of 13 roots per cutting; 200 ppm IBA-24 hour soak rooted 100% with 61 roots per cutting. Girdling 30 days before taking cuttings increased rooting from 55 to 100% and hastened rooting to 28 days versus about 90 days. In a Brazilian study, 10" long cuttings rooted 97% and gave best shoot and root development.

Summer firm wood should root as well but no literature was found.

TISSUE CULTURE: Shoot tips were successfully cultured on MS medium with 0.18 mg/l NAA, 0.1 mg/l BA, and 0.03 mg/l GA. Shoots proliferated and were rooted on MS plus 0.5 mg/l NAA plus 0.5 mg/l IBA and successfully transplanted to the greenhouse [*HortScience* 17(1):86-87 (1982)]. See also Tissue culture of economically important plants. Proc. of Int. Symp. Singapore, 1981. Rao (ed) p. 505 (1982).

Ficus pumila Creeping Fig

CUTTINGS: Easily rooted from June to September using current season's growth (4" long), 1000 ppm IBA-solution, well drained medium. Rooting approaches 90 to 100%. The mature form roots about 75% with the same treatment. IAA and NAA are not as effective as IBA. This is an easy plant to root and forms aerial roots (root-like holdfasts) with which it cements itself to structures. See PIPPS 28:306 for more information than is needed to successfully root the plant.

Firmiana simplex Chinese Parasol Tree

SEED: A rather exotic tree with green stems and large tropical-appearing leaves. Only hardy in Zone 7 and south. There is basically nothing in the literature on the species. Senior author collected shriveled seeds from campus plants (Georgia) with sporadic but good germination occurring over an 8 week period. Seed that was stratified did not germinate any better. Perhaps fresh seed would germinate uniformly.

Fontanesia fortunei Fortune Fontanesia

SEED: Collect and sow immediately. Some cold stratification may help.

CUTTINGS: Unusual plant with a bamboo-like effect. Hardy to −17°F. Good screening and massing plant. July, untreated, sand, rooted 100% in 8 weeks. June, 5 ppm IAA-24 hour soak, rooted 53% in 3 weeks. Higher IAA concentrations were less effective. October, untreated, did not root; 20 ppm IBA-24 hour soak improved rooting.

Forestiera neomexicana New Mexican Forestiera

SEED: Fresh seed germinates without pretreatment.

CUTTINGS: Late July (Colorado), 5 to 8" long, rooted 80% in sand; 35% in peat: perlite. 960 ppm Ethrel increased rooting to 100% in peat: perlite.

Forsythia Forsythia

Popular yellow-flowered, spring blooming shrubs that are almost ridiculously easy to propagate from softwood or hardwood cuttings. Cuttings taken from May through September root readily. The 4-lobed yellow flowers are followed by dehiscent capsular fruits with brown seeds. Seeds require a brief cold stratification (1 to 2 months) and often put down a radicle in the cold.

Forsythia europaea Albanian Forsythia

SEED: Seed sown after extraction germinated 56%, 72% after one month cold stratification, and 86% after 3 months cold.

CUTTINGS: May be more difficult than most F × *intermedia* types. June-July, 1000 ppm IBA, well drained medium, mist, should root in 4 to 5 weeks.

Forsythia × *intermedia*
(F. *viridissima* × F. *suspensa* var. *fortunei*)

SEED: Seed of 'Spring Glory' was given 3 months cold stratification and radicles developed during stratification. Virtually every seed germinated and it would have been fun to line out and observe for flower differences but time and space limitations precluded this effort.

CUTTINGS: Exceedingly easy from May to September using 1000 to 3000 ppm IBA-talc or solution. A hormone is not required but the quality of the root system is definitely improved over the control. Hardwood cuttings root well. Senior author sampled hardwood cuttings in early January (Georgia), 1.0% IBA-solution, outside cold frames with bottom heat, peat: perlite or bark with 60% rooting in 12 weeks. January cuttings rooted best; those taken later broke bud and food reserves were shunted to growing shoots. Rooted cuttings were transplanted into bark medium in 6" containers and fertilized regularly. They made excellent growth.

'Arnold Giant', late June (Wisconsin), perlite, bottom heat and mist. In 6 to 7 weeks, 79% rooting occurred. This cultivar is considered difficult to root but is one of the most flower bud hardy forms, which along with F. *ovata*, flowered profusely in Madison, Wisconsin after severe winters.

TISSUE CULTURE: 'Spectabilis' buds from long shoots are collected in February, March, disinfested, and grown on MS medium. Leaves (shoots) are produced and these are cultured on MS plus 1 mg/l IAA and 1 mg/l BA. Shoots and roots are produced and plantlets are established in soil.

Forsythia mandschurica 'Vermont Sun'

CUTTINGS: A selection by the University of Vermont with excellent flower bud hardiness (−25 to −30°F). Rooting studies at Vermont indicated that June, 3000 ppm IBA-talc, vermiculite: perlite medium, mist, rooted 85% in 8 to 12 weeks.

Forsythia ovata Early Forsythia

SEED: Sown without pretreatment germinated 74%; 82% after one month cold; 64% after 3 months cold.

CUTTINGS See F. × *intermedia*. Late June (Wisconsin), perlite, bottom heat and mist. In 6 to 7 weeks, 55% rooting occurred.

Forsythia suspensa and varieties *atrocaulis*, *fortunei* and *sieboldii*

SEED: One or two months cold stratification.

CUTTINGS: Basically like F. × *intermedia*. June cuttings benefit from a hormone as rooting is hastened (3 weeks) and percentage may be increased over untreated controls.

Forsythia viridissima Greenstem Forsythia

SEED: Fresh seed germinated 48%. A one to two month cold treatment would improve the response.

CUTTINGS: May through September. July, untreated, rooted 100%. 'Bronxensis', a low growing cultivar, has been described as difficult to root. Senior author, summer, 0, 1000, 5000, 10,000 ppm IBA-solution and 1000 ppm NAA-solution, peat: perlite, mist, cuttings rooted 100% in all treatments. In fact, the control was as good as hormone treatments.

Fothergilla gardenii Dwarf Fothergilla

SEED: Six months warm/3 months cold stratification resulted in good germination. The two-valved capsules ripen in September and should be collected before they dehisce, placed in large paper sack or cardboard box with a lid. After a time, a strange popping sound (like corn) will be evident as the seeds are ejected from the capsules and hit the side of the vessels. The seeds are shining black. Senior author has never observed great quantitites of fruit especially on isolated shrubs. Cross pollination may be necessary for good fruit set.

CUTTINGS: Exceedingly easy member of the Hamamelidaceae to root and keep alive after rooting. Senior author has rooted this species in late July, 4000 ppm IBA-solution, sand: perlite, mist, with 100% success. Plants can be left in the flat and overwintered or potted, fertilized and usually will continue to grow thus assuring survival. Inspect the rooted cuttings before transplanting and if the terminal is still elongating then transplant. If the leaves appear mature size and an end bud is evident simply overwinter in a cool greenhouse, pit house or suitable structure.

Fothergilla major (*F. monticola*) Large Fothergilla

SEED: The best seed information comes from Mr. Alfred Fordham, Propagator, at the Arnold Arboretum [see *Arnoldia* 31:256-259 (1971)]; seeds are somewhat difficult to germinate for they exhibit a double dormancy and pretreatment must be accomplished in two stages, seeds require warm fluctuating temperatures followed by a period of cold. *Fothergilla major* seeds have required exceptionally long warm periods with 12 months being optimum; after warm treatment they should be placed at 40°F for 3 months.

CUTTINGS: Wood taken from suckers or root cuttings yielded good results, best with bottom heat; untreated cuttings taken when shrubs were in flower rooted 67% in sandy soil in 60 days; June cuttings set in sand: peat rooted 67% without treatment, and 100% in 42 days after treatment with 200 ppm IAA/24 hours; for some reason the two species treated here have been listed as difficult to root; senior author has a "love affair" with these shrubs and would like to describe some of his propagation adventures [see Dirr. *Horticulture* 40(12):38-39. (1977), Fothergillas: A garden aristocrat]; softwood cuttings can be readily rooted when collected in June, July and as late as August; they should be treated with IBA-quick dip and placed in peat: perlite under mist; one experiment, late July [see *The Plant Propagator* 24(1):8-9, 1978], *F. major*, involved 0, 1000, 2500, 5000 and 10,000 ppm IBA-quick dip treatments and 1000 ppm NAA-quick dip; all cuttings, even the controls, rooted 80 to 100%; the greatest numbers of roots among IBA treatments were produced with the highest IBA levels but surprisingly the NAA treatment resulted in the greatest number of roots. I have found that if the cuttings are still growing when rooted they can be transplanted, if not the following handling practice should apply: often after rooted cuttings are transplanted they enter a dormancy from which they never recover; the problem can be avoided if the cuttings, when rooted, are left in the flats and hardened off; the flats of dormant cuttings are transferred to cold storage, which is maintained at 34°F; in February or March the flats are returned to a warm greenhouse and when growth appears the cuttings are potted. Cuttings can also be maintained in polyhouses, pit house or other suitable structure.

TISSUE CULTURE: Plants are now available through a commercial tissue culture operation.

Franklinia alatamaha Franklinia

SEED: Collect capsules in October-November before they split (look for suture lines opening), dry inside, shake out small brown seeds, sow immediately and good germination will ensue. If allowed to dry a one month cold stratification is recommended. Dr. Darrel Apps, Longwood Gardens, indicated that some germination takes place immediately but best germination occurs after a cold period.

CUTTINGS: Quite easy. One report noted rooting them in a jar of water shaded from the sun. Sounds a bit like *Coleus* propagation. Senior author has rooted the species numerous times using June, July, August cuttings, 1000 ppm IBA-solution, peat: perlite, mist, with 90 to 100% success. Overwintering can be a problem but if rooted early enough the cuttings continue to grow and have no trouble surviving. Excessive moisture in the rooting or growing medium can result in decline. A wilt caused by *Phytophthora cinnamoni* can be serious.

A large grower uses the following prcedure to achieve 85% plus success. Mid-July (New Jersey), from stock plants pruned hard every 3 years, 5 to 6″ long with 5 or 6 top leaves cut in half, peat: perlite, styrofoam, sandy soil medium (peat: perlite works as well), mist, shade, outside, directly stuck in peat pots (to avoid transplanting), root in 3 to 4 week and gradually weaned from mist over the next 3 to 4 weeks. Overwintered in heated pit house covered with opaque plastic and maintained at 38°F night temperature.

Fraxinus Ash

Large trees with polygamo-dioecious, non-showy flowers in early spring before the leaves. *F. ornus* is a notable exception for the showy white flowers appear after the leaves in May-June. The fruit is a canoe-paddle-shaped samara that ripens in October. Collect seeds when the samaras turn brown, dry and store in sealed containers under refrigeration. Seed has been stored with no loss of viability for 7 years. A cut test is wise to determine soundness of seed. Cuttings have proven virtually impossible to root even from young trees. Recent breakthroughs in tissue culture offer hope for the future.

Fraxinus americana White Ash

SEED: Fall plant, 2 to 3 months cold or 1 month warm/2 months cold stratification. Most growers fall plant. There is evidence for embryo dormancy as well as an impervious seed coat. Possibly seed collected early and sown immediately would germinate well in spring. The embryos, which may be immature, require a period of afterripening after which exposure to low temperature satisfies the dormancy requirement. *F. americana* requires only a cold period although the literature is conflicting on this point.

GRAFTING: Cultivars are usually field budded on seedling understock in late July and August. Bud take is good and subsequent growth the following year can range from 8 to 12′ in height. To keep the budded trees growing, regular fertilization, irrigation and spray programs are necessary. Once a terminal bud is set, linear growth essentially stops. New cultivars of F.

americana have proliferated in recent years.

TISSUE CULTURE: Has been accomplished with seedling explants but no cultivars have been successful to date.

Fraxinus angustifolia Narrowleaf Ash

SEED: Not well known in the U.S. but sometimes encountered in Europe. Seed requires only cold stratification to satisfy dormancy requirements. Fall planting will suffice.

Fraxinus excelsior European Ash

SEED: Immature embryo and impermeable seed coat necessitate warm/cold stratification. According to English literature, seed picked green and sown will germinate the following spring. There are, however, comments that counter this contention. If dry seed is fall planted, it will germinate during the second spring. Seeds planted in soil in September in a cold frame, a mulched and board-covered frame, and a 70°F greenhouse were examined in mid-March of the following year. In all situations the immature embryos grew to the full length of the seed. Simply by visual inspection, the researchers reported the seeds should have germinated. However, this did not occur. By early June, only those seeds that received afterripening and cold germinated. Seeds maintained in the greenhouse did not germinate. Obviously growth of the embryo alone does not always bring about changes necessary for germination.

GRAFTING: In England, chip-budding (*q. v.*) has proven highly successful. Numerous cultivars exist and some can be topworked (i.e. 'Pendula') to produce a cascading effect. Both pot grafting in winter and mid-spring grafting in the field are/have been practiced. A Canadian nursery grafts in the field but mentions the importance of leaving a bud on the top of the understock (called a drawing bud). The scion is two-budded at least. *Fraxinus pennsylvanica* seedlings have been used as an understock in the United States for F. *excelsior* cultivars. 'Aurea' was shield and chip budded with 100% success on F. *excelsior* rootstocks in mid-August. Earlier and later, chip budding was more successful than shield budding.

Fraxinus holotricha

SEED: Fall plant or cold stratification. This is not a well-known species and 'Moraine' is the best known cultivar.

Fraxinus nigra Black Ash

SEED: Immature embryo which requires a period of warm stratification (2 months) followed by cold stratification (3 months).

GRAFTING: See F. *excelsior*. 'Fallgold', a seedless, disease-free, yellow fall-coloring selection, can be budded on green ash understock and, no doubt, the species.

Fraxinus ornus Flowering or Manna Ash

SEED: Fall plant or 2 to 3 months cold stratification.

Fraxinus oxycarpa 'Raywood'

SEED: Popular tree in California for the wine-red fall color. Fall plant or 2 months cold stratification. 'Raywood' must be budded onto seedlings of the species or possibly F. *angustifolia*, with which it is closely allied.

Fraxinus pennsylvanica Green Ash

SEED: Variety *lanceolata* is included in F. *pennsylvanica*. Generally, fall planting or 2 to 3 months cold stratification will suffice. Seed given 2 months warm/3 months cold germinated 92%.

GRAFTING: Cultivars are budded on seedling F. *pennsylvanica* in July-August.

Fraxinus quadrangulata Blue Ash

SEED: Generally considered to have an immature embryo and warm followed by cold stratification is required. One study reported excellent germination with only 2 months cold stratification.

Fraxinus uhdei Shamel Ash

SEED: No pretreatment necessary. Freshly planted seed germinated in high percentages.

Gardenia jasminoides Cape Jasmine

SEED: No pretreatment required. In one test, seed of G. *jasminoides* 'Fortuniana', 'Mystery' and 'Radicans' germinated 83, 83 and 43% in 2 to 4 weeks.

CUTTINGS: Easy to root. 3000 ppm IBA works well. June-July to March cuttings can be used. Actually about any time of year is acceptable. Well-drained medium and mist or polytent are necessary. Rooting takes 3 to 6 weeks. A hormone is not necessary but hastens rooting. April, untreated, rooted 56%; 88% with IBA treatment in 7 to 8 weeks. The cultivars 'Radicans', 'Mystery', etc., are also rooted easily. Although one reference stated rooted cuttings are hard to transplant, this is generally not the case.

TISSUE CULTURE: 0.3 to 1 mg/l BA and 1 mg/l IAA promoted axillary shoot development from nodes of greenhouse stock plants. Rooting occurred in 3 weeks with 1 mg/l IAA plus 2 g/l activated charcoal. Plants were established and flowered normally. J. *Plant Physiol.* 116(5):389-407 (1984). See also *The Plant Propagator* 29(3):13-14 (1983).

Gaultheria procumbens Checkerberry

SEED: Collect fruits in fall through winter; remove seeds from mealy matrix by rubbing through screen or maceration in water. Seeds can be directly sown and will germinate although a short cold period may improve germination.

CUTTINGS: Usually handled by dividing pieces of "sod" in late winter or early spring. The sod is divided into small plantlets with roots, potted in an organic medium and allowed to develop. No fertilizer is required until roots reach soil: container interface and then only small quantities. The rooting of March cuttings was improved by the use of IAA and NAA. Since no one will ever make their fortune with this plant the first method suffices nicely.

Gaylussacia brachycera Box Huckleberry

SEED: Cleaned seed planted in mid-August germinated 22% in 10 weeks. 3 months cold stratification did not improve germination. One month warm/followed by 1 or 2 months cold stratification resulted in 80% and 96% germination, respectively.

CUTTINGS: Late August (Boston), 8000 ppm IBA-talc, sand: peat, mist, rooted 100% in 12 weeks. Cuttings taken in

September, November, December and January also rooted well. From senior author's experience, August, 8000 ppm IBA-solution, sand: perlite, mist, rooted 20%. 'Amity Hall North' rooted 80% with same treatment. November appears to be the *best* month to collect cuttings although success is possible virtually year-round. A light wound facilitates a dense root system.

Gelsemium sempervirens Carolina Yellow Jessamine

CUTTINGS: June-October (Georgia), 3000 ppm IBA, peat: perlite, mist, root 90 to 100%. Rooted cuttings transplant readily and continue to grow. Probably could root the plant anytime of the year. 'Pride of Augusta' (double-flowered) roots readily also.

Genista Genista, Woadwaxen

Generally low growing broom-like plants with yellow flowers and at times very difficult to separate from *Cytisus*. The fruit is a small pod that should be collected in color transition, dried, and seeds extracted and stored. In general, seed has a hard, impermeable coat and 30 minute acid or 30 minute hot water (190°F) soak will work. Acid treatment is the best treatment and this might be followed by a 24 hour cold water soak (seed imbibition).

Genista germanica German Woadwaxen

SEED: 30 minute acid treatment produces good germination.

CUTTINGS: Late July (Boston), 8000 ppm IBA—talc plus thiram, rooted 100% with good root systems. 'Prostrata' also rooted readily from summer softwoods.

Genista hispanica Spanish Woadwaxen

SEED: 30 minute acid treatment.

CUTTINGS: June, 2½" long, no hormone, well drained medium, root 90-95% in 8 to 9 weeks. They can be transplanted after rooting and overwintered in a cool greenhouse.

Genista lydia

CUTTINGS: This and the above species are planted extensively in Europe. The propagation requirements are similar.

Genista pilosa Silkyleaf Woadwaxen

SEED: Untreated seed did not germinate. 18-hour hot water soak resulted in 4% germination; 30 minute acid treatment resulted in 98% germination. Best results are obtained by a combination of acid/hot water soak (18 hrs).

CUTTINGS: September-October, untreated, rooted 90%. Late August (Boston), 3000 ppm IBA-talc, rooted 100% in 3 to 4 weeks.

Genista tinctoria Dyer's Woadwaxen

SEED: Same treatment as described under G. *pilosa*. Work in Massachusetts showed 15% germination of untreated seeds at 70°F. Hot water and acid treatments did not greatly improve response. Cold stratification for 3 months at 41°F increased germination to 70%.

CUTTINGS: Treat like above. Variety *plena*, late September (Boston), 8000 ppm IBA-talc, sand: peat, rooted 87% in 4 weeks. Although June-July are good months to take cuttings, August-September appear as good. This is the most popular species in the United States.

Ginkgo biloba Ginkgo

SEED: A warm/cold stratification is required as the embryo is immature and needs to develop before being subjected to cold temperatures. The "fruits" should be collected in October, the pulp removed and the seed warm stratified for 1 to 2 months followed by 1 to 2 months cold. Fresh seed can germinate and one report noted 29% germination, with one month cold stratification inducing 62%. If seed is collected and sown outside good germination should take place in spring.

CUTTINGS: Ginkgo is not difficult to root from cuttings and even material from old trees roots reasonably well. Most literature points to July as the optimum month for taking cuttings. Senior author experienced good success in June (Illinois), 4 to 6" long, 30-year-old tree, 8000 ppm IBA-solution, peat: perlite, mist, with rooting in 7 to 8 weeks. Other reports support these results. Late July (Boston), 8000 ppm IBA-talc, mist, rooted 100% in 7 weeks. Late June, terminal shoots elongating, 8000 ppm IBA—talc plus thiram, rooted 100% in 4 to 5 weeks. Spur cuttings (older wood), same treatment as above rooted 80%. The cuttings produced new shoots and grew well. Although the root system is rather coarse, rooted cuttings transplant well and can be overwintered without any problem.

For some reason *Ginkgo* is usually not produced from cuttings for commerce. There is no doubt the rooted cuttings are slow to grow off compared to grafted or budded trees. Tennessee nurseryman commented that rooting was no problem and cuttings were overwintered in place. The next season (not transplanted) they did not grow (height increase). The second summer after rooting they grew and ranged from 6 to 24" in height. A New Jersey nurseryman noted the same response but mentioned that early rooted and transplanted cuttings grow off better than those rooted later. One grower said hormone (IBA) was not beneficial, another insisted (with data) that 8000 ppm IBA-talc was best.

Cultivar differences exist and one grower reported 'Autumn Gold' rooted well and 'Princeton Sentry' did not.

GRAFTING: Cultivars are budded on seedling understocks in summer. Whip and cleft grafts have been used for pot grafting in January-February.

TISSUE CULTURE: See Z. *Pflanzenphysiol.* 85:61-69 (1977).

Gleditsia triacanthos var. *inermis* Thornless Common Honeylocust

SEED: Acid treatment for 1½ to 2½ hours is necessary. Seeds sown without pretreatment germinated 5%; 29% in hot water, 62% after 1½ hours sulfuric acid, and 98% after 2½ hours sulfuric acid. Honeylocust has no internal dormancy, strictly external (hard seed coat), and once the coat is ruptured by any means water is imbibed and germination can proceed. The seed is used in plant propagation classes to demonstrate the use of acid scarification. 0, 1, 2, 3, 4, 5 hours acid treatment are followed by sowing with germination occurring 4 weeks later. Students then evaluate the germination responses.

Honeylocust is so widely used in the midwest as to be disgusting. Seed collection involves collecting pods in October, drying and thrashing seed. In a large parking lot on the Illinois campus, honeylocust was the only tree. The pods would drop to the ground, disintegrate under cars with bullet-like brown seeds rolling everywhere. One Illinois nurseryman would come in late fall and sweep the parking lot to the tune of 100 pounds of cleaned seed.

CUTTINGS: Difficult (maybe impossible) to root. The thorny form (juvenile) was propagated by dormant hardwood or root cuttings. Thornless clones (mature), 'Millwood' and 'Calhoun', were exceedingly difficult from hardwood cuttings and did not regenerate from root cuttings. Commercially, no one produces the tree from cuttings. Root cuttings, however, have been reported to work quite successfully. Senior author observed numerous shoots developing from roots where plants had been dug in the field and successfully rooted them.

GRAFTING: Budding as early as June-early July in Ohio; early June in Tennessee. Budwood should be green. T-bud is standard procedure. Interestingly, the bud piece has multiple buds and 2 or 3 may break. This necessitates removing all but one. Honeylocust grows rapidly and a 7 to 8′ (10′) whip is produced in a single season.

Gordonia lasianthus Loblolly-bay

SEED: Since flowers occur from May to October (Georgia) the capsules ripen over a long time period. Collect when brown, dry, extract small seeds and sow immediately. No dormancy exists and seeds germinate readily.

CUTTINGS: Easily rooted from June-July-August, 3000 ppm IBA, peat: perlite, mist, 90 to 100%. Transplants readily after rooting. Rooted 100% in summer in a high humidity chamber. Senior author has taken cuttings in March with 100% rooting. *Gordonia axillaris* also roots easily.

Gymnocladus dioicus Kentucky Coffeetree

SEED: Seeds are the size of a small jawbreaker and 2000 times harder. The impermeable seed coat is the only hindrance to germination. In one study, 0, 2, 4, 8, 16, 32 hours concentrated sulfuric acid treatment resulted in 7, 93, 100, 95, 82 and 87% germination, respectively. Two to 4 hours acid treatment is ideal, the seeds imbibe water quickly when sown and germinate uniformly. Species is dioecious and female trees produce large leather-hard pods that abscise differentially from late fall into winter (spring). Collect pods from ground and basically run through some type of flailing device. The hard coated seeds will not be hurt. Seed can be dried and stored almost indefinitely.

CUTTINGS: Root cuttings will work if taken December through March. In fact, this may be the only effective way to vegetatively propagate the tree. No success with budding was reported. A nurseryman mentioned digging coffeetree and failing to fill the holes. Soon suckers were evident at the root/soil interface. These shoots could possibly be rooted. The best approach might be root pieces ½″ diameter, 2 to 3″ long, placed horizontally in sand: peat.

GRAFTING: Interestingly, a globe form has been described. Grafting was used to propagate the form but no specifics were given. Chip budding may be the best way to proceed.

Halesia Silverbell

Southeastern United States natives, refined in character and grace, delicate in flower, and of the first order for garden use. Bell-shaped white flowers appear before or as the leaves emerge in April-May. The 2 to 4-winged drupe ripens in October and can be collected, dried and stored under refrigerated conditions.

Halesia carolina (H. *monticola*) Carolina Silverbell

SEED: The best procedure is fall planting and waiting until the second spring for germination. Seedlings are precocious and may flower when only 4′ high although flowering usually occurs later. Senior author and others have fooled with warm/cold stratification combinations with minimal success. In one study, non-pretreated seed did not germinate. 9 months warm/3 months cold resulted in 20% germination. The normal recipe is 2 to 4 months warm/2 to 3 months cold. Senior author followed this with no success. Seeds were still good (determined by a cut test) and were put back in cooler for 2 additional months. General germination occurred finally. A Tennessee graduate student extracted embryos from green and brown seeds. He found that both grew. However, embryos extracted from seed that had been stratified germinated better and came up in four days. Fine stands can be obtained with fall planting and patience for two years. Good paper related to seed germination: *Contributions Boyce Thompson Institute* 4:27-37 (1932).

CUTTINGS: Not difficult to root but surprisingly very few propagators take advantage of this fact. May, June, July, August (Illinois), 2500 or 10,000 ppm IBA-solution, peat: perlite, mist, rooted 80 to 100% with profuse root systems. Cuttings were easily transplanted and grown. The earliest rooted survived the best. The species can be container-grown in small sizes and a two-year-old plant (including rooting year) should average 3 to 4′ high. The pink forms (var. *rosea*) can also be rooted easily. Actually, 1000 ppm IBA-solution works well. June might be the best month to root cuttings.

GRAFTING: No recorded instances but the pink form might be pot grafted on seedlings of the species. Summer budding should also work.

TISSUE CULTURE: See *HortScience* 20(3):592 (Abst. 504). WPM plus 1 to 2.5 mg/l BA promoted excellent shoot proliferation. Optimum of 10 to 14 shoots in a 9-week period.

Halesia diptera Two-wing Silverbell

SEED: Takes two years to germinate if fall planted. 3 months cold/6 months warm/4½ months cold — radicles developed in cold.

Halesia parviflora Small Silverbell

SEED: Seed should be handled like H. *carolina*. Best to fall plant and wait in line.

CUTTINGS: Senior author has attempted to root this and H. *diptera* with no success. September cuttings of H. *diptera*, and February of H. *parviflora* have not rooted.

Hamamelis Witch-hazel

Shrubs with yellow to red, often fragrant flowers in January-March and in one species, October-November. The 2-valved dehiscent capsules ripen in October and eject two shiny black seeds. Capsules should be collected in the yellow stage before they open, dried in a closed container, and the seeds will be ejected. The seeds can then be easily screened. Seed can be stored dry for one year under refrigeration without loss of viability.

Hamamelis × intermedia (*H. japonica* × *H. mollis*)

As now known, a whole group of shrubs with pale and bright yellow to orange and copper to red flowers in February-March on robust, often 15 to 20′ high shrubs. 'Arnold Promise', 'Diane' and others set reasonable quantities of fruit and their offspring will yield interesting color forms.

SEED: Not practiced to any degree since the cultivars need to be propagated vegetatively to insure trueness-to-type. If one is interested in securing seedlings the best choice would be fall planting with germination the second spring. Three months warm/3 months cold induced good germination. Another report noted 12 months warm/3 months cold was effective.

CUTTINGS: Many researchers and nurserymen have worked on rooting techniques. Results have been variable but generally 'Arnold Promise' and others have been rooted successfully at 80% plus. The biggest factor is overwintering and, after rooting, cuttings should not be disturbed. Senior author has never successfully overwintered 'Arnold Promise' although rooting success has been excellent. June-July, 1.0% IBA-solution, peat: perlite, mist is best. Firm wood has rooted well and cuttings of 'Arnold Promise' taken in early September (Georgia), treated as above, rooted 90%. The stock plants were 5-year-grafts that were pruned heavily each year, irrigated and fertilized. Cuttings are now routinely taken in August, rooted in individual cells (36/flat, bedding plant style) in about 6 weeks, removed from mist and hardened off outside for another 3 to 4 weeks under saran cloth. When outside temperatures dropped to 40°F the plants were put in an opaque poly house but losses were high. One percent IBA-solution appears to be an ideal concentration for successful rooting. 8000 ppm IBA-talc has been used frequently with good success. To avoid root disturbance, place cuttings in pots, cell packs, or 2¼ by 2¼ by 5″ bottomless rose-type pots. A hormone is necessary and improves rooting percentage as well as root quality.

GRAFTING: *Hamamelis virginiana* is the preferred understock but in the authors' minds anything but cutting propagation is foolhardy. Understocks sucker profusely and may overgrow the scion. The average gardener or nurseryworker will have a difficult time telling the suckers of the understock from the actual growth of the scion. Pot grafting (side or whip graft) in January-February or summer is practiced. T-budding is also used in August and as late as October (England). 90% take was reported on H. × *intermedia* 'Jelena', 'Pallida', and H. *mollis* 'Gold Crest'. Other grafting techniques (principally Great Britain) include chip grafting in late spring using dormant budwood collected in January. A chip graft is identical to chip budding except the plants are placed in a closed case or polytent. Any witch-hazel species can serve as rootstock for any other. Availability may be the key. *Distylium racemosum*, an evergreen species, has also been used as an understock but subsequent growth was slow. Supposedly, if the scion is grafted onto the hypocotyl of the seedling (H. *mollis*), no suckers develop.

TISSUE CULTURE: Although not perfected both H. *vernalis* and H. *virginiana* are in production in one commercial laboratory (U.S.). Shoot tips of H. × *intermedia* were cultured on MS medium plus 1% polyvinylpyrrolidone. Best results occurred with shoot tips taken in July. *Acta Horticulturae* 54:101-104 (1975).

Hamamelis japonica Japanese Witch-hazel

SEED: Variety *arborea*, 7 months warm/3 months cold stratification induced good germination. 12 months warm/3 months cold might be best. Fall plant and wait two years.

CUTTINGS: Not well known in the United States but certainly a worthy garden plant. Several cultivars are available. In general, this species is more difficult to root than H. × *intermedia* cultivars. Softwood, late May (Ireland), 0, 2000 and 8000 ppm IBA-talc, peat, mist, rooted 37, 62 and 64%, respectively. 'Flavopurpurascens', late June (Boston), 0, 3000, and 8000 ppm IBA-talc rooted 8, 0, and 8%, respectively.

GRAFTING: Variety *arborea*, 'Flavopurpurascens' and others are grafted or budded as described under H. × *intermedia*.

Hamamelis mollis Chinese Witch-hazel

SEED: Every reference pointed a different direction but warm/cold stratification is a necessity or fall plant and germination occurs the second spring. Three months warm/3 months cold stratification resulted in 88% germination. Additional warm stratification did not improve germination. One report noted that seeds planted in the fall, although taking 2 years to germinate, grew 3′ high in a single season.

CUTTINGS: Not a particularly difficult species to root and some of this genetic delight has been transferred to H. × *intermedia* cultivars. Mid-June (Boston), 0, 8000 and 10,000 ppm IBA-talc, sand: perlite, mist, rooted 10, 60 and 60%, respectively, in 5 to 6 weeks; heaviest roots occurred with 1% IBA treatment. Softwood, late May (Ireland), 8000 ppm IBA-talc, Captan soak, mist, rooted 90% in peat, 50% in 2 peat: 2 sand, and 55% in 2 sand: 1 peat. Again, Ireland, 0, 2000, 4000, 8000 ppm IBA-talc, peat, mist, rooted 75, 62, 87, and 100%, respectively. The researchers in Ireland commented that H. *mollis* was subject to heavy overwintering losses (50%). The procedure for handling H. *mollis* overwinter is similar to that described under H. × *intermedia*. Variety *brevipetala*, late June (Boston), 8000 ppm IBA-talc plus thiram, sand: perlite, mist, rooted 100% and were not disturbed, overwintered in rooting flat and all survived.

GRAFTING: See H. × *intermedia*.

Hamamelis vernalis Vernal Witch-hazel

SEED: The easiest from seed. An interesting study showed that freshly sown seed did not germinate; 70% germination after 3 months cold stratification; 75% after 3 warm/3 cold; 81% after 4 warm/3 cold; 85% after 5 warm/3 cold. Fall planting will also be rewarded.

CUTTINGS: This is the easiest species to root and keep alive. Senior author was perennially looking for good flowered forms in the midwest. Standard rooting procedure included early June cuttings, 1000 ppm IBA-solution, sand, mist, with 70 to 80% rooting in 8-12 weeks. Generally, H. *vernalis* breaks bud sooner than H. *virginiana* and can be collected and stuck sooner. Early May (Ireland), rooted 80% in 8 weeks and all survived the winter. Late May, rooted 75% in 8 weeks and 66% survived winter.

Senior author has observed continued growth after rooting which may contribute to overwintering success (through carbohydrate buildup). In fact, end buds are often not set in September (Georgia) and linear growth is still occurring. The species grows fast under good nutrition and cultural practices and rooted cuttings should make a good 2 to 3′ plant in a full growing season. Senior author prunes twice in growing season to produce a well-branched plant. This might be tried with other witch-hazels.

Not everyone has the same degree of success. One report noted 43% rooting, late June, 8000 ppm IBA-talc, in sand: styrofoam. Untreated cuttings, early summer, rooted 72% in sand in 17 weeks at 85 to 90% humidity, 40% at 65 to 70%. May (Ireland), 2 to 3″ long, from juvenile mother plants, remove basal leaves, base immersed in Captan, 8000 ppm IBA-talc. 2 peat: 1 sand, bottom heat (72°F), covered with clear plastic (no mention of mist), 75% rooting in 6 to 8 weeks. Winter survival of hardened rooted cuttings was improved by repotting in 1 peat: 3 sand medium and providing a slow release fertilizer.

GRAFTING: Not practiced to any degree with this species although several cultivars exist. This is the worst species from a suckering standpoint and grafting is really not suitable unless one is interested in two-tone shrubs. However, rooted cuttings in May (Ireland) could be used by August of the first year, although cuttings were best after 15 months from sticking. H. *vernalis* proved a better rootstock than H. *virginiana* and was readily compatible with H. *mollis*.

TISSUE CULTURE: Embryo explants through callus culture have produced embryoids. *HortScience* 12:389. Abst. 62 (1977).

Hamamelis virginiana Common Witch-hazel

SEED: Three months warm/3 months cold stratification. Fall plant with germination occurring the second spring. Also fruits collected in late August, seed planted in early October, germinated 90% the first spring. Seed sown without pretreatment did not germinate and 5 months cold stratification did not help. The key is breaking the leather-hard coat down to allow imbibition and the germination processes to proceed. Work in England (PIPPS 34:334-342, 1984) showed that 2 months warm/2 cold/2 warm/4 cold produced 88% germination; 2 warm/1 cold/½ warm/4 cold produced 84%.

CUTTINGS: Senior author's H. *vernalis* recipe has produced 2 to 5% rooting. Cuttings should be collected as early as possible. An Illinois nurseryman said this species was easy to root. Late June-early July (Ohio), untreated, sand, rooted 92% in 11 weeks. July, untreated did not root; 200 ppm IBA-24 hour soak rooted 33%. Work in Ireland (the best authors have read) indicated 75% rooting in peat, 60% in 1 peat: 2 sand, and 55% in 2 sand: 1 peat with late May cuttings, 8000 ppm IBA-talc, Captan soak, mist. Late May (Ireland), rooted 100% in 5 weeks and 70% survived the winter. Senior author collected cuttings from 15 to 20-year-old plants and maturity no doubt was a significant factor in rooting failure. The following example illustrates this point. Four-year-old seedlings forced in the greenhouse served as stock plants. The cuttings were taken in early April, 1% IBA-talc or 1% NAA—talc and rooted 93 and 100%, respectively, in 7 weeks. However, late June, 3000 and 8000 ppm IBA-talc rooted 5 and 0%, respectively. Stock plants maintained by renewal pruning will give better results than a 20-year-old plant on the edge of a clearing.

GRAFTING: See under H. × *intermedia*.

TISSUE CULTURE: Embryoid explants regenerated roots and small leaves. *HortScience* 12:389. Abst. 62 (1977).

Hamamelis postscript

The literature indicated emphatically that early cuttings rooted best and may produce new growth that assists with carbohydrate accumulation and increased winter survival. One of the more interesting studies the authors read was a thesis concerned with *Hamamelis*. Monthly timing studies were conducted on H. *japonica*, H. *mollis*, H. *vernalis*, and H. *virginiana*. H. *japonica* rooted best in May (32%), June (18%), July (10%) (northern Delaware); H. *mollis* best in May (39%), June (57%), and July (53%); H. *vernalis* best in April (42%), May (62%), June (79%), July (72%), August (44%), and H. *virginiana* best in May (49%), June (51%), July (32%). It is rather obvious that June is the best month. Tissue culture might prove exceedingly beneficial with this group.

Hedera Ivy

Ivy is so easy to root from cuttings that recipes are not really warranted. Flowers occur on mature wood in September-October and are followed by blackish drupes. The fruits are poisonous.

Hedera canariensis Algerian Ivy

CUTTINGS: Not common in the United States but used in coastal southeastern U.S. and western U.S. Larger leaf than H. *helix* but exhibits same juvenile/adult pattern. Cuttings can be taken any time of year but are best after first flush of growth (extremely soft) has firmed to a degree. 1000 ppm IBA-solution or talc, well drained medium, mist, with 90 to 100% rooting in 4 to 5 weeks.

Hedera colchica Colchis Ivy

CUTTINGS: Considerably more cold hardy than the above and just as easy to propagate. Many cultivars exist and are very prevalent in English gardens.

Hedera helix English Ivy

SEED: The fleshy outer coat must be removed. In one study, uncleaned seed sown immediately or stratified did not germinate. Cleaned seed sown at once germinated 70%. Stratification may not be absolutely necessary but a brief cold period is good insurance. The seedlings look like the juvenile form (maple leaf shape) even though the seed came from the adult stage.

CUTTINGS: About any time of year, June-July through October, might be ideal. Senior author has brushed snow off English ivy, taken cuttings, and rooted them successfully. Untreated cuttings root well but 1000 to 3000 ppm IBA-solution or talc hastens, unifies and results in a more profuse root system. Single leaf bud cuttings, approx. 2″ long, 3000 ppm IBA-15 second dip, rooted in high percentages.

TISSUE CULTURE: Internodes formed greatest callus on MS plus 0.1 or 2 mg/l BA and 0.1, 2 or 10 mg/NAA. After callus initiation, best callus development occurred at 10 or 50 mg/l BA plus 0.1 or 1.0 mg/l NAA [*Gartenbauwissenschaff* 46(3):116-119 (1981)]. Shoot apex explants led to rooted shoot tips with no shoot proliferation [*J. Amer. Soc. Hort. Sci.* 95:398-402 (1970)]. Stem internodal explants from mature plants through callus culture produced somatic embryos [*Z. Pflanzenphysiol.* 92:349-353 (1979), *Plant Physiol.* 63(5):138 (1979)]. Embryonic hypocotyl explants produced adventitious shoots and roots directly [*Planta* 145:205-207 (1979)].

Helleborus orientalis Lenten Rose

SEED: The seeds are borne in profusion in May (Georgia) and June (Boston). The fruit is a follicle and should be collected before the suture opens and the seeds abscise. The black seeds look like small BB's. Seeds require 2 months cold stratification or should be sown as soon as collected in late spring, mulched, kept moist and germination occurs in spring.

One grower collects seed as soon as ripe, provides 2 to 3 months cold stratification and then sows the seed outside in October. Germination occurs the following April. Self-sown seedlings appar in abundance in spring but never during summer or fall although the seed has been in the ground as early as late May (Georgia). German work with a Helleborus hybrid showed that fully imbibed seeds on filter paper in petri dishes germinated in high percentages after maintenance at 68°F for 2½ months followed by 41°F for 2½ months. Seed treated with GA alone did not germinate.

CUTTINGS: Although not a "true" woody plant, the leaves are evergreen and the foliage and flower beautiful. Division is the principal method of vegetative propagation and can be conducted successfully in fall or spring. The soil should be high in organic matter, moist and well drained. Shade is also a necessary cultural prerequisite. Cultivars exist and this is the preferred way to reproduce them.

Hibiscus syriacus Rose-of-Sharon

SEED: 5-valved capsules ripen in fall and can be collected, dried, and the seed removed. There are no pregermination requirements. Self-sown seedlings are often a nuisance attesting to this species' rather "weedy" nature.

CUTTINGS: June-July, 1000 ppm IBA-solution, peat: perlite, mist, root readily. Cuttings can be rooted in late summer or winter. Hardwood cuttings are successfully rooted at one large nursery by insertion into prepared soil outside in October-November. No hormone is used and rooting averages 80%. Most research indicated that a hormone is beneficial. A few examples illustrate this fact. July, untreated, sand, rooted 65%; 96% with 8000 ppm IBA-talc in 8 weeks. July, untreated 52%; 100% with 50 ppm IBA-soak, 6 hours, in 5 weeks. Terminal hardwood cuttings, October, untreated, rooted 100% in 9 to 10 weeks. However, basal or middle cuttings in comparison with tip, are improved by a hormone.

Notice in the above studies the *time* of rooting. Hardwoods take considerably longer than softwoods (July), however, softwood cuttings treated with an IBA-soak rooted much faster than those treated with IBA-talc. This is due to the IBA absorption in the proper concentration to induce early and uniform rooting.

Hippophae rhamnoides Sea Buckthorn

SEED: The drupe-like fruit has a fleshy coat that should be removed by maceration. Fruits ripen in September-October and persist through winter. Seed can be fall planted or given 3 months cold stratification. Germination percentages are generally high. The species fixes atmospheric nitrogen and is able to colonize infertile sandy soils.

CUTTINGS: A dioecious species and female forms are desirable because of the handsome orange fruit. A male (1 in 6) is necessary to effect good pollination. Stem cuttings are considered difficult but one report noted 76% rooting with 3000 ppm IBA-talc from late July cuttings. Rooting took 4 to 5 weeks. In Russian work, terminal cuttings from 7 to 10-year-old stock plants collected at the beginning of growth in spring rooted well. Other Russian work with girdling and application of IBA or IAA paste facilitated excellent rooting. After 25 days, 49 roots were present on treated shoots; while girdled, without hormone, averaged 2.8 roots. Root cuttings offer an alternative method. This species suckers profusely; a good indication that it will regenerate from root pieces.

Hovenia dulcis Japanese Raisintree

SEED: Interesting tree but not well known in commerce. Seed requires about two hours acid scarfication and will then germinate readily in 2 to 3 weeks. Seed sown without acid treatment produced a few seedlings in 9 months. Two hour acid-treated seed germinated uniformly in 16 days.

Hydrangea Hydrangea

Shrubs or small trees with magnificent flowers and totally insignificant capsular fruits containing numerous minute winged seeds. Most species are propagated from cuttings.

Hydrangea anomala subsp. *petiolaris* Climbing Hydrangea

SEED: Most hydrangea seed germinates without pretreatment but 2 to 3 months cold stratification induces best germination of this species. Seeds should be germinated in flats in a greenhouse. An interesting anomaly occurred when untreated seed was planted in a fog house structure. Germination was quite good.

CUTTINGS: Not exactly akin to the other hydrangea species when it comes to ease of rooting. Senior author has tried many times with minimal success (5%). Optimum time to secure cutting wood is late spring/early summer before stems turn brown. Treat with 8000 to 10,000 ppm IBA, peat: perlite, mist, rooting takes 5 to 7 weeks.

In one interesting study, terminal, mid-June (Boston), 8000 ppm IBA-talc plus thiram, peat: perlite, mist, 75°F bottom heat, rooted 80% in 5 to 6 weeks. The cuttings showed interesting root development patterns with roots developing from the base of the cuttings as well as internodal roots that coincided with the aerial root-like holdfasts. Most plants with aerial root formation (*Hedera*, *Euonymus*, etc.) root readily but that is not the case with this plant.

One nursery (Rhode Island) roots late June cuttings in sand under mist outdoors. Roots do not appear until September and cuttings must be protected against bark splitting. Rooted cuttings are lifted, flatted in peat: perlite, and placed in a 40°F cold storage unit. In March they are potted and maintained in a cold frame until the following spring.

Overwintering can be a problem and the earlier cuttings are rooted the better the chance for survival. In England, soft cuttings ¾ to 3″ long are taken in April-May, 3000 ppm IBA-talc, peat: sand, mist, 65 to 70°F bottom heat with 80% success in 4 to 5 weeks. Cuttings are transplanted after weaning from mist and by fall are 6″ high. They are overwintered at 45°F with excellent survival. Direct sticking into containers (2¼ by 2¼ by 5″ or other rooting containers) would avoid transplanting and root disturbance.

Hydrangea arborescens Smooth Hydrangea

SEED: No pretreatment; seeds are small and should be sown in flats in the greenhouse.

CUTTINGS: Softwoods are easy to root. June-July, 1000 ppm

IBA-solution or -talc, peat: perlite, mist, root in 3 to 4 weeks in high percentages. Might be better to root H. *arborescens* early to insure additional growth and overwintering success. This species also suckers or forms colonies and can be divided in late winter or early spring and containerized or field planted. Hardwood cuttings are used commercially, usually in November. An Illinois nurseryman maintains fertilized stock blocks and cuts the long canes in fall, saws them into 8″ lengths with a bansaw, plants them about 6½″ deep and has good rooting by spring. In another study, hardwood cuttings benefited from a 1000 ppm NAA treatment (75% rooting); untreated cuttings rooted 38%. 'Annabelle', a superior form, is now extensively grown by the nursery trade. Summer softwoods are used primarily. Senior author used 5000 ppm K-IBA on early July cuttings with 100% rooting in 3 weeks. Early December hardwoods, 3000 ppm IBA-talc, open bench (cool top, bottom heat) or polytent, rooted 100%.

Hydrangea macrophylla Bigleaf Hydrangea

SEED: No pretreatment necessary.

CUTTINGS: May, June, July, terminal, 1000 ppm IBA-solution or talc, peat: perlite, mist, root 90 to 100% in 3 to 5 weeks. Cuttings can be taken virtually any time of year...soft, firm or hard and will root with ease. Leaf bud cuttings are also successful. In several studies, July untreated or IAA-soaked rooted 100%. October cuttings benefited from hormone treatment.

A large Rhode Island nurseryman rooted this species in outdoor polytents with good success. Mid-July cuttings, 8000 ppm IBA-talc, soil: perlite, no mist but well moistened cuttings and medium, shaded poly. Rooted cuttings are not disturbed and allowed to overwinter in place. Also, 100% rooting has been obtained from softwood cuttings under high humidity systems (fog). Indiana study indicated 5000 to 10,000 ppm IBA-talc, sand, mist, produced good rooting of several florist cultivars.

TISSUE CULTURE: Meristems were grown into young plants on modified MS or Knop media (French reference). Shoot tip explants produced axillary bud proliferation. Plant cell and tissue culture: Principles and applications. Ohio State Univ. Press, Columbus. Sharp et al. (eds), pp. 441-452 (1979). *HortScience* 19:717-719 (1984).

Hydrangea paniculata Panicle Hydrangea

SEED: No pretreatment required.

CUTTINGS: May, June, July, terminal, 1000 ppm IBA-solution or talc, well-drained medium, mist, 90 to 100% rooting in 4 to 5 weeks. It is almost impossible not to succeed with this plant. Like H. *macrophylla* it can be rooted from soft, firm or hardwood cuttings. A hormone in most published work was shown to be beneficial. Untreated June/July cuttings rooted 20/40%; 80/100% with IBA treatment. One grower reported that 8000 ppm IBA-talc was too strong for summer softwoods. December hardwoods also benefitted from hormone treatment.

Hydrangea quercifolia Oakleaf Hydrangea

SEED: No pretreatment necessary, fresh seed germinated profusely in two weeks.

CUTTINGS: Probably the second most difficult species to root after H. *anomala* subsp. *petiolaris*. July, untreated, rooted 10%, 100% with 4000 ppm IBA-talc in 5 to 6 weeks. Overwintering can be a problem with the species and rooted cuttings should not be disturbed. May, June, July, terminal (perhaps reduce leaf area), 3000 to 5000 ppm IBA-talc or solution, well drained medium, mist, provides the best recipe.

Root cuttings, February-March, ¼″ by 2″ long, pot vertically, greenhouse (possibly under polytent).

Several beautiful cultivars are known and these must be propagated by stem or root cuttings. Senior author used 5000 ppm K-IBA, early June terminal cuttings, perlite or peat:perlite, mist, greenhouse, 100% success. Cuttings were transplanted, given Osmocote 18-6-12 and continued to grow.

Hypericum St. Johnswort

Wonderful shrubs with rich green to bluish green foliage and delightful yellow flowers of various sizes. Most flower in spring-summer (and into fall). The fruit is a dehiscent capsule with small petunia-ish seeds. Collect capsules in September-October, dry inside, extract seeds. Senior author has germinated seeds of 30 species with 100% success. Seed has no pregermination requirements and germinates 2 to 3 weeks after sowing. Authors can never remember a question in the Plant Propagator's Question Box relative to the rooting of cuttings. All are rooted easily from June-July to October cuttings, terminal, 1000 ppm IBA-talc or solution (alcohol will burn soft cuttings), peat: perlite, mist, root in 4 weeks, transplant readily and will continue to grow if fertilized. Hardwood cuttings (England) are also used.

The following species have been propagated from seed or cuttings by senior author. Included are general observations on performance.

Hypericum androsaemum Tutsan

Grown from seed (Georgia); all died in landscape from wilt-type disease. Rather pretty red changing to black fleshy capsule. Flowers are small and insignificant. A weed in Europe.

Hypericum calycinum Aaronsbeard St. Johnswort

One of the most popular ground cover species with 2½″ diameter bright yellow flowers. Foliage will discolor (purple-brown) in cold climates. Used extensively in Pacific Northwest. An English report noted that this species does not do well in containers although it is commonly grown that way in the U.S. Many ground cover growers root this and other hypericums in cells or individual pots. In senior author's experiences, excessive moisture in rooting media or landscape situations will result in death or decline. This species flowers on old wood and if stems are killed during winter, even though new shoots develop from crown, no flowers appear (based on Georgia observations).

Hypericum frondosum Golden St. Johnswort

Southeastern native found on rocky cliffs, etc., with handsome foliage and flower. 'Sunburst' has 2″ diameter bright golden yellow flowers on a compact 4′ by 4′ shrub with rich bluish green foliage. Wood needs to be firm (July) before taking cuttings (senior author's observation). This cultivar withstood −20°F in Illinois.

Hypericum kalmianum Kalm St. Johnswort

Hypericum prolificum Shrubby St. Johnswort

Probably the two hardiest species. Flowers average about one inch in diameter and are borne in great profusion in June-July. Both grow 3 to 4′ high and wide. Stem damage was not observed on these species even after exposure to −20°F.

Hypericum × moseranum Moser's St. Johnswort

Hybrid between H. *patulum* and H. *calycinum* growing 1½′ tall. 2 to 2½″ diameter flowers appear in July through October on growth of the season. Used extensively in Europe; not common in U.S. 'Tricolor' is a rose-white variegated leaf form that does not possess the vigor of the species.

Hypericum olympicum Olympic St. Johnswort

Amazing gray-green leaf form with 1½-2″ diameter bright yellow flowers. Interestingly, injured in Boston, but flowered on new growth. Flowered one year in senior author's Georgia garden but literally melted out in summer heat.

Hypericum patulum 'Hidcote'

Common form that can be treated as a herbaceous perennial from Zone 5 to 8. Large, 3″ diameter golden yellow flowers appear on new growth of the season. Senior author always cuts shrub to within 6″ of ground in March, fertilizes and by early June (GA) has a wealth of rich foliage and flowers. Apparently, this cultivar sets no fruit which may explain its long flowering period into October (GA). Senior author conducted an experiment with this form using different hormone formulations. Untreated cuttings [mid-September (Georgia), 5 to 6″ long, terminal cuttings, peat: perlite, mist, rooting evaluated after 4 weeks] rooted 95% but had an average of 20 roots/cuttings; 3000 ppm IBA—solution or 1000 ppm NAA plus 2000 ppm IBA-solution produced 100% rooting and an average of 50 roots/cutting. See *The Plant Propagator* 29(3):6-7 (1983).

Iberis sempervirens Candytuft

SEED: No pretreatment necessary. Seeds germinate readily.

CUTTINGS: June-July-August, 1000 to 3000 ppm IBA-talc, root readily under mist. Wood should be firm. No doubt, cuttings taken later in season also root well.

Idesia polycarpa

SEED: Macerate mealy fruits, remove seeds and sow. Seed requires no pretreatment. Germination is heavy as most seeds are sound. Senior author has raised thousands of seedlings. In fact, whole fruits were planted and germination still proceeded uniformly.

CUTTINGS: Early August, probably young plants, wound, 8000 ppm IBA-talc, bottom heat, rooted 92% in 4 weeks. Late July (Amherst, MA), untreated, sandy soil, rooted 40%; 76% after 25 ppm IBA-16 hour soak.. Cuttings would be the logical way to proceed since the species is dioecious with the female producing long pendent racemes of bright red fruits. Authors do not know whether cuttings from mature (fruiting) trees root easily. Certainly worth the try!

Ilex Holly

Evergreen or deciduous trees with insignificant creamy flowers on male and female plants. Drupaceous fruits are often spectacular and range in color from white, yellow, orange, red to black. Each fruit contains one or more hard-coated "seeds" called a pyrene. The pyrenes should be removed by maceration from the mealy flesh. Germination of holly seed is not easy and may take 2 to 3 years. In general, the hard impermeable seed coat and an immature embryo contribute to dormancy. Authors have covered numerous references and queried a number of growers and breeders relative to the best germination procedures. The ultimate answer is patience. Sow the seed in a suitable medium in a flat, moisten, place in plastic bags, leave on shelf in a room (warm temperature 60 to 80°F), shaded bench, etc., and wait until seedlings emerge. At this time, remove bag or covering, place under normal germination conditions. Senior author followed this approach with *Ilex × attenuata* 'Foster's #2', *Ilex cornuta* 'Burfordii' and others. Eighteen months after placement in medium the first seedlings emerged.

Cuttings are the preferred method but all species and cultivars do not root with equal facility. I. *aquifolium*, I. × *altaclarensis*, I. *latifolia*, etc., are more difficult than I. *cornuta*, I. *crenata* and others. None, however, should be put on the difficult-to-root category like *Chionanthus virginicus*. The literature is absolutely crammed with references relating to cutting propagation of *Ilex*. The following general recipe is assembled from many sources. Late June to February with perhaps late August to November ideal, terminal cuttings with firm wood (soft cuttings do not respond well), many growers use one-year-old wood with a portion of two-year-old wood, fatter (larger) cuttings are better than skinny ones; hormone is important and 1000 to 3000 ppm IBA-solution or talc for easy to root types, 1 to 2% IBA for I. *opaca*, I. *aquifolium*, etc.; wound on the hard-to-root types is beneficial, double wound is recommended; medium should be well drained, peat: perlite, peat: sand and others work well; 70 to 75°F bottom heat is beneficial; mist or polytent are necessary. Many references referred to high humidity during the rooting process and its importance. Rooting takes place in 4 to 8 weeks (longer for certain types) and rooted cuttings can be successfully transplanted after rooting or left in place.

Ilex × altaclarensis (I. *aquifolium* × I. *perado*) Altaclara Holly

CUTTINGS: Not extensively grown in the U.S. although 'Camelliifolia', 'James G. Esson', 'J. C. Van Tol' and 'Wilsonii', all female clones, are certainly handsome broadleaf evergreen hollies. Cuttings should be treated as described under *Ilex*. High hormone (1 to 2% IBA-talc) is necessary.

Ilex aquifolium English Holly

SEED: Difficult. Fall plant and wait 2 to 3 years or follow approach in the *Ilex* discussion. A 12 month warm/3 month cold produced 3% germination. Embryos incubated on agar medium plus sucrose required 55 days to reach germination size [JASHS 100:221 (1975)].

CUTTINGS: Numerous cultivars exist. The procedure is basically as described. One California grower uses 4″ long terminal cuttings, strips all but two upper leaves, September through February or early March, fungicidal treatment, 7000 ppm IBA-10 second dip, sand: peat or 2 perlite: 1 peat in flats under mist in greenhouse with 65°F bottom heat. Rooting takes 10 to 12 weeks and 15 weeks is considered ideal for heavy root system production. Percentage rooting is seldom less than 100% and 90% is considered low. Nurseryman maintains stock trees under high nutritional status. Late November (Oregon), 5-6″ long, with 4 to 5 leaves, 2-sided, 1½″ long wound, perlite, 75°F bottom heat in plastic greenhouse with 50°F minimum air temperature, 12 hour soak in 50 ppm IBA-water or same plus 22 ppm boron. Rooting was consistently 95% or greater for both treatments. Root number and length were generally better with the boron treatment. English nurserymen have conducted considerable research with the species and cultivars. Much of their work indicated late August to December as the ideal time for taking cuttings. The rest of the procedures are standard except they root under poly, double glass, and mist.

TISSUE CULTURE: Zygotic embryo explants led to direct embryogenesis from cotyledons [*Phytomorph*. 21:103-107 (1972)]. Seed embryo explants led to direct embryogenesis [*Proc. Holly Soc. Amer.* 54:5-6 (1977); *Z. Pflanzenphysiol.* 89:41-49 (1978)].

Ilex × *aquipernyi* (*I. aquifolium* × *I. pernyi*)

CUTTINGS: Largely as described for *I. aquifolium*. Several good cultivars are known. Early October, 1% IBA-talc, resulted in good rooting.

Ilex × *attenuata* 'Foster's #2' (*I. cassine* × *I. opaca*)

SEED: Senior author germinated hundreds of seedlings from Foster's #2. Interestingly, *all* look like *I. opaca* in leaf morphology. None have reached the fruiting stage so their future performance should be interesting. It took 18 months for the first seedling to show after placement in warm stratification. Senior author is not convinced most hollies require a cold period to germinate. The under-developed embryo requires an extended period of after-ripening and once this has occurred it appears the seed can germinate. None of the seeds were exposed to any cold.

CUTTINGS: Much like *I. opaca* in requirements. Late September, October, November are prime times according to one nurseryman. Senior author rooted this form in early March, 8000 ppm IBA-quick dip, wound, peat: perlite, mist with 85% success.

Ilex cassine Dahoon

CUTTINGS: Best handled like *I. aquifolium*. One report noted that it was not difficult to root.

Ilex ciliospinosa

CUTTINGS: Early April, 8000 ppm IBA-talc plus thiram, poly, rooted 92%. Late November, same treatment, rooted 90%.

Ilex cornuta Chinese Holly

SEED: Handle like *I. attenuata* 'Foster's #2' (*q. v.*). In this case, germination was not as good.

CUTTINGS: The Chinese hollies are not difficult to root. Authors doubt seriously whether anyone is rooting the species since cultivars 'Burfordii', 'Carissa', 'Dazzler', 'Dwarf Burford', 'Rotunda' and others dominate in the market place. In general, the cultivars are easy to root. Senior author has taken cuttings of 'Burfordii', 'Dwarf Burford', 'D'Or' as late as April (GA), 1000 to 3000 ppm IBA-solution, peat: perlite, mist, with 90% plus rooting in 6 to 8 weeks. Cuttings can be transplanted easily, fertilized and may make two or more growth flushes during the year. A large grower takes cuttings of Burford and others after the growth flush has hardened. Two to 3 flushes can be expected and might coincide with June, August, October if spring is early (south) or July, September, November, if late. One of the keys to success is firm wood, at least as firm as possible after each flush has ceased elongation. These cuttings can be rooted under poly and shade in the field, mist or any suitable structure. A hormone is definitely warranted. Rooting may take place in 2 to 3 weeks with 4 being normal.

Authors have always been of the opinion that hollies prefer a well drained medium and the only controlled study that more or less confirmed this was the following: Burford was rooted in 1 bark: 1 sand (20%), 2 bark: 1 sand (13%), sand (8%), 1 peat: 1 perlite (65%), 1 peat: 1 sand (58%). The results pretty much speak for themselves. An Alabama nurseryman uses 3750 ppm IBA plus 750 ppm NAA in alcohol for 'Rotunda'.

GRAFTING: 'Burfordii' has been utilized as an understock for many *Ilex* species. Where root rot is problematic, Burford could be used successfully as an understock especially for *Ilex aquifolium* cultivars. Unfortunately, Burford, especially in larger sizes, is somewhat difficult to transplant. A good discussion on holly grafting is presented in *The Plant Propagator* 16(4):11-15 (1970).

TISSUE CULTURE: Embryo explants produced callus and/or direct embryogenesis. *Plant Physiol.* 49:31. Abst. 174 (1972).

Ilex crenata Japanese Holly

SEED: Black fruits should be treated as described under *Ilex*. Clean, plant seed and wait. This species does not take as long as *I. opaca* and others. Seed of 'Convexa' germinated 76% without pretreatment, 90% after 3 months cold stratification. Two months warm/3 months cold stratification produced 90% germination.

CUTTINGS: There are numerous cultivars varying in shape and foliage characteristics. Most root easily from cuttings handled like *I. cornuta*. The species and cultivars grow in flushes and cuttings should be taken as each flush hardens. 1000 to 3000 ppm IBA-solution is ideal. Senior author has collected cuttings as late as March with good success. Rooting percentages should approach 100%. 'Highlight' has been a disappointment with an average around 10%. 'Rotundifolia' rooted best with 5000 ppm IBA-dip. Senior author noted a range of rooting responses from cuttings taken in early March (GA), 8000 ppm IBA-dip, peat: perlite, heat, mist, with 'Compacta' rooting 100%, 'Helleri' — 60% and 'Hetzii' — 40%.

Ilex decidua Possumhaw

SEED: Patience. Follow recipe described under *Ilex*.

CUTTINGS: This is a deciduous species and is not particularly easy to root and two great growers describe two seemingly opposite methods for successful cutting propagation. Early July (southern Indiana) as growth begins to harden. The proper degree of wood maturity is critical. 5 to 6" long, 7500 ppm IBA-solution, peat: polystyrene in flats, greenhouse, mist, with rooting occurring in 4 to 6 weeks. Rooted cuttings are transplanted to 2½ peat pots, carried in polyhouses with a combination of mist and fog, and wintered at 33 to 35°F. The plants are carried under lath for another season and planted out in the 3rd year.

Another method involves 8 to 10" cuttings after 2 or 3 frosts or by mid-October (southern Illinois), leaves removed, 8000 ppm IBA-talc, peat: perlite, 72 to 75°F bottom heat, in a polyhouse, the beds enclosed in a polytent with rooting occurring in 4 to 6 weeks. Root growth is slow and top growth occurs about the same time. If cuttings are taken too late, top growth starts before rooting and all is lost. Plants are maintained in these beds until April-May, then transplanted to one-gallon containers containing 5 bark: 1 sand medium and fertilized.

All cultivars do not root with the same facility and 'Byer's Golden' (yellow-fruited) is whip-grafted onto established I. *decidua* understocks. In a controlled study testing media effects, peat: perlite was superior to sand, bark and their combinations. Interestingly, a yellow-fruited form treated with 8000 ppm IBA-talc plus thiram rooted 100%.

Ilex glabra Inkberry

SEED: Same treatment as others but this species may germinate faster. Fresh seed germinated 57%. Stratification treatments reduced germination.

CUTTINGS: Much like I. *cornuta* and I. *crenata* in rooting requirements. Senior author has rooted firm wood, 1000 ppm IBA-solution, peat: perlite, mist, with 90 to 100% success. 'Compacta' appears easier to root than species. One nurseryman roots the species from late August-early November with excellent success. Again, wait until growth firms before taking cuttings. Rooted cuttings should be 12 to 15" (18") high by their second season. July to February-March cuttings can be rooted successfully. Senior author has succcessfully rooted this species from July to early March using firm wood. Alabama, sand, mist, rooted 41, 96, 95 and 86%, respectively, after treatment with 0, 3000 ppm IBA-talc, 8000 ppm IBA-talc or Chloromone. Best root systems occurred at 8000 ppm IBA-talc.

Ilex integra

SEED: 5 months warm/3 months cold produced good germination.

CUTTINGS: December, untreated, did not root; 100% rooting with 1.0% IBA-talc in 8 weeks.

Ilex × koehneana 'Wirt L. Winn'

SEED: 5 months warm/3 months cold produced good germination. Fall planted seeds would take 2 years to germinate.

CUTTINGS: Somewhat difficult. Senior author has had little success. Suspect it requires high hormone, wound, etc., as described for I. *aquifolium* (*q. v.*). December, untreated, rooted 20%; 100% after treatment with 4000 ppm IBA-talc. Early March (GA), wound, 8000 ppm IBA-quick dip, peat: perlite, mist, rooted 40% in 9 weeks.

Ilex latifolia Lusterleaf Holly

CUTTINGS: Generally difficult to root. Alabama report indicated late June, 5-6" long, 1 sand: 1 vermiculite, mist, 70°F bottom heat, dipped 10 seconds in a mixture of 5 ppm Bayer (aspirin) plus 125 ppm IBA plus 1000 ppm NAA rooted 80%. Commercial nurseryman (Alabama) uses 10,000 ppm IBA plus 1500 ppm NAA in isopropyl alcohol with good success.

Ilex × meserveae Blue Hollies, Meserve Hybrids

SEED: Most of these cultivars are patented and cannot be reproduced vegetatively unless growing rights are granted by holder of patent. Each plant also carries a royalty fee. One way to circumvent this might be the production of seed-grown material from the female members of the clan with subsequent selection of superior types. Authors abhor the number of cultivars already on the market and believe this would be a good way to keep the confusion in high gear.

CUTTINGS: This group of hybrids between I. *rugosa* (hardy parent) and I. *aquifolium* resulted in 'Blue Boy', 'Blue Girl', 'Blue Prince', 'Blue Princess', 'Blue Angel', 'Blue Maid' and 'Blue Stallion'. New cultivars with I. *rugosa* as one parent include 'China Boy', 'China Girl', and 'Dragon Lady'. All can be rooted from cuttings treated as described for I. *cornuta*. August to November is possibly the best time to root cuttings; however, in Georgia work, cuttings have been rooted 80% and above year-round as long as firm wood and a hormone are used.

Ilex 'Nellie Stevens'

CUTTINGS: Easy to root, 3000 to 5000 ppm IBA from August to March promotes good rooting. Senior author has used 8000 ppm IBA-alcohol dip on early March cuttings in peat: perlite, mist, with 80% success in 9 weeks. In 1986, cuttings were taken June 3, 5000 ppm IBA-solution, peat: perlite, mist with 100% rooting in 6 weeks.

Ilex opaca American Holly

SEED: See general discussion under *Ilex*. It is rather obvious as one walks through the southern woods that birds do a good job disseminating seed, for seedlings in all shapes and sizes are everywhere. This species takes a long time to germinate. If fall planted, seed may germinate over a 2 to 3 year period. 12 months warm/3 months cold gave reasonable germination.

CUTTINGS: Follow directions under I. *aquifolium*. Several points should be emphasized. Wounding is definitely beneficial. In early December senior author received hardwood cuttings of 'Jersey Knight' and 'Jersey Princess', a one-inch wound was applied to one side, 3000 ppm IBA-solution, peat: perlite, mist, 75°F bottom heat. In 6 weeks a number of cuttings had rooted with roots originating primarily from wounded area creating a one-sided rooted cutting. A double wound may be the best way to proceed with this species. Cuttings can be taken from late August to February with optimum period falling between September-November. Early January, NAA and NA-acetamide, rooted 95 to 100% in 6 to 7 weeks compared to 30% for untreated cuttings. The use of a hormone is essential for good rooting. Early October (Illinois), 3 photoperiods × 3 IBA concentrations, peat: perlite, mist, rooted 90% with or without hormone under natural and 8 hr. light photoperiod. 1 or 2% IBA-quick dip improved quality of rooting but not percentage.

Clonal differences in rooting are known but, in general, with attention to detail rooting can be successful for most clones. One reference to a 100-year-old American holly noted the cuttings rooted 90 to 95% every year. A smart propagator will select good foliage and fruiting types that are easy to root.

Ilex pedunculosa Longstal Holly

SEED: See *Ilex*. Apparently quite difficult to germinate.

CUTTINGS: One report in literature indicated this species was difficult to root but senior author has had excellent success. July/August and February (Illinois), 1000 ppm IBA-solution, peat: perlite, mist, rooted 80% or greater. Summer wood should be firm. Early summer, untreated, sand, rooted 96% in 16 weeks. Late December, untreated, sand: peat, 83% rooting in 23 weeks; IBA soak caused injury.

Ilex pernyi Perny Holly

SEED: See *Ilex*.

CUTTINGS: Variety *veitchii*, early summer, sand, rooted 76% at 85 to 90% humidity; 64% at 65 to 70%. November, untreated, rooted 54%; 100% after 60 ppm IBA-24 hour soak. Senior author has not had good success.

Ilex rugosa Prostrate Holly

SEED: See *Ilex*.

CUTTINGS: Handsome, rough veined, dark green leaf species that has been used as the cold hardy parent in the production of the "Blue" hollies (*I.* × *meserveae*). Unfortunately, it does not appear to be heat tolerant. Senior author has had good success with November cuttings. Another researcher had 100% success with untreated November cuttings. Rooting took 5 months. In same experiment, cuttings treated with 100 ppm IBA-18 hour soak rooted 100% in 3 months.

Ilex serrata Japanese Winterberry

SEED: Three months cold stratification produced good germination in 7 days. Without pretreatment, only one seed germinated in 7 months.

CUTTINGS: See *I. verticillata* for details. June after growth starts to harden is the ideal time. Late July, 8000 ppm IBA-talc plus thiram, peat: perlite, mist, rooted 93%. The white and yellow fruited forms are also easily rooted. Rooted cuttings present no overwintering problem.

Ilex sugerokii Sugeroki Holly

CUTTINGS: Mid-August and late November, rooted 20 and 40%, respectively, when treated with about 1.0% IBA-talc.

Ilex verticillata Common Winterberry

SEED: Senior author provided three months cold stratification and had 23% germination. The same treatment in a later experiment did nothing. Five months warm/3 months cold stratification produced 74% germination.

CUTTINGS: Perhaps the easiest of the deciduous hollies to root. If in doubt as to the best approach for *I. amelanchier*, *I. laevigata*, *I. longipes*, *I. montana* and others follow the directions described here. Senior author has had excellent success with June-July cuttings, terminal, 1000 ppm IBA-solution, peat: perlite, mist, 90 to 100% rooting in 6 to 8 weeks. The same procedure used for June (Indiana) *I. decidua* cuttings is also used successfully for cultivars of this species. Plants 10 to 30" tall can be expected from these June cuttings. In two years, heavy fruited, well branched plants are ready for sale. The definitive study (PIPPS 24:454) looked at soft and hardwood cuttings. Hardwood rooted about 14%. Mid-June cuttings rooted 82% in peat and 67% in peat: sand. 7500 ppm IBA-talc proved optimum. Mist and polytent were equally effective.

Ilex vomitoria Yaupon

SEED: Germinate quicker than *I. opaca* but still require same patience as outlined under *Ilex*. Thirty-minute acid treatment and direct sowing produced 63% germination. Untreated seed germinated 11%. Various warm/cold treatments resulted in 28% germination.

CUTTINGS: Reasonably difficult. Senior author has had no better than 50% success with October-November cuttings, 3000 ppm IBA-solution, peat: perlite, bottom heat or no bottom heat, mist. A female plant of the species and 'Nana' were used. For southern nurserymen this has been a difficult plant to produce consistent rooting year to year. One report (Alabama) mentioned close to 100% success with October cuttings of the species and 'Pendula' in poly houses. All cuttings were taken from young, healthy, vigorous container plants. Interestingly, early March cuttings of 'Nana', 'Pendula', 'Stokes Dwarf' and a heavily pruned plant of the species, 8000 ppm IBA-quick dip, peat: perlite, mist, rooted 30, 20, 30 and 100%, respectively, in 9 weeks. November (GA) cuttings, hormone, perlite, mist, rooted quite well. This species may be alcohol sensitive and the K-salt of IBA can be used to possible advantage.

Ilex yunnanensis Yunnan Holly

SEED: Five months warm/3 months cold produced good germination.

CUTTINGS: Foliage reminds of *I. crenata* but the fruits are red. Easily rooted from October-November cuttings, 3000 ppm IBA-talc, sand: peat, mist or poly. Late December, untreated, rooted 80% in 26 weeks; 80% with 50 ppm IBA-24 hour soak in 16 weeks.

Illicium anisatum Japanese Anise-tree
Illicium floridanum Florida Anise-tree

SEED: Fruit is a star-like aggregate of follicles that contain small brown seeds. Collect in fall, dry and shake out seeds. Seed supposedly requires no pretreatment. At Callaway Gardens (Georgia), *I. floridanum* has seeded in to the point of weediness. Senior author has planted seed of *I. floridanum* immediately after extraction from the fruits but germination did not occur.

CUTTINGS: July (Athens), 3000 ppm IBA-dip, leaves cut, bark, mist, rooted 95% in 6 weeks. September, same treatment, rooted similarly. Cuttings should be firm.

Illicium parviflorum Small Anise-tree

CUTTINGS: The three evergreen species are rooted easily from June-July to November (Georgia) and into March-April. Avoid extremely soft new growth. Senior author has rooted the above two but most work was conducted with *I. parviflorum*. Terminal cuttings taken in early July and late August (Georgia), 3000 ppm IBA-solution, peat: perlite, mist, rooted 100% in 4 to 6 weeks. They transplant easily and make excellent growth in a pine bark medium when properly fertilized. Root cuttings will also work as plants tend to sucker, especially *I. floridanum*.

Indigofera Indigo

Low growing shrubs that are almost unknown outside of botanical gardens. Rather pretty pinkish or rosy purple flowers appear in June-July on the growth of season. *I. kirilowii* may be knocked to the ground in cold winters but regenerates and flowers. Fruit is a pod with small hard seeds. Impermeable seed coat is the only barrier to successful germination. Stem and root cuttings work well and division is effective if a few plants are needed.

Indigofera amblyantha Pink Indigo

SEED: Hot water (190°F), 24-hour soak induced good germination.

CUTTINGS: Mid-December, untreated, sand: peat, rooted 18%; 64% after treatment with 50 ppm NAA-20 hour soak. IBA did not prove as effective as NAA.

Indigofera gerardiana Himalayan Indigo

SEED: Hot water (190°F) soak produced 75% germination.

CUTTINGS: June-July root readily, 1000 ppm IBA-talc or solution, well drained medium. Division of parent plant works for small numbers.

Indigofera incarnata Chinese Indigo

SEED: Hot water soak induces good germination.

CUTTINGS: Late July, 4000 ppm IBA-talc plus thiram, rooted 100%. Late June cuttings of variety *alba*, 3000 ppm IBA-talc, rooted 100% in 2 to 3 weeks. June-July-August cuttings root with ease.

Indigofera kirilowii Kirilow Indigo

SEED: Hot water soak.

CUTTINGS: Late June, 3000 or 8000 ppm IBA-talc, 100% rooting.

Itea ilicifolia (evergreen) Hollyleaf Sweetspire

CUTTINGS: Not grown in U.S. to any degree, perhaps in Pacific Northwest. Cuttings in September, wound, 8000 ppm IBA-talc, well drained medium, root 80 to 90%. This Chinese species has a holly-like leaf and long racemes of greenish white flowers in late summer. It is so different from the following species that its familial relationship is hard to believe.

Itea japonica Japanese Sweetspire

SEED: No pretreatment necessary and seeds (extremely small) should be sown lightly on well drained medium. Fruit is a dehiscent capsule and should be collected in October, dried inside on paper, and the dust-like seeds extracted.

CUTTINGS: 'Beppu' ('Nana') is the common form in cultivation in the U.S. A small (3'), almost evergreen (Georgia) shrub but decidedly deciduous to the point of herbaceousness at −10°F. June to November, 1000 ppm IBA-solution, peat: perlite, mist, root 100% in 3 to 4 weeks. It is a suckering shrub and can be divided if a few plants are needed. Authors doubt that there is a time of year when plant cannot be rooted successfully.

Itea virginica Virginia Sweetspire

SEED: No pretreatment necessary. Seeds are exceptionally small and young seedlings lack vigor. Careful attention to watering must be provided after germination.

CUTTINGS: As above. One report noted early July, untreated, rooted 100% in 4 weeks with or without treatment. Senior author collected late August cuttings from wild populations, carted them around in a plastic bag for a day until the leaves looked like wilted lettuce, treated them with 3000 ppm IBA-solution, peat: perlite, and had 100% rooting. This is a fantastically easy plant to root. Collect cuttings when growth is firm. Cuttings collected in November rooted. Hardwood cuttings will root. Root sytems are exceptionally profuse. Cuttings will continue to grow in rooting bench.

Jasminum Jasmine

Shrubs, often tender, with white, pinkish or yellow flowers. Some species have fragrant flowers but the hardier types J. *floridum*, J. *mesnyi*, and J. *nudiflorum* do not. Fruit is a berry which senior author has not observed on plants in cultivation. Softwood cuttings root easily.

Jasminum floridum Showy Jasmine

SEED: Little information is available but one reference noted that fresh seed germinated upon sowing.

CUTTINGS: It is more difficult not to root this plant than to root it. June to October terminal cuttings, 1000 ppm IBA-solution, peat: perlite, mist, root 90 to 100% in 3 to 4 weeks. Root systems are profuse and transplanting is no problem. Rooted cuttings continue to grow after transplanting.

Jasminum mesnyi Primrose Jasmine

CUTTINGS: See J. *floridum*.

Jasminum nudiflorum Winter Jasmine

CUTTINGS: Reports indicate this species may be more difficult than J. *floridum*, but senior author has rooted the species using same format described under J. *floridum*. Late June, 3000 ppm IBA-talc, rooted 100%. July, 1000 ppm NAA-talc, rooted 78%; 100% after treatment with 1000 ppm naphthaleneacetamide-talc. A hormone generally improves rooting of this species although English literature indicated it was not necessary. Summer softwoods rooted 90 to 100% without treatment. Hardwood cuttings in November also work.

TISSUE CULTURE: CRSAS III 293(6):343-346 (1981).

Jasminum officinale Common White Jasmine

CUTTINGS: Use same recipe described for J. *nudiflorum*.

TISSUE CULTURE: Leaf and stem section explants led to direct regeneration of plants; stem apex explants through callus culture led to shoot regeneration. *Compt. Rend. Acad. Sci. Paris* 288D:323-326 (1979).

Juglans Walnut

Trees with monoecious flowers. Fruits (nuts) covered by a thick, indehiscent husk. Fruits can be collected in fall after they abscise from the trees. Husks may be mechanically removed or allowed to soften and removed by hand. Seeds can be stored for long periods at high humidity and 34 to 38°F. Seeds of most species possess a dormant embryo and fall sowing or 3 to 4 months cold stratification are necessary.

Juglans cinerea Butternut

SEED: Fall plant or 3 to 4 months cold stratification.

Juglans nigra Black Walnut

SEED: Fall plant or 3 to 4 months cold stratification.

CUTTINGS: Etiolated shoots from juvenile plants, girdled, 1% IBA in lanolin in a 4/5″ long band above girdle, covered with aluminum foil, severed from plant, rooted 55% (average) from April to June [*The Plant Propagator* 19(2):13-14 (1973)]. Kansas researcher rooted cuttings from seedling plants with some success. Cuttings from older trees did not root as well. Anyone wildly interested in black walnut propagation should see *The Plant Propagator* 18(3):4-8 (1972).

GRAFTING: Most cultivars are bench grafted on seedling understock. Shield or T-budding might be tried.

Juglans regia English Walnut

SEED: Fall plant or one to five months cold stratification. The cold requirement is probably considerably lower than that of *J. nigra*.

CUTTINGS: Softwoods have been rooted, especially those used for rootstocks. The rooted cuttings cannot be disturbed after rooting. Survival is poor.

GRAFTING: One nurseryman fall plants the seed, undercuts seedlings in spring to get branched root system, T-buds in late August with good success. Budwood is taken from current season's growth on trees that were irrigated.

TISSUE CULTURE: Endosperm explants through callus culture produced adventitious shoot buds and roots. Plant tissue culture 1982, Proc. 5th Int. Cong. Plant Tiss. Cell Cult., Japan. Jap. Assn. Plant Tissue Cult., Tokyo, Fujiwara (ed.), pp 111-112 (1982). Cotyledon explants through callus culture produced roots only [*HortScience* 17:195-196 (1982)]. Seed explants gave rise to multiple shoot formation from seedlings; subcultured shoot explants through shoot tip culture produced shoot proliferation [*HortScience* 17:591 (1982)].

Juniperus Juniper

Invaluable needle evergreens displaying cosmopolitan personalities that allow them to survive where other landscape plants succumb. Most are dioecious or occasionally monoecious trees, shrubs, or ground covers that "flower" in spring with the berry-like cones ripening the first or second (occasionally third) season. Each cone contains 1 to 4, to 12 seeds. Collect ripened fruits in late fall or winter, clean by maceration, dry seeds and store in sealed containers under refrigeration.

An expanded discussion of juniper propagation is necessary because of their importance to the nursery industry and the tremendous volume of literature. Simply stated, there is no ubiquitous recipe for all junipers. Some root easily; others with difficulty. Many upright types must be grafted. Cutting propagation is broken down by category.

CONDITION OF STOCK PLANT: A vigorous, healthy plant is superior to an overgrown plant that has been setting for 10 years unattended in a corner of the nursery. Major growers stress that cuttings taken from their container grown plants are superior. They root in higher percentages and have better root systems.

TIMING: Generally, cuttings should be taken after several hard frosts or freezes. Major container juniper growers take cuttings in December to March. References point to any time from July through April. The data presented for summer rooting indicated that percentages were not as high as for late fall-winter rooting. It is truly a mixed bag as far as timing is concerned. November through February is considered optimum. Cuttings of *J. horizontalis* 'Plumosa' rooted 65% in August, 52% in September, 91% in October, 100% in December, 96% in February, 33% in April and 2% in June. A large grower noted that the best time to take cuttings from a physiological standpoint was when they had stopped growing. In one study, junipers were rooted year-round but best rooting occurred between November and December with 4000 ppm IBA-dip. *J. chinensis*, *J. communis*, *J. horizontalis*, *J. sabina* and *J. squamata* were easier to root than *J. scopulorum* and *J. virginiana*.

HORMONE: The standard recommendation throughout the literature is 3000 ppm IBA-solution or talc for easy-to-root types, 8000 IBA-solution or no hormone to 4.5% IBA-talc. Based on published data, a hormone definitely serves to improve rooting. The following hormones and concentrations are used by the largest and most successful juniper growers in the world: 600 to 5000 ppm IBA-solution; 3000, 8000, 3.0%, 4.5% IBA-talc or 2000 ppm IBA + 1000 ppm NAA-solution; 5000 ppm IBA-solution; 1870 ppm to 1.0% IBA-solution; 1.6% IBA-talc; 3000 to 8000 ppm IBA-talc; 3.0 to 4.5% IBA-talc. About all this indicates is that everyone does it differently. Virtually all nurserymen/researchers agree that a hormone is beneficial. The reason for the disparity of rates is that easy-to-root types like *J. horizontalis* would be treated with a low hormone concentration, difficult-to-root types like 'Maneyi', 'Blaauw', *J. virginiana* cultivars, etc., with higher levels. There are several reports that indicate greener cuttings (July-August) require greater hormone concentrations than those taken later.

Senior author has observed alcohol burn on juniper cuttings, especially those taken in summer or early fall, although damage (basal burn) has occurred on *J. conferta* (one of the most susceptible) in January through March. A recent study [*The Plant Propagator* 29(2):8-10 (1983)] with *J. chinensis* 'Hetzii' confirmed and quantified this observation. 5000 ppm to 2.0% IBA in 50% ethanol or prolonged exposure to 50% ethanol (2 minutes) caused basal end necrosis (burning) and reduced rooting. KIBA (water soluble formulation) at similar IBA concentrations and exposure times did not damage cuttings and resulted in slight increases in percent and number of roots. Rooting ranged from a low of 18% (1250 ppm IBA in 50% ethanol-2 minute dip) to 86% (5000 ppm KIBA-water-2 minute dip).

CUTTINGS: SIZE/CONDITION/WOUNDING/BOTTOM HEAT/ROOTING TIME: Four to 6" long cuttings with a tinge of brown (mature) wood seem to be preferred. If actively growing July-August cuttings are taken this type of wood is not possible. Cuttings are stripped one-third to one-half. This, in effect, provides a wound which appears to assist in juniper rooting. In general, stockier, larger cuttings root better than small spindly ones. Bottom heat (66-68°F) is used by some growers and not by others. It should be advantageous if cuttings are rooted outside in cold frames or poly houses in ground beds where the air temperature is rather cool. Rooting time varies from 6 to 8 weeks to 12 weeks and possibly longer. Senior author noted that *J. conferta* (January, February, March) rooted quickly and appeared to require less heat than other conifers. In a controlled study using outdoor ground beds with and without bottom heat with two media (bark or peat: perlite), *J. conferta* rooted under all conditions (better and faster with bottom heat) while leyland cypress in the non-heated frames did nothing (not even callus). It may be that junipers have a lower heat requirement for the root initiation process to proceed.

Senior author worked for a nurseryman during college who took all his juniper cuttings in late summer, provided IBA-talc treatment, sand, on a shaded greenhouse bench, syringed when needed and waited patiently for 3 to 4 months for rooting. He had good success with a wide range of junipers.

MEDIUM: The choice of media is about like the hormone rate...variable. Sand, sand: peat, bark, sandy soil, peat: perlite, vermiculite, etc., have been used successfully. In a comparative study, vermiculite, sand: peat, soil: peat: perlite proved better than sand or German peat.

WATER/SANITATION: Mist in summer months with hand syringing or controlled mist in fall-winter months. Observations indicate junipers do not require a great amount of water during the rooting process. Keep medium moist, but not wet. Do not leave a continual film of moisture on cuttings. Water tends to bead in axils of the needles and the opportunity for fungal invasion (*Phomopsis*) is excellent. Some growers dip all the cuttings in a Benlate-Captan solution before preparing them for rooting. Again, it is a mixed bag as far as those that do and don't. The large growers practice preventative medicine.

AFTERCARE: Cuttings can be rooted in place and allowed to go through the winter and transplanted into containers in spring. The better the root system, the more successful the transplanting. Junipers present no special problems as far as aftercare. Proper cultural and fertility practices yield a high quality juniper.

GRAFTING: A general overview is more logical than to treat the subject under each species. Many upright forms of *J. chinensis*, *J. virginiana* and *J. scopulorum* are grafted because they do not root easily. Choice of understock has been argued for 34 years in the Plant Propagator's Proceedings with *Juniperus chinensis* 'Hetzii' the clear leader followed by seedling *J. virginiana*, the cultivar 'Skyrocket' and a strange juniper called *J. pseudocupressus*. A large nursery noted 'Sky Rocket' was an excellent understock because it is straighter and more graftable and resists fungal diseases better than 'Hetzii' and *J. virginiana*. 'Hetzii' or 'Hetz Columnaris', however, are the most popular. Understocks should be 1/5 to 1/3" diameter, cleaned 3 to 4" from soil of needles, etc., tops evened up. Process can run from December-February. The potted understock is brought inside and evidence of root growth is a signal to start. Understocks are sprayed with a fungicide prior to grafting. Scionwood is collected (like cutting wood) from vigorous stock plants, is 5 to 6" long with 1 to 2 " brown wood at base, lower 2" are cleaned, soft growth at top is removed, scions dipped in fungicidal solution, side-grafted, wrapped and allowed to heal in greenhouse bench. 25% of grafts are healed in 2 weeks, most are healed after 5 to 6 weeks. Watch for new growth on scion as an index to successful healing. Grafts can be placed under shade or moved to containers for growing on.

Juniperus chinensis Chinese Juniper

SEED: Impermeable coat and embryo dormancy necessitates warm/cold period. 60 minutes acid followed by 3 months cold stratification resulted in good germination.

CUTTINGS: Many of the upright cultivars are considered difficult to root and an attempt is made here to point out specific requirements. Most information relative to degree of difficulty was derived from Plant Propagator's Proceedings. For most *J. chinensis* types follow the recommendations for cuttings described under *Juniperus*. There are 99 ways to skin a cat and 999 ways to propagate a juniper. One interesting comment concerns the use of bottom heat on *J. chinensis* cultivars and subsequent callus formation with little root initiation. Recommended they be stuck in September (Canada) in an unheated greenhouse without bottom heat.

'Ames' — Difficult to root from cuttings, most effectively grafted.

'Armstrong' — See 'Pfitzeriana'.

'Blaauw' — Grafted on 'Hetzii'; very difficult to root. California grower reported good success with fall, 6 to 12" long pencil thick cuttings and noted these rooted better than smaller 2 to 4" long cuttings. These cuttings were rooted under mist, outside, and treated with 3000 ppm IBA-solution. Canadian work showed this cultivar susceptible to rot when in the greenhouse under mist (winter). 50 to 60% rooting was obtained with terminal 6 to 8" long cuttings, 8000 ppm IBA-talc under mist indoors.

'Blue Vase' — Easy to root; can be rooted in outside beds with success.

'Fairview' — Difficult to root; one reference mentioned 33% rooting. Usually grafted on 'Hetzii'.

'Hetzii' — The "universal" understock for most junipers. Suspected parentage is *Juniperus virginiana* 'Glauca' × *J. chinensis* 'Pfitzeriana'. Easily rooted from cuttings using the procedures described. 5 to 8" long cuttings, 8000 ppm IBA-talc, rooted 84 to 100% in controlled studies. The only cultivar mentioned that did not do well on 'Hetzii' was 'Keteleeri'. If the parentage is true, this may explain why *J. virginiana* and *J. chinensis* cultivars can be grafted successfully. In a timing study, March was the worst month with October the best.

'Iowa' — See 'Ames'.

'Kaizuka' ('Hollywood') — Often considered difficult to root. In Oregon, Oct. 1-Nov. 15 or January represent best times to take cuttings, the first 2 to 3 weeks in October being ideal. Hormones are 2000 ppm IBA plus 1000 ppm NAA-solution followed by 8000 ppm IBA—talc treatment (double-dipping, if you will). Some growers stick the cuttings, take them out and remove callus, redip in hormone (8000 ppm IBA-talc) and supposedly rooting is quite good. Rooting will approach 65 to 75%. The entire process starts in October and may end the following October. In a Florida study, late January cuttings, Benlate/Captan treatment, 3000 ppm IBA-talc, 74 to 78°F bottom heat, rooted 92% in 6 to 7 weeks. Non-bottom heated cuttings did not root in this same time frame. Interesting Chinese study utilizing 300 ppm Naligonate 6 hr. soak that produced 77% rooting.

'Keteleeri' — An upright form that is difficult to root and usually grafted. One nurseryman obtained 33% rooting, late January, 8000 ppm IBA-talc, sand, polytent. In Holland, Keteleer cuttings taken in fall, 2 to 3" long, sand: peat, and treated with a variety of hormones rooted 0 to 80%. 1000 ppm IBA + 500 ppm NAA-solution gave 52% rooting. 2000 ppm NAA-solution or 10000 ppm IBA + 5000 ppm NAA-solution resulted in no rooting.

'Maneyi' — Like 'Ames' difficult to root, a seedling from *J. chinensis* var. *sargentii*. 50% rooting reported from cuttings taken in early September (Chicago) and placed in greenhouse.

'Mountbatten' — Upright form, roots with difficulty although a Canadian nurseryman reported September cuttings rooted 70%, unrooted but callused cuttings when restuck brought total to 90%.

'Obelisk' — Upright form, early August (Canada), sand, cold frame, one-third strength Chloromone, good rooting in 9 weeks.

'Pfitzeriana' — The granddaddy of juniper cultivars, long a staple in the nursery/landscape profession. Over the years many cultivars have arisen as branch sports. Pfitzer roots readily if handled as described under juniper cuttings. One thing is evident from the literature that not everyone obtains the same or even good rooting. October-November cuttings rooted 85 to 93%, the highest rooting obtained with a wound and no hormone. Summer cuttings have been successful in Texas. Rooting takes 10 to 12 weeks. A partial list of pfitzeriana sports include 'P. Aurea', 'P. Compacta', 'P. Dwarf', 'P. Glauca', 'P. Glauca Compacta', 'P. Hill's Blue', 'P. Kallay's', 'P. Moraine', 'P. Nana', 'P.

Nelson's Compact', etc. Pfitzer, late January, 3000 ppm IBA-talc, peat:sand, 70°F bottom heat, rooted 40% from heel cuttings, 56% without a heel, 81% without a heel but with a light wound in 8 weeks. Several nurserymen have alluded to the positive effect of wounding. Roots emerged from the cut surfaces. In Alabama, Pfitzer has been rooted directly in the field in September-October. Cuttings are laid in furrows, covered, sheared back to 3 to 4" and allowed to root.

'Plumosa Aurea' — Heel cuttings, 1.5% IBA-talc, mist, 75 to 80°F bottom heat, rooted 80%.

'Robusta Green' — Upright type, October (Oregon), 2000 ppm IBA plus 1000 ppm NAA-solution gives best results.

var. *sargentii* — Considered somewhat difficult. November-December (Maryland) with rooting occurring in 8 to 10 weeks. October (Oregon), best success.

TISSUE CULTURE: Pollen explants produced callus. *Compt. Rend. Acad. Sci. Paris* 297D:651-654 (1974).

Juniperus communis Common Juniper

SEED: Fall plant and seeds germinate the second spring. Two to 3 months warm/3 plus months cold will short circuit the fall planting method.

CUTTINGS: Almost circum-global in distribution in temperate latitudes. Species and cultivars not amenable to successful culture in southern climates (zone 8 and south). Many cultivars have been selected and most root without difficulty. One reference described the rooting process as "easy but slow". 'Suecica' (Swedish) is easily rooted. In fact, 12 to 15" cuttings taken in winter (January-February) root readily in 5 to 6 weeks. A large grower reported that most forms of J. *communis* are easily rooted and cited specifically var. *depresssa* (also the golden and vase-shaped forms of this variety), var. *montana*, 'Hibernica', 'Suecica' and 'Suecica Nana'. 3000 to 8000 ppm IBA-talc or solution are ideal for these forms. October-November appears to be the best time, although as discussed under J. *chinensis* there is wide latitude in rooting times.

Many of the forms ('Hibernica', 'Suecica', etc.) are treated as hardwood cuttings and stuck directly in rows in the field without any special pretreatment.

TISSUE CULTURE: Pollen explants produced callus. *Compt. Rend. Acad. Sci. Paris* 279D:651-654 (1974).

Juniperus conferta Shore Juniper

CUTTINGS: Possibly the easiest juniper to root although many nurserymen will debate this point. Senior author has taken cuttings in June-July and January-March with excellent success. The alcohol quick dips have damaged the soft summer cuttings. 'Blue Pacific' and 'Emerald Sea' are the preferred forms because of rich blue-green foliage and more compact growth habit. In southern California, mid-December to mid-January cuttings root best, fungicidal drench, 3000 ppm IBA-talc, peat: perlite, mist, greenhouse, maintain 80°F heat on bottom of flats, 86% rooting in 5 to 6 weeks.

Although the California grower uses high bottom heat, senior author believes this is a waste. Cuttings in unheated beds collected in January, February, 1.0% IBA-solution, peat: perlite or bark, rooted 80 to 90% in 12 weeks outside in cold frames covered with milky plastic.

Juniperus davurica Dahurian Juniper

CUTTINGS: Not well known in north but used extensively in Southeast. Handsome dark green (tinge of blue), largely scale-like foliage. Has been successfully rooted from summer, fall and winter cuttings. 'Parsoni' is the form in cultivation.

Juniperus excelsa 'Stricta' Spiny Greek Juniper

CUTTINGS: Easily rooted as described under *Juniperus*. A poor plant and cannot compete with many of the upright types.

Juniperus horizontalis Creeping Juniper

SEED: Probably should be treated like J. *chinensis*.

CUTTINGS: Possibly along with J. *conferta* the easiest juniper to root. Some nurserymen do not use a hormone on the species and cultivars. Late summer, fall and winter propagation has been successful. This species may be injured by high hormone concentrations. 8000 ppm IBA-talc or lower is recommended. 'Blue Rug', January (North Carolina), terminal 8" long cuttings, needles removed from basal 2", 9 different hormone treatments, all cuttings rooted 100%; greatest root number occurred on 5000 ppm IBA-50% isopropyl, Chloromone and 8000 ppm IBA-talc. IBA in alcohol caused basal necrosis.

To authors' knowledge no cultivar is difficult to root and if percentages fall below 75% then one should backtrack through the system to find the glitch.

Juniperus procumbens Japgarden Juniper

CUTTINGS: Not particularly difficult to root if treated as described under *Juniperus*. Varying degrees of success: 46% up to +90%. A southern nurseryman takes cuttings in August-September, but best results are achieved with December-February cuttings. 'Nana' roots better than the species. In summer, 3000 ppm IBA-talc is used, 8000 ppm IBA-talc in winter, peat: perlite: sand medium. Other growers affirm the January-February time frame. Most experienced propagators listed excessive water as a limiting factor in successful propagation of the species. Water only when necessary. K-salt of IBA which is water soluble is advantageous in preventing basal burn of cuttings and improving rooting percentages.

Juniperus sabina Savin Juniper

CUTTINGS: Follow general procedures outlined under *Juniperus*. Most forms are low growing or spreading forms and propagate without difficulty. The recipes given for *Juniperus horizontalis* apply to this species and the cultivars. Variety *tamariscifolia* (Oregon), July-August, October-November, 8000 ppm IBA-talc plus Benlate.

Juniperus scopulorum Rocky Mountain Juniper

SEED: Handle exactly as described under J. *communis*. Dormancy is caused by impermeable seed coat and dormant embryo.

CUTTINGS: Most of the cultivars are considered difficult-to-root and are grafted. A California nurseryman reported success with varieties of the species using fall and winter, 6 to 12" long pencil-thick cuttings, 3000 ppm IBA-solution, peat: pumice: sawdust, 2" diameter plastic tubes, mist, outside. A combination of 960 ppm Ethrel plus 2500 ppm IBA-solution resulted in 35 and 50% rooting after 8 to 9 weeks in sand and peat: perlite, bottom heat, mist, respectively, from 5 to 8″ long, late July (Colorado) cuttings.

Interestingly, the break even point for rooting cuttings compared to grafting is 25% for this species and the cultivars. A large grower (California) roots 'Cologreen', 'Gray Gleam', 'Pathfinder', 'Table Top', 'Wichita Blue', 'Welchii', and 'Wintergreen'. Only 'Cupressifolia Erecta' and 'Tolleson's Weeping' are produced exclusively by grafting. The grafted plants grow much faster the first season which is probably attributable to the better developed root system. Cuttings are taken in late November (minimum temperatures at this time, 35 to 45°F), 3″ long with hardwood at base, heel cuttings preferred, foliage removed from bottom inch, dipped in fungicidal solution, 9 perlite: 1 peat in flats, outdoor under mist in full sun with bottom heat (60 to 65°F for first 6 weeks; 70 to 75°F thereafter). Bottom heat at least in this particular operation is essential to root J. *scopulorum*. Cuttings are rooted sufficiently and can be transplanted 5 to 6 months after sticking. Hormones are specific for each cultivar. 'Cologreen' — 3000 ppm NAA-solution, 40% rooting; 4.5% IBA-talc resulted in 74% rooting. 'Gray Gleam' — 6000 ppm IBA-solution, 52% rooting. 'Pathfinder' — 8000 ppm IBA-solution, 62% rooting; higher IBA levels have an adverse effect. 'Table Top' — 1.6% IBA-talc, 63% rooting. 'Wichita Blue' — 8000 ppm IBA-solution, 57% rooting. The key, according to this nurseryman, is finding the ideal (best) hormone and concentration. 'Moonglow' (Oregon) — July-August, October-November, 2000 ppm IBA plus 1000 ppm NAA-solution.

GRAFTING: Graft if you must but heed the nurseryman's advice under cuttings. The description under *Juniperus* applies to this species and the upright forms of J. *chinensis*.

Juniperus squamata 'Meyeri' Meyer's Singleseed Juniper

CUTTINGS: To authors' knowledge the species is not represented in cultivation. 'Meyeri' is the most common cultivar although 'Blue Star' and 'Blue Carpet' are gaining. Senior author has always considered this an easy form to root. However, there is evidence to counter this contention. Work in Boskoop assessing the best time reported 61% rooting in late August with lows of 10 and 19 in early and late February, respectively. Other work indicated 40 to 50% rooting. In Oregon, mid-July to mid-August and early October to late January proved the optimum time to root 'Blue Star' and 'Prostrata'. 'Blue Star' received 8000 ppm IBA-talc: Benlate treatment; 'Prostrata', a double dip of approximately 2000 ppm IBA plus 1000 ppm NAA solution followed by 8000 ppm IBA-talc: Benlate treatment.

Juniperus virginiana Eastern Redcedar

SEED: Fall plant or provide 2 to 3 months cold stratification.

CUTTINGS: Upright and spreading types occur within the framework of this species. Upright types are difficult to root; spreaders easier. One nurseryman takes October, November, December cuttings with rooting occurring 8 to 14 weeks later. In general, very few specific references to rooting J. *virginiana* cuttings occur in the literature. The ones that do are not very positive. July-August (Tennessee), 1.6% IBA-talc, sand, mist, outside, full sun, rooted 5 to 10% ('Burkii') and 50% ('Densa Glauca'). A combination of a fungicide (Captan) and IBA produced a positive response on 'Canaertii' and 'Hillii'. No other specifics were given. Late January (Connecticut), 8000 ppm IBA-talc, sand, bottom heat, polytent, 75% rooting of 'Canaertii'. Cuttings came from 4- to 6-year-old plants. Several nurserymen described their success/failure in rooting 'Canaertii' and other J. *virginiana* clones. 40% rooting on 'Canaertii' was obtained with 1% naphthaleneacetamide; all other rooting compounds resulted in 0 to 20%. The root sysems on these cuttings may be sparse and the resultant plants do not grow off as fast as grafted plants. One grower reported an average of 5 to 8 roots per cutting.

'Grey Owl', a spreading form, has been rooted most successfully from late August to late October. Tests in Holland showed 85 to 100% rooting at these times, 1 to 14% from November to March. This work was carried out over a number of years and the results were consistent. 'Kosteri' is also rooted easily from early December cuttings. Cuttings have also been stuck directly in the field (Alabama, fall) with good success. 'Skyrocket' is easily rooted (percentage as high as 96% is reported). Italian work reported that the natural tendency for rooting 'Skyrocket' is low (4% in January and February) but application of 5×10^{-4} M IBA improves rooting to 70%.

GRAFTING: The preferred understock would seem to be J. *virginiana* (see discussion under juniper) but 'Hetzii', 'Skyrocket' and 'Pseudocupressus' (J. *virginiana* form) are used with excellent success. For many of the upright types grafting is the only logical approach with success ranging from 85 to 95% and higher. No doubt, someone will unlock the secret of rooting J. *virginiana* cultivars and grafting will become passé.

Kalmia Mountain-laurel

Beautiful evergreen shrubs with spectacular flowers in May-June. Fruit is a capsule that, like rhododendron, should be collected before the seeds dehisce in fall. The dust-like seeds should be treated as described under *Calluna*.

Kalmia angustifolia Lamb-kill

SEED: Sow on surface of milled sphagnum, place under mist or plastic, seed germinates readily. Some indication that limited cold stratification may improve germination.

CUTTINGS: Late November (Boston), 8000 ppm IBA—talc plus thiram, peat: perlite, bottom heat, polytent, rooted 100% in 11 weeks. 'Hammonasset', a selected form, late summer-early fall, 3333 ppm IBA plus 1666 ppm NAA-solution, mist or plastic, rooted satisfactorily. Vegetative shoots with no flower buds root more readily than shoots with flower buds.

Kalmia cuneata Whitewicky Kalmia

SEED: 1000 ppm GA_3, 24-hour soak effectively breaks the dormancy.

Kalmia hirsuta Sandhill Kalmia

SEED: Dormancy is most effectively broken under humid conditions for 10 min to 19 hours between 140 and 194°F; the higher the temperature the shorter the treatment period required.

Kalmia latifolia Mountain-laurel

SEED: 1.4 million seeds to the ounce; germinate without any special pretreatment but percentage germination of freshly harvested seed can be enhanced by 50% by 200 ppm GA_3-12 to 24 hour soak. Seed stored for 6 months requires no GA treatment. Best germination at 72°F; higher temperatures reduce germination. Seedling growth is often slow and 200 ppm GA foliar spray stimulates elongation. Seed is long-lived and when kept dry and refrigerated remained viable for 20 years. See J. *Amer. Soc. Hort. Sci.* 96(5):668 (1971).

CUTTINGS: Wow! Some say yes; others forget it. Some say a hormone is necessary; others say it makes no difference. One absolute entity is that tissue culture has solved most of the propagation problems on a commercial scale. Senior author has rooted a few cuttings but success has been mediocre. A commercial concern in Oregon tried about every conceivable combination of media, hormone, time, etc., and came away with the following recipe: January best, although late October acceptable, cuttings taken from young plants. Twice the rooting from young compared to old plants. Benlate dip, double wound, mixture of 5000 ppm IBA, 5000 ppm NAA, 175 ppm Borax solution at half strength, 4 fir-4 cedar sawdust, 1 perlite:1 peat in greenhouse benches 6″ deep with 73 to 75°F bottom heat and mist. January cuttings root 60 to 75%, depending on cultivar by June, are transplanted, and the unrooted cuttings restuck in flats with mist without bottom heat. By fall these have rooted.

This species takes a long time to root (4 to 5 months) and growers mention the advisability of using a high humidity chamber rather than mist because of excess moisture.

One east coast grower reported October-November is the best time to take cuttings. Mid-October, 8000 ppm IBA-talc rooted 95% in January but only 60% produced new growth. He used a polytent, peat mix and bottom heat. In a "mini"-test with an impossible to root clone, Chloromone resulted in 100% rooting.

A significant paper (PIPPS 27:479) addressed propagation of K. *latifolia* by cuttings. Pertinent points to ponder were the use of a polytent, peat: perlite, 70°F bottom heat, cuttings from current season's growth as soon as hardened, double wound, patience for 4 to 6 months. Best hormones included IBA plus NAA solutions at various strengths or 2, 4, 5-TP-talc treatments. Other reports indicated 1000 ppm 2, 4, 5-TP in talc produced the best rooting of *Kalmia* but the highest percentage was only 39. In this same study, September was the best month to root cuttings. In Clemson tests, 10 sec-dip in 10,000 ppm IBA plus 5000 ppm NAA plus 2, 4, 5-T at 1000 ppm produced best rooting. Success on a number of cultivars ranged from 30 to 100%.

The problem with fall cuttings is the lack of cold and subsequent lack of bud break. December-January cuttings have sufficient cold to break and grow. Clonal differences in rootability are pronounced. Over a six year period 'Ostbo Red' averaged 34% and 'Pink Surprise' 76%.

Juvenility has been shown to be an important factor in rootability. The younger the stock plant, whether seedling or cutting grown, the better the rooting. Cuttings of one clone rooted 30% from a 15-year-old original plant; 31% from an 8-year-old rooted cutting; and 94% from a one-year-old rooted cutting. The same relationship holds for seedlings: 21% from 4-year-old; 33% from 2-year-old; 89% from one-year-old seedlings.

About 25 named selections are known and, as mentioned, root in varying degrees. 'Pink Charm', October (Connecticut), rooted 82% in 5 peat: 2 perlite, 70 to 75°F bottom heat, polytent. 'Goodrich', 'Ostbo Red' and 'Stillwood' are difficult to root (less than 50%); 'Pink Surprise', 'Nipmuck' and 'Quinnipiac' are relatively easy (70 to 95%).

GRAFTING: Bench grafting in December-January is practiced using side, splice or cleft graft. With tissue culture, grafting should be obsolete.

TISSUE CULTURE: At least three commercial laboratories are producing cultivars in abundance. For specifics see PIPPS 30:421-427.

Kalmia polifolia Bog Kalmia

SEED: No pretreatment necessary. See K. *latifolia* for specifics.

CUTTINGS: Semi-hardwood, August, 8000 ppm IBA-talc, shaded cold frame, rooted well.

Kalmiopsis leachiana Oregon Kalmiopsis

SEED: Treat as described under *Calluna*; one investigator [*The Plant Propagator* 18(1):11-13 (1972)] reported excellent germination from seeds sown on milled sphagnum.

CUTTINGS: October (Oregon), 3 to 5″ long, 72°F bottom heat, 18 hours light/day, polytent, 92% rooting in 3 peat: 1 sand treated with 3 parts 8000 ppm IBA-talc: 1 part Captan 50. All cuttings survived transplanting.

Kalopanax pictus Castor-aralia

SEED: Small black drupe with a fleshy coat that ripens in September-October. Clean seeds by maceration, dry, and either fall plant and wait two years for germination or provide 5 months warm/3 months cold stratification.

Soaking seeds for 30 minutes in sulfuric acid may suffice for the warm period. Controlled studies produced the following results: 5 warm/3 cold — 77%; 6 warm/3 cold — 47%.

CUTTINGS: Root cuttings, as soon as frost is out of ground, dig large roots, cut into 3 to 4″ pieces, stick vertically with proximal end (closest to stem) upright in the medium in a cool greenhouse or suitable structure preferably with bottom heat. Stem cuttings are difficult if not impossible to root at least from mature trees.

Kerria japonica Japanese Kerria

SEED: An achene that senior author has not observed on cultivated plants. No specifics were found regarding propagation.

CUTTINGS: Sinfully easy. From May-June-July to October and probably as hardwoods in winter. 1000 ppm IBA-solution, peat: perlite, mist, rooting takes place in 2 to 4 weeks at 90 to 100% rates. This procedure also applies to 'Picta' and 'Pleniflora'. Rooted cuttings are transplanted easily and continue to grow. The plant tends to sucker and if a few plants are desired simply carve off a chunk. Best time to divide is late winter or early spring before growth begins. One nursery reported good success with hardwood cuttings in the greenhouse during the winter months. The species and cultivars lend themselves to outdoor propagation in ground beds. Although a hormone is not needed the quality of the root system is better when a low level IBA solution or talc formulation is used. Late August (Boston), 3000 ppm IBA, sand, rooted 100%. 'Picta', July (Boston), 3000 ppm IBA-talc plus thiram, rooted 100%.

Koelreuteria Goldenraintree

Medium-sized trees with rich green compound pinnate (bipinnate) leaves, golden yellow flowers in large panicles during June-July (August-September on one species) and interesting inflated capsules. The black seeds, often two per carpel (six per fruit), are about the size of small peas. Collect when capsule changes to brown but before opening, dry and flail out seeds. Seed purity is high (95%) and seeds can be stored in sealed containers under refrigeration for long periods. Dormancy is controlled primarily by hard seed coat and perhaps by limited embryo dormancy.

Koelreuteria bipinnata Bipinnate Goldenraintree

SEED: Flowers in August-September (Georgia), pink capsules ripen in October-November. Seedlings appear everywhere the following spring. Suspect dormancy is related to seed coat and fall planting will suffice. An acid treatment for 30 to 60 minutes would probably work.

CUTTINGS: Root cuttings may work but to authors' knowledge have not been tried with this species.

Koelreuteria paniculata Panicled Goldenraintree

SEED: After treatment with sulfuric acid for one hour, seed was sown or given one or two months cold stratification. The first treated seeds germinated in 113 days. Seeds in cold germinated in bag. Senior author does not believe the seeds are doubly dormant (as described in some literature). Make the seed coat permeable and some germination will proceed. An early paper in PIPPS reported 30 minute sulfuric acid soak followed by 78 days cold stratification was superior to other treatments. Mechanical scarification followed by cold stratification for 3 months produced 91% germination 10 days after planting. It was suggested that the embryos are in various states of maturity; hence, the erratic (non-uniform) germination if seeds are only scarified. The stratification period allows for embryo development [*The Plant Propagator* 25(2):6-8 (1979)]. 'September' is a late flowering (August-September), vigorous-growing form that comes largely true from seed.

CUTTINGS: February-March, treat as described for root cuttings; root shoots that develop. Procedure described for *Albizia julibrissin* should work.

Kolkwitzia amabilis Beautybush

SEED: A bristly capsule that should be collected in September-October, dried and seed extracted. No pretreatment necessary as fresh seed will germinate soon after sowing.

CUTTINGS: Late May-early June (New York), light hormone, well drained medium, good rooting. Many growers root beautybush in outside beds with mist during the summer months. One grower reported that cuttings taken after mid-June (New York) produce callus and few roots. In many cases a hormone overrides callus production and induces good rooting. Several growers note that cutting wood should be soft. Late June (Missouri), mist under lath shade, rooted 50% with note that cuttings were taken too late. Rooting takes 3 to 5 weeks. One grower takes cuttings when new growth is long enough to make a cutting (3 to 4″). June, untreated, rooted 4%, 92% with 80 ppm IBA-10 hour soak. July, untreated, did not root; 100% with 60 to 100 ppm IBA-4 hour soak. October cuttings, untreated or IBA-treated did not root well. Hardwood cuttings have not been commercially successful on this species.

+ ***Laburnocytisus adamii***

CUTTINGS: Interesting graft-chimera with yellow and pink flowers as well as leaves of both parents on the same tree. Late November (Boston), 1% IBA-talc, rooted 40% in 10 weeks. Firm wood in July might root better but quantitative work has not been conducted with this plant.

GRAFTING: Bench grafted in winter using side graft on seedlings of *Laburnum anagyroides*.

Laburnum Goldenchain

Splendid small trees that literally drip with yellow rain in May. The 2″ long pod contains small, hard-coated black seeds. Collect in late August to October, dry, extract seed. Percentage of sound seed is high. Seed can be stored dry for two years without loss of viability.

Laburnum alpinum Scotch Laburnum

SEED: Dormancy is caused by impermeable seed coat which must be ruptured to allow imbibition of water. The following data indicate the essentiality of breaking the seed coat. No treatment — 0% germination; hot water — 18%, one hour sulfuric acid — 29%; two hours sulfuric acid — 68%. In the highest treatment only one sound seed was found in the ungerminated seeds indicating the acid treatment was extremely effective.

GRAFTING: Forma *pendulum* is bench grafted on seedlings of the species in winter using side or whip graft. In Europe, T-budding is practiced in July-August on two-year-seedlings. The buds make six feet of growth in the first growing season.

Laburnum anagyroides Common Laburnum

SEED: A hot water soak produced 24% germination, 30 to 60 minutes sulfuric acid produced the best germination.

CUTTINGS: July, basal cut at node, untreated, rooted 100%. February (England), 6 to 9″ long, 8000 ppm IBA-talc, wound, lined out in open ground, little growth made during rooting year; transplant in fall and during next season train to a single leader.

Laburnum × watereri Waterer Laburnum

SEED: Pods collected in late July (Boston), seed coat was soft (thin green membrane), divided into two lots: as is or coats punctured with a needle; in five days each group germinated uniformly. Hard seed is another matter and 2 hour sulfuric acid treatment resulted in 100% germination.

CUTTINGS: Interestingly, not a great amount of information relative to rooting. Leaf bud cuttings, untreated, early summer, rooted 80%. 'Vossii' hardwood cuttings rooted better with hormone treatment and Captan dip. Senior author during undergraduate days at Ohio State collected soft summer cuttings from root-sprouts with 100% success. Other information indicated juvenile cutting material roots "better" than older material.

GRAFTING: 'Vossii' is bench grafted on seedlings of *L. anagyroides*. Any of the species would serve as a suitable understock.

Lagerstroemia Crapemyrtle

Shrubs or small trees with gorgeous summer flowers of white through red. Fruit is a 6-valved dehiscent capsule that ripens in October. The capsules are held upright so all seeds do not fall out when the dehiscence occurs. Senior author collected fruits in January in grocery sacks, dried (in same sacks), shook like a cocktail, removed small winged seeds. Seeds require no pretreatment and germinate within 2 to 3 weeks of sowing. A brief cold period may unify germination. Seed should be stored in sealed containers after extraction and sown in flats when days are lengthening in spring. In a single season (Georgia), it is possible to have seedlings in flower.

Lagerstroemia indica Common Crapemyrtle

SEED: Discussed under *Lagerstroemia*. The only reference authors found indicated germination was sparse when seeds were planted without pretreatment; one month cold stratification induced excellent germination. However, senior author can find nothing to support the necessity for cold stratification. Great fun to grow crapemyrtles from seeds since the variation in seedling populations is so great.

CUTTINGS: Easily rooted from soft and hardwood materials. Senior author has rooted the plant year round. June, July, August, 1000 ppm IBA-solution, peat: perlite, mist, root 90 to 100% in 3 to 4 weeks and continued to grow in rooting bench. These rooted cuttings are easily transplanted and will continue to grow. Cuttings taken later in the season will root but often experience overwintering problems. It is difficult to properly acclimate succulent plants for the October-November freezes.

Hardwood cuttings are routinely used by many commercial firms. The thicker the cutting the faster the rooting response: 3/8 to 1/2" diameter cuttings rooted better than 1/4" diameter cuttings. November-December collected and planted or collected, stored (bedded in outside) and spring planted. Rooting is about 50%. Senior author collected hardwood cuttings in early January, early February and early March. Best rooting occurred with February (43%) cuttings, bottom heat, in both peat: perlite or bark. The rooted cuttings, when transplanted, make excellent growth during the season. A large commercial crapemyrtle producer (Alabama) sticks hardwoods from September to December with 80% resultant success.

Dwarf crapemyrtles were assessed for rootability from hardwood and softwood cuttings. Late March (Arkansas), different stem diameters, hormones, media, outside and under mist. Larger diameter cuttings rooted best outside; small and large inside under mist. Hormone was helpful. Best rooting was 40% in greenhouse. June softwoods under mist in greenhouse rooted 100% in 4 weeks regardless of hormone. However, hormone increased root length and dry weight. August cuttings did not root in as high percentages as June cuttings.

Senior author conducted a June (Georgia) study with 'Tuscarora', 'Natchez' and 'Muskogee' using no hormone and 1000 ppm IBA-solution. Although percentages were the same, root quality (number and length) was infinitely superior with hormone treatment. It is good insurance to use a hormone. Cuttings taken later in season (August-September) did not root as well as earlier cuttings.

For small numbers from hardwood tissue, senior author lays hardwood cuttings (late November-December) horizontally on peat: perlite under mist. Small shoots develop and when one to two inches long they are given a 1000 ppm IBA-quick dip and rooted in peat: perlite. This process is superior to hardwood cuttings placed vertically in medium which root indifferently over a long time period at rates of 30 to 50%. The return is simply one-to-one whereas a hardwood cutting placed horizontally may produce 4 to 7 shoots (depends on number of buds) all of which are rooted easily.

Root cuttings offer another alternative but should be the third approach after soft and hardwood cuttings. Many years small plants are killed to the ground yet multiple shoots appear. A hole produced when a balled and burlapped plant is removed will quickly have numerous sprouts around sides. Suckers which develop from roots can essentially overgrow an established plant. All this observational evidence supports the idea that root cuttings will produce plants. Best to dig them in March and plant in sand: peat in greenhouse.

Larix Larch

SEED: Cone-bearing trees of cold climates with deciduous needles. Monoecious with small cones ripening in fall. Collect in October, dry in heated room, remove seeds with a shaker. Seed can be stored in sealed containers under refrigeration for 3 or more years with little loss of viability. Seed, in general, will germinate without pretreatment although a one to two month cold stratification may hasten and unify germination.

CUTTINGS: Difficult; *Larix decidua* — European larch, L. *kaempferi* — Japanese larch, L. *laricina* — Tamarack, have been rooted from seedling material but this serves little purpose. Softwood cuttings of L. *griffithii*, L. *laricina*, and L. *sibirica* have been rooted in low percentages with 8000 ppm IBA-talc. The more juvenile the stock plant the better the chance for success. Extended photoperiod will cause buds to break and grow. Might be worthwhile to approach rooting much like that practiced on *Acer rubrum*. Doubtful cuttings will ever become significant for the propagation of the larches. A blue needle form in Connecticut, late August, from 10-year-old tree, terminal, heavy wound, 2% IBA-talc, sand, mist, 41% rooted by December but with tremendous variation in root quality from a single root to 10 roots/cutting. Length varied from 1/8" to 3". No mention was made of survival.

GRAFTING: The selected cultivars, usually weeping types, are bench grafted (whip and tongue) on established understocks in the winter. *Larix decidua* serves as a good understock.

TISSUE CULTURE: Cambial explants produced callus [*For. Abst.* 38:3330 (1975)]. Strobili discs through callus culture produced adventitious shoots (from one tree) [*Can. J. Bot.* 60:1357-1359 (1982)]. Stem and leaf explants produced callus [*For. Abst.* 39:809 (1978)].

Larix × *eurolepis* Dunkfeld Larch

CUTTINGS: Summer and winter cuttings rooted in high levels from young seedlings. Rate of rooting was hastened with IBA treatment but final percentages were similar. *Silvae Genetica* 28:5-6.

Lavendula angustifolia Common Lavender

SEED: No pretreatment necessary although one nurseryman provides a cold period (length not given) before sowing seed in early February (New York). Senior author has germinated seed quite successfully without pretreatment. Seeds soaked in 200 ppm GA_3 and kept at 50 or 68°F for 16 hrs, and 86°F for 8 hours germinated greater than 90%.

CUTTINGS: In England, late summer, semi-hardwood cuttings are rooted in outdoor frames containing soil: peat: sand. No special treatment is afforded the cuttings. June (Ireland), outside frames, root 100% in 3 to 4 weeks. Summer cuttings root readily in sand without treatment. Senior author had problems with excessive moisture in rooting bench. Best to keep on dry side during rooting. In one study, IBA dip at 150 ppm improved rooting of two cultivars 35 and 20% over untreated control.

TISSUE CULTURE: See *Ann. Bot.* 45:361-362 (1980).

Ledum groenlandicum Labrador-tea

SEED: Treat like *Calluna*; no pregermination requirement.

CUTTINGS: Broadleaf ericaceous evergreen with rather pretty white flowers in May-June. Mid-October (Boston), 8000 ppm IBA-talc or 4000 ppm IBA-talc plus thiram, rooted 67 and 100% in 10 weeks, respectively, with excellent root systems. A polytent with a peat: perlite medium and perhaps bottom heat would be the best system. Mid-December cuttings rooted 97% with 4000 ppm IBA-talc, peat: perlite, polytent in 10 weeks.

Leiophyllum buxifolium Box Sandmyrtle

SEED: No pretreatment required; handle as described for *Calluna*. Sow on surface of medium and maintain under lights.

CUTTINGS: A handsome, dainty, white flowered broadleaf ericaceous evergreen. Mid-October (Boston), 3000 ppm IBA-talc, rooted 100% in 8 weeks. July-August cuttings also root well using 1 to 1½" long material, peat: sand, bottom heat. The best procedure is the tent method described above for *Ledum groenlandicum*. New York work indicates short side shoots from near the main branch terminals in October root well. These, when pulled off, yield heel cuttings.

Leitneria floridana Corkwood

SEED: Respectable germination after three months cold stratification.

CUTTINGS: Mid-July cuttings rooted 100% in 8 weeks after treatment with 4000 or 8000 ppm IBA-talc plus thiram under mist. Root cuttings can also be used. The species suckers and a clump can be divided if several plants are needed. Good plant for moist or wet soil areas.

Lespedeza bicolor Shrub Bushclover

SEED: Seed that was directly sown after harvest in fall germinated well. The seed coat probably becomes hard with drying and a hot water or short (8 to 15 minute) acid treatment would help. Seeds have been stored as long as 20 years and maintained viability.

CUTTINGS: June-July root readily; 1000 ppm IBA—solution or talc, peat: perlite, mist, with 80 to 100% rooting in 4 weeks.

Lespedeza thunbergii Japanese Bushclover

SEED: Probably the same as for *L. bicolor*. Authors have not observed fruit set on this species.

CUTTINGS: According to one report, October cuttings, untreated, did not root; 66% in sandy soil after treatment with 100 ppm IAA-18 hour soak. Senior author collected mid-June cuttings, 1000 ppm IBA-solution, peat: perlite, mist, with 100% rooting in 4 weeks. Rooted cuttings were transplanted and continued to grow.

Leucothoe Leucothoe

Deciduous or evergreen shrubs with urn-shaped white flowers in spring followed by dehiscent capsules containing small dust-like seed in fall. Collect before capsules open, dry and remove seed. Seed germination should be handled as described for *Calluna*.

Leucothoe axillaris Coast Leucothoe

SEED: Direct sow; follow procedure described for *Calluna*.

CUTTINGS: Mid-October (Ohio), 4 to 5" long terminal, leaves stripped from basal one inch, 3000 ppm IBA-talc, 5 peat: 2 coarse perlite: 2 fine perlite: 1 sand, 65 to 70°F bottom heat, benches in greenhouse, manually operated mist, weekly dusting of Captan, rooting occurs in 6 to 8 weeks and rooted cuttings are transplanted.

Leucothoe fontanesiana Drooping Leucothoe

SEED: No pretreatment required. Handle like *Calluna*.

CUTTINGS: Procedure described for *L. axillaris* applies to this species. November and December, 3000 ppm IBA-talc, rooted 100% in 10-11 weeks. Senior author has rooted the species in June, 1000 ppm IBA-solution, peat: perlite, mist. July, untreated, peat: sand, rooted 100% in 12 weeks. IBA treatment only slightly hastened rooting. Cuttings can be taken from June to December (latest reported date) and probably later. Key is healthy cutting material, well drained medium, and a 10 to 12 week wait. They can be transplanted after rooting and, if fertilized, will initiate growth. Virtually every reference reported excellent rooting (100% in most cases) with or without a hormone.

Leucothoe grayana

SEED: No pretreatment required.

Leucothoe keiskei Keisk's Leucothoe

SEED: No pretreatment required.

Leucothoe populifolia Florida Leucothoe

SEED: No pretreatment necessary.

CUTTINGS: Used in southern gardens and has much to recommend it. Should be handled like *L. fontanesiana*. One research report noted that IAA, IBA, and NAA produced a slight increase in rooting over untreated cuttings but did not give specifics. Senior author collected mid-January cuttings, IBA, peat: perlite, mist, with 70% rooting. Cuttings are easily transplanted.

Ligustrum Privet

Deciduous or evergreen small trees and shrubs with handsome foliage, earthy commonness, white heavy scented flowers and bluish to blackish drupes that ripen in fall and often persist through winter. The seeds should be removed by maceration from the pulp, dried and stored in sealed containers under refrigeration. Fresh seed of most species germinates without cold stratification. Stored seeds of *L. vulgare* require 2 to 3 months cold to induce germination. As a graduate student at the University of Massachusetts, senior author collected fruits of *L. amurense* and sowed one lot while cleaning the pericarp from the other before sowing. The second lot germinated like beans indicating no embryo or seed coat dormancy. It also indicated that the fruit wall contained inhibitors or was impermeable to water.

Ligustrum amurense Amur Pivet

SEED: Fresh seed germinates readily; dry or stored seed should be given one to two months cold stratification to insure uniform germination.

CUTTINGS: June, July, August, 1000 ppm IBA-solution or talc, well drained medium, mist, ±90% rooting in 4 to 5 weeks. All privets are easy to root and many growers root them outside in ground beds with mist during summer or as hardwood cuttings placed in rows in the ground during fall. One interesting report noted that NAA may be more effective on privets than IBA. Untreated August cuttings rooted 84%; cuttings sprayed with a hormone, 98% in 5 weeks.

Ligustrum × ibolium Ibolium Privet

CUTTINGS: July, rooted 76% without treatment; 87% in 5 weeks after 50 ppm IBA-6 hour soak. Untreated September cuttings rooted 100% in 2½ weeks.

Ligustrum japonicum Japanese Privet

SEED: As described; abundant fruits are set in southeastern states. This species is not as weed-like as L. *lucidum*.

CUTTINGS: Anytime from June (Georgia) to March based on senior author's experiences. 1000 to 3000 ppm IBA-solution or talc, peat: perlite, mist, good success. Literature reported 95, 92, 86% rooting from June, July, November cuttings plus hormone, respectively. IBA was definitely beneficial and in July and November cuttings almost doubled rooting percentages compared to control. Included here are 'Recurvifolia', 'Rotundifolium', 'Texanum', etc.

Ligustrum lucidum Waxleaf (Glossy) Privet

SEED: Fruits heavily; fresh seed requires no pretreatment.

CUTTINGS: Basically as described for L. *japonicum*. This species is rooted without difficulty. Rooted 100% in a high humidity chamber. NAA may be more effective on this species.

Ligustrum obtusifolium Border Privet
L. o. var. *regelianum* Regel's Border Privet

SEED: Regel's border privet has been grown from seed and the original selection is probably not what is offered to the public.

CUTTINGS: Basically as described under L. *amurense*.

Ligustrum ovalifolium California Privet

SEED: Fresh seed germinates readily.

CUTTINGS: Essentially as described for L. *amurense*. June, August and September hormone-treated cuttings rooted better than untreated cuttings. This applies to the species and cultivars. IBA and NAA produced good results. 1000 ppm solution or talc would suffice.

Ligustrum sinense Chinese Privet

SEED: No pretreatment required.

CUTTINGS: Root readily; in the South, 'Variegatum' is a popular form and is rooted easily as described for L. *amurense*.

Ligustrum × vicaryi Golden Vicary Privet

SEED: Senior author germinated a few seedlings and noticed yellow foliage in the beginning stages. Might be fun to look for superior types. This would be a good choice if yellow foliage is desired.

CUTTINGS: Exceedingly easy; often rooted outdoors in summer in ground beds under mist.

Ligustrum vulgare European Privet

SEED: 83% germination on freshly planted seed. Literature indicated that dried seed might require one to two months cold stratification. One report noted that some of the varieties came relatively true-to-type from seed but did not mention which ones.

CUTTINGS: Root easily. Hormone appears to improve response.

GRAFTING: Privet seedlings have served as an understock for grafting *Syringa vulgaris*. This practice has largely been discontinued. One of the principal problems was the suckering of the understock.

Lindera angustifolia Narrowleaf Spicebush

SEED: A Chinese shrub with rose-red autumn foliage that turns brown and persists on the plant throughout the winter. 3 months cold stratification produced 50% germination, 4 months — 83%.

CUTTINGS: Softwoods taken early and treated like those of L. *obtusiloba* will root.

Lindera benzoin Common Spicebush

SEED: Yellow flowers are dioecious and appear from March to May. Fruit is a red drupe and ripens in August or September. The fresh fruit should be macerated and the pulp floated off. A pound of cleaned seed contains 4500 seeds. The seed loses viability soon after maturity, but storage at low temperatures may prolong viability. Pregermination treatments have indicated the possibility of both warm and cold stratification requirements. In one study, 1 month warm stratification followed by 3 months cold stratification produced good germination. In another study, 3.5 months cold stratification produced the best results. In nursery practice the seed can be fall sown and mulched over winter or spring sown after stratification. Senior author provided 3 months cold stratification to seeds collected and cleaned in late October with resultant 85 to 90% germination. Seedlings continue to grow and may be 2′ or larger the first season.

Lindera obtusiloba Japanese Spicebush

SEED: The authors suspect that a cold stratification period of about 3 months is required. Unfortunately, the species is dioecious and plants in cultivation are of one sex. The only data presented indicated that fresh seed germinated 3%; 57% after 3 months cold.

CUTTINGS: Early June cuttings (Boston), 8000 ppm IBA-quick dip, sand:perlite (1:1, v/v) and mist produced 50% rooting. 4000 ppm NAA-solution produced similar results.

Liquidambar formosana Formosan Sweetgum

SEED: This species differs from the following in having a 3-lobed leaf and bristly fruits. The species is self-incompatible and a single tree on the Georgia campus sets abundant fruit but no viable seed. Another seedling is required to effect cross pollination. The absolute dormancy requirements are not known but

2 weeks to one-month cold stratification would probably hasten and unify germination.

GRAFTING: Superior fall coloring types have been described. Graft or bud on seedlings of the species and possibly on L. *styraciflua*. Follow the procedures described under L. *styraciflua*.

Liquidambar styraciflua American Sweetgum

SEED: Fruiting heads fade from green to yellowish green or yellow to brown at maturity. After harvest, the fruits should be spread to dry until they open and release the seeds. A pound of cleaned seed contains about 82,000 seeds. Seeds should be stored at 10 to 15% moisture content in sealed bags at 35 to 40°F. Viability will be maintained for at least 4 years. Sweetgum seed has a shallow dormancy and cold stratification for 1/2 to 3 months improves germination if spring sown. 78% germination occurred after 2 months cold stratification; the same after 3 months. Fall sown seed should be mulched. There is considerable variation in hardiness from different seed sources.

CUTTINGS: 'Worplesdon', early June, 8000 ppm IBA-talc, mist, peat:perlite, rooted and survived well. Cuttings of 'Gumball', June, 4000 ppm IBA-solution, mist, rooted readily. Overwintering may be a problem.

GRAFTING: Cultivars are vegetatively propagated by budding on seedlings of the species. One year seedlings are lined out and grown a full year with budding taking place the second season at the end of June (Ohio) when the bark slips. Tie with rubber strips for 10 to 15 days, then remove. The understock is cut back to 2′ above the bud at the first sign of swelling. Tie bud with tape close to the stem and later stake for straight stem. Bench grafting on 2-year potted seedling of L. *styraciflua* using a whip graft in early spring under glass is conducted in England.

TISSUE CULTURE: Lateral buds are surface sterilized by washing in a dilute soap solution, 30 seconds in 5% Amphyl, 20 minutes in 20% laundry bleach plus 0.1% Tween 20, a second 5 minutes in 20% laundry bleach, and 4 rinses in sterile water. The propagation medium used was Woody Plant Medium (WPM) with 0.2 mg/liter BA. Explants on the Linsmaier-Skoog medium turned brown. Shoots were rooted on WPM with 0.5 mg/liter IBA. Cultures were incubated at 77°F with a 16-hour photoperiod under 60 microE m^{-2} sec^{-1}. See PIPPS 33:113 (1983).

Liriodendron tulipifera Yellow Poplar, Tuliptree

SEED: The fruit is a green cone-like aggregate of samaras, eventually turning brown. A pound of cleaned seed contains about 14,000 seeds. There is great variation in the number of full seeds but the percentage is usually low (10%). Dried samaras can be stored in sealed containers at 36 to 40°F for several years without loss of viability. The seed requires cold stratification for 2 to 3 months. Fresh seed sown in the fall will germinate the following spring. Mulch seed beds after sowing [*The Plant Propagator* 20(3):14 (1974)].

CUTTINGS: Difficult from mature plants. One report noted that July cuttings, made with basal cut ½″ below node, rooted 52%. Juvenility is probably an important factor if this species is to be rooted in high percentages.

GRAFTING: 'Aureo-marginatum', 'Fastigiatum' and other cultivars are whip grafted on pot grown seedling stock of L. *tulipifera*. Perhaps chip budding might serve as an alternative during the summer months.

Lonicera Honeysuckle

The number of honeysuckle species and cultivars is enormous and includes about 180 species of deciduous and evergreen shrubs. Many species are cultivated for their attractive flowers and fruits. They are easy to grow, propagate readily from seeds and cuttings, and are generally free of serious insects and diseases.

SEED: The ornamental berries are translucent, orange, red, purple, blue or black in color, and mature in the summer or early fall. Because honeysuckle species hybridize freely, seed should be collected from isolated plants. Depending on the species, each fruit may contain several to many small seeds. Seed extraction is accomplished by maceration in water. Typical average seed yields per pound vary from about 140,000 for L. *tatarica* and L. *maackii* to 326,000 for L. *involucrata*. Heit reported little loss in viability with L. *tatarica* after 15 years with dry seed stored at 34 to 38°F. Other species probably react the same.

Except for L. *canadensis* and L. *dioica*, all species show some dormancy. In most species, embryo dormancy is responsible. Representative of the embryo dormant group would be ornamental species such as L. *tatarica* and L. *maackii*. Cold stratification (41°F) for 30 to 90 days is recommended for embryo dormant species. For species with seedcoat dormancy, cold stratification should be preceded by warm stratification (68 to 86°F). See *Amer. Nurseryman* Nov. 15 p. 12, 1967; and Aug. 15 p. 8, 1968.

CUTTINGS: Softwood, June, July, August, root with ease under mist. For example, vine types such as L. × *brownii*, L. × *heckrottii* 'Gold Flame', L. *japonica* 'Halliana', L. *japonica* 'Aureo-reticulata', L. *periclymenum*, and L. *sempervirens* root readily with nodal tip cuttings or double leaf bud cuttings using 3000 ppm IBA-talc. IBA in the range 1000 to 3000 ppm hastens and unifies rooting. Cuttings often continue to grow after rooting [*Gardener's Chronicle* 180(1):17 (1976)].

Shrub honeysuckles, including L. *alpigena*, L. × *bella*, L. *caerulea*, L. *fragrantissima*, L. *korolkowii*, L. *maackii*, L. *maximowiczii*, L. *morrowii*, L. *nitida*, L. *pileata*, L. *tatarica*, and L. *xylosteum* root easily from softwood under mist without hormone. However, 1000 to 3000 ppm IBA-talc or solution improves and unifies the rooting.

Hardwood cuttings of shrub types, such as L. × *xylosteoides*, root readily. Make 8″ cuttings in the fall as soon as the wood is ripe and plant 7″ deep into a sandy soil. Mulch fall planted hardwoods [*The Plant Propagator* 15(3):10 (1969)].

Layering, although not necessary with present propagation methods, was mentioned in the literature with a broad number of species [*Arnoldia* 13(4):25 (1953)].

Loropetalum chinense Chinese Loropetalum

SEED: Fruit is a two valved capsule like *Hamamelis*. There are no reports of successful germination in the literature. Authors suggest a 3 month warm/3 month cold period.

CUTTINGS: Senior author rooted cuttings about 80% in late July, with 3000 ppm IBA-quick dip, peat:perlite, and mist. The rooting medium should not hold too much moisture or the cuttings will rot. Ideally, cuttings should be firm at the base and direct stuck in rooting containers or left in the flat until spring. Overwintering has been a problem for some growers. In general, do not transplant after rooting. Hardwood cuttings, late December, 4000 ppm IBA-talc and thiram, rooted 75%. The use of a warm bench or polytent during the winter months might be beneficial.

Lycium chinense Chinese Wolfberry

SEED: Fresh seed germinates without pretreatment.

Lyonia mariana Staggerbush Lyonia
Lyonia ovalifolia Tibet Lyonia

SEED: All germinate without pretreatment. Treat like *Rhododendron* or *Calluna*.

Maackia amurensis Amur Maackia

SEED: Collect pods in fall, dry inside, remove small, brown, hard-coated seeds and store dry under refrigeration. The seed can be soaked in hot water and allowed to cool for 24 hours or acid scarified for 60 minutes. Untreated seed germinated 5%; 94% after a hot water soak; and 96% after a one hour acid treatment. M. *chinensis* is similar and could probably be treated the same.

CUTTINGS: Root cuttings collected in the fall and planted outside are successful. M. *chinensis* responds similarly [*Gardeners Chronicle* 169(3):22 (1971)].

× *Macludrania hybrida*

CUTTINGS: Mid-July, 8000 ppm IBA-talc + thiram, rooted 90% in 8 weeks and all survived.

Maclura pomifera Osage-orange

SEED: The fruit is a large, glabrous compound fruit (syncarp of drupes) which ripens in September and October and averages 4 to 5″ in diameter. Seeds can be extracted by macerating the fruit and floating off the pulp. An easier method is to ferment fruits for several months before extraction or store in a pile over winter out-of-doors which yields easily macerated fruit. Seeds from fruits stored out-of-doors germinate readily. One pound of clean seed contains approximately 14,000 seeds. Fresh seed exhibits a slight dormancy that can be overcome with stratification at 41°F for 30 days. A two day water soak has alleviated dormancy and permitted germination to occur. Average germination is about 60%. Seed can be stored dry for at least 3 years at 41°F.

CUTTINGS: Softwood, mid-June (Kansas), or hardwood cuttings in mid-winter (January) root best. Hormone treatments improve rooting. For softwoods use 5000 or 10,000 ppm IBA, mist and sand medium. 100% rooting can be achieved in 40 to 50 days. Hardwood (4 to 6″), late January, previous season's growth, cool greenhouse, bottom heat (68°F), and 5000 to 10,000 ppm IBA-quick dip will root. Cover the hardwoods with an opaque poly-covering until more light is needed for shoot development. Rooted in 6 weeks. These rooted cuttings produced about 30″ of new growth in a single season which is about twice as much as a 2-year-old rooted softwood. See PIPPS 30:348 (1980) and *The Plant Propagator* 30(1):6 (1984).

Root cuttings taken in fall and early winter are also possible.

GRAFTING: Fall sown seed produces the best understock for budding. Fall sown seed will produce seedlings large enough to bud by August of the following year or the following May. Best trees are produced from August budding.

Magnolia Magnolia

The magnolias belong to the Magnoliaceae, which is composed of about 85 species native to the Americas and Asia. The genus includes deciduous and evergreen types that are valued primarily for their showy flowers but also foliage and fruit. *Magnolia grandiflora*, southern magnolia; M. *kobus*, kobus magnolia; M. × *soulangiana*, saucer magnolia; M. *quinquepeta*, lily magnolia; M. *stellata*, star magnolia; and M. *virginiana*, sweetbay magnolia are the most important landscape species.

SEED: The major reason for raising magnolias from seed is to provide understock. Although most species can be propagated from seed, they are slow to flower and differ from the parent. In the United States, M. *kobus* and M. *acuminata* are widely used as understocks. Other species often raised from seed include M. *grandiflora*, M. × *soulangiana*, M. *stellata* and M. *virginiana*.

The cone-like (aggregate of follicles) fruits of magnolia vary from 2 to 6″ in length and are composed of several to many, 1- to 2-seeded follicles. The fruits ripen from late summer to mid fall. At maturity the orange-red to red outer seed coat (aril) is fleshy and oily, and the inner seed coat is hard. Seed should be harvested when the fruit is fully mature, that is when the follicles are opening to expose the seeds. If all the follicles are not open, the fruit should be placed in a warm room to complete ripening. Collecting the fruits early is important because the seeds, once they emerge from the follicle, will often dehisce.

Optimum germination occurs with the fleshy pulp removed. The pulp is removed by maceration. Soak in water for a couple of days with daily water changes, macerate by squeezing and lightly rubbing over a screen, and then float off the pulp. Many investigators have reported that the fleshy aril contains inhibitors that delay germination. Excessive drying after maceration can lead to loss of viability. Seed size varies considerably with different collection years and geographical origins.

Magnolia seed can be kept for several years with little loss of viability when stored in sealed containers at 32 to 41°F. Some magnolia species lose viability if stored over winter at room temperature. Magnolia seeds exhibit embryo dormancy which is overcome by cold stratification at 32 to 41°F for 2 to 4 months. Fresh seeds show extreme variation in seedling production potential. It is believed that improper or incomplete fertilization of the embryo is the cause. In nursery practice, magnolia seeds can either be fall sown with heavy mulching or stratified and sown in late winter or spring. Ample evidence exists that suggests the embryos of many *Magnolia* species are immature (morphologically dormant) and require a period of after-ripening during which embryos reach germinable size. In senior author's work, freshly cleaned seeds of *Magnolia grandiflora* were sown in flats in a warm greenhouse and germinated about 75% four months later.

CUTTINGS: Cutting propagation is the preferred method of asexual propagation. Many species and cultivars can be cutting propagated. Cutting wood selected from young, previously rooted stock plants provides the best source of propagation material. The importance of juvenile stock plants cannot be overemphasized. A large north Alabama nursery cuts stock plants of *Magnolia* × *soulangiana* to the ground in late winter to induce new shoot growth. Fungicides, such as Benlate, are often included in combination with the rooting hormone and as a spray while in the propagation bench. Care must be exercised in the transplanting process because roots are large, fleshy and often sparse. Senior author has rooted cuttings of many deciduous and evergreen species. In general, the cuttings should be between the soft and semi-hardwood condition. The terminal bud should be developing with the last leaves close to full size. Magnolia is responsive to higher hormone levels and 5000 to

10,000 ppm IBA-talc or solution (and higher) are used. Medium should be well drained and a uniform mist provided. In summer, bottom heat is not necessary but can be used. For *Magnolia grandiflora*, high bottom heat (75 to 85°F) has been recommended but senior author has found it not absolutely necessary. Magnolias do not resist transplanting after rooting and will make significant root growth. Roots are large, coarse, fleshy and usually not present in great numbers (3 to 7 per rooted cutting). Care should be exercised in the transplanting process.

Wounding may be beneficial and is probably helpful, however, many species root well without the procedure.

Not all magnolias root with equal facility. *Magnolia heptapeta* (*M. denudata*), *M. acuminata*, *M. cordata*, *M. grandiflora* and large tree type species are difficult. Like *Syringa*, *Magnolia* species and cultivars have a narrow "window of rootability". For most taxa, June and July are optimum months. However, *Magnolia grandiflora* roots best in August-September from firm-wooded cuttings. There is some evidence that NAA is more effective than IBA on *M. grandiflora*. It is possible this relationship extends to other species. Magnolias take 6 to 8 or 10 to 12 weeks for complete rooting, depending on species or cultivar.

GRAFTING: Grafting or budding provide reliable methods of propagating magnolia species and cultivars that are difficult to root from cuttings. The grafting and budding of magnolias has been recently reviewed by Treseder (*Magnolias*. 1978. Faber and Faber Limited, London). The type of magnolia used for understock does not appear to be important as incompatibilities are not a problem. The understock should be 3 to 4-year-old seedlings or rooted cuttings, and hardwooded at the base where the grafting will be carried out.

There is no evidence that closely related species produce better ultimate plants and the choice depends on the understock available. *M. hypoleuca* and *M. kobus* are widely used, with *M. acuminata* often used in the United States. In England, *M. sargentiana* var. *robusta* is often used. Other species, including *M. salicifolia*, *M. sieboldii*, and *M. tripetala* have also been mentioned. The type of graft utilized does not affect the success of the operation and the side and veneer grafts are commonly used [PIPPS 3:113 (1953)].

Grafting is possible from mid-August until growth begins in the spring. Some species, however, showed reduced success after February. Potted stock is best from mid August until November. Bare-root stock is used after November. Stock that is hardened at the base, that which has brown bark, is best. The same for the scion, with side spur type growth preferred. Late summer grafts are best side grafted and waxed after taping to avoid the entry of water that will lead to failure. Reduce the leaf surface by half on the scion. Place the grafts in a grafting case and cover the pots with moist peat. Regular syringing together with shading and controlled ventilation is required to prevent leaf wilting. After a few weeks hardening off begins.

Chip-budding can be practiced any time of the year when material is available. The buds should be tied and waxed or covered with polyethylene wrap or parafilm. To achieve rapid callusing, transfer to humid atmosphere at 70 to 80°F for 2 to 3 weeks. The stocks are headed back at bud break.

T-budding is extensively practiced in the orient. In late summer seedlings are budded and wrapped with polyethylene. The following spring the grafts are headed back.

TISSUE CULTURE: Success has proven elusive for *Magnolia* species and cultivars. One commercial laboratory has successfully cultured 'Elizabeth' and one or two others. Apparently the time in culture before proliferation takes place is extended and multiplication rates are not high. A Woody Plant Medium with 2 mg/l BA and patience would be a good starting point. Senior author has tried a *Magnolia grandiflora* clone with no success. Phenolic "bleeding" is a problem and perhaps with numerous subcultures the problem could be arrested.

Magnolia acuminata American Cucumber Tree or Cucumber Tree Magnolia

SEED: *Magnolia acuminata* is widely used as an understock for grafting in North America. This magnolia is usually self incompatible with its own pollen and seldom fruits in isolation, however, self-compatible clones are known. Seeds can be cleaned, dried and stored for over a year at 32 to 41°F. One pound of clean seed contains approximately 5,000 seeds. *Magnolia acuminata* exhibits embryo dormancy which is overcome by moist stratification at 32 to 41°F for 3 months. Germination percentage is variable but can average 80%. Freshly cleaned seed germinated 17%; 73% after 3 or 4 months cold stratification. Fall and spring (after stratification) sowing are practical in the nursery trade. After germination the seedlings require ½ shade the first summer. Seedlings grow fast and make handsome small trees [*Arnoldia* 41(2):36 (1981)].

GRAFTING: Can be chip-budded in August. The buds should be completely covered with polyethylene plastic. Often used as understock for grafting other magnolias because of tolerance to varied soils and climates. *Magnolia grandiflora* has been successfully grafted on this species.

Magnolia ashei Ashe Magnolia

SEED: Three to 4 months cold stratification at 40°F has proven best.

Magnolia cordata

SEED: Fresh seed germinated 6%; 72% after 5 months cold stratification.

Magnolia cylindrica

SEED: Five months cold stratification at 40°F has proven best.

CUTTINGS: Late July, 8000 ppm IBA-talc plus thiram, sand: peat, rooted 70%.

Magnolia fraseri Fraser's Magnolia

SEED: Propagation by seeds is currently the only method available for reproducing this species. One pound of seed contains approximately 4,500 seeds. Fraser's magnolia germinates about 80% after 3 months cold stratification.

Magnolia grandiflora Southern Magnolia

SEED: See *Magnolia* introduction and *M. acuminata* for general comments on seed harvesting and storage. *Magnolia grandiflora* is highly self-fertile and produces large amounts of seed. Average number of cleaned seeds per pound is about 6,000. Seed requires 3 months at 40°F or can be cleaned and sown without cold stratification with germination occurring in 4 months. The delayed germination is related to an immature embryo and the subsequent long after-ripening process. Results have been variable probably due to condition of embryos. The seeds will germinate without cold stratification [*Amer. Nurseryman* 162(9):38-51 (1985)].

CUTTINGS: Cutting propagation is the preferred method for this species. The importance of bottom heat, sanitation and the use of fungicides has been stressed. One report mentioned the use of Clorox treated metal flats, for the control of pathenogenic microorganisms. The months of July and August (September) are the optimum for cutting propagation, although reports have noted successful cutting propagation into November. Take terminal cuttings, 4½ to 8″ long, treat with 3000 or 5000 ppm IBA-talc and place under mist with bottom heat (75 to 80°F). NAA also has proven effective when applied at a concentration of 10,000 ppm in 50% alcohol. Rooting of 'Bracken's Brown Beauty' was significantly improved with NAA compared to IBA at comparable concentrations. Wounding has been reported to promote rooting and the leaves are not reduced in size. A range of rooting media, including vermiculite, perlite, sand, sand and perlite, have been used. When using high bottom heat (80°F) it is important to keep the cuttings well watered. Fungicides, such as Captan, have been used as a cutting dip prior to sticking followed by a fungicide spray every 2 weeks. A soluble fertilizer application at the time of callus formation and rooting also has been reported to improve subsequent cutting development. Roots are brittle and coarse, and care is necessary to prevent root damage which will retard cutting growth. The above procedure also works for 'Claudia Wanamaker', 'Margaret Davis', 'Hasse', and 'Shady Grove #4, 5 and 6'. 'Victoria' roots better in November. 'Edith Bogue', late November (Boston), 10,000 ppm IBA, plastic tent, rooted 100% in 10 weeks. 'Gallissoniensis'' roots better from apical cuttings between August and November than median or basal cuttings. Rooting conditions include 20,000 ppm IBA-quick dip and mist [PIPPS 17:261 (1967); 25:88 (1975); 31:620 (1981); 33:622 (1983)].

Senior author has found the following recipe effective for a number of *Magnolia grandiflora* clones:

Collect 4 to 6″ long, firm-wood cuttings with end bud set, August-September, leave 2 to 3 top leaves, 10,000 ppm NAA-50% ethanol, 5-second dip, horticultural or coarse perlite at least 4″ deep, mist, polytent covering the media and mist, 53% shade. Rooting takes 10 to 12 weeks. The two most important ingredients are the NAA and perlite. In 1986, rooting was not as good as previous years. At time of the evaulation, the cuttings that were alive were retreated with 10,000 ppm NAA-solution. Six weeks later, rooting had doubled over the initial results averaging 50 to 70% for five different clones. For all but one clone, the cutting wood was collected from mature trees in North Florida that had experienced one of the worst droughts in the history of the southeast. Another key in success is the enclosed polytent which traps the heat and keeps the medium warm. See Chapter II, Cutting Propagation, for photographs of *Magnolia grandiflora* cuttings rooted with this system.

GRAFTING: The usual procedure (California) is to graft onto seedling understock in 1 gallon containers. The understock stem diameter must be between 2/5 to 1/2″. In September or October, the understock is pruned to 24″ for uniformity. In November, after some cold, the understock is brought into the greenhouse and the side branches and lower leaves are removed, then placed in a grafting tent with a layer of moist peat on the bottom. The understock is sprayed with Physan and allowed to dry before grafting. The scion wood, 8 to 10″ terminal stem ends, is stripped of leaves and disinfested in a bath of Physan (200 ppm). Scions can be placed in plastic bags and stored for 3 to 4 days at 38 to 40°F before use if necessary. A side graft is made with longitudinal cuts 1″ in length. The graft is wrapped with wide rubber strips from top to bottom and sprayed with Benlate. In 2 weeks the first signs of growth occur and the grafts are removed when 1 to 2 new leaves are fully expanded. After 2 weeks, spray weekly on an alternating schedule with a fungicide such as Benlate. Harden-off by placing on an open bench and mist as needed for 2 weeks and then move to a shaded area and continue misting for an additional 2 weeks [PIPPS 30:92 (1980)].

T-budding in the spring or fall is possible when controlled greenhouse environments are not available or stock is limited. Seedling understock is grown in containers and must be 1/4″ or larger to accommodate the large bud. Because the bark is thick it is necessary to cut deeply into the rootstock and remove the wood from the shield for good cambial contact. After bud insertion wrap with a budding rubber strip and remove 1/3 to 1/2 of the rootstock. The budding rubber is removed after 3 to 4 weeks and the final cut on the understock is made 4 weeks later. In fall budding, the final cut is not made until new growth starts in the spring [PIPPS 31:616 (1981)].

Magnolia heptapeta (*M. denudata*) Yulan Magnolia

SEED: Possibly self-incompatible and rarely sets seed but has been found to readily accept pollen from other magnolias. Seed, therefore, may be of hybrid origin. It is the female parent of the M. × *soulangiana* race of magnolias. Seeds should be cold stratified 3 months at 32 to 41°F to overcome embryo dormancy.

CUTTINGS: The best time for cutting propagation is at the end of June or beginning of July (New Jersey) just as the first flush of growth is ending. Terminal cuttings are used with all but the top 2 to 3 leaves removed. The hormone used has varied from 3000 to 8000 ppm IBA-talc, 4.5% IBA plus Benlate to a mixture of 4% IBA + 4× Cutstart (1:1) + 1/16 by volume of fungicide such as Captan. Wounding may be beneficial. Media used include either sand or sand:peat (1:3, v/v) with bottom heat at 70 to 75°F, and mist. Rooting occurs in 6 to 10 weeks. Early June cuttings (Boston), wound, 1% IBA, rooted 85 to 100% in 13 weeks; 3000 and 8000 ppm IBA-talc produced 20 and 50%, respectively. The use of high hormone is important for most magnolias. This particular species is considered one of the most difficult to root from cuttings. As mentioned in the introduction, timing is critical [PIPPS 30:374 (1980); 32:89 (1982)].

GRAFTING: *Magnolia heptapeta* is often grafted or budded on seedlings of M. *kobus* or M. × *soulangiana*.

Magnolia hypoleuca (*obovata*) Whiteleaf Japanese Magnolia

SEED: Cold stratification at 40°F for 4 and 5 months produced 96 and 92% germination, respectively. A 3 month cold stratification resulted in 32% germination. The improved success with the longer cold period might relate to the maturation of the embryo.

Magnolia kobus Kobus Magnolia

SEED: Imported seed of this species from Japan failed to germinate possibly because of drying during shipping. This species is used as a major understock for grafting in Asia. Average number of cleaned seeds per pound is about 4,300. Cold stratification (40°F) for 3 months facilitates best germination [*Amer. Nurseryman* Sept. 15, (1975); PIPPS 23:285 (1973)].

CUTTINGS: Late June to early July is reported to be the best time. A heavy wound plus 2% IBA-talc are used. Place in a greenhouse with mist. Rooting takes about 5 weeks. Early June (Boston) cuttings, 8000 ppm IBA-talc plus thiram, rooted 50%. In another study, cuttings were taken from a mature tree (flowering stage M. *kobus* var. *borealis*), late June, 2% IBA and rooted 80%. Untreated cuttings did not root. See *Plant Propagator* 29(1):4 (1983); PIPPS 24:390 (1974) and 30:374 (1980).

Magnolia × loebneri Loebner Magnolia

CUTTINGS: A hybrid group of great beauty that includes the cultivars 'Merrill' and 'Leonard Messel'. Senior author has rooted 'Leonard Messel' 100% from late July (Boston) cuttings, 4000 ppm IBA-quick dip, peat:perlite, and mist in 8 weeks. The wood should be firm and the end bud set; no wound is necessary. Also included here are 'Spring Snow' and 'Ballerina'. 8000 to 10000 ppm IBA-solution produces good results with this group.

Magnolia macrophylla Big Leaf Magnolia

SEED: Seeds should be stratified for 3 months at 35 to 45°F as soon as cleaned. One report, however, indicated that dried seed, planted immediately, germinated 44%; seed stratified for 3 months at 40°F germinated no better.

CUTTINGS: *Magnolia macrophylla* is not propagated by cuttings because of leaves that can range in size up to 3′ long. It is also one of the large pith species which do not propagate readily from cuttings.

GRAFTING: In addition to budding on seedlings of the same species, M. *tripetala* and M. *officinalis* can also be utilized. McDaniel, however, recommended seedlings of the same species as the best understock. He noted that because of the thick pith and thin bark the usual tree grafting methods were not feasible. He utilized a thin deep-bud taken from last year's well developed stems if budding was done in the spring. Stock should be 1/2″ or larger. For August to early September budding, buds can be taken from current season's wood. After budding, wrap the bud with polyethylene plastic film for about 3 weeks.

Magnolia officinalis

SEED: Three months cold stratification induces the best germination.

Magnolia quinquepeta (M. *liliflora*) Lily Magnolia

SEED: This species may be self incompatible.

CUTTINGS: This species has been rooted with and without mist. In the non-mist method, cuttings 3 to 6″ long with the soft terminal removed and leaves reduced by 1/2 are used. IBA at 8000 ppm plus Benlate is applied to wounded cuttings. Media included sand or sand:peat which are drenched with Benlate prior to sticking. Cuttings are well watered and covered with sash and lath. Keep well watered the first 2 weeks and apply weekly sprays of Benlate. Do not dig until the cold period is satisfied. In February pot and place in the greenhouse and then to the field. This technique may prove useful for other smaller leaved magnolias. 'Nigra', mid July (Boston), 10,000 ppm IBA, mist, rooted 88% with heavy root systems in 5 weeks [PIPPS 28:570 (1978)].

Magnolia salicifolia Anise Magnolia

SEED: A 3 month cold period induces the best germination.

CUTTINGS: Late June (Boston), wound, 10,000 ppm IBA, rooted 90% with heavy root systems in 5 weeks. Wounding was beneficial. Untreated (no hormone) cuttings did not root.

Magnolia sieboldii Siebold Magnolia

SEED: Handle as described in the introduction.

CUTTINGS: Canada, 8000 ppm IBA-talc, wound, peat moss: perlite:sand, rooted 93 to 100% under various supplemental carbon dioxide and light treatments. Root numbers ranged from 6.5 to 10.3 per cutting indicating strong root systems. *Magnolia sinensis* and *Magnolia wilsonii* might also be rooted this way.

Magnolia × soulangiana Saucer Magnolia

SEED: This species is of hybrid origin (M. *heptapeta* × M. *quinquepeta*) and the numerous cultivars are vegetatively propagated. This species is self-fertile and produces seed. In Europe it is widely used as an understock. Average number of cleaned seeds per pound is about 2,700. 3 to 6 months stratification is recommended. Again, the longer period may be necessary for seeds in which the embryo has not fully developed.

CUTTINGS: Take 4 to 6″ cuttings from vigorous field stock or container plants, remove the tip and wound. Cutting wood of 'San Jose', 'Lennei', 'Lennei Alba' and 'Rubra' is ready when the terminal snaps out. Coarse sand and perlite or pumice (1:1 v/v) have been used with high bottom heat (75 to 78°F) plus mist. IBA from 8000 ppm to 1.6% is recommended. Misting is critical with the objective to keep the cuttings from wilting — no mist is applied at night. After sticking use a Benlate spray and continue as needed until leaf drop. Do not overcrowd the cuttings because *Botrytis* can be a problem. Pot after rooting and overwinter in a cool greenhouse with enough heat to prevent freezing. Pot as new growth commences or plant out in the spring. Commercial growers maintain stock blocks by cutting plants to the ground in late winter to induce vigorous growth. These "juvenile" cuttings are much easier to root than those collected from older plants [PIPPS 25:79 (1975); 31:619 (1981)].

'Amabilis' — early August, 8000 ppm IBA-talc plus Captan, rooted 75% in 6 weeks. 'Candolleana' — mid-July, 8000 ppm IBA-talc, rooted 50%. 'Speciosa' — early September, wound, 8000 ppm IBA-talc, rooted 75% in 9 weeks. 'Verbanica' — as with 'Speciosa', rooted 76% in 9 weeks.

LAYERING: Layering was widely used with this species before cutting propagation. In late June to early July (Alabama) stems about 4′ high with a diameter of 3/8 to 5/8″ are used for layering. As low as possible a tongue cut is made half way through the stem. Sphagnum moss is inserted to keep the cut apart and also carry the hormone. The hormone-containing moss is prepared by mixing 4 tablespoons hormone (8,000 ppm IBA-talc) into a 10 qt. pail of well moistened sphagnum moss. Prepare a container (18″ high) with no bottom and fill with well rotted sawdust. It is important to keep the sawdust moist during the rooting process. The layers are well rooted by August and at that time are pruned back to within 6 to 8″ of the top of the mound. Sever and transplant into the field in November [PIPPS 14:67 (1964)].

Magnolia stellata Star Magnolia

SEED: The star magnolia is self-fertile and produces seed. Average number of cleaned seeds per pound is about 5,300. A 2 to 3 month cold stratification (40°F) produces the best germination.

CUTTINGS: Cuttings root readily in June-July. 1.0% IBA-solution, wound, sand or peat:perlite, mist, are ideal. Six to 9" cuttings are collected in early morning to avoid wilting. Bottom heat at 68 to 78°F is recommended. Early June (Boston), wound, 10,000 ppm IBA-talc, rooted 100% in 5 weeks. Untreated cuttings did not root while lower IBA concentration (3000 ppm) resulted in 40% rooting. Late July cuttings, wound, 8000 ppm IBA-talc plus thiram, rooted 94%. The var. *rosea* ('Rosea'), late June, 2% IBA-talc, rooted 40% in 5 weeks. Untreated cuttings did not root. The hybrid 'Little Girl Series' (M. *stellata* × M. *quinquepeta*), also roots readily from cuttings. 'Ann', late July (Boston), 4000 ppm IBA quick dip, peat:perlite, mist, rooted 100% in 8 weeks. Cutting wood should be firm with the end bud set [PIPPS 13:170 (1963); 32:89 (1982)].

Leaf bud cuttings also have been rooted successfully from early July cuttings in a plastic tent. Sharp sand, bottom heat (72 to 75°F) and 1000 ppm IBA-talc, were employed with a fungicide drench.

GRAFTING: Use a side graft with M. × *soulangiana* and M. *kobus* seedlings serving as understock.

Magnolia tripetala Umbrella Magnolia

SEED: This species is self-fertile. One pound of seed contains 7,500 seeds. Cold stratify (35 to 41°F) for 1½ to 2 months; a germination percentage of greater than 80% is common. Numerous fruits are often produced with ripening (rose-red blush) evident as early as late August into September.

Magnolia virginiana Sweetbay Magnolia

SEED: One pound of seed contains an average of 7,500 seeds. Cold stratify (35 to 41°F) for 1 to 2 months; a germination rate of 80% is common [*Arnoldia* 41(2):36 (1981)].

There are two varieties: (1) *virginiana* — which is largely deciduous and highly compatible with its own pollen; and (2) *australis* — which is more evergreen and incompatible with its pollen. M. *virginiana* has been reported to hybridize with M. *grandiflora*, M. *obovata* (*hypoleuca*), M. *macrophylla*, and M. *tripetala*. McDaniel recommends the cultivars 'Havenei' and 'Mayer' as good seed sources. 'Havenei' when pollinated with either M. *grandiflora* or M. *macrophylla* produces apomictic seed [PIPPS 20:199 (1970)].

CUTTINGS: Root well from young trees, however, with maturity rooting response drops. Early July softwoods, 5000 to 8000 ppm IBA-talc, peat:perlite (1:1, v/v), rooted up to 64%. Timing and hormone need to be worked out. Cuttings from mature trees are not easily rooted [*The Plant Propagator* 30(4):11 (1984)].

× *Mahoberberis aquicandidula*

CUTTINGS: An intergeneric cross between *Mahonia aquifolium* and *Berberis candidula*. In general, cuttings of × *Mahoberberis* crosses should be taken after several hard freezes and treated in the same manner as *Mahonia*. Early December stem cuttings, 3000 ppm IBA-talc, sand, rooted 93% in 10 weeks. Cuttings have also been rooted in September.

× *Mahoberberis aquisargentiae*

CUTTINGS: An intergeneric hybrid between *Mahonia aquifolium* and *Berberis sargentiana*. When treated as × M. *aquicandidula*, 100% rooting resulted. Cuttings taken earlier did not root as well. A wound did not prove beneficial. September and November cuttings root well.

× *Mahoberberis neubertii*

CUTTINGS: An intergeneric cross between *Mahonia aquifolium* and *Berberis vulgaris*. Late November stem cuttings, 3000 ppm IBA-talc, rooted 44% in 8 weeks.

Mahonia Grapeholly

Fruit is a fleshy, bluish berry with multiple seeds. Collect fruits in May-Summer, macerate, dry and store seeds under refrigeration. Timing is very important since birds will consume fruits especially on early ripening species like M. *bealei* and M. *japonica*. Mahonias hybridize freely and fruits should be collected from isolated plants.

Mahonia aquifolium Oregon Grapeholly

SEED: The blue-black berry ripens August to September and may persist until December. Collect when ripe and clean by maceration and sow. Embryo excision tests have indicated the possibility of an immature or improperly developed embryo. Maximum germination is obtained with 4 months warm stratification (68°F) followed by 4 months cold stratification (36 to 40°F). The warm period permits embryo maturation to occur.

CUTTINGS: Propagated from both leaf bud and stem cuttings. Conflicting reports regarding the best time to collect cuttings. In one study, cuttings taken at the time of the first heavy frost rooted in 5 weeks. Cutting wood selected earlier blackened on the bottom, while cuttings taken later took 3 to 4 months to root. Use 8000 ppm IBA-talc and place under mist. In other studies (see later) late summer cuttings rooted well. In milder climates (England) the cuttings are stuck in double-glazed frames outside with soil warming cables holding the temperature at 68°F. Mid-August (Boston), 3000 ppm IBA-talc, rooted 84% in 8 weeks; late November, 3000 ppm IBA-talc, rooted 74% in 7 weeks. Several nurserymen have noted that November is an ideal time to take cuttings. See *The Plant Propagator* 26(3):11 (1980).

'Compactum' is reported to be more difficult to root. Terminal cuttings root best when harvested in late September (California) but can be taken into November. Avoid green wood cuttings. Selection of proper wood appears to be more important than hormone used. Mahonia cuttings do not tolerate heat while rooting. In California, rooting takes place in a lath house with shade (50%) on gravel beds under hourly watering provided by 'Spray Stakes'. 'Compactum' requires a well aerated medium such as peat and perlite (1:10, v/v) — other mahonias can tolerate less. No bottom heat is used. The cuttings are stuck in flats after being treated with 8000 ppm IBA-talc plus Benlate. Benlate sprays are also applied during rooting. Other propagation schemes report soaking the cuttings in fungicidal solution followed by regular Benlate/Captan sprays. Rooting commences in about a month. Avoid stress from heat, bright sun, wind or dryness which will cause leaves to redden and drop.

Mahonia 'Golden Abundance' (hybrid between M. *aquifolium* and M. *amplectens*) roots best if taken in November when the cuttings are firm and beginning to turn slightly reddish brown. Treat with 8000 ppm IBA-talc and place under mist in a peat:perlite medium (1:1, v/v).

Mahonia bealei Leatherleaf Mahonia

SEED: This species benefits from a limited period of cold stratification (1 to 2 months) to improve the rate and uniformity of germination. Typically fruits are collected in late spring

(Georgia), cleaned and sown immediately with excellent germination. Seedlings are 2 to 3" high by September.

CUTTINGS: The hybrid 'Arthur Menzies' has been successfully rooted from leaf-bud cuttings taken at the end of June to late July. Other times may also work. Two-thirds of each leaf is removed before dipping the cutting in 8000 ppm IBA-talc plus Benlate (5:1, v/v). A coarse river sand and peat moss (3:1, v/v) medium is drenched with Truban (1 tablespoon/3 gallons) to reduce stem rot. Mist plus bottom heat (70 to 72°F) are used. Rooting takes place in 7 to 8 weeks.

GRAFTING: *Mahonia bealei* can be grafted on M. *aquifolium* utilizing a side graft in mid-January. After grafting, the pots are plunged in moist peat moss to a point slightly above the graft union. The bench is covered with plastic for 5 weeks followed by hardening-off.

Mahonia japonica Japanese Mahonia

SEED: Germination pretreatments are slight, but the species benefits from a limited period of cold stratification, 1 to 2 months, to improve the rate and uniformity of germination. Probably very similar to M. *bealei*.

CUTTINGS: Leaf bud cuttings taken in December and January (England) have been successful. Wound, 2% IBA-talc, sphagnum peat moss medium in a heated greenhouse (65 to 68°F). Wet the cuttings and cover with clear polyethylene. The poly cover is removed once a week and left off overnight to air. Spray foliage lightly and cover the next day.

Terminal or leaf bud cuttings, October to November, without hormone, sphagnum peat:perlite (2:1, v/v), bottom heat (65°F), and mist also have been successful. This same procedure has been utilized to root M. *x media* 'Charity', 'Winter Sun' and 'Lionel Fortescue'. 'Charity' terminal or leaf bud cuttings, October to November, without hormone, sphagnum peat:perlite (2:1, v/v), bottom heat (65°F), mist, have also rooted [NGC 188(2):33 (1980)].

Mahonia lomariifolia

SEED: Does not require pretreatment to germinate.

CUTTINGS: Refer to M. *japonica*.

Mahonia nervosa Cascades Mahonia

SEED: 3 to 5 months cold stratification.

CUTTINGS: Late November, polytent, sand, 8000 ppm IBA-talc + thiram, rooted 100% in 15 weeks with excellent root systems.

Mahonia pinnata Cluster Mahonia

CUTTINGS: Terminal or leaf bud cuttings, October to November, no hormone, sphagnum peat:perlite (2:1, v/v), mist, root readily. Late November, 8000 ppm IBA-talc, rooted 40%. If leaf drop is a problem, collect cuttings from August to September.

Mahonia repens Creeping Mahonia

SEED: 3 months cold stratification.

CUTTINGS: Not the easiest plant to root, mid-August, 3000 ppm IBA-talc, rooted 18% in 6 weeks; late November cuttings same treatment rooted 22%; early December, 8000 ppm IBA-talc, peat:vermiculite, rooted 100%.

Mahonia wagneri

SEED: 6 months cold stratification.

CUTTINGS: Late October, 3000 ppm IBA-talc or 8000 ppm IBA-talc, rooted 50 and 100%, respectively, in 6 weeks.

Malus Crabapples (Apples)

Crabapples are an outstanding group of small flowering trees for landscape planting. They are chiefly valued for their flowers and fruits but also for foliage and growth habit. The actual number of types, both species and hybrids, is extensive and probably hovers around 700. Crabapples tend to be cross fertile and freely hybridize. Practically all flowering crabapples are propagated by asexual methods such as grafting, budding or cuttings. Commercially, crabapples are usually field-budded in August or bench grafted during the winter months. Cutting and tissue culture propagation are assuming increased commercial importance.

SEED: Seed is extracted from the fruit by macerating in water, floating off the pulp and screening out the seed. Small quantities can be place in plastic bags, the pulp softened and then macerated or passed through a screen. Seeds are about 1/16" long and rich brown. Seeds dried to a moisture content of less than 11% and stored in sealed containers at 36 to 50°F have been stored for over 2 years without loss of germination capacity. Crabapple seed displays embryo dormancy which is overcome by cold moist stratification for 1 to 4 months. Stratified seed germinates in 30 to 60 days. In nursery practice unstratified seed can be sown in late fall or stratified seed can be sown in the spring. See *The Plant Propagator* 14(4):7 (1968).

Seeds of M. *pumila* (*communis*), and possibly some of the crabapples, will start to sprout once dormancy is broken — even at 36°F. Such sprouting can be safely inhibited by freezing the seed to 23°F after dormancy is broken.

Malus florentina, M. *hupehensis*, M. *sargentii*, M. *sikkimensis* and M. *toringoides* will reproduce true from seed [PIPPS 5:84 (1955)]. Selected forms of other species, such as M. *sargentii* and M. *hupehensis*, should be asexually propagated. Apomictic species of *Malus* produce more uniform seedlings than do sexually reproduced seedlings. Some of those apomictic species may be useful as rootstocks. The following asiatic species are apomictic and breed true even when pollinated by other species: M. *hupehensis*, M. *toringoides* and M. *sikkimensis*. M. *tschonoskii* produces uniform seedlings only if isolated from other crabapples.

CUTTINGS: *Softwood*. May or early June softwoods can be effectively rooted in commercial quantities (80-100%) [*The Plant Propagator* 22(4):4 (1976)]. Later cuttings root but generally not as high. Malus taxa rooted include M. *x atrosanguinea*, M. *baccata* var. *mandschurica*, M. *floribunda*, M. 'Hopa', M. *hupehensis*, M. *purpurea* 'Eleyi', M. 'Selkirk', and M. *sieboldii* var. *zumi* 'Calocarpa' and numerous others. Wide variances in rooting are observed among crabapple taxa and specific recommendations must be developed for each crabapple taxon. A longer time is required to produce a tree suitable for sale when propagated by cuttings. Cuttings 4 to 6" from current season's growth, 2500 to 10,000 ppm IBA-solution, peat:perlite, mist, rooted in high percentages. Cuttings can be transplanted, fertilized lightly and will continue to grow. Several nurserymen are producing many cultivars from cuttings. See *The Plant Propagator* 22(4):4-5 and PIPPS 27:427-432.

Hardwood. In general, crabapples are very difficult to root from hardwood cuttings. Apple rootstocks, however, have been propagated from hardwood material. This technique is most logically accomplished in regions with mild winter temperatures. Hardwood cuttings of apple are harvested fall or late winter, the basal ends are treated with root-promoting chemicals (IBA 2500 to 5000 ppm). After hormone treatment the cuttings are bundled together and placed in damp packing material (peat moss and sand) contained in insulated bins with bottom heat (64 to 70°F) for about 4 weeks. The tops are left exposed to the cool temperatures. Cuttings are transplanted before bud growth commences which is about the time roots begin to emerge [*The Plant Propagator* 26(1):6 (1980)].

GRAFTING: Crabapples are primarily propagated by budding or grafting.

Budding. Budding is generally the preferred method of propagation for crabapples and apple fruit trees because much larger 1- and 2-year plants are possible. The understock can be commercial apple seedling understock (generally 'Jonathan' and 'Delicious'), crabapple seedlings such as M. *baccata*, M. 'Hopa' and 'Dolgo' or a number of clonal rootstocks such as the Malling rootstocks. Late summer and fall are considered the best times to bud crabapples. Seedling understock for budding should be growing vigorously. Best results are normally secured where scion wood is of such size and vigor that the buds can be removed free of wood beneath the shield. Removal of the seedling portion should be accomplished just as the bud breaks dormancy. The cut should be a 45° angle 1/8 to 1/4" above the bud. For some varieties the removal of understock shoots, staking and scion pruning are necessary for the production of well-shaped trees.

A valid reason for not using clonal rootstocks in the past was that most ornamental crabapples were shown to be sensitive to virus infections from the clonal rootstocks. Sources now are available free from known viruses and methods are available to minimize reinfection. Results indicate that natural spread of infection is not a hazard in apples. The clonal rootstocks should be budded 12" high so they can be planted 10" deep for added support. Clonal roots include:

1) 'Malling 27' — very dwarfing and produces a bush-like tree; may be useful for ornamental crabapples.
2) 'Malling 9' — dwarfs apples to 30% of standard size; poor bud take has been reported with this clone, which may be caused by latent viruses; trees grafted on 'Malling 9' are very precocious, bear very early fruit and need staking.
3) 'Malling 26' — popular among commercial fruit growers because of excellent dwarfing and fruiting; produces a well anchored and free standing tree.
4) 'Malling 7' — most tested and valuable semi-dwarfing rootstock; compatible with all cultivars tested; well anchored and produces precocious trees.
5) 'Malling Merton 106' and '111' — semi-vigorous; 'MM 106' is known to be susceptible to *Phytophthora* (collar rot) and needs well drained soil.

Chip Budding. Theoretically it can be accomplished anytime of the year but is usually confined to the period April to early September with the most common time July to August with dormant buds from the current season's growth. The use of polyethylene film was noted as an essential requirement. Rubber strips, plastic strips or twine were reported to be inferior.

Whip Grafting. Whip grafting has been chosen over budding by a number of propagators because it eliminates the "dog-leg" problem often associated with budding. In late January to early February, the grafts are made, wrapped with wax string and placed in cold storage until planting time. The grafts can be kept in cold storage for 3 months without loss or excessive callus. The scion should be the same size as the rootstock and contain 2 to 3 buds. Some propagators dip their rootstocks and scion wood in a fungicide solution before grafting to control pathogen problems during storage. An alternate method is to pack the grafts in boxes with moist shingle-toe or long fiber sphagnum moss over the roots, and then place at 70°F to get as much callus as possible before buds swell. After callusing the grafts are placed in a cooler at 36 to 38°F. Inspect weekly until potted into 3" pots and grow in a greenhouse until danger of frost is past.

Root grafting is also practiced with crabapples. Straight root pieces from seedlings are whip grafted during winter as noted above. The root piece is cut into several sections and each piece is used with a relatively long scion. A longer time is required to produce saleable plants by this technique.

TISSUE CULTURE: Rapid multiplication of crabapples (M. *sieboldii* var. *zumi* 'Calocarpa', M. × *purpurea* 'Eleyi', M. 'Almey' and 'Hopa') has been achieved in vitro. Proliferation of shoot tips was induced on MS medium containing benzyladenine (BA) at 1 and 2 mg/liter. Transferring the shoots to MS plus NAA at 0.1 or 0.2 mg/liter resulted in good root development [*HortScience* 17:191 (1982) and 19:227 (1984)].

Zimmerman micropropagated 9 cultivars of apple ('Delicious', 'Northern Spy', 'Nugget', 'Ozark Gold', 'Spartan', 'Spuree', 'Rome', 'Stayman', and 'Summer Rambo') from actively growing shoot tips on a Murashige and Skoog (MS) medium supplemented with 1.0 mg BA, 0.1 mg IBA, 0.5 mg GA, and 30,000 mg sucrose. Initiation was on a liquid medium and proliferation on an agar based medium. Shoot tips were washed in a detergent solution for 5 minutes, 1/2 strength calcium hypochlorite solution + Tween 20 for 20 minutes, and then rinsed in 2 changes of sterile water. Zimmerman reported improved rooting (80+%) of in vitro shoots of *Malus* when placed for 3 to 7 days in darkness in liquid medium containing 43.8 mM sucrose plus 1.5 microM IBA. Raising the temperature from 77 to 86°F was beneficial [JASHS 110:34 (1985)].

In vitro multiplication of 'Malling 7' rootstock was obtained on MS medium with salts at 1/2 strength, 0.5 mg BA, and 0.6% agar. Rooting was on 1/3 strength salts, 0.27% agar, and 2 mg IBA [*HortScience* 15:509 (1980)].

Meristem Grafting. Meristem grafting can be utilized to recover virus-free plants. Seeds, after sterilization with sodium hypochlorite solution, are sown in moist, autoclaved vermiculite. A mixture of 4.5 g finely ground vermiculite and 25 ml deionized water is placed in 25 × 250 mm tubes to germinate the seeds. Before grafting the seedling rootstock is transferred to a sterile petri dish. The meristem is grafted directly to the hypocotyl of the germinating seedling after decapitating below the cotyledonary node. A shoot apex from a disinfected stem tip is severed and transferred to the decapitated rootstock. The cut surface of the decapitated seedling is sufficiently moist to achieve adherence between the shoot apex and rootstock. All manipulations are carried out in a clean air hood to minimize airborne contamination. The grafted seedlings are transferred to sterile culture tubes (25 x 150 mm) containing 10 ml Hoagland's solution. A filter paper support with a hole in the center is utilized to support the grafted plant. Successful grafts are transferred to soil [PIPPS 29:393 (1979)].

Melia azedarach Chinaberry

SEED: The chinaberry is a short-lived deciduous tree from Asia that has become naturalized in the southern United States. Easily propagated from seed. Fruits are picked after the leaves have fallen and either macerated to remove the fruit pulp or the entire fruit is planted. Fruits can be stored dry for at least one year without a loss of seed viability. A pregermination treatment is not required. Seeds require 1 to 3 months to germinate. In the nursery, seeds can be sown immediately after harvest or in the spring. 'Umbraculiformis' comes relatively true-to-type from seed.

CUTTINGS: Root cuttings will work quite well on the species and 'Umbraculiformis'. Follow procedures described in Chapter 2.

Menispermum canadense Common Moonseed

SEED: The common moonseed is a native plant grown for its attractive foliage. Fruits can be harvested from September through November and macerated to remove the pulp from the seed. A pound of seed contains 7,600 seeds. The seeds do not require a pregermination treatment. Cold stratification at 41°F for 2 months, however, will speed germination. In nursery practice, the seed can be sown immediately in the fall or stratified and spring sown.

DIVISION: The common moonseed is capable of spreading by underground stems and may be increased on a limited scale by division.

Metasequoia glyptostroboides Dawn Redwood

SEED: Male and female cones are borne on the same tree. Cones should be collected when the scales naturally begin to separate — earlier harvest may lead to difficulty in opening of the cones. Cones are ready for collection by late October into November, about the time leaves fall. Seed can be stored dry in air-tight containers at 34 to 40°F. Fresh seeds will germinate without pretreatment, however, some work indicates a short cold period (1 month) improves, unifies and hastens germination. Germination may begin 5 days after planting. Seedlings are very susceptible to damping-off fungi which are minimized by using a pasteurized germination mix. Young seedlings thrive in high humidity [*The Plant Propagator* 6(4):7 (1980); *Arnoldia* 37:59 (1977)].

CUTTINGS: One of the easiest conifers to root and can be propagated from either softwood or dormant hardwood cuttings. Dormant cuttings are harvested in late December or early January (California), cut into 4″ lengths and washed with chlorine water (15 ppm) followed by a fungicide solution. Cuttings are placed in clear plastic bags and stored at 45°F for 1 month. After cold storage, they are given a 3,000 ppm IBA-quick dip, placed in perlite:peat moss (9:1, v/v) in outdoor beds or on an open bench, mist, bottom heat (about 65 to 72°F) and low air temperature. In 4 months, 90% have rooted. Cuttings have a brittle root system and can be stuck in individual pots to minimize disturbance during transplanting. Shade when budbreak occurs [PIPPS 32:327 (1982)].

Softwoods have been successfully rooted from late June and early July cuttings. 3000 ppm IBA-talc, mist or polyethylene structures. The approximate rooting time is 7 to 8 weeks. Mid-August (Boston), 8000 ppm IBA-talc, sand, rooted 93% in 10 weeks. Rooted cuttings should not be disturbed after rooting and are best overwintered in flats and transplanted in spring.

Michelia figo Banana Shrub

CUTTINGS: Easy to root from May to fall at 90 to 100% with 3000 ppm IBA-quick dip, peat:perlite, and mist. Rooted cuttings are easily transplanted and grow vigorously. Cutting wood should be firm. This is an evergreen species and, no doubt, could be rooted year-round.

Mitchella repens Patridgeberry, Twinberry

SEED: A ¼″ diameter,red berry-like drupe. Seeds need to be removed from fruit before stratifying. Fresh seed germinated reasonably well but 3 months at 40°F produced better germination.

CUTTINGS: Early November (Boston), 8000 ppm IBA-talc plus thiram, sand:peat, polytent, rooted 100% in 10 weeks. An evergreen ground cover in the wild. Tends to root in contact with moist soil.

Morus alba White Mulberry

SEED: Fruit should be harvested as soon as ripe to avoid losses from birds. One hundred pounds of fruit yield 2 to 3 pounds of seed which averages 235,000 seeds per pound. Macerate the fruits immediately and soak in water to remove pulp and empty seeds. Air dry before storage or use. Subfreezing temperatures (-10 to +10°F) are recommended for storage. Germination of fresh seed may vary because some seed will contain dormant embryos and impermeable seed coats. Seed collected in early July, cleaned immediately and sown, germinated 75%. Germination is improved by cold stratification at 33 to 41°F for 1 to 3 months. Properly pretreated seed can be sown in spring. Seedlings require half-shade for a few weeks after germination.

CUTTINGS: Easy to root, especially from summer softwoods. June and July are optimum months for rooting. 'Pendula' responds similarly. Mid-July cuttings, 8000 ppm IBA-talc, sand, rooted 100% in 3 weeks with heavy root systems. Hardwoods also root. Root cuttings taken in fall and early winter are also possible [*The Plant Propagator* 14(4):4 (1968)].

GRAFTING: *Morus alba* 'Pendula' is grafted on a standard of *M. alba* var. *tatarica* to produce a small weeping tree. Triangling has proved to be a better form of grafting because the understock is larger than the scion. Grafting is accomplished outside (Canada) when the frost starts to go out of the ground but not during freezing weather. Tie with raffia and cover the wound with an emulsion. The scion should contain 3 buds.

'Pendula' can also be greenhouse grafted during the winter. Understock is dug in the fall, wrapped in burlap, pruned for grafting and placed in an area where it will receive a normal cold period. In February the understock is brought into the greenhouse and allowed to start growth. Standard grafting techniques are employed with high humidity.

Bareroot grafting with a splice graft also has been utilized. Lift the understocks in the fall, layer in boxes and place in cold storage. After grafting, callus at 55 to 60°F.

Morus alba has been reported to form an incompatible union when grafted on *M. nigra*.

TISSUE CULTURE: Buds for tissue culture can be collected from August to April. Surface sterilize the buds in 4% calcium hypochlorite for 30 min. and after removing the bud scales, the apex, including several immature leaves, is dissected out and placed on a modified MS medium. The MS medium is supplemented with (in mg/liter):glycine, 2.0; nicotinic acid, 0.5; HCl-thiamine, 0.1; HCl-pyridoxine, 0.5; inositol, 100; Na-FeEDTA, 32; and BA, 1. Agar at 0.4-0.5% and sucrose or fructose at 3% were also added. Dormant buds do not develop when sucrose is used but when fructose is present shoot development occurs from buds taken any time of the year. Multiplication is achieved with the same medium. Shoots (2/5" long) transferred to an MS medium plus 0.1 mg NAA initiated roots in 10 days.

Leaf explants derived from either subcultured shoots or primary leaves of seedlings in culture have generative capacity when cultured on the above BA containing medium. Leaves from plant material grown on a BA-free medium do not produce any buds. The generative capacity is restricted to the transition zone between the mid-rib and petiole. Removal of the petiole is necessary for bud initiation.

Morus indica has been tissue cultured from axillary vegetative buds from mature trees when surface sterilized and cultured on MS medium supplemented with kinetin (1 mg/liter), NAA (0.5 mg/liter) and 0.8% agar. Each explant developed into a single shoot which also develops roots. Plantlets were easily transferred to soil or divided and multiplied. Refer to M. *alba* for additional details.

Morus rubra Red Mulberry

SEED: Fresh seed sown at once produces good germination. When germinated under laboratory conditions seed requires artificial light and alternating 68 to 86°F temperatures.

Myrica cerifera Southern Waxmyrtle

SEED: The fruit is a small, 1/8 to 1/4" diameter globose drupe enclosed in a grayish white waxy coating. Germination is hindered by the waxy coating. Dewaxing is achieved by vigorous rubbing across a screen. Dewaxed fruit yields about 84,000 seeds per pound. Seeds germinate readily without a pretreatment based on Georgia work. However, there is evidence that shows a 1 to 3 month cold stratification period is necessary. Cleaned seed sown at once germinated 4%; 2 months stratification at 40°F produced heavy germination. Gibberellic acid (24 hours at 500 ppm) was reported to enhance the germination rate to 76% in one study with fresh seed. The wax must be removed if water is to be imbibed. See *HortScience* 12:565 (1977); *Arnoldia* 43:20 (1983).

CUTTINGS: Semi-hardwood cuttings collected in mid-August and treated with a 1 to 1.5% IBA-quick dip rooted 90%. A peat:perlite medium (1:1, v/v) in outdoor mist beds with a 47% shade covering was used. May through August cuttings have rooted best. A comparison with talc formulations of the same IBA concentration showed the liquid quick dip to be more effective. *Myrica pumila* cuttings should be treated like M. *cerifera*. In Georgia work, timing (May-August) was more important than hormone. Cuttings taken during winter rooted at extremely low percentages. During peak months rooting averaged over 80%.

Southern wax myrtle can be propagated from root cuttings. Use 1½ to 2" long pieces from thick young roots.

Myrica gale Sweet Gale

SEED: The sweet gale needs only a period of cold for germination as the fruit does not have a wax coating. Seeds germinate well after 3 months stratification at 40°F. Light was reported to be essential for germination. Treatment of fresh seeds with gibberellic acid for 3 to 20 hours increased germination rate.

CUTTINGS: Sweet gale is capable of shoot regeneration from root cuttings. Use 1.5 to 2.0" long pieces from thick young roots.

Myrica pensylvanica Northern Bayberry

SEED: Seed is a small round drupe, enclosed in a waxy coating, which ripens in October and persists on the plant until late winter or early spring. The wax coating is removed by rubbing over a screen. For long-term storage (10 to 15 years), seeds should be dewaxed, air dried in a heated room and placed in sealed, moisture proof containers at 34 to 38°F. Embryo dormancy can be overcome by 3 months stratification at 40°F after wax removal. Cold stratification is ineffective if wax remains. Gibberellic acid is not effective on dewaxed druit. Fruits dewaxed and scarified showed some response to gibberellic acid [*Contrib. Boyce Thompson Instit.* 4:19 (1932)].

CUTTINGS: Propagation by cuttings is difficult. Stem cuttings, mid-June, 5000 ppm IBA-quick dip, rooted 53% after 8 weeks. Use a quick draining medium and minimal mist. Timing appears quite critical on this species. Root cuttings are possible [*The Plant Propagator* 23(2):14 (1977)].

Nandina domestica Heavenly Bamboo

SEED: Fruit is a true berry containing 2 seeds. Seeds should be removed from fruit by maceration. Conflicting reports regarding the germination requirements of this species exist. The seeds exhibit a delayed germination due to the extremely slow rate of development of the embryo. At the time when the fruit is collected the embryo is still in a rudimentary form. Embryo development reportedly takes place whether the seed is stored moist or dry, at high or low temperatures. Stratification at a warm temperature for several months followed by cold stratification for several months speeds embryo development. A most interesting characteristic is the germination during a definite season — late fall or early winter (October-December). Seeds held in cold storage for 9 to 10 months germinate as well as those sown immediately after collecting. However, according to Arnold Arboretum results, seeds germinate without any treatment. Cleaned and freshly sown yielded 65% germination. Two months at 40°F resulted in 15% germination and 3 months 30%. A north Alabama nurseryman sows seed in September and has good success.

CUTTINGS: Dwarf nandina cultivars including 'Purpurea' ('Atropurpurea Nana') and 'Harbour Dwarf' are cutting propagated. A large Alabama nursery uses 1250 ppm IBA + 1500 ppm NAA solution with excellent success. 'Purpurea' is propagated by tip cuttings taken from late spring to mid fall (Alabama). Once the cuttings have hardened-off, which is indicated by a reddening of the foliage, they become more difficult-to-root. The foliage should show lots of green and as much as possible should be left. Stem tissue should have stiffened, be pinkish in color and the cut made just above where the wood turns brown. Dip in fungicide (Captan at 2 lb/100 gal) for 10 to 15 minutes, and trim to 1½ to 2½". Treat with IBA (1250 ppm for very soft and 1870 ppm for stiffer), place in pine bark:Canadian peat moss:perlite (3:2:3, v/v/v) in 4" pots, 50% shade and mist. The addition of 6 pounds Osmocote (18-6-12), Micromax minor elements and dolomitic lime (3:1:5, v/v/v) to each cubic yard of mix improved establishment. The propagation benches are sprayed weekly with Benlate and Captan. A major problem with *Nandina* is securing sufficient cutting wood. In general, cuttings root readily and grow off vigorously [PIPPS 33:624 (1983)].

'Harbour Dwarf' is treated the same as 'Purpurea', except that the number of leaflets is reduced. Once stuck under mist, 'Harbour Dwarf' is extremely sensitive to drying. 'Harbour Dwarf' can be propagated by separation of suckers which are readily produced by this cultivar. Initially treat the suckers as cuttings until established. Root cuttings will work [*The Plant Propagator* 14(4):4 (1968)].

TISSUE CULTURE: In vitro establishment and multiplication of *Nandina domestica* 'Purpurea', 'Harbour Dwarf', and other cultivars from lateral and terminal buds have been reported on a modified Gamborg B5 medium supplemented with (in mg/liter): thiamine-HC1, 0.4; myo-inositol, 100; NAA, 0.1; BA, 1.0; activated charcoal 1,500; agar, 8,000; and sucrose, 30,000. Iron was supplied according to the MS medium. Cultures were maintained under 24 microE $m^{-2}s^{-1}$ light intensity for 16 hours at 78 ± 4°F. Shoot multiplication occurs on the same medium minus charcoal, with rooting on 1/3 MS plus activated charcoal. After rooting the plantlets are established in a sphagnum peat moss and perlite mix. Several New Zealand and North American nurseries are producing a number of cultivars through tissue culture. Some 'Atropurpurea Nana' is virus infected and develops puckered-blistered leaves. Tissue culture techniques have been used to elimiate viruses from nandina. When the virus is removed the leaves appear smooth. Tissue culture will result in a glut of *Nandina* cultivars because of the ease with which they are produced [*HortScience* 18:304 (1983)].

Neillia sinensis Chinese Neillia

SEED: No clear cut recommendations were found, however, some cold is apparently required to facilitate germination.

CUTTINGS: Easy to root from softwoods in summer with 1000 ppm IBA-quick dip and mist. Summer cuttings, 8000 ppm IBA-talc, high humidity in shaded greenhouse root readily. Late August cuttings, 8000 ppm IBA-quick dip, peat:perlite, mist, rooted 80%.

Nemopanthus mucronatus Mountain-holly

SEED: The red berry-like fruits, which ripen in summer, are 1/4 to 1/3″ diameter. Seeds are doubly dormant, requiring a 5-month warm stratification and 3-month cold stratification. A close relative of *Ilex* with an immature embryo. Time may be the best aid to germination.

CUTTINGS: Late June (Boston), 8000 ppm IBA-talc, mist, rooted only 18%.

Nerium oleander Oleander

SEED: Fruits are 5 to 7″ long capsules containing hundreds of seeds. Seeds germinate without pretreatment. Collect the fruit in the fall before capsules split naturally. Oleander knot, a bacterial disease, can be a problem when using seed propagation. Seed should be collected only from disease-free capsules as the pathogen is passed on to seed from infected pods [*The Plant Propagator* 14(1):8 (1968)].

CUTTINGS: Summer cuttings root easily; collect in late July or August (mature wood), 3000 ppm IBA-talc, and mist. Oleander is extremely easy to root from cuttings and can probably be rooted any time the wood is firm.

Neviusia alabamensis Snow-wreath

CUTTINGS: Softwood, June-July, 1000 ppm IBA solution, peat:perlite, mist, root in high percentages. If a few plants are required simple division of the parent plant works well.

Nyssa aquatica Water Tupelo

SEED: The dark reddish purple drupe ripens from September to October and fruit drop occurs from October to November. The ripe fruit should be run through a macerator and the pulp floated off. A pound of cleaned seed contains about 450 seeds. Seeds exhibit moderate dormancy. See *N. sylvatica* for pretreatment conditions.

Nyssa sinensis Chinese Tupelo

SEED: The need for a pregermination requirement is not known, however, the authors feel that the seed probably exhibits embryo dormancy and should be treated like *N. sylvatica*.

CUTTINGS: Softwoods from plants forced in a greenhouse root readily. Refer to *N. sylvatica*. See *Gardeners Chronicle* 165(1):29 (1969).

GRAFTING: Bench grafting on *N. sylvatica* seedlings established in pots is successful. Refer to *N. sylvatica* for details.

Nyssa sylvatica Black Tupelo

SEED: The bluish black drupes ripen from late September through early October. Although the removal of the pulp does not appear to be essential, the fruit is generally run through a macerator. Black tupelo averages 3,300 cleaned seeds per pound. *Nyssa* seed exhibits variable embryo dormancy and benefits from cold stratification. Some seed lots require as little as 1 month while others benefit from 4 months. Cold stratification for 3 months at 40°F has generally produced excellent germination. The seed may be fall sown in the north but spring sowing after cold stratification is recommended. Root pruning is required to stimulate a well-branched root system [*The Plant Propagator* 10(4):5 (1964)].

CUTTINGS: Bring stock plants into a greenhouse and force in late winter. Softwood (1 to 1½″), 8000 ppm IBA-talc, sand, mist, bottom heat (70°F), rooted 90 to 100%. Cuttings can be transplanted in 6 weeks. If properly managed, they will grow into early fall and may reach 2½ to 3′. Protect from freezing the first winter. See *The Plant Propagator* 14(4):11 (1968).

GRAFTING: Establish seedlings of *N. sylvatica* in pots. Graft in mid winter with a 2″ scion using a whip graft. Reduce stock to 1½″, wax, and place on open bench or in closed case at 55 to 60°F. Do not water but spray with a fine mist 2 times a day. Once union has healed and buds are breaking, apply adequate water.

TISSUE CULTURE: See *Ornamental Plants* 1986: *A Summary of Research*. Res. Circ. 289. Ohio Agr. Res. Dev. Center. 1986. p. 27.

Orixa japonica Japanese Orixa

SEED: The genus *Orixa* consists of a single species native to Japan, South Korea, and China. It is a deciduous shrub about 10 feet tall that is valued for its foliage. The inconspicuous flowers are unisexual, with the sexes on separate plants. Fruits are composed of 4 flattened carpels joined at their base, and each contains one black seed. A 3 month cold stratification (40°F) produced 92% germination; untreated seed germinated only 2%.

CUTTINGS: Early to mid-June, 1000 to 5000 ppm IBA quick-dip and mist [PIPPS 34:603 (1984)].

Osmanthus

This genus is composed of evergreen shrubs and trees with fragrant flowers native mainly to eastern and southeastern Asia, and North America. They belong to the olive (Oleaceae) family and number about 30 species. Most cultivated types have male and bisexual flowers. The fruit is a drupe.

Osmanthus americanus Devil-wood, American Tea-olive

SEED: Senior author has had no success with seed germination. 3 months cold stratification did not stimulate germination. The seed is enclosed in a bony, hard endocarp which apparently must be broken down. Possibly warm and cold stratification periods will prove successful.

CUTTINGS: Not the easiest plant to root. Senior author has tried many times without success. However, a Georgia nurseryman has had success taking August-September cuttings from young plants. A strong juvenility factor comes into play with this species. One report noted early December hardwoods, 8000 ppm IBA-talc + thiram, peat:perlite, polytent, rooted 80%. Early December (Georgia), 8000 ppm K-IBA solution, perlite, bottom heat, mist, rooted 75% in 8 weeks.

Osmanthus delavayi Delavay Tea-olive

SEED: Seeds are reported to be difficult and slow to germinate.

CUTTINGS: Stocky, 3 to 4″ long side shoots root best. Treat with 8000 ppm IBA-talc and place under mist. From 50 to 70% will be rooted in 6 weeks. August to September nodal cuttings with heel, 8000 ppm IBA-talc, wound, peat:sand (1:1, v/v), bottom heat (70°F), also root. Cuttings have a tendency to drop leaves [NGC 180(7):26 (1976) and 188(2):33 (1980)].

Osmanthus × fortunei Fortune's Tea-olive

SEED: A hybrid between *Osmanthus fragrans* and *Osmanthus heterophyllus*. May be a male or sterile as senior author has never observed fruit set even on 75-year-old plants.

CUTTINGS: Wood should be firm and in Georgia work a hormone made no difference. Early June, 0, 5000 ppm P-ITB, 5000 ppm IBA, 2.0% P-ITB in solution, peat:perlite, mist, rooted 57, 43, 30, 47%, respectively, in 6 weeks. Root systems in all treatments were profuse. Cuttings that had not rooted at the time of initial evaluation were restuck and another evaluation 5 weeks later (11 total weeks) showed 67, 80, 67, and 93%, respectively.

This species shows two distinct types of leaf morphology. A spiny-margined leaf associated with young plants; an entire margin with older plants. The literature reported that the adult, entire-margined leaf form, if propagated vegetatively, would maintain this condition. In Georgia work this did not prove true. Cuttings collected from 75-year-old plants on campus were rooted, transplanted and the leaves on the new flush of spring growth had spiny margins. The spines were not as abundant as the typical juvenile form. Senior author suspects that a measure of juvenility (because of proximity to root system) was reintroduced into these cuttings.

Osmanthus fragrans Fragrant Tea-olive

SEED: No definitive work was discovered but the same comments relative to O. *americanus* probably apply.

CUTTINGS: Certainly the most fragrant of the tea-olives. Unfortunately, reserved for Zone 9 and warmer. A lovely orange-flowered form ('Aurantiacum') is known. Treat cuttings as described under O. × *fortunei*. Rooting will take 10 weeks.

Osmanthus heterophyllus (*ilicifolius*) Holly Osmanthus

SEED: Seeds are reported to be difficult and slow to germinate.

CUTTINGS: Late summer, 3000 ppm aqueous solution of the K-salt of IBA for 20 seconds, sphagnum peat moss:perlite (1:1, v/v), mist, and 16 hour photoperiod, rooted 75%. A pretreatment with a NaOH solution (pH 10.5) for 10 minutes was reported to increase the rooting to 100%. Cuttings can be fall rooted (California) in outdoor bottom-heated (65 to 70°F) beds covered with polyethylene. 'Purpureus', mid-November, 3000 ppm IBA-talc, sand, rooted 67% in 4 months. In another study, cultivars were rooted from cuttings in August and September using 8000 ppm IBA-talc, peat:perlite (2:1, v/v), mist. Prone to leaf drop problems and fungicide sprays may be beneficial. Excessive moisture in the rooting bed will cause leaf drop. A polytent approach during the fall and winter months is probably best.

Ostrya carpinifolia European Hophornbeam

SEED: Has an internal dormancy and requires cold stratification: see O. *virginiana* for details.

Ostrya virginiana American Hophornbeam

SEED: The nutlets are contained in small sacs (hop-like in shape) in a catkin. Collect when the color is pale greenish brown. When fully ripe, the strobile falls apart and the seed can be separated from the chaff by screening. Ten pounds of fruit will yield approximately 2 pounds of seed and each pound averages 30,000 seeds. Seeds have a dormancy that is difficult to overcome. In one study, seeds sown immediately after collection failed to germinate; seed given 3 months cold stratification germinated 2%; 3 months warm followed by 3, 4 or 5 months cold produced 81, 92 and 92% germination. In nursery practice the seed can be fall sown after harvest or spring planted after warm and cold stratification treatments.

Seeds harvested while immature have been reported to germinate 100%. The seeds should be harvested and sown just as the endosperm is going into the mealy stage (mid-August in Des Moines, Iowa) and before the seed coat becomes hard [*Amer. Nurseryman* 76(11):22 (1940)].

Oxydendrum arboreum Sourwood

SEED: Sourwood can be seed propagated by methods identical to those used for *Rhododendron* and *Calluna*. Seeds germinate without pretreatment. Soon after fall harvest the seeds are sown on the surface of flats filled with a fine milled sphagnum:vermiculite (1:1, v/v) and misted. The flats are covered with plastic, taking care to keep it off the surface, and then placed under continuous lighting supplied by fluorescent lights. Germination occurs within 2 weeks and the seedlings grow fast. At the 2 to 3 leaf stage transplant the seedlings, being careful not to check growth. Two-foot high plants may be obtained in 7 to 9 months [*The Plant Propagator* 24(2):13 (1978)].

CUTTINGS: Softwoods (short side shoots) made with a heel rooted 80% when taken in late July, 90 ppm IBA soak, sand:peat (1:1, v/v) and mist. In general, cuttings are difficult to root and the above is the only report in the literature that indicated significant success. Senior author tried repeatedly with no success. Significant variation in habit, flower and fall color occur to warrant vegetative propagation. Many nurserymen tried with little success.

Pachysandra axillaris Chinaspurge Pachysandra

CUTTINGS: Early August cuttings root readily under mist (see P. *terminalis*).

Pachysandra procumbens Allegheny Pachysandra

CUTTINGS: Allegheny pachysandra is not difficult to propagate from cuttings and either IBA or NAA are suitable hormones. Leafy cuttings, early August, 4000 or 8000 ppm IBA solution or 4000 ppm NAA-solution, mist, perlite:sand (1:1, v/v), rooted 100% after 10 weeks. Cuttings should be firm [*The Plant Propagator* 25(2):9 (1979)].

DIVISION: Propagation by division is possible but slow.

Pachysandra terminalis Japanese Pachysandra

SEED: In one study by the junior author, freshly harvested seed germinated only 20% after 30 days; scarified seeds germinated 90% in 34 days.

CUTTINGS: Very easy from cuttings that are taken after the spring flush of growth has hardened. Although a hormone treatment is not necessary, 1000 to 8000 ppm IBA-talc improves rooting. A wide variety of methods have been utilized and a few are presented here: 1) *Mist* — Mid to late summer, sand:peatmoss (1:1, v/v), mist, root readily. A hormone application, such as 1000 ppm IBA, improves root quality. 2) *Polytent* — Outdoor polytent propagation in mid-July is also possible. Treat the cuttings with a hormone (1000 ppm IBA-talc), place in outdoor ground beds, cover with 4 mil polyethylene, and shade (75%). Cuttings will be rooted in 4 to 5 weeks and can be hardened-off. Sterilize the beds with methyl bromide prior to sticking the cuttings. 3) *Thermoblankets* — Japanese pachysandra has been summer propagated under thermoblankets. The blanket is laid directly on the cuttings and then covered with white plastic. Cuttings should be checked weekly and watered when necessary. 4) *Leafless underground stems* — Stolons can be cut into 1½″ long sections (each with a bud) and will root and develop a shoot. Early spring is the best time when maximum stolon activity is occuring. Plants propagated by this method will take longer to produce a saleable plant.

Paeonia suffruticosa Japanese Tree Peony

SEED: Tree peony seed has a double dormancy and requires warm stratification followed by cold stratification. The seeds are potted, given warm stratification (65 to 75°F) for 3 months to allow the hypocotyl/root axis to grow and then a cold stratification (41°F) for 3 months to break epicotyl dormancy. Following cold stratification move to warm conditions (65 to 70°F) for epicotyl growth.

GRAFTING: Tree peony scions are grafted onto herbaceous roots toward the end of August. Herbaceous hybrids that do not produce adventitious buds are preferred. Included in this group are cultivars such as 'Early Windflower', 'Red Charm', 'Requiem', 'White Innocence', 'Mons. Jules Elie', 'Charles White' and 'Krinkled White'. The roots are dug a few weeks before the actual grafting and allowed to dry until rubbery. At this stage the root piece will not split away from the scion. Roots are cut into 4″ long sections and triangling is used as the grafting method. One-year-old scion wood is used with only one bud per scion. Leaves are reduced but left on the scion. After tying with a rubber strip, all exposed surfaces are covered with a wound dressing. Grafts are plunged into a shaded cold frame with the bud exposed and checked weekly. A fungicide spray is helpful for controlling *Botrytis*. By November the grafts have knitted and the leaves are removed. Keep the frame closed and shaded until spring. Leave the grafts in the frame until the following fall and then plant very deep to encourage scion rooting [PIPPS 21:387 (1971) and 32:512 (1982)].

Direct planting of the grafts into outdoor beds specially prepared with generous amounts of well rotted manure and peat moss to improve texture is also utilized. With this method the scion has 2 to 3 buds and the grafts are covered with 2″ of soil followed by 2″ of wood chips. A layer of black plastic is placed over the chips and in November the black plastic is covered with 1′ of straw. In early spring the straw and black plastic are removed. *Paeonia lutea* is similarly treated but the grafting operation is later because the buds mature later.

Division is possible. All stems are cut off several inches above the ground in fall and then the plant is dug and divided. Divide the plant in such a way that several buds are present on each division [*Arnoldia* 29:25 (1969)].

Stem layering works but it take several years to root. Air layers are also possible using poly-film.

TISSUE CULTURE: Tree peony embryos have been successfully cultured in vitro. A modified Linsmaier-Skoog medium was used with 6g per liter of agar. Peony embryos develop rapidly with cotyledons appearing and the root axis growing 3 to 4″ in 6 to 8 weeks. At this time a cold period (4 to 6 weeks) is required to produce the first true set of leaves [PIPPS 26:272 (1976)].

Parrotia persica Persian Parrotia

SEED: No quantitative published literature is available on seed propagation of this species. Most plants do not produce fruits under cultivation. Senior author has observed only one fruiting tree. Warm stratification for 5 months and then 3 months at 41°F is recommended [*The Plant Propagator* 30(4):9 (1984)].

CUTTINGS: This is one of the easier members of the Hamamelidaceae to root. Cuttings collected in June-July that have firmed at the base or when the terminal has ceased to grow (up to August) will root readily if given a 1000 to 3000 ppm IBA quick-dip. In one study, July cuttings treated with a rooting hormone and placed in sand:perlite under mist rooted readily. Best rooting was obtained with IBA-talc plus thiram (100%) and 24-hour soaks in 400 and 800 ppm K-salts of IBA (85 to 95%). Senior author, during sabbatical at Arnold Arboretum, collected cuttings from a 97-year-old tree, 4000 ppm IBA quick-dip, peat:perlite, mist with 100% rooting. Overwintering and survival of the rooted cuttings is recognized as a problem. Allow the cuttings to go dormant and overwinter in rooting flats. Transplant in the spring after new growth has started. In many cases if the cuttings continue to grow after rooting they prove relatively easy to overwinter. This species is easily managed after rooting by applying a light application of Osmocote 18-6-12 or liquid fertilizer. Cuttings will push out new shoots. Cuttings rooted early in the season produce more uniform shoot growth in response to fertilizer application. See PIPPS 26:296 (1976).

GRAFTING: *Parrotia persica* 'Pendula' is grafted on species understock produced from cuttings. It can be whip grafted onto a potted rootstock in the early spring or top worked using a side graft onto a standard under glass in the early spring. *Parrotia persica* has been grafted on *Hamamelis virginiana* seedling understock. This is not advisable because of suckering from the understock and latent incompatibility problems.

Parrotiopsis jacquemontiana

SEED: Like other members of the Hamamelidaceae, seeds are difficult to germinate. Seed has double dormancy that requires a long warm stratification followed by a cold stratification. Treat like *Fothergilla major*. Seeds provided 8 months warm/3 months cold germinated poorly. See PIPPS 26:296 (1976).

CUTTINGS: Mid-summer (August), 4000 to 8000 ppm IBA-talc and thiram, rooted 100% with heavy root systems in 8 weeks. They should be left to overwinter in the flats into which they were rooted and transplanted the following spring after new growth starts. Might be treated like *Parrotia* after rooting.

GRAFTING: *Parrotiopsis jacquemontiana* has been grafted to *Hamamelis virginiana* in July and August using a side veneer graft. Removal of the leaves enhances scion survival and reduces disease incidence. Grafts are placed under a polytent or double glass. This combination is not the best because of understock suckering and possible incompatibility problems.

Parthenocissus henryana Silvervein Creeper

SEED: See P. *quinquefolia*.

CUTTINGS: Single leaf bud cuttings, May to July, 3000 ppm IBA-talc, peat:perlite (1:1, v/v), mist, root readily [NGC 185(21):14 (1979)].

Parthenocissus quinquefolia Virginia Creeper, Woodbine

SEED: The bluish black berries should be collected in October and macerated to remove the fleshy pulp. The seed coat is thin and extraction should be done carefully to prevent damage. Seeds have a dormant embryo and cold stratification (41°F) for 2 months is recommended. Seed can be fall sown for natural stratification or spring sown after cold stratification.

CUTTINGS: Softwoods during June, July or August, mist, sand:peat, rooted 100% without hormone treatment. Single leaf bud cuttings with 3000 ppm IBA-talc root readily. Select cuttings without tendrils as no buds form at nodes with tendrils [*The Plant Propagator* 26(1):15 (1980)]. Hardwood cuttings also can be rooted.

Parthenocissus tricuspidata Japanese Creeper, Boston Ivy

SEED: See P. *quinquefolia*. In one study, seeds sown at once or provided 2, 3 or 6 months cold stratification at 40°F germinated 11, 92, 89 and 92%, respectively. Even one month cold stratification stimulated heavy germination.

CUTTINGS: Softwoods, 8000 ppm IBA-talc, mist, sand:peat (1:1, v/v), root readily. Single leaf bud cuttings, May to July, 3000 ppm IBA-talc, also root. Select cuttings that do not have tendrils as no buds form at nodes with tendrils. Hardwoods of this species probably can be rooted. 'Veitchii' roots from hardwood cuttings.

GRAFTING: In England, the Japanese creeper is grafted by some propagators because a saleable plant is produced faster. The rootstocks are P. *quinquefolia* seedlings or rooted hardwood cuttings. Hardwood cuttings provide a better length of stem on which to graft and have a more fibrous root system. 'Veitchii' is the major scion grafted. Scion wood must be selected with care as no axillary buds are formed at nodes where tendrils occur. Two types of grafts are used: (1) a whip graft with hardwood cuttings, and (2) a cleft graft when there is a spongy hypocotyl and a smaller scion. January and February are the major grafting months. After grafting, pot and place in a shaded closed case at 65 to 68°F. Below 65°F the grafts do not form good callus. Air daily for 30 to 40 minutes and provide a Benlate spray every 10 to 14 days. When 1 to 2" of new growth has developed and a small "button" of callus forms, the grafts are hardened [*The Plant Propagator* 26(1):15 (1980)].

Paulownia tomentosa Royal Paulownia

SEED: Seeds are the main method of propagation. The two-valved, dehiscent capsules should be collected in the fall before they split and seed dispersal occurs. The seeds have no pretreatment requirements, although light may be necessary. Sow thinly on the surface of the medium and cover lightly. Seeds germinate 90% in 3 weeks. Seedlings grow rapidly and because of large leaves occupy a large area. Sow in spring to coincide with increasing daylength.

CUTTINGS: Root cuttings have been successful. The plant tends to sucker profusely and root cuttings would produce more plants than any nurseryman could sell in a lifetime [*Gardeners Chronicle* 169(3):22 (1971)].

TISSUE CULTURE: Nodes from mature tissue of greenhouse grown plants were cultured on a modified (1/2 strength salts) Murashige-Skoog (MS) supplemented with 1.0 mg BA and 0.1 mg NAA per liter. Explants were disinfested by a 20-second dip in 70% ethanol followed by a 10-minute soak in 0.5% sodium hypochlorite and 3 rinses in autoclaved water. Shoots rooted in 7 to 10 days on 1/2 strength MS salts plus either 0.5 or 1.0 mg IBA/liter or quick-dipped (15 seconds) in the K-salt of IBA (500 or 1000 ppm), then placed under mist. Explants were cultured under conditions of 16 hours of light (42 micromoles $s^{-1}m^{-2}$) and 75 to 82°F [*HortScience* 20:760 (1985)].

Paxistima canbyi Canby Paxistima

CUTTINGS: Propagates readily from cuttings taken after the spring growth has hardened and until later into winter. Summer cuttings root the fastest. Rooting conditions include 8000 ppm IBA-talc, mist, sand:peat medium. Cuttings will root 80 to 100% in 6 weeks. Senior author used late July (Illinois) cuttings, 1000 ppm IBA, peat:perlite, mist, with good success [PIPPS 17:316 (1967); *The Plant Propagator* 17(4):3 (1971)].

Paxistima myrsinites Myrtle Paxistima

CUTTINGS: Early April (Boston), 8000 ppm IBA-talc, rooted 67%. Anytime new growth has firmed (Canada) until bud break in spring, 8000 ppm IBA-talc, well drained medium, bottom heat polytent in winter, mist anytime, 97% rooting with heavy root systems in 8 weeks [PIPPS 35:293 (1985)].

Perovskia atriplicifolia Russian-sage

CUTTINGS: Summer softwoods, June into August, 1000 ppm IBA-quick dip, well drained medium, mist, root in 10 to 14 days. As soon as rooted the cuttings should be removed from mist because excess moisture leads to deterioration. Collect cuttings early since flowers appear in late summer and good cutting wood is difficult to locate. Cuttings continue to grow after removal from the rooting bed.

Phellodendron amurense Amur Corktree

SEED: The black, 1/3 to 1/2″ diameter drupes remain attached to the tree after leaf fall. Fruits should be macerated to remove the pulp and seeds stored dry until sowing. Cleaned seeds will range in number from 26,800 to 48,000 per pound. The seed germinates without pretreatment and, if sown indoors, do so in late winter or early spring so the seed will germinate with the lengthening days. Seed can be sown outdoors in the late fall or spring. If seed is stored for a period of time, 2 months cold stratification is advisable [*Arnoldia* 14:25 (1954)].

CUTTINGS: The Amur corktree can be propagated by root cuttings. Dig in the fall and store during the winter in moist sand or sphagnum. Shoot cuttings taken with a heel in July have been reported to root. Details were not available.

Phellodendron chinense Chinese Corktree

SEED: This species has the largest fruit the authors have observed on the cultivated species. Two to 3 months cold stratification produced 93% germination; one month only 48%; and fresh seed germinated just 17%.

Phellodendron japonicum Japanese Corktree

SEED: No pretreatment is necessary.

Phellodendron lavallei Lavalle Corktreee

SEED: Seed requires no pretreatment for germination and is the usual method of propagation. See P. *amurense*.

Phellodendron sachalinense Sakhalin Corktree

SEED: Seed requires no pretreatment for germination.

Philadelphus Mockorange

Philadelphus contains about 65 deciduous flowering shrub species and numerous cultivars. Some of the more popular species include P. *coronarius*, P. x *lemoinei*, P. *microphyllus*, P. x *nivalis*, and P. x *virginalis*. *Philadelphus* have 4 petaled white or yellowish flowers which is their most outstanding feature. Fruit is a four-valved dehiscent capsule containing numerous seeds. Collect in late summer, dry inside and store seeds under refrigeration.

SEED: Seed from *Philadelphus lewisi*, P. *coronarius*, and P. *virginalis* require cold stratification plus light for germination. A 20 to 40 day treatment at 32 to 39°F satisfies the cold requirement. In general, seed of most species requires no pregermination treatment [JASHS 91:742 (1967)].

CUTTINGS: Summer softwoods, June and July, root readily (100%) with 1000 ppm IBA-solution, peat:perlite (1:1, v/v), mist. Although cuttings will root without hormone treatment, a better root system is produced with hormone treatment. Cuttings will continue to grow after transplanting.

Hardwoods, harvested and stuck in fall or early spring, root easily. Treat with 2500 to 8000 ppm IBA and insert 8″ cuttings 7″ deep into a sandy soil in the fall or as early in the spring as possible. Fall plantings should be mulched to prevent heaving [*The Plant Propagator* 15(3):10 (1969); PIPPS 34 (1984)].

Layering, although not necessary with present propagation methods, was mentioned in the literature for a number of species [*Arnoldia* 13:25 (1953)].

Photinia x fraseri Fraser Photinia, Red-tip Photinia

SEED: A red, berry-like pome that ripens in October and persists through winter. Usually not abundantly produced on this species but occasional plants are heavily fruited. Senior author collected fruits in early February, macerated and removed pulp and debris. Clean seeds were sown immediately or provided cold stratification. Untreated seeds germinated in high percentages in 4 weeks indicating cold stratification was not required. There is also the possibility that the seeds had self-stratified while on the plant. If in doubt provide 1 to 2 months cold stratification. This procedure can be applied to *Photinia glabra* and P. *serrulata*.

CUTTINGS: Many growers experience difficulty rooting the species but there is no reason for this. Cuttings can be rooted anytime the wood is firm. The key is high IBA concentration (10,000 ppm IBA in solution), a well drained medium and mist. Rooting takes about 4 weeks. Several southern nurserymen direct stick cuttings in gallon containers in the field in full sunlight and provide overhead mist. Several researchers have thoroughly discussed the procedures necessary to successfully root the species. See *Proc. Southern Nurs. Assoc. Res. Conf.* 28:224-226, 227-228 (1983).

Photinia x *fraseri* is a great plant with which to conduct rooting studies. Without hormone treatment, rooting is virtually nil. If the hormone or plant growth substance is worth its salt, this plant will respond. Also, it is virtually impossible to injure the cuttings by various carries such as alcohol and dimethylsulfoxide. The cuttings withstand excessive moisture and do not decline or rot during the experimental period.

Photinia koreana Korean Photinia

SEED: Three months cold stratification produced good germination.

Photinia parvifolia Littleleaf Photinia

SEED: Two to 3 months cold stratification produced good germination.

CUTTINGS: Early August, 3000 ppm IBA-talc, sand:perlite, mist, rooted 75%.

Photinia villosa var. ***laevis*** Smooth Oriental Photinia

SEED: Two months cold stratification produced good germination.

Photinia villosa var. ***maximowicziana*** Veinyleaf Oriental Photinia

SEED: Fresh seed germinated 12%; 2 months cold stratification produced 92%.

Phyllodoce glandulifera Cream Mountainheath

SEED: Should be handled like *Calluna*.

Physocarpus opulifolius Common Ninebark

SEED: The fruit is an inflated follicle that turns brown when ripe. Seeds germinate without pretreatment.

CUTTINGS: Softwoods, 3000 ppm IBA-talc, mist or polytent, sand:peat, root readily. Easily rooted from June-July softwoods, 1000 ppm IBA-solution, well drained medium, mist. Hardwoods in December, sand:peat:perlite (6:4:3, v/v/v), in a greenhouse or field [*The Plant Propagator* 15(3):10 (1969)].

Picea Spruce

The spruces represent an interesting group of tall, symmetrical, conical evergreen trees. The genus includes nearly 40 species which are largely restricted to the colder regions of the Northern Hemisphere. Numerous cultivars occur within selected species (P. *abies*, P. *glauca*, P. *pungens*) but their availability is limited in the nursery trade.

Seeds of most *Picea* species germinate without pretreatment, but cold stratification has been used for a few species. Cutting propagation is possible in isolated cases but percentages are usually low. Grafting often is used with cultivars of P. *abies*, P. *pungens* and other species. A side graft is used on seedlings of P. *abies* or the same species.

Spruces are adapted to a wide range of diverse environments and a vast number of individual, ecological and geographical variations exist. The importance of seed provenance cannot be over stressed. For example, it is known that the provenance of P. *omorika* can affect the shape of the tree and the susceptibility to frost. Also, with P. *pungens* var. *glauca* it is important to select a seed source that will give a high percentage of glaucous-blue colored seedlings. When ranges of compatible species overlap, natural hybridization is common. Artifical hybrids between widely separated species have been reported when species are planted together. The selection of seed source, therefore, is second in importance only to selection of the species to be grown.

Spruce cones mature in one year and open on ripening to shed seed in the autumn and winter. Therefore, spruce cones must be collected immediately on ripening to avoid seed loss. Seeds are generally mature before the cones change to the ripe colors and can be harvested earlier. *Picea* seed may lose viability if not extracted promptly from the cones and seed is sensitive to adverse storage conditions. Cones can be air dried for a few weeks or in a convection kiln for 6 to 24 hours at 100 to 120°F. Separate the wings and chaff from the seeds and store in sealed containers at 33 to 38°F with a moisture content of 4 to 8%. Spruce seeds have been stored for periods of 5 to 17 years. All spruce species appear to be fairly similar in longevity characteristics and storage requirements [*Arnoldia* 37:62 (1977)].

Seeds of most species germinate promptly without a presowing treatment. A sowing rate of 400 to 500 seeds per square yard appears to be optimum. Benefits are gained from soil sterilization and fungicidal treatments of the seed to control damping-off. Spring or fall sowing out-of-doors is utilized for most of the commonly grown *Picea* species. However, fall sowing late enough to ensure no germination is often preferred because it ensures more uniform spring germination. If no cold stratification is given, then all seeds prior to sowing should be soaked in water for up to 24 hours in order that imbibition takes place. The raising of seedlings under glass or polyethylene structures can be practiced to germinate expensive seed or to produce grafting understock faster.

When germinated under laboratory conditions, P. *engelmannii*, P. *mariana*, and P. *omorika* require artificial light. They germinate rapidly and completely within 7 to 12 days at a normal temperature range. P. *glauca*, P. *orientalis*, P. *rubens*, and P. *sitchensis* require artificial light and alternating temperatures of 68 to 86°F.

Blue Colorado spruce, dragon spruce, Engelmann spruce, Koyama spruce and tigertail spruce are not generally recommended for fall sowing. They germinate rapidly at a lower temperature and do not need cold stratification for maximum germination. The following spruces perform well with late fall sowing; P. *brewerana*, Brewer; P. *bicolor*, Alcock; P. *glauca*, white; P. *glehni*, Sakhalin; P. *jezoensis*, Yeddo; P. *mariana*, black; P. *omorika*, Serbian; P. *orientalis*, oriental; P. *rubens*, red; P. *sitchensis*, Sitka; and P. *spinulosa*, Sikkim [*Amer. Nurseryman* Sept. 15, p. 10 (1967) and April 15, p. 12 (1968)].

Though many nurseryman consider P. *abies* the most satisfactory understock for all taxa, there appears to be a wide latitude in the selection of understock for *Picea* clones. Two periods of the year are used for grafting: August and September or December, January and February. Advantages gained by grafting in August and September include: 1) less risk of "flooding of the union" because excessive sap flow is less likely to occur, and 2) improved scion growth the following year. Scions are side-grafted on established understocks potted one growing season in advance of their use. The plants are then plunged in peat moss to a depth that covers the union. The following graft combinations have survived a significant number of years at the Arnold Arboretum. Root stock - scion compatibilities include: *Picea abies* with *Picea abies*, P. *glauca*, P. *jezoensis*, P. *mariana*, P. *orientalis*, P. *pungens*, P. *purpurea*, P. *rubens*; P. *glauca* with P. *abies*, P. *aurantiaca*, P. *mariana*; P. *omorika* with P. *obovata*; P. *pungens* with P. *abies*, P. *aurantiaca*, P. *pungens*. See PIPPS 20:225 (1970) and 28:219 (1978); *Arnoldia* 37:62 (1977).

Picea abies Norway Spruce

SEED: The cone of P. *abies* is brown when ripe and matures from September to November with seed dispersal occurring from September to April. It is best to harvest the cones just before they ripen. A pound of seed averages 64,000 seeds. The seed requires no pre-sowing treatment, although reference has been made to a 3-week cold stratification for more uniform germination. The seed may benefit from a 24-hour water presoak before spring planting. The germinative capacity is 80%.

CUTTINGS: Rooting is difficult and a challenge. Nutrition, cutting type, season, medium and hormonal treatments are important variables. A controversial topic concerns the time of year that the cuttings should be taken. There have been a number of studies and what proves best for one propagator may not be successful for another. Girouard (Canada), in a seasonal study with 47-year-old plants, found that cuttings should be taken in the spring (April to May), just before or during bud opening, or in mid-autumn (late September to mid-November) after the plants have been subjected to cool temperatures. Other researchers have noted better rooting in June and July, and mid-September until mid-April. Perlite is recommended as the rooting medium although peat moss:sand (1:1,v/v) has also been utilized. Cuttings collected from juvenile trees root best and root within a 15 week period, whereas mature cone-bearing trees require

20 to 23 weeks. Terminal and lateral cuttings give different types of growth after rooting. Lateral cuttings produce a plagiotropic (spreading) habit of growth and also develop into a more bushy form. Terminal cuttings produce the natural orthotropic (upright) growth habit. Position or location in the tree crown additionally affects the ease of rooting. Cuttings from the lower 1/3 of the crown root in higher percentages that those from the upper 1/3 of the crown.

Reports on the effectiveness of rooting hormones vary from no promotion (and possibly inhibition) to promotion. Girouard noted that cuttings not treated with auxin were among those with the best rooting, survival, and shoot development. Enright, however, found that IBA increased rooting in direct proportion to the amount of IBA added. Cuttings were placed under mist during the daylight hours and incandescent lamps were used to provide long days. See *The Plant Propagator* 14(2):5 (1968), 19(2):16 (1973), 20(1):20 (1974), 21(3):9 (1975).

Cutting propagation of P. *abies cultivars.* Cuttings of numerous Norway spruce cultivars with dwarf characteristics can be rooted in commercial percentages. There are as many recommendations relative to rooting P. *abies* cultivars as there are people rooting them. The following procedure was developed by Iseli and Van Meta for the rooting of the following cultivars: 'Gregoriana Parsoni', 'Little Gem', 'Macronata', 'Nidiformis', 'Ohlendorfii', 'Pumila', 'Pygmaea', 'Remontii', 'Repens', and 'Sherwoodii' ('Sherwood's Multinomah') [*The Plant Propagator* 26(3):9 (1980)]. Cuttings from containerized stock plants maintained at optimum nutrition levels are best. The collection of cuttings begins in late July (Oregon) and continues until mid-September. Larger cuttings (5 to 6") root as successfully as small cuttings (1 to 2"). Once prepared the cuttings are dipped for 5 seconds in a 10 to 1 solution of water and Dip 'N Grow, and placed in coarse perlite to ameliorate the deleterious effects of overwatering. Mist is supplied until the cuttings are mature. This usually requires a few weeks of misting and care is necessary to ensure that the needles are always moist without overwatering the medium. Bottom heat is beneficial for speeding the time to rooting but not rooting percentage. Foliar feeding of the cuttings begins once root initiation is observed. After the cuttings have well developed roots, they are moved outside to harden-off and then potted into small containers. Kelly conducted research trials with dwarf spruce cultivars ('Barryi', 'Capitata', 'Decumbens', 'Dumosa', 'Maxwellii', 'Microsperma', 'Nidiformis', 'Pseudo Maxwellii', 'Pumila Glauca', 'Repens' and 'Tabuliformis') and found that with summer cuttings, all taxa except 'Repens' showed increased rooting up to mid-August [PIPPS 22:238 (1972)].

Considerable variation exists in rooting "witches' brooms." Cuttings from forty, 7-year-old "witches' broom" clones, collected in mid-November, 1000 ppm IBA-talc, peat:sand (1:1, v/v), showed 9 clones rooting 100%, 31 clones rooting 50-75%, 9 clones not rooting. Refer to *Pinus* introduction for additional information on "witches' brooms" [PIPPS 35:555 (1985)].

GRAFTING: Refer to the spruce introduction for general comments. *Picea abies* is the preferred understock. Understocks should be established for one year in pots for good results. Mention, however, has been made relative to potting up seedling understocks in October and November, and establishing them in a greenhouse to stimulate root growth before use. Seedling understock is grown for 2 years in the seedbed prior to lifting in the spring of the third year and potting in 2½" pots. Winter or early spring is the preferred time for grafting. Immediately prior to or just as root growth starts are considered the best stages [*The Plant Propagator* 9(1):11 (1962)].

Terminals or terminal shoots from side branches are used as scion wood. The understock is headed back when grafting, however, it is important to keep plenty of top growth on the rootstock. Needles are scraped from the lower half of the stem using a sharp knife, being careful not to damage the scion. A side or side veneer graft is used and care is taken not to cut too deeply into the scion when preparing the graft.

Aftercare of the grafts is very important, particularly during the first 6 weeks. The tops must be prevented from desiccating. This can be accomplished by syringing, light shade or covering with white poly. Two methods of aftercare are practiced. The finished grafts can be placed in a greenhouse bench with the pots plunged in moist medium (sand, perlite, peat moss or any combination) deep enough to cover the graft union. Bottom heat of 65 to 75°F is desirable with the top air temperature at 50 to 60°F. After callusing has progressed to assure a "good take", the medium temperature is lowered to 55 to 65°F. Older literature discusses the use of soft cotton twine and covering the scion with paraffin wax.

Alternatively, the finished grafts can be placed under a plastic tent and healed in sphagnum peat with the union uncovered. The grafts are kept on the dry side until callusing has commenced and then gradually ventilated over the next 3 weeks. After callusing has occurred along the entire length of the union, the grafts are ready for hardening-off. Some growers prefer not to cut the understock back completely until spring planting because they believe a stronger union develops with spruce.

Picea brewerana Brewer Spruce

SEED: The immature cone is green and changes to brown or black when ripe. Cones ripen September to October and seed dispersal occurs at the same time. Brewer spruce averages 61,000 seeds per pound of cleaned seed. The seed germinates without pretreatment. A germinative capacity of approximately 50% can be expected. This species is considered rare and would be best germinated under controlled conditions (such as a greenhouse) and transplanted into individual pots. In nursery practice, the Brewer spruce is spring sown.

GRAFTING: The time taken to produce this plant for market from seed, with the typical weeping characteristics, is considerable. Grafting results in a plant bearing the foliage and weeping habit of the mature species [*The Plant Propagator* 14(2):19 (1968)]. P. *abies* is the preferred understock. Understocks are grown in the seed bed for 2 years, and then potted into 3" pots and grown for an additional year. In early February the rootstocks are brought into a cool greenhouse to stimulate root action. Keep the stocks on the dry side. Scions, pencil thick, with the lower needles removed are side grafted and 1/3 of the stock is removed before placing on an open bench. The union is bound with tape or rubber ties and waxed. Care is necessary to prevent drying but watering is kept to a minimum and shade applied on sunny days. When callus development is evident, the stock is reduced by 1/2 and careful watering and shading applied. The temperature should not exceed 65°F. When the apical bud breaks, the understock is reduced to a few inches. Gradually increase ventilation to harden off. An alternative method is bare-root grafting. The seedlings are grown as above for 2 years then planted 4" apart for one more year. In early

February the seedlings are lifted, boxed in peat and root growth is stimulated. After grafting, the plants are boxed for callusing and treated as above [*Gardeners Chronicle* 165(1):29 (1969)].

Picea engelmannii Engelmann Spruce

SEED: The immature cone is green and changes to brown when mature. Cones ripen in August and September with seed dispersal occurring in September and October. A pound of cleaned seeds averages 135,000 seeds. Seeds require no pretreatment for germination and have a germinative capacity of about 70%.

Picea glauca White Spruce

SEED: Immature cones are green and change to pale brown when ripe. The white spruce averages 226,000 seeds per pound. Seeds lack a pregermination requirement but a period of cold stratification probably unifies and shortens the time required for germination. In nursery practice the seed is fall sown but could be spring planted after cold stratification.

Expensive seed of rare forms, such as *Picea glauca* 'Densata', can be germinated under laboratory conditions and transplanted into individual pots. 'Densata' seed is sown in flats containing either fine perlite or vermiculite. The seedling flats are covered with plastic, placed in a germination chamber lighted with fluorescent tubes and the temperature maintained at 75°F. Germination begins in a few days and at this point the plastic is removed. The flats are misted several times per day. Within 7 to 10 days the seedlings are large enought to transplant.

CUTTINGS: Girouard (Canada) reported the successful rooting of cuttings from 11-year-old plants. Cuttings were taken in early May from lateral shoots of the previous growing season on the lower third of the tree. The cuttings were trimmed to 3" and inserted half their length into shredded sphagnum peat:sand (1:1, v/v) with bottom heat (68°F). Light was reduced by 50% with mist provided during daylight hours. A 70% rooting success was obtained after 5 months. The var. *albertiana* has been rooted from July cuttings. Heeled cuttings are put in sand with bottom heat and no mist or hormone. Cuttings will be rooted in 4 to 5 months. See *The Plant Propagator* 23(3):5 (1977).

Picea glauca 'Conica' can be raised from softwoods, semi-hardwoods and hardwoods [*The Plant Propagator* 24(2):8 (1978)]. Containerized stock can be brought into the greenhouse in February with a temperature around 50°F to promote bud break. Suitable cutting material will be available late April to early May. Softwoods from outdoor stock plants treated with 8000 ppm IBA-talc and placed under intermittent mist also root. Semi-hardwood and hardwood cuttings from current season's growth are taken after becoming sufficiently hardened in July and placed in a shaded cold frame or plastic tunnel. IBA (8000 ppm) in talc or 1% quick dip is recommended. Upright and side shoots work equally well and any quick draining medium such as sand is suitable. The rooted cuttings are left in the cold frame until the following spring. Eighty to 85% rooting can be expected. Plants are grown on in 50% shade.

GRAFTING: Selected forms of the white spruce are grafted on the species or P. *abies*. Scion wood is taken (December to February) and side-grafted on understock potted one year in advance. Refer to *Picea* introduction and P. *abies* for additional information. See PIPPS 8:98 (1958).

Picea jezoensis Yeddo Spruce

SEED: The immature cones are green and change to brown when ripe. A pound of cleaned seed contains 184,000 seeds. This species is reportedly difficult to germinate and the only available information states that a 5 month cold stratification gave some germination.

Picea koyamai Koyama Spruce

SEED: The immature cone is green and changes to brown when mature. A pound of cleaned seeds contains about 100,000 seeds. The seed has no pregermination requirement and in one study germinated 94%.

Picea mariana Black Spruce

SEED: Germinates in fair percentages without any pretreatment, but a period of cold unifies and shortens the time required for germination.

Picea omorika Serbian Spruce

SEED: The immature cones are bluish black and dark brown when ripe. The Serbian spruce averages 140,000 cleaned seeds per pound. Although the seed does not require pretreatment for germination it may benefit from 3 months stratification at 40°F. The seed has a reported germinative capacity of about 87%. *Picea omorika* is a rare species and might be germinated under laboratory conditions and transplanted into individual pots.

GRAFTING: The Serbian spruce is normally grafted December through February on established understock with a side-graft. Fall grafting, September to December, is feasible but maximum survival and growth require the grafts be chilled during this period. P. *abies* and *glauca* can serve as understocks. Refer to P. *abies* for additional information.

Picea orientalis Oriental Spruce

SEED: The immature cone is reddish purple and changes to brown when mature. A pound of cleaned seed averages 74,000 seeds. The seed requires no pretreatment for germination and an average germinative capacity of 60% can be expected. Late fall sowing of this species is possible, however, since it is a rare species it might be germinated under controlled greenhouse conditions (refer to P. *glauca* for general details on this procedure).

CUTTINGS: 'Aurea Compacta' roots in low percentages when treated with 8000 ppm IBA + Benlate (8:1, w/w) (refer to rooting the dwarf cultivars of P. *abies*). 'Nana' has been reported to root at a low percentage (27%) from summer (August) cuttings. See PIPPS 25:81 (1975).

GRAFTING: Refer to *Picea* introduction and P. *abies* for grafting details. *Picea abies* is a suitable understock.

Picea pungens Colorado Spruce

SEED: The cone is green when immature, becoming pale brown when ripe. Cone ripening occurs in the fall with seed dispersal in fall and winter. Colorado spruce averages 106,000 cleaned seeds per pound. Seed requires no pretreatment and has a germinative capacity of about 80%. Fall planting is not recommended because the seeds may germinate at low temperatures, with resultant winter-kill. Spring planting is practiced and cold stratification at 34 to 41°F for 1 to 2 months is recommended to unify and shorten the time required for germination. With P. *pungens* var. *glauca* it is important to select a seed source that will give a high percentage of glaucous-blue colored seedlings. Heit reported that the most consistent blue color was produced by seed from the Manti-Lasal Forest in Utah at the 9000-foot elevation, the San Juan National Forest in Colorado, and the Kaibab National Forest in Arizona.

CUTTINGS: Selected blue forms of P. *pungens*, such as var. *glauca*, have been rooted from cuttings. The cuttings are cut in June with a heel of older wood, treated with 3000 ppm IBA-talc, and placed in sand without removal of the needles. Rooting (80%) occurs in about 8 weeks. Success rate increases by taking cuttings from rooted cuttings or lateral shoots that come off the one-year old wood. Terminal cuttings are not good rooters. Proper root structure and leader development require special attention. After rooting, the cuttings have 1 or 2 roots and when transplanting to flats, the root system is pruned and root pruned again at field planting to stimulate lateral root development. Staking is required to develop a symmetrical plant form. In another study, the cultivars 'Fat Albert', 'Globosa', 'Hoopsii', 'Iseli Fastigiata', 'Iseli Foxtail', 'Prostrata Blue Mist', 'R.H. Montgomery', 'Select', 'St. Mary's Broom', 'Swartzii', 'Thompsonii', 'Thume', and 'Walnut Glen' were rooted in commercially acceptable numbers. Rooting is higher from robust container-grown plants under high nutrition. January to February (Oregon) terminal cuttings, 1 to 10 dilution Dip 'N Grow, very coarse perlite, bottom heat (65°F), 49% shade, and mist during the day (required to keep foliage moist), but dry going into the night. Roots appear in May and weekly feeding starts with transplanting in September. See PIPPS 31:440 (1981); *The Plant Propagator* 27(1):5 (1981).

GRAFTING: Blue spruce cultivars ('Glauca Compacta', 'Glauca Pendula', 'Hoopsii', 'Koster', 'Moerheimii', 'Montgomery' and 'Procumbens') are traditionally grafted on P. *abies* or P. *pungens* seedling understocks. *Picea glauca* has also been utilized. Refer to the introductory comments on spruce grafting and grafting comments under P. *abies* for details.

Picea rubens Red Spruce

SEED: The immature cones are green and change to brown when mature. A pound of cleaned seed contains 139,000 seeds. Two months cold stratification induces good germination.

Picea sitchensis Sitka Spruce

SEED: The immature cones are yellow-green and change to brown when mature. A pound of cleaned seed contains about 210,000 seeds. No pretreatment is required and seed sown fresh gives good germination, however, a period of cold unifies and hastens germination.

Picea smithiana Himalayan Spruce

SEED: The immature cones are green and change to brown when ripe. Cones ripen in October and November with seed dispersal occurring at that time. A pound of cleaned seed contains about 34,000 seeds. Seed has no pregermination requirements and gives a good germination percentage.

Picrasma quassioides India Quassia-wood

SEED: Closely related to *Ailanthus*. The fruits are drupes that ripen in October. Seeds require 3 months cold stratification.

CUTTINGS: Root cuttings will work.

Pieris floribunda Mountain Pieris

SEED: The seed has no barriers to germination. If sown in late winter or early spring they will germinate and grow with the lengthening days. Seed capsules are gathered just prior to opening in fall, dried, and the fine seeds screened from the capsules. A seeding medium consisting of pure milled sphagnum moss works well. See *Calluna* or *Rhododendron* for details [PIPPS 17:236 (1967) and 34:524 (1984)].

CUTTINGS: *Pieris floribunda* is often considered difficult to root. Cuttings are treated similar to rhododendrons: wound heavily on opposite sides and treat with 8000 ppm IBA-talc + thiram (15%) or Benlate (10%). Sphagnum peat moss and horticultural grade perlite (1:1, v/v) with bottom heat at 75°F are ideal. Cuttings taken in early July to late March (Massachusetts) rooted well (70 to 80%). Summer cuttings are placed under mist and winter cuttings under polyethylene. Fall hardwoods have been rooted using 2, 4, 5-TP at 5000 ppm. October cuttings placed in a Nearing frame rooted 70% when lifted the following August. Rooting conditions include wounding, a soak for 18 hours in IBA at 375 ppm and peat:sand (1:1, v/v). 'Millstream', a compact growing form, has been rooted (70 to 80%) when treated with 8000 ppm IBA-talc plus Benlate (10%) or 2, 4, 5-TP at 5000 ppm. Non-flowering shoots should be used [PIPPS 27:495 (1977)].

'Brouwer's Beauty' (P. *floribunda* × P. *japonica*) rooted readily (75 to 100%) from December cuttings. Treat with a 50 to 400 ppm IBA soak for 24 hours, 10,000 ppm IBA dip, or 5000 ppm NAA dip and place in peat:perlite (1:1, v/v) in a bottom heated polytent. Non-flowering cuttings should be used [*The Plant Propagator* 30(3):2 (1984)].

Pieris 'Forest Flame'

CUTTINGS: A hybrid between P. *formosa* var. *forrestii* 'Wakehurst' and P. *japonica*. Terminal cuttings, August to September, 8000 ppm IBA-talc, sphagnum peat:perlite (2:1, v/v), bottom heat (65°F), mist, root readily. Excellent results have also been obtained from December cuttings, but important to check mist frequency to avoid leaf drop. See NGC 188(2):33 (1980).

Pieris formosa and ***P.f.*** var. ***forrestii*** Chinese and Himalayan Pieris

CUTTINGS: Terminal cuttings, August to September, 8000 ppm IBA-talc, sphagnum peat:perlite (2:1, v/v), bottom heat (65°F), mist, root readily. Excellent results have also been obtained from December cuttings, but important to check mist frequency to avoid leaf drop.

Pieris japonica Japanese Pieris

SEED: Capsules are gathered when mature (brown) in fall, dried, and the seeds screened from the capsules. The seeds have no dormancy and can be planted as soon as harvested. Seeds germinate in 2 to 3 weeks. See *Rhododendron* for details. See PIPPS 33:571 (1983).

CUTTINGS: *Pieris japonica* and cultivars are easily (90 to 100%) propagated from softwoods and greenwoods during July and August. Terminal cuttings harvested from container or field stock are best. Trim to 4 to 5", remove the lower leaves, 3000 ppm IBA-quick dip or IBA-talc at 5000 to 8000 ppm. Wounding is sometimes helpful. The rooting medium consists of peat moss:perlite (1:1, v/v), and the cuttings are placed under mist or in a polytent. Mist is applied as needed with Benlate and Manzate D application while the cuttings are under mist. Rooting is slow, often taking 6 or more weeks. Reduce the mist gradually after rooting. Sensitive to excess moisture in the rooting medium, and various pathogens induce stem and root rots.

Cuttings harvested in mid-October and late November also root satisfactorily (100%). Treat with 3000 ppm IBA-talc and place in benches with bottom heat (65 to 75°F), mist as needed, and apply fungicides weekly.

Pieris phillyreifolia

CUTTINGS: Mid-November, wound, 8000 ppm IBA-talc or 5000 ppm 2, 4, 5-TP, peat:perlite (1:1, v/v), closed case, bottom heat (70°F), rooted readily. Softwoods from the above rooted cuttings that were treated the same also rooted with good results [*The Plant Propagator* 23(4):11 (1977)].

Pinus Pine

The genus *Pinus* constitutes by far the most important of the coniferous genera with about 90 species and numerous varieties and cultivars. Pines are distributed throughout the Northern hemisphere and range in distribution from the artic circle to Guatemala, North Africa and Malayan Archipelago. Forty-one species are native to the United States.

SEED PROPAGATION: Pines are monoecious with the male and female cones (strobili) borne separately on the same tree. Female cones are generally located in the upper part of the tree crown and male cones are found in the lower part. After pollination the female cones begin a slow development. At the end of the first growing season they are 1/8 to 1/5 the length of mature cones. Fertilization occurs in spring or early summer about 13 months after pollination and cone development proceeds rapidly. Cones and seeds mature rapidly during late summer and fall of the second year. The mature cone is very woody and is composed of thick, closely-packed scales. Cones of most species open shortly after ripening. Serotinous cones of the so-called "fire" pines (*P. attenuata*, *P. contorta* var. *latifolia*, *P. muricata*, *P. pungens* and *P. radiata*) disperse their seeds only after the intense heat of a forest fire. Some species combine both opening features (*P. banksiana*).

The importance of collecting seed from the proper source cannot be stressed too strongly. Seed provenance is very important in determining ornamental characteristics and the ability of a species to grow and survive in a given environment. Cones should be collected as soon as ripe, since seeds are shed promptly from open cones, and immediately dried to avoid mold and internal heating. Some species, after initial air drying, may require additional heat in a kiln or a heated shed (130°F). Serotinous cones have been opened by dipping in boiling water. Seeds are separated from open cones by screening.

Pine seeds are variable in longevity and sensitivity to environmental factors during storage. Maturity of cones and handling during collection, extraction and cleaning contribute to the keeping quality of seeds during storage. Storage at high humidity and fluctuating temperatures is detrimental to keeping quality. Seeds of the hard serotinous coned pines, such as Bishop, jack, lodgepole, knobcone and Monterey, have good keeping qualities, and cold storage may not be necessary if kept for only a few years with moisture content below 10%.

Most of the common hard pines, such as Aleppo, Austrian, Mugo, Jelecote, Japanese black, Japanese red, pitch, ponderosa and Scotch can be held for up to one to two years without cold storage, provided the moisture content is below 10%.

Seeds of the southern pine group, loblolly, longleaf, shortleaf, and slash should be stored at temperatures of 35°F or below, with moisture contents ranging from 8 to 12%.

The soft or 5 needled pines, such as Swiss stone pine, limber pine, sugar pine, and eastern white pine are more difficult to store. Swiss stone and sugar pines can be stored for several years at subfreezing temperatures. White pine can be stored at 32 to 38°F with 8% moisture for 10 years.

Germination behavior varies widely depending on the species and seed lot. Seeds of many species germinate satisfactorily without pretreatment, but cold stratification unifies and shortens the time required for germination, especially if the seeds have been stored. Stratification is accomplished by first soaking the seed for 1 to 2 days in water and then placing in a moist medium (sand and peat moss) at about 40°F for the appropriate time.

In temperate regions, pine seeds can be sown in fall or spring. However, it is now common practice to spring sow nondormant species. Dormant seeds are spring sown after cold stratification.

When germinated under laboratory conditions, *P. banksiana*, *P. nigra*, *P. mugo* var. *mughus*, *P. ponderosa* var. *scopulorum*, *P. rigida*, and *P. sylvestris* require artificial light and germinate rapidly and completely within 7 to 12 days at a normal temperature range; *P. densiflora*, *P. echinata*, *P. elliottii*, *P. taeda*, *P. thunbergiana*, *P. virginiana*, and *P. wallichiana* require artificial light and 68 to 86°F alternating temperatures; *P. flexilis*, *P. glabra*, *P. leucodermis*, and *P. strobus* require 21 to 28 days cold stratification followed by artificial light at a temperature of 68 to 86°F for 14 to 28 days.

When fall sown, the extremely dormant species, such as *P. armandii*, *P. cembra*, *P. koraiensis*, *P. lambertiana*, *P. monticola*, *P. parviflora*, and *P. pumila* benefit from a few months' warm stratification prior to winter. Pines which benefit from late October or early November sowing include *P. glabra*, *P. leucodermis*, *P. thunbergiana*, and *P. strobus*. Fairly late fall sowing with good success is obtained with *P. densiflora*, *P. echinata*, *P. flexilis*, *P. wallichiana*, *P. ponderosa*, *P. resinosa*, *P. rigida*, *P. taeda*, and *P. virginiana*. See *Amer. Nurseryman* Jan. 15, p. 12 (1967); Sept. 15, p. 10 (1967); Oct. 15, p. 14 (1967).

VEGETATIVE PROPAGATION: Pines as a group are very difficult to root from cuttings, an exception being *P. radiata*, and vegetative propagation is usually accomplished by grafting [*Arnoldia* 37:65 (1977)].

Most pines are grafted in January and February. The understock should be the same species or a closely related species. As a general rule, the number of needles per fascicle can be used to gauge grafting compatibility. For example, 5-needled pines are always compatible with other five-needled pines. There are exceptions, for example, *P. bungeana* and *P. rigida* (both three needled pines) are not compatible. It is a general rule that pines in the same subgenus are more than likely to be compatible.

The following graft combinations have survived a significant number of years at the Arnold Arboretum. *Pinus nigra* with *P. densiflora*, *P. heldreichii*, and *P. nigra*, varieties and cultivars; *Pinus resinosa* with *P. densiflora*, *P. heldreichii*, *P. heldreichii* var. *leucodermis*, *P. nigra* var. *caramanica*, *P. ponderosa* 'Pendula', *P. resinosa* cultivars; *Pinus strobus* with *P. aristata*, *P. ayacahuite*, *P. bungeana*, *P. cembra*, and 'Stricta', *P. cembroides*, *P. flexilis*, *P. holfordiana*, *P. hunnewelliana*, *P. koraiensis*, *P. parviflora* varieties and cultivars, *P. peuce*, *P. pumila*, *P. strobus* varieties and cultivars, and *P. wallichiana* and 'Zebrina'; *Pinus sylvestris* with *P. densiflora*, *P. densiflora* 'Globosa', *P. mugo*, *P. mugo* 'Prostrata', and *P. sylvestris* varieties and cultivars.

Understocks are potted well in advance of grafting time to permit good establishment. The recommended practice is to use dormant, sturdy, straight-stemmed seedlings between 1/8 and 1/4" diameter. After potting, the understocks are placed in a frame, bed, or bench, with 50% shade and grown until grafting time. In early winter (January and February), after the dormancy requirement is satisfied, the understocks are brought into the greenhouse. Rootstocks are grafted after 3 to 4 weeks when they show root activity.

The most widely used graft is the side graft although a veneer is sometimes used. To prepare the stock plants for grafting, needles are removed from the stem for several inches above the soil line. A shallow downward cut 1¼" is made through the bark and cambium and slightly into the wood. A deep cut will produce poor callusing. The scion and rootstock are bound with rubber strips and then placed in a heated greenhouse bench with the pots plunged into moistened medium to a depth sufficient to cover the graft union. The medium may be sand, peat moss, perlite, or any combination. Initially the temperature of the medium should be kept in the range of 65 to 75°F for about 4 to 6 weeks. Ambient air temperature should be maintained at 50 to 60°F. After callusing has progressed, the medium temperature is lowered to approximately 55 to 65°F. The tops are supplied with sufficient moisture by syringing, shading the greenhouse lightly, or by covering with white polyethylene. Daily lifting of the poly is recommended to introduce fresh air. After callusing is evident along the entire length of the union, the grafts are ready for hardening-off. The understock is reduced by half with the remainder removed after scion growth is well advanced. In the spring, the grafts are planted with the union below the soil after removing the grafting ties.

Grafting pines out-of-doors has also been accomplished in mild climates, such as Washington State, in August to October. The 2½ to 3½" scions should be firm, semi-mature wood taken from current season's growth with all needles, except one inch at the top, removed. A side graft is used and the scion is placed on the north side. After grafting, the union area is covered with sawdust mulch which helps to protect from frost and wind conditions. After 20 days, 30 to 40% of the stock top is removed, and the following spring, just before growth starts, the remainder is removed, leaving a 3/4" stub.

"WITCHES' BROOMS": The term "witches' broom" translates directly from the German word *Hexenbesen*. Both parts of the German word are found in English as *hex*, meaning to bewitch, and *besom*, a bundle of twigs bound together to form a sweeping implement. "Witches' brooms" have been found on many species of woody and non-woody plants, and have been shown to result from mutations (bud sports), fungi, bacteria, viruses and the stimuli of feeding mites and insects. Increasing interest in dwarf and slow-growing conifers has given added importance to "witches' brooms" of genetic (mutation) origin. Some dwarf conifers that originated as vegetatively propagated "witches' brooms" are: *Picea abies* 'Maxwellii' and 'Tabulaeformis', *Pinus nigra* 'Hornibrookiana', and *Pinus sylvestris* 'Beauvronensis'. "Witches' brooms" produce seeds that yield plants which are generally 50% normal and 50% dwarf [*Arnoldia* 27(4):29 (1967)].

Dr. Sidney Waxman, University of Connecticut, has a major research program on the development of new forms of dwarf conifers from "witches' brooms". He has over 20,000 seedlings mainly from brooms on two *Larix* species, one *Picea* species, six *Pinus* species, and one *Tsuga* species. Cuttings from "witches' broom" seedlings exhibit clonal variation in rooting. Cutting propagation results are found under the individual species. See PIPPS 33:500 (1983).

Pinus albicaulis Whitebark Pine

SEED: Similar to *P. flexilis*; refer to for seed germination requirements.

Pinus aristata Bristlecone Pine

SEED: The immature cone is green to brownish purple and changes to chocolate brown when ripe. Cones ripen in August and September and seed dispersal occurs at the same time. Ripe cones require 2 to 8 days air-drying to open. Bristlecone pine averages 21,500 cleaned seeds per pound. Fresh seed requires no pretreatment, however, stored seed is variable and may require one month stratification at 33 to 41°F. This pine germinates over a wide range of temperatures and at colder temperatures than most pines. A germinative capacity of approximately 90% can be expected. Seed has been stored 9 years and still germinated 50%. See *Amer. Nurseryman* 127(2):14 (1968).

Pinus armandii Armand Pine

SEED: Fresh seed germinates without pretreatment but 3 months at 40°F unifies germination.

Pinus balfouriana Foxtail Pine

SEED: In one study, fresh seed germinated 87%, with no improvement when seed was stratified for 2 or 3 months at 40°F.

Pinus banksiana Jack Pine

SEED: The immature cone is green and changes to tawny yellow to brown when ripe. The cone ripens in September and seed dispersal occurs at that time. Ripe cones require 2 to 4 hours of kiln drying at 150°F for opening. Jack pine averages 131,000 cleaned seeds per pound. Seed can be stored 10 years and still yield a 50% germination rate. Fresh or stored seeds require no pretreatment or at most one week of cold stratification. A germinative capacity of approximately 70% can be expected. Seedlings from a prostrate form of Jack pine were found to be prostrate or strongly reclining. Dwarf forms also can be grown from seeds produced by "witches' brooms" [*Arnoldia* 27(4-5):29 (1967)].

CUTTINGS: Considerable variation exists in rooting from "witches' brooms." Cuttings from four, 6-year-old "witches' broom" seedlings, collected in mid-March, 5000 ppm K-IBA-quick dip, granular styrofoam:sawdust (1:1, v/v), rooted 0, 40, 60 and 80%, respectively [PIPPS 35:555 (1985)].

GRAFTING: Jack pine is not easily grafted.

Pinus bungeana Lacebark Pine

SEED: The immature cone is green and changes to light yellowish brown when ripe. Use floating technique to check for seed viability; sound seeds will sink. No pregermination requirements exist and seed germinates readily upon sowing.

Pinus cembra Swiss Stone Pine

SEED: The immature cone is greenish violet and changes to purplish brown when ripe. Cones ripen in August and September and the seeds are not dispersed until the detached cone disintegrates. Therefore, the cones must be mechanically broken up to free the seeds. Swiss stone pine averages 2,150 cleaned seeds per pound. Seeds should probably be floated to determine soundness. The seed coat is thick but not impervious to water as once thought. Fresh and stored seeds require 3 to 9 months cold stratification. Some lots of *P. cembra* also may have immature embryos and require a warm stratification (2 to 3 months at 70 to 80°F) preceding the cold stratification for maximum germination. Late fall sowing has given poor germination.

Fall sown seed in August or early September has given good results because this species benefits from 2 to 3 months warm moist pretreatment. Shade seedlings for 2 years. *Pinus cembra*, because of its lengthly and complex seed dormancy problems, together with its difficulty in seed extraction, is a good candidate for vegetative propagation. See *The Plant Propagator* 21(4):7 (1975)]

GRAFTING: *Pinus cembra* can be grafted on P. *strobus* using side or veneer grafts. It is commonly reproduced this way to maintain particular characteristics.

Pinus contorta Lodgepole Pine

SEED: Fresh seed will germinate without pretreatment.

Pinus densiflora Japanese Red Pine

SEED: The dull tawny yellow to brown cone ripens in August and September with seed dispersal occurring in September and October. Ripe cones require 3 to 4 days air drying for opening. Placing the cones in boiling water for 30 seconds may improve opening. Japanese red pine averages 47,000 cleaned seeds per pound and can be safely stored for 2 to 5 years. This is the smallest-seeded of the oriental pines. Seed requires no pretreatment for germination and can be spring sown with a germinative capacity of 80+%. The seed of the Japanese red pine is sometimes mixed with the Japanese black pine. Seedling stem color of the red pine in pinkish red, and green in the black pine.

GRAFTING: 'Umbraculifera', 'Oculis-draconis', 'Pendula' and other cultivars are grafted on seedling P. *densiflora* using a side or veneer type graft [PIPPS 18:255 (1968)].

Pinus sylvestris is also listed as an understock [NGC 188(21):31 (1980)].

Pinus echinata Shortleaf Pine

SEED: The immature cone is green and changes to light brown or dull brown when ripe in October to November. Seed dispersal occurs at the same time. Mature cones require 48 hours kiln drying at 105°F for opening. Shortleaf pine averages 46,000 cleaned seeds per pound. Seeds can be stored for 35 years. Fresh seed requires 0 to 15 days of cold stratification, while stored seeds require ½ to 2 months pretreatment. A 90% germination rate can be expected. In nursery practice seeds are pretreated and spring sown.

Pinus flexilis Limber Pine

SEED: The immature cone is green and changes to brown when ripe in August to September. Seed dispersal occurs September to October. Ripe cones require ½ to 1 month air-drying for opening. *Pinus flexilis* seeds are variable in size, averaging 2,300 to 5,400 cleaned seeds per pound. Seed can be stored for 5 years. Most seed sources have shown some embryo dormancy. Fresh or stored seed requires ½ to 3 months pretreatment and has a germinative capacity of approximately 80%.

Pinus halepensis Aleppo Pine

SEED: The cone ripens the third year and is yellowish brown or reddish when ripe. Seed germinates readily without pretreatment. The variety *brutia* appears to benefit from a cold treatment of 3 months. Optimum temperature for seed germination is 68°F.

Pinus koraiensis Korean Pine

SEED: The mature cones are yellowish brown when ripe in September and seed dispersal occurs in October. Korean pine averages 820 cleaned seeds per pound and can be stored for 3 years. Cold stratification for 3 months is required for germination. In some seed lots embryos may be immature and require warm stratification (2 months, 70 to 80°F) before cold stratification. In one study, seed scarified followed by immediate sowing germinated 70%, scarification plus 2 months at 40°F produced 90% germination, and fresh seed (untreated) germinated only 40%.

Pinus leucodermis Balkan Pine

SEED: The number of seeds per pound averages about 20,000. The seed is extremely dormant for a hard pine and requires 40 to 60 days cold stratification if spring sown. Early fall sowing is recommended.

Pinus monophylla (P. *cembroides* var. *monophylla* Singleleaf Pinon Pine

SEED: The seeds are large and lots have run only 900 to 1350 per pound. Seeds are not dormant but have been found sensitive to warm temperatures over 72°F during germination in the laboratory. Very early spring sowing when the soil is fairly cool has given best results.

Pinus monticola California Mountain Pine, Western White Pine

SEED: Fresh seed requires no pretreatment. Fresh seed germinated 90%.

Pinus mugo Mugo Pine

SEED: The immature cones are violet purple and turn tawny yellow to dark brown when ripe. Cones ripen about October and seed dispersal occurs in November and December. Ripe cones require 48 hours of kiln drying at 120°F for opening. Mugo pine averages 69,000 cleaned seeds per pound and seeds can be stored for 5 years. Fresh or stored seed requires no pretreatment for germination and a germinative capacity of 45 to 80% has been reported.

Extreme variation in growth habits of Mugo pine occurs. For the most compact, slowest-growing plants, obtain seeds collected from the true dwarf trees (shrubs) growing on the high Tyrolean Alps of central Europe.

CUTTINGS: In one study, *Pinus mugo* var. *mugo*, early June (Connecticut), current season's growth when needles were ½ to ¾ length from young plants less than 8-years-old, peat:coarse sand (1:1, v/v), IBA + Benlate (1 part 8000 ppm IBA-talc + 1 part 5% Benlate), plastic cover and mist, rooted 74% in 6 to 8 weeks [*The Plant Propagator* 22(3):9 (1976); PIPPS 30:372 (1980); *HortScience* 9:350 (1974), 12:270 (1979)].

Pinus nigra Austrian Pine

SEED: The immature cones are yellowish green and change to light brown when ripe. Cones ripen September to November and seed dispersal occurs in October and November. Ripe cones require 24 hours of kiln drying at 115°F for opening. Austrian pine averages 26,000 cleaned seeds per pound and the seed can be stored for 10+ years. Fresh seed has no pregermination requirements, however, stored seed is variable and may require up to 2 months of cold stratification. The germinative capacity averages greater than 85%. In nursery practice, if fall sown, no pretreatment is given. However, spring sown seed is cold stratified for 1 to 1½ months before sowing.

Pinus parviflora Japanese White Pine

SEED: The brownish red cones ripen in September and seed dispersal occurs in November. Ripe cones require 5 to 15 days air-drying to open. Japanese white pine averages 3,900 cleaned seeds per pound. Fresh and stored seeds require 3 months cold stratification for germination. Germinative capacity is 80%. Sow outdoors in August or early September because this species benefits from a 2 to 3 month warm period.

Pinus peuce Balkan Pine

SEED: The immature cones are green to yellow and turn tawny yellow to light brown when ripe. Cones ripen in the fall and seed dispersal occurs at the same time. Balkan pine ranges from 8,000 to 10,000 cleaned seeds per pound. The seed has a rather tough, hard seed coat. Seeds seem to have a partially impermeable seed coat and dormant embryo. Some lots have responded better with 30 minutes of acid scarification before cold stratification. The seed is variable in its cold requirement. Fresh seed may require up to 2 months cold stratification, and stored seed from 2 to 6 months. A practical cultural solution may be to sow the seed in late spring, mulch to prevent drying and carry through to the second spring when complete germination should take place.

Pinus ponderosa Ponderosa Pine

SEED: The immature cones are green to yellow green and become yellow brown to brown when ripe. Cones ripen August to September and seed dispersal occurs at the same time. Ripe cones open after 4 to 12 days air-drying or 3 hours of kiln drying at 120°F. Ponderosa pine averages about 12,000 cleaned seeds per pound. Fresh seed has no pregermination requirement, however, stored seed may require 1-2 months of cold stratification. The seed can be stored for approximately 15 years and has a germinative capacity of approximately 70%. In nursery practice, the seed is given a presowing cold stratification period and spring sown.

Pinus pumila Japanese Stone Pine

SEED: Excellent germination resulted from 3 months warm stratification at 65 to 85°F followed by 3 months cold stratification at 40°F. An acid scarification might eliminate the need for the warm period. This is one of the handsomest 5-needle pines and tremendous variation is evident in seedlings.

Pinus resinosa Red Pine

SEED: The immature cones are green and change to brown when ripe. Cones ripen August to October and seed dispersal takes place in October and November. Ripe cones require 9 hours of kiln drying at 130°F for opening. Red pine averages 52,000 cleaned seeds per pound and has been stored successfully for 30 years. Fresh seed has no cold requirement for germination, but stored seed requires 2 months cold stratification. A germinative capacity of 75 to 80% can be expected. Optimum temperature for germination is 77°F. In nursery practice, red pine is either fall or spring sown.

Pinus rigida Pitch Pine

SEED: The immature cones are green and change to yellow brown when ripe. Cones ripen in fall and seed dispersal occurs in the fall. Many cones can remain closed for several months to years. Soaking ripe cones in water (24 hours) and then drying at room temperature reportedly causes cone opening. Pitch pine averages about 62,000 cleaned seeds per pound. The seed has been successfully stored for 11 years. Fresh seed has no pregermination requirement but stored seed may need 1 month cold stratification. The germinative capacity averages 70+%. In nursery practice the seed is spring sown after pretreatment.

Pinus strobus Eastern White Pine

SEED: The immature cones are green and change to brown when ripe. Cones ripen August to September and seed dispersal occurs at the same time. Ripe cones respond to kiln drying for 4 to 12 hours at 130°F for opening. Eastern white pine averages 26,000 cleaned seeds per pound and has been successfully stored for 10 years. Fresh and stored seeds require 2 months cold stratification. The germinative capacity averages greater than 90%. In nursery practice, the seed is fall sown or spring sown after 2 months cold stratification. In one study, fresh seed germinated 2%; 28% after 1 month cold stratification, and 88% after 3 months cold stratification. If fall sown, the seed should be sown in early fall during October.

White pine "witches' brooms" can be propagated by seed. Only female cones have been found on brooms and seed production occurs by cross pollination with normal trees. Usually 50% of the seedlings are abnormal. The abnormal types will vary in form, texture, size, needle length and color, and branch density. This allows for selection of new and unusual dwarf forms [*Arnoldia* 27(4-5):29 (1967)].

CUTTINGS: There is considerable variation in rooting response among different clonal trees. In one study, late December cuttings from the lower part of a 12-year-old tree, peat:perlite (1:1, v/v), mist during daylight, cool greenhouse (50 to 60°F), bottom heat (78°F), rooted 100, 63, and 50% when treated with 2500 ppm IBA + 2500 ppm NAA + 25% Captan; 5000 ppm IBA + 25% Captan; and 5000 ppm IBA, respectively. In the same study, cuttings from a hedge planting rooted 30% when 5000 ppm IBA + 25% Captan was applied [*The Plant Propagator* 26(4):9 (1980)]. Cuttings from 10 seedlings derived from a single "witches' broom" clone, late December, 4000 ppm IBA-talc, coarse perlite, rooted 100, 100, 100, 100, 100, 80, 60, 40, 40 and 0%.

Pinus sylvestris Scotch Pine

SEED: The immature cones are green and change to varying shades of brown when ripe. Cones ripen September to October and seed dispersal occurs from December to March. Ripe cones require kiln drying for 10 to 16 hours at 120°F or air drying for 3 to 7 days for opening. Scotch pine averages about 75,000 cleaned seeds per pound. The seed has been successfully stored for 15 years. Fresh seed has no pregermination requirements, however, stored seed is variable and may require ½ to 3 months cold stratification. The germinative capacity generally averages greater than 80%. In nursery practice the seed is fall sown or spring sown after a pretreatment of up to 2 months. This species exhibits extreme variation in growth habit and needle color traits from different seed sources.

CUTTINGS: There is considerable variation in rooting response among different clonal trees. In one study, cuttings from a 9 to 10-year-old tree, late January, peat:perlite (1:1, v/v), mist during day-light, cool greenhouse (50 to 60°F), bottom heat (78°F), gave rooting percentages of 90, 80, and 40% with treatments of 2500 ppm IBA + 2500 ppm NAA; 2500 ppm IBA + 2500 ppm NAA + 25% Captan; and 5000 ppm IBA + 25% Captan, respectively [*Physiol. Plant.* 35:66 (1975)].

Pinus tabulaeformis Chinese Pine

SEED: Number of seeds per pound averages about 14,000. Fresh seed germinates without pretreatment. Late fall or spring sowing is recommended.

Pinus taeda Loblolly Pine

SEED: The immature cones are green and change to light brown to reddish brown when ripe. Cones ripen September to October and seed dispersal occurs from October to December. Ripe cones are subjected to kiln drying for 48 hours at 105°F for opening. Loblolly pine averages about 18,000 cleaned seeds per pound. The seed can be stored for more than 9 years. Fresh and stored seed has a 1 to 2-month cold stratification requirement for germination. In nursery practice, a 1 to 2-month cold pretreatment is provided before spring sowing.

Pinus thunbergiana Japanese Black Pine

SEED: The immature cones are purple and change to reddish brown when ripe. Cones ripen October to November and seed dispersal occurs in November to December. Ripe cones are sometimes given a boiling water treatment (up to 30 seconds) to facilitate opening before air drying for 5 to 20 days. Japanese black pine averages 34,000 cleaned seeds per pound and has been stored successfully for 11 years. Fresh seed has no pregermination requirement, however, stored seed may require 1 to 2 months cold stratification pretreatment. The seed has a germinative capacity in the range of 75 to 85%. In nursery practice, the seed is fall or spring sown after 1 to 2 months cold stratification. Fall sowing produces better seedlings.

Pinus virginiana Virginia Pine

SEED: The immature cones are green and change to reddish brown or dark brown when ripe. Cones ripen September to November and seed dispersal occurs in the fall. Many cones remain closed for several months to years. A 2 hour kiln drying at 170°F is recommended to help opening. Virginia pine averages 55,000 cleaned seeds per pound and the seed has been stored successfully for 5 years. Fresh seed is variable in its requirement for a pregermination treatment but will germinate without any pretreatment. Some seed lots may require 1 month cold stratification. Stored seed requires 1 month pretreatment before planting. The germinative capacity varies from 65 to 90%. In nursery practice the seed is either fall or spring sown after 1 month cold stratification.

Pinus wallichiana (*griffithii*) Himalayan Pine

SEED: Immature cones are green and change to light brown when ripe. Cones ripen August to October and seed dispersal begins in September. Himalayan pine averages 9,400 cleaned seeds per pound. Fresh seed is slightly dormant and a short cold stratification of 15 days is recommended. Stored seed may require ½ to 3 months cold stratification for uniform germination. The germinative capacity averages about 65%. In nursery practice the seed is late fall sown or spring sown after pretreatment.

Pistacia chinensis Chinese Pistache

SEED: Seed propagation is the most common method of growing this species in commercial nursery production. A dioecious species with the bluish fruits (drupes) occurring only on female trees. The fruits ripen in September and October and are soaked for 2 or 3 days to allow the pulp to lightly ferment and slip from the seed. After maceration the heavier seed settles to the bottom and the lighter seed (hollow) and pulp can be floated off.

Germination improves dramatically when seed is stratified at 40°F for 2 months — from 60 to 90% of stratified seeds germinate compared to 0 to 24% for nonstratified. A common practice is fall planting with germination occurring the following spring. *Pistacia* seedlings tend to suffer considerably from shock in transplanting. The key to success is good seed. Invariably red fruits are hollow so collect only blue fruits. Seedlings, unfortunately, tend to grown irregularly [PIPPS 32:497 (1982)].

CUTTINGS: Juvenility is a very important factor in rooting. Softwoods from one-year-old seedlings root 50 to 90% using May or June cuttings, 10-second dip in 5000 or 10,000 ppm IBA, coarse sand and mist. Cuttings from a 2-year-old budded clone failed to root under similar conditions. Hardwood cuttings from the same clone also did not root. This species has been difficult to root from mature trees. Apparently, juvenility is lost at a young age.

GRAFTING: Budding is more widely practiced than grafting, however, success is variable and uncertain. Experience has shown the importance of using vigorous budwood from current season's growth. Pruning plants back insures a good supply of vegetative buds. Opinion differs as to the best time of year to bud. T-budding in August, however, is more successful than spring budding. Superior fall coloring clones are known but their propagation has proven elusive. Continued research with grafting and cutting techniques is necessary. See PIPPS 10:287 (1960) and *The Plant Propagator* 9(2):11 (1963).

Pittosporum tenuifolium Tawhiwhi Pittosporum

SEED: Seed germinates without pretreatment.

CUTTINGS: In one study, 'Atropurpureum', 'Garnettii', 'Silver Queen', 'Tresederi' were rooted from December and January (England) cuttings, 6000 ppm IBA-talc, grit:sand (1:1, v/v), bottom heat (70°F), and mist at an 80% success rate in 8 to 12 weeks. 'Abbottsbury Gold' and 'Garnettii' were rooted from June and July cuttings taken from indoor stock or small shoots on previously rooted crops with peat:perlite (2:1, v/v), 8000 ppm IBA-talc, and mist. Avoid the use of thin and weak cuttings. Remove from mist as soon as possible. A fungicide soak before sticking and during propagation has been recommended. See *The Plant Propagator* 17(3):12 (1971); NGC 188(2):33 (1980).

Pittosporum tobira Japanese Pittosporum

SEED: Possibly like P. *tenuifolium*

CUTTINGS: An environmentally tough broadleaf evergreen shrub but hardy only to 0 to 5°F. 'Variegata', 'Wheeler's Dwarf' and the species can be rooted year-round as long as the wood is firm. Use 1000 to 3000 ppm IBA-talc or solution, well drained medium, mist.

Platanus × acerifolia London Planetree

SEED: This species is a hybrid between P. *orientalis* × P. *occidentalis* and must be asexually propagated to maintain the desirable bark, foliage, fruit and disease resistance. However, common practice has been to propagate the species from seed. The resultant trees have lost (during segregation) some of the original traits, such as anthracnose resistance. Fresh seed may germinate but in low percentages. Two to 3 months cold stratification improves germination. London planetree is monoecious with two, 1 to 1½" diameter, rounded syncarps of achenes. The fruits ripen about November and seed collection can be accomplished into spring because the fruit persists. Fruits should be dried in ventilated bins in thin layers until broken apart. During extraction some method of dust removal needs to be supplied and workers should wear masks. Cleaned seed will average between 150- and 190-thousand seeds per pound. For storage longer than one year, seeds should be dried to 10 to 15% moisture and stored in air-tight containers at 20 to 38°F.

CUTTINGS: Softwoods harvested after the shoots have begun to mature can be rooted. The highest rate of success is achieved with cuttings taken from young, vigorous stock plants cut back annually during the dormant season. Cuttings are placed in sand:peat (1:1, v/v) or peat:perlite (1:1, v/v) under mist (3 seconds/2.5 minutes). Leaves can be reduced by ½ to increase sticking density. Untreated and those treated with 8000 ppm IBA-talc rooted best with the control giving the highest percent. Leaf abscission and stem necrosis occur with IBA treatments above 8000 ppm. By 45 days, the cuttings are well rooted.

Myers and Still [*The Plant Propagator* 25(3):9-11 (1979)], June softwoods (Kansas) from a 45-year-old tree and a 20-year-old tree, control, 8000 ppm, 1.6, 3.2% IBA-solution, best rooting occurred with the control and was 100% for 20-year-old tree and 95% for the 45-year-old tree; IBA at the two lowest concentrations was effective but not better than control; NAA proved valueless.

In mild climates (England), P. × *acerifolia* can be propagated from hardwoods inserted in the field in autumn. A success rate of about 70% is reported. Cuttings are taken from current season's growth and are 6 to 9″ long. A cutting with 4 buds is best and thin terminal growth should be avoided. Stool block and pollarded trees produce the best cuttings. Tree age may be important and results suggest that trees older than 20 years root less readily. The top cut should be just above a bud and the lower cut just below a bud. Insert the cuttings shortly after leaf fall and no more than 1 inch should be above the soil surface. In early spring, any cuttings heaved by frost should be reinserted. Provide shade until the new growth has hardened off.

Hardwood cutting propagation (England) utilizing a modified heated bin technique also is possible. The bin contains equal parts peatmoss and sand with a temperature at the cutting base of about 60°F. Cuttings are collected in January and made to a length of at least 1′. In mid-March, they are lifted and field planted. Although control (0 hormone) and IBA (1250 ppm) rooted equally well, field survival for IBA-treated cuttings was low. Two-node hardwood cuttings taken in January and inserted into rockwool blocks have been reported to root in 6 weeks. Bottom heat is maintained at 60 to 65°F.

GRAFTING: 'Bloodgood' is summer budded on seedling understock. Other named clones like 'Liberty' and 'Columbia' are treated likewise. All can also be rooted from cuttings.

Platanus occidentalis American Planetree, Sycamore

SEED: Percentage of sound seed is variable and may be low. The fruit ripens in November and seed dispersal occurs February through April. While pregermination treatments are not required, there is evidence cold stratification improves germination. Shading is beneficial during the first month after germination.

CUTTINGS: Follow procedures described under P. × *acerifolia*.

Platanus orientalis Oriental Planetree

SEED: The fruit ripens in September and October and can be harvested at that time. Refer to P. × *acerifolia* for additional comments. Oriental planetree will average 140,000 cleaned seeds per pound. Pregermination treatment is not required and spring planting is best. Evidence indicates that 2 months cold stratification improves germination.

CUTTINGS: Late November (Boston), 8000 ppm IBA + Captan, sand, open bench (cool air temperature), bottom heat, rooted 83% in 7 weeks. This species is easier to root than P. *occidentalis* and has passed this characteristic to P. × *acerifolia*. In another study, late May (Boston), 3000 to 8000 ppm IBA-talc + thiram averaged 83% rooting.

Platanus wrightii Arizona Planetree

SEED: Requires no pretreatment and will germinate upon sowing.

CUTTINGS: Cuttings from young plants root readily without hormone treatment.

Podocarpus macrophyllus Yew Podocarpus

SEED: Not a great deal of information available, however, Fordham [*Arnoldia* 37(1):68, 1977] recommended sowing seed in late winter. Seeds germinate in good percentages but slowly. This species may have the same problem as *Taxus*. Seeds are covered with a fleshy aril and should be collected in fall-winter, cleaned and stored or planted.

CUTTINGS: Southern nurserymen root the species and var. *maki* from firm-wood cuttings in late summer or fall with 3000 to 8000 ppm IBA-talc, well drained medium and mist or polytent. Rooting is slow. The approach for this species is much like that used for *Taxus*.

Polygonum Polygonums

SEED: Require no pretreatment.

CUTTINGS: A large group of vines and suckering, herbaceous perennials that, although not woody, are found in gardens. Softwood in late spring-summer, root cuttings and division will work. *Polygonum aubertii*, P. *baldschuanicum*, P. *cuspidatum* and var. *compactum* are common in America. Numerous other garden forms are used in Europe.

Poncirus trifoliata Hardy-orange

SEED: Collect small 1-2″ diameter fruits (hesperidiums) in fall, split and remove seeds. Seeds can be dried and stored for long periods under refrigeration. Although literature reports the seed will germinate without pretreatment, the rate is sporadic. Senior author has planted seeds immediately after extraction and germination occurred over a 3 month period. Seed stratified for 1 month at 41°F germinated uniformly in 3 to 4 weeks. Seed quality is high and germination averages 95%. Polyembryony is often evident with several shoots developing from a single seed. For some reason, the presence of albinos is quite frequent. Seed for citrus grafting is direct sown into 3″ propagation tubes in early spring, grown until early summer and then potted into 3 quart pots for autumn budding.

CUTTINGS: Softwood cuttings taken in summer rooted after treatment with a 50 ppm IBA, 17 to 24-hour soak. Late October, failed to root without treatment, but rooted 76% after a 50 ppm NAA-24 hour soak.

GRAFTING: The hardy orange is used commercially as a dwarfing understock for citrus. In commercial propagation, T-budding is used. Decline of navel orange trees on trifoliate orange rootstocks has occurred at 15 to 20-years of age. Trees on some rootstocks, such as 'Argentina', 'Dryder 60-2', and 'Christiansen' trifoliates, were more severely and regularly affected. 'Rich 16-6' trifoliate rootstock has not exhibited this decline.

Populus Poplar

The poplars are a group of large, often fast-growing, deciduous trees. There are about 30 species widely distributed in North America, Europe, North Africa and Asia. In addition, there are numerous hybrids and named cultivars. Hybridization has been reported among almost all poplar species. Flowers are produced in catkins on leafless shoots in the spring, and except for minor deviations, all species are classified as dioecious.

SEED: The fruit is a 3 to 4-valved dehiscent capsule and the time of ripening is species dependent and often highly variable within a species. Wide variation between individual trees has been reported for P. *deltoides*, P. *grandidentata*, and P. *tremuloides*. A safe criterion for time of fruit collection is when a small percentage of the capsules are beginning to open.

The seed is surrounded by a conspicuous tuft of white, cottony hairs which enables dissemination by the wind. Small quantities of seed can be cleaned by rubbing over a wire screen. The most efficient method of freeing poplar seeds is with an air stream in combination with 20, 40, and 60 mesh screens. The range of cleaned seeds per pound is highly variable and those given by the *Seeds of Woody Plants in the United States* include: P. *deltoides* var. *deltoides*, 350,000 to 1,250,000; P. *deltoides* var. *occidentalis*, 250,000 to 479,000; P. *grandidentata*, 3,000,000+; P. *heterophylla*, 141,000 to 165,000; P. *tremula*, 2,660,000; and P. *tremuloides*, 3,600,000. Seed longevity under natural conditions has been reported to vary from 2 weeks to one month and 2 to 3 years after proper storage. Pre-storage drying immediately after collection is essential for successful storage. A moisture content of 5 to 8% improves viability and germination of stored seed. After air drying for 4 days store in a sealed container at 41°F.

Fresh seeds germinate within a few hours. Seed is not covered and requires a water-saturated seedbed. The need for adequate moisture can be met by shallow irrigation ditches or fine-spray sprinkler systems. Seedlings are very susceptible to drying, heavy rains, and damping-off type fungi [*Amer. Nurseryman* Nov. 15, p. 12 (1967)].

CUTTINGS: Softwood and hardwood cuttings of most poplars root readily with the exception of gray poplars and aspens. White poplar (P. *alba*), black poplars, cottonwoods (P. *deltoides*, P. *nigra*), and their hybrids can be propagated by hardwood cuttings and less commonly by sets. Cuttings are usually 10 to 15″ long pieces of dormant 1-year-old stems. Sets are dormant stems, 4 to 10′ in length, and 1 or 2-years old. Cuttings are commonly obtained from stool beds. In Europe, many propagators feel that cuttings should be cut at an angle with the top cut immediately above a bud. Cuttings are made anytime in the winter and stored (37 to 38°F) until used. There is no advantage to early insertion. In England, cuttings are usually stuck by March. Cuttings are planted either flush with the soil or with 2 or more buds above the soil. Planting space varies widely. In June the plants are pruned to one good shoot. In cuttings from easy to root poplars, roots develop from preformed root initials and 90 to 100% rooting is common. Difficult-to-root species, such as aspens, lack preformed root initials. A rich, well-drained soil produces the best plants.

Gray poplar, P. × *canescens*, is a mixture of natural hybrids between P. *alba* and P. *tremula* and few clones root well from hardwoods. P. *tremula*, European aspen, and P. *tremuloides*, quaking aspen, are similarly difficult or impossible from hardwoods. These plants propagate readily from root suckers, root cuttings or layers. Gray and aspen poplars can be propagated by rooting softwoods from root pieces. In February, 2 to 4″ long root pieces are placed in moist peat moss or similar medium in a greenhouse. By the end of March the root pieces have produced soft shoots that root readily. Species difficult to root from hardwood may be readily rooted under mist with bottom heat. The aspens and the *Leucoides* poplars root least readily from softwoods, while the black (*Aigeiros*) and balsam (*Tacamahaca*) poplars are easiest to root. Cuttings are selected from vigorous stock plants and treated with IBA-talc preparations. See *The Plant Propagator* 18(4):6 (1972) and 28(2):7 (1982).

GRAFTING: The grafting and budding of poplars have received little attention. Nelson recorded three combinations of white poplar and aspen, and the successful combination of P. *tremula*/P. *trichocarpa* from two different sections. The latter combination indicates that intersectional compatibility is present. Compatibility and growth of P. *tremula* 'Erecta' on a broad range of poplar rootstocks also has been examined by Ronald and Cumming. Understocks tested included: Brooks #4 and #6 hybrids, P. × *canadensis* 'Serotina de Selys', P. *deltoides* 'Dakota', P. *songarica*, P. *nigra* 'Thevestina', P. *acuminata* and P. *tremuloides*. Only Brooks #4 and #6, 'Thevestina' and 'Dakota' clones, and P. *tremuloides* proved compatible for two or more years. Brooks #4 and #6 proved best because they readily root from hardwoods and do not sucker. T-budding in late August or whip and tongue grafting in the winter are possible.

TISSUE CULTURE: Cambium tissue from 1-year-old terminal branches and shoot tips were used to initiate callus tissue from which whole plants were regenerated from the following species: P. × *euroamericana*, P. *nigra*, P. *tremula* and P. *canescens*. The best growth was obtained on Linsmaier and Skoog medium supplemented with NAA (2 mg/liter), BA (1 mg/liter), and L-arginine (100 mg/liter). Callus was initiated under continuous light (fluorescent and incandescent, 10,000 lux) and a temperature of 77°F. The Linsmaier and Skoog medium supplemented with BA at 0.4 mg/liter supported the initiation and growth of leafy shoots from the callus. Undifferentiated callus was placed in agar plates at 77°F, 10,000 lux and 16 hours of light. Best development of roots was achieved on shoots transferred into a sterile mixture of perlite:sand (3:1, v/v) containing 1/2 strength Wolter and Skoog medium without cytokinin, and 0.2 mg/liter of NAA [*Biologia Plantarum* 16:316 (1974)].

The difficult-to-propagate poplar species, P. *alba*, P. *alba* × *glandulosa*, P. *alba* × *tremula*, P. *canescens*, P. *tremula*, and P. *tremuloides*, have been propagated in vitro by enhanced axillary branching of shoots. Apical and axillary buds from adequately chilled formant shoots, actively growing shoots, and excised root sucker shoots can serve as explants. Buds are rinsed in ethanol and then surface sterlized in 0.15 to 0.3% sodium hypochlorite and 0.05% Tween 80. This is followed by 3 sterile distilled water rinses. Excised buds 1 to 3 mm are cultured on a solid agar medium using a modified MS salts medium (adenine sulfate, 20 mg; lysine, 100 mg). Refer to table for individual hormone requirements for each species. Cultures were incubated at 77°F under fluorescent lights (1000 to 3000 lux) with a 16 hour photoperiod. See PIPPS 28:255 (1978).

Callus production is reduced in the rooting medium by the elimination of lysine, inositol and adenine sulphate from the basal medium. Good root development is obtained in 10 to 14 days. The plantlets are removed from culture, sprayed with a fungicide and placed under mist for 7 to 14 days.

Hormone requirements (mg/liter) of *Populus* species for in vitro propagation.

Species	Stage 1	Stage 2	Stage 3
		concentration in mg/liter	
P. alba	BA 0.2	BA 0.2 NAA 0.02	IBA 0.2
P. alba x *glandulosa*	BA 0.5	BA 0.5 NAA 0.02	IBA 0.5 NAA 0.1
P. alba x *tremula*	BA 0.5, 1	BA 0.5 NAA 0.02 BA 0.5 NAA 0.05	IBA 0.2, 0.5
P. canescens	BA 0.5, 1	BA 0.5 NAA 0.02	IBA 0.2, 0.5
P. tremula	BA 0.5	BA 0.5 BA 0.5 NAA 0.02	IBA 0.5 NAA 0.1
P. tremuloides	BA 0.5	BA 0.5 BA 0.5 NAA 0.02	IBA 0.5 NAA 0.1

Populus nigra 'Italica', P. 'Flevo' (*P. deltoides* x *P. nigra*) and *P. yunnanensis* were similarly propagated with dormant buds serving as the explant source. A modified MS medium similar to the above was used and the following hormone treatments were used (mg/liter): Stage 1, BA 0.2, NAA 0; Stage 2, BA 0.1, NAA 0.02; Stage 3, BA 0.1, NAA 0.01 [PIPPS 26:340 (1976)].

Potentilla fruticosa Bush Cinquefoil

SEED: Seeds require no pretreatment.

CUTTINGS: Softwood, 1000 ppm IBA, peat:perlite or suitably well drained medium, mist, root 100%. Very sensitive to excess water and the mist should be reduced as soon as rooted. Rooting takes place in 3 weeks and cuttings should be transplanted or removed from mist. Cuttings will continue to grow after rooting.

TISSUE CULTURE: See *HortScience* 17:190 (1982).

Prinsepia sinensis Cherry Prinsepia

SEED: The orange-red to red drupe ripens in July, August and September. Macerate the fruit to remove the pulp. Fresh or dry-stored seeds with intact endocarp germinated satisfactorily, although slowly, and chilling did not increase germination. GA_3 (100 to 500 ppm) hastened germination of seed in the endocarp but did not increase final germination. Optimum germination temperature is 60°F; 77°F gave no germination. Germination is inhibited by light. The shoot must be chilled after germination to bring about normal shoot elongation. Work at the Arnold Arboretum showed that 2 months cold stratification produced 74% germination; untreated seed germinated 38%.

CUTTINGS: Softwoods, early July, root readily if taken before growth hardens. Treat with 3000 to 8000 ppm IBA-talc and place under mist. Harder cuttings taken in mid-August rooted 100% with 8000 ppm IBA-talc.

Prinsepia uniflora Cherry Prinsepia

SEED: Cleaned seed germinated 30%; only 15% after 6 months cold stratification. Somewhere in the 1 to 2 month range lies the best treatment.

CUTTINGS: Not the easiest plant to root. Early July, 3000 or 8000 ppm IBA-talc, rooted 55 and 69%, respectively, but only after cuttings were removed and the callus removed and the cutting retreated with hormone. Early August, 8000 ppm IBA-talc + thiram, mist, rooted 40%; variety *serrata*, mid-July, 100% rooting. Senior author has had no luck with this species. It appears that the cutting wood must be firm.

Prunus Almond, Cherry, Nectarine, Peach, Plum, Apricot

Prunus flowers are bisexual, and occur singly or in clusters. The fruit normally is a one-seeded drupe that is fleshy, except in almond, and contains a bony stone or pit. Fruits are collected when mature which facilitates cleaning. Cleaning is accomplished by maceration and floating off or screening out the pulp. Fermentation has been used to soften fruit, but it is risky since it may reduce germination if the temperature gets too high or continues too long.

There is general agreement that excessive drying is detrimental. However, what is excessive for most species has not been defined. *Prunus armeniaca* can be dried to 8% while *P. avium* varies from 9 to 11%. If sown or stratified immediately, seeds need not be dried. Seed used within a few months should only be surface dried and stored at 33 to 41°F. When stored for one or more years it is necessary to dry further at room or lower temperatures. Dried seed is stored in sealed containers at 33 to 41°F.

Prunus seeds have embryo dormancy. Good germination can be obtained with cold stratified seed of most species. The stony endocarp is permeable to water.

Prunus americana American Plum

SEED: Seeds may take 2 years in the field if fall planted or one year if cold stratified prior to outplanting. Other times they do not require this; it depends on the seed lot. Although satisfactory results have been obtained by sowing whole fruits, it is generally desirable to clean the seed of pulp. Seeds can be dried to 6% moisture. For controlled stratification with fresh seed, 5 weeks warm stratification at 60 to 70°F followed by a cold stratification at 41°F for 3 to 6 months is satisfactory. It is important to examine the seed periodically and sow when seed coat splitting and radicle emergence begins [PIPPS 28:106 (1978); *Contrib. Boyce Thompson Instit.* 4:39 (1932)]. Best results with stored seed are obtained by placing dry seed in aerated water for 2 weeks and planting by mid-July. The seed germinates the next spring.

CUTTINGS: Not the easiest plant to root; late June cuttings, 3000 or 8000 ppm IBA-talc, wound, did not root; late January (hardwoods), 8000 ppm IBA-talc, sand, rooted 7% in 6 months.

Prunus armeniaca Apricot

SEED: One to 1½ months cold stratification (41°F) is required for germination [*The Plant Propagator* 12(4):10 (1966)].

CUTTINGS: Difficult from cuttings.

GRAFTING: Cultivars are T-budded or chip budded on seedling understocks of the species, peach (*P. persica*), and Myrobalan plum (*P. cerasifera*) (*Plant Propagation*, Hartmann and Kester). Apricot is the preferred stock. Fall budding is generally used, however, spring and June budding are possible. Bench grafting on species seedling understock has been successful. A whip and tongue graft is used, grafts are wrapped with cloth grafting tape, and callused for 10 days at 65°F before placing in cold storage until spring planting.

TISSUE CULTURE: P. *armeniaca* 'Canino' has been micropropagated on a Lloyd and McCown medium supplemented with a 2 mg/liter 2iP [*HortScience* 19:229 (1984)]. Explants were derived from grafted plants forced in a greenhouse. The cultures were kept in a chamber with a light intensity of 40 umol $s^{-1}m^{-2}$ in a 16-hour light period. There was almost no axillary bud break and the propagation process was based mainly on the segments of the shoots that were again cut and placed on new medium. The rooting medium contained 0.5 mg/liter NAA and the cuttings were wounded before insertion. Rooting reached 70 to 90% in 2 weeks.

Prunus avium Sweet Cherry

SEED: The fruit ripens in June-July. Optimal storage occurs under cold, dry conditions (*Amer. Nurseryman* Nov. 15, p. 12, 1967). Warm storage at a high moisture content for only a few months is harmful. Difficulty in obtaining satisfactory germination of sweet cherry is often experienced and appears to be dependent on the year and inherent genetic characteristics. Joley (California) recommended a 3 week warm-moist stratification at 70°F followed by cold stratification at 40°F for 12 to 15 weeks before sowing [*The Plant Propagator* 13(4):10 (1967)]. Once dormancy has been broken, seeds will sprout in the cold. Seeds, after stratification, can be stored at 23°F to prevent germination [PIPPS 29:205 (1979)].

Freshly extracted seed may be planted immediately with after-ripening and warm stratification, plus cold stratification, taking place in the field. This is the simplest method. The second method applies to stored seed. The seed should be stored at 32 to 38°F with a moisture content of 7 to 10%. The stored seed is soaked for a week, with alternating changes of water, and then stratified in moist sand. Four to 6 weeks of warm stratification at 60 to 70°F is followed by 5 months of cold stratification at 41°F. When most of the seed coats have cracked the seeds are planted [*The Plant Propagator* 18(4):8 (1972)].

CUTTINGS: *Prunus avium* 'F 12/1' roots from leaf-bud cuttings in mid-April when treated with a 1000 ppm IBA-quick dip. Firm cuttings are taken from greenhouse-grown stock plants and placed under mist.

GRAFTING: Sweet cherry is conventionally late summer or early fall T-budded on P. *avium* and P. *mahaleb* seedlings, or on rooted cuttings of 'Stockton Morello' (P. *cerasus*) and 'Colt'. P. *mahaleb* is widely used as a drought-resistant rootstock and gives trees slightly smaller than those on sweet cherry, but it is not completely compatible with all cultivars. Seed collected from virus indexed trees should be used [PIPPS 72:352 (1972)].

TISSUE CULTURE: Four cultivars were propagated in vitro. The best basal medium for bud establishment was Knop's solution and for bud proliferation was Murashige-Skoog's medium. BA at 1 mg/l and IBA at 1mg/l were optimum for both bud establishment and bud proliferation. Increased rooting efficiency was obtained with basal shoot wounding and treatment with 0.5 mg/l NAA. Plantlets transferred to soil developed naturally [*HortScience* 17:192-193. (1982)].

Prunus besseyi Sand Cherry

SEED: The purplish black fruit ripens from July to September. Cleaned seeds per pound average about 2,400. The seed requires cold stratification (41°F) for 3 months [*The Plant Propagator* 21(4):5 (1975)].

CUTTINGS: Softwoods, June to mid-August root readily [*The Plant Propagator* 21(4):5 (1975)]. Soak the subterminal 3-node, 2-leaf cuttings in a Benlate solution (1000 to 6000 ppm) for 15 minutes, air dry for about 10 minutes, treat with 1000 or 8000 ppm IBA-talc, perlite, under mist [PIPPS 28:63 (1978)]. Junior author has rooted cuttings 100% with the above method. The best time to harvest the cuttings is in the morning to avoid wilting. Wilting is detrimental possibly because the cuttings contain prunasin, a cyanogenic glycoside, that breaks down during wilting and releases cyanide.

Prunus × blireiana Blireiana Plum

CUTTINGS: A hybrid between P. *cerasifera* 'Atropurpurea' and a double form of P. *mume*. Refer to P. *cerasifera* for details on softwood and hardwood cuttings. Readily propagated by layering and one of the easiest by this method.

GRAFTING: Summer T-bud or chip-bud on seedling or rooted cutting understocks. P. *cerasifera*, 'St. Julien A', 'Myrobalan B' and 'Brompton' are suitable. *Prunus spinosa* also has been shown to be compatible. Bush trees can be produced by whip and tongue grafting onto field lined rootstocks in early spring, using a 3 to 4 bud scion.

Prunus campanulata Taiwan Cherry

SEED: Refer to general introduction for information on collection, cleaning and storage. A 3 month cold stratification treatment induces good germination.

Prunus caroliniana Carolina Cherrylaurel

SEED: Carolina cherrylaurel is primarily propagated by seed. The black fruit is inconspicuous when ripe and persists into winter. A 1 to 2 month cold stratification or fall planting satisfy dormancy requirements. In the southeast, the species is weed-like and numerous seedlings appear in waste areas.

CUTTINGS: Softwood to semi-hardwood, June to September, mist, root 85 to 90% by late August. 3000 to 8000 ppm IBA talc or solution improve rooting response. Cuttings can also be rooted in fall and winter.

Prunus cerasifera Myrobalan Plum

SEED: The reddish fruit matures in July and August. Refer to the introduction for details on collection, cleaning, and storage of *Prunus* seed. A pound of cleaned seed contains approximately 1000 seeds. Myrobalan plum seed requires warm (15 days) and cold (6 months) stratification treatments. However, another source suggested 3 months at 41°F for optimal germination [*The Plant Propagator* 12(4):10 (1966)]. Stratified seed should be planted as early in the spring as possible. It is best if a high proportion of the seed has cracked stones and the radicle has not emerged.

CUTTINGS: A number of the purple leaf forms are propagated by hardwood cuttings in England [*Gardeners Chronicle* 171(19):14 (1972)]. Success depends on the production of suitable stem material with a high capacity to root. This is induced by the production of vigorous stems produced on a hedge system of severely pruned plants. The cuttings are taken at leaf fall or in early spring, 5000 ppm IBA-quick dip, peat:sand (1:1, v/v), heated bin (70°F). Late fall is the preferred season because of better establishment. In England, fall cuttings also are stored in a peat:sand mix without any heat in a barn and put outside in February [*The Plant Propagator* 26(1):6 (1980)].

Softwood cuttings taken in mid July, treated with a hormone, placed in sand under mist will root [PIPPS 33:39 (1983)]. Many nurserymen propagate by summer softwoods. Cuttings should be firm, 3000 to 8000 ppm IBA-talc or solution, well drained medium, mist, root 70 to 90% in 4 to 6 weeks. Cuttings taken from container-grown plants under high nitrogen status do not root as well as those from plants grown under moderate to low fertility. Mound layering works for the colored-leaf forms such as 'Hessii' and 'Nigra'; an effective method for these plants [*Gardeners Chronicle* 166(15):18 (1969)].

GRAFTING: Purple forms are T-budded or chip budded in summer on seedling or rooted cutting understocks. P. *cerasifera*, 'St. Julien A', 'Myrobalan B' and 'Brompton' are suitable. Whip and tongue grafting onto field-lined rootstocks in early spring, using a 3 to 4 bud scion, is possible.

TISSUE CULTURE: Actively growing shoot tips proliferated in vitro on a Linsmaier and Skoog medium containing 0.5 mg/liter BA. Actively growing shoot tip explants were surface-sterilized in 0.5% sodium hypochlorite containing a few drops of detergent for 15 minutes and rinsed twice in sterilized water. Cultures were maintained at 77±2°F and light was provided by Cool White fluorescent tubes at 4.4 klx for 16 hours per day. An average of 6 shoot tips was obtained after a 4-week culture period. Microcuttings rooted in vitro in the presence of 0.5, 1.0 and 2.5 mg/liter IBA and rooting increased when cultures were incubated in the dark for 1 week prior to illuminated incubation [J. *Hort. Sci.* 61(1):43 (1986); *HortScience* 17:190 (1982); J. *Amer. Soc. Hort. Sci.* 107:44 (1982); *The Plant Propagator* 32(3):5 (1986)].

Prunus cerasus Sour Cherry

CUTTINGS: The cultivars 'Stevnsbaer', 'Kelleriis 14' and 'Kelleriis 16' are easily rooted from cuttings. Late June, 1000 ppm IBA, mist. Bottom heat is maintained at a minimum of 70°F. Greenhouse forced plants rooted easily without IBA treatment [PIPPS 32:182 (1982)].

GRAFTING: The rootstocks are either P. *avium* or the P. *avium* clone, 'F 12/1'.

TISSUE CULTURE: Buds with dormancy satisfied were established on a Boxus and Quoirin medium with Murashige and Skoog (MS) micro elements. Multiplication medium was Jones medium without phloroglucinol. The multiplication rate was 4x in 3 weeks. To initiate roots the microcuttings were placed on 1/2 strength MS medium with 1 mg/liter IBA for 6 to 8 days and then transferred to the same medium without IBA. Bud explants were surface-sterilized in 80% isopropanol for 30 minutes and rinsed three times in sterilized water. Cultures were maintained at 79±4°F and light was provided by Gro-Lux lamps at 600 uW/cm in the visible range for 16 hours per day [*Scientia Hortic.* 19:85 (1983)].

Prunus × cistena Purpleleaf Sand Cherry

CUTTINGS: Softwoods, 1000 to 3000 ppm IBA-talc solution root readily. Fall hardwoods, 8″ long, 2500 ppm IBA-quick dip, planted outdoors, root. Easy to root from softwoods but cuttings should be firm at the base [PIPPS 34:540 (1984)].

GRAFTING: *Prunus tomentosa* and P. *nigra* are used to produce P. × *cistena* standards. 'Brompton', 'Myrobalan' and 'St. Julien' seedlings have been reported to be compatible [PIPPS 17:303 (1967)].

Prunus cyclamina Cyclamen Cherry

SEED: A lovely, rich pink flowered cherry that is the rival of the best Sargent, Yoshino and Higan types. A 3 month cold stratification treatment produces optimum germination [*Arnoldia* 40(3):146 (1980)].

CUTTINGS: Cuttings treated with IBA at 5000 to 10,000 ppm form heavy callus but no roots. NAA at 5000 to 10,000 ppm resulted in high mortality [*Arnoldia* 40(3):146 (1980)].

GRAFTING: Grafting has been successful using P. *avium* as the roostock. Make the graft as close to the soil level as possible.

Prunus dulcis (P. amygdalus) Almond

SEED: The brownish fruit is harvested in August-October. A pound of seed contains about 180 seeds. Cold stratification for 2 months at 41°F or 3 months at 32°F yielded good germination percentages (90%) [SWP-US; *The Plant Propagator* 13(2):10 (1967)].

CUTTINGS: Stem cuttings have given only slight success in rooting.

GRAFTING: Almond cultivars are T-budded in fall, spring or June on almond or peach seedlings, 'Marianna 2624' cuttings, or almond-peach hybrid seedlings or cuttings. In California, 95% of the almond trees are propagated on peach seedlings. Almond seedlings are satisfactory in deep, well drained soils. Seeds of the bitter types, or selected cultivars such as 'Texas', are used. 'Marianna 2624' is not satisfactory with all almond cultivars (*Plant Propagation*, Hartmann and Kester). 'Texas' and related cultivars form a compatible union while 'Nonpareil' and related cultivars show graft incompatibility. Research by the junior author has implicated cyanogenic glycosides as the possible cause of the incompatibility.

TISSUE CULTURE: Dormant shoot buds collected from December to February proliferated in vitro on a modified Knop's macroelement mineral solution, microelements and organic supplements of Murashige and Skoog medium, FeEDTA, 1 mg/liter BA, 2% sucrose, and 0.8% agar. The pH was adjusted to 5.9. Cultures were maintained at 77°F and light was provided at 2.2 to 4.3 klx on a 16 hour photoperiod [*HortScience* 12(6):545 (1977)].

Actively growing shoot tips from peach-almond hybrids proliferated in vitro on Anderson's salts containing 2.5 mg/liter BA, 0.1 mg IBA, 0.1 mg BA, 100 mg myo-inositol, 0.4 mg thiamine, 30 grams sucrose, and 7 grams agar. Explants were surface-sterilized in 0.5% sodium hypochlorite containing a few drops of detergent for 5 minutes and rinsed three times in sterilized water. Cultures were maintained at 77±2°F and light at 3 klx for 16 hours per day. Microcuttings rooted in vitro in the presence of 1 mg/liter IBA [*The Plant Propagator* 29(4):13 (1983)].

Prunus glandulosa Dwarf Flowering Almond

SEED: Senior author has never observed fruit set on the single or double form. Fruit is described as a 2/5″ diameter, red, subglobose drupe. Suspect 1 to 2 months cold stratification would prove optimum.

CUTTINGS: Easily rooted from summer softwoods with 1000 ppm IBA-solution, peat:perlite or well drained medium, mist.

Prunus 'Hally Jolivette' — Hally Jolivette Cherry

SEED: 'Hally Jolivette' is the result of a cross between P. *subhirtella* × P. × *yedoensis* back-crossed on P. *subhirtella*. Fruit set has never been observed by the authors.

CUTTINGS: New growth, 8000 ppm IBA-talc, sand, mist, rooted 72%. Senior author has rooted this hybrid using softwood, 1000 ppm IBA-quick dip, peat:perlite, and mist. One of the easiest cherries to root. Extremely easy to root and will produce a 24 to 30″ and larger plant in the second growing season.

TISSUE CULTURE: Actively growing shoot tips of 'Hally Jolivette' proliferated in vitro on a Murashige and Skoog medium containing 200 mg/liter casein hydrolysate, 1.0 mg/liter BA and 0.1 mg/liter NAA. Shoot tip explants were surface-sterilized in 0.5% sodium hypochlorite for 15 minutes and rinsed twice in sterilized water. Cultures were maintained at 75 to 80°F with a photosynthetic photon flux density of 29 to 42 umol/sec/m^2 provided by Gro-Lux fluorescent lamps. A 500-fold increase in shoot tips was obtained after 25 weeks culture. Microcuttings rooted in vitro in the presence or absence of 0.1 mg/liter NAA and under nonsterile conditions in a sphagnum peatmoss:perlite medium [*HortScience* 18(2):182 (1983)].

Prunus × *hilleri* 'Spire'

CUTTINGS: Softwood cuttings taken when large enough to handle, 8000 ppm IBA-talc, mist, peat and sand (2:1, v/v), 70 to 75°F bottom heat, rooted and made growth the first year.

GRAFTING: 'Spire' is whip grafted on bareroot stock of P. *avium* [*New Gardeners Chronicle* 184(12):35 (1978)].

Prunus × *incam* 'Okame' — Okame Cherry

SEED: Of hybrid origin and must be asexually propagated to maintain desirable characteristics.

CUTTINGS: Okame cherry roots easily from softwood cuttings. Six-inch cuttings are taken from mid- to late June (Pennsylvania) and treated with 1000 ppm IBA-talc plus thiram and placed under mist. The cuttings are well rooted within 4 weeks. Terminal cuttings yield plants with the best upright form; lateral require pruning to form a strong leader [*Arnoldia* 45:23 (1985); PIPPS 33:482 (1983)].

Prunus incisa Fugi Cherry

CUTTINGS: Propagation is readily accomplished (over 90%) with cuttings taken at the beginning of June, 8000 ppm IBA-talc, mist, and bottom heat at 75°F [*Gardeners Chronicle* 169(21):44 (1971); *The Plant Propagator* 22(2):11 (1976)]. Rooting occurs in 25 days and cuttings taken in early June often will produce a second flush of growth. Early July, 8000 ppm IBA-talc, rooted 85% in 4 weeks [*Gardeners Chronicle* 169(21):44 (1971)]. Overwinter in a coldframe above freezing.

GRAFTING: Whip and tongue grafting, using a 3 to 4 bud scion, on established field-lined understock in late winter is used in England.

Prunus laurocerasus Common Cherrylaurel

SEED: Cold stratification for 2 to 3 months. Untreated seed did not germinate. Radicles will emerge during cold stratification. Senior author has germinated seeds of 'Otto Luyken' and some of the progeny displayed the compact habit of the parent but the largest percentage, after 3 years, appeared similar to the species.

CUTTINGS: Cherrylaurel can be rooted anytime the growth is firm from June to July through April. 1000 to 3000 ppm IBA-talc or solution, peat:perlite, and mist. Rooting takes place in 4 to 6 weeks. Senior author has successfully rooted the species, 'Otto Luyken', 'Schipkaensis' and 'Zabeliana' with the above procedure. Mid-November (Boston), 3000 ppm IBA-talc, rooted 97%. Cuttings of 'Otto Luyken' may have a tendency to develop excess callus [*New Gardeners Chronicle* 180(7):26 (1976)].

In another study, September through November cuttings (Georgia), broad spectrum fungicide treatment (Captan or Benlate and Manzate), 2000 ppm IBA + 1000 ppm NAA-quick dip, plastic tunnel + 50% shade, regular fungicidal sprays, were found to root well by spring. The moisture needs to be checked regularly [PIPPS 33:547 (1983)].

Common cherrylaurel can be rooted during winter under thermoblankets. Six inch terminal cuttings, November, 8000 ppm IBA-talc, inserted in containers placed directly on heating pads; quarter-inch microfoam was then laid directly on top of the cuttings with all the edges coming in direct contact with the gravel base. White co-polymer was next placed over the microfoam and sealed. Temperature in the medium was maintained at 80°F until the cuttings were rooted. After rooting the temperature was reduced to 45°F.

Prunus lusitanica Portuguese Cherrylaurel

SEED: Treat as described under P. *laurocerasus*.

CUTTINGS: See P. *laurocerasus*.

Prunus maackii Amur Chokecherry

SEED: The fruit is black when ripe. The seed is doubly dormant and requires 4 months warm stratification followed by 3 months cold stratification [*Arnoldia* 46(2):25 (1986)].

CUTTINGS: Softwoods, late June (Boston), 8000 ppm IBA-talc, sand:perlite, mist, root in high percentages. Basal cuttings, same treatments, rooted 100%. Best not to disturb after rooting and overwinter in a refrigerated storage or polyhouse. Hardwoods have not been successful.

GRAFTING: Bench grafting onto seedlings of P. *avium* or P. *serrulata* has been successful. The whip and tongue grafts are tied with rubber budding strips and placed in a medium of damp peat moss on a greenhouse bench, making certain to cover the union. Bottom heat of 70°F is necessary.

Summer budding on seedlings is successful. Cut back the understock the following spring.

Prunus maritima Beach Plum

SEED: 2 to 3 months cold stratification. Fresh seeds did not germinate; seeds given 3 months cold stratification germinated 39%.

CUTTINGS: Mid-July cuttings, 8000 ppm IBA-talc plus thiram, mist, rooted 60%.

Prunus nigra Canada Plum

SEED: 3 months cold stratification.

Prunus padus European Bird Cherry

SEED: Untreated seed does not germinate. A 3 to 4 month cold stratification period produces good germination.

CUTTINGS: Heel and nodal, 5000 ppm IBA-quick dip, cold frame, October, no basal heat, root [*The Plant Propagator* 26(1):6 (1980)].

GRAFTING: Cultivars are T-budded during summer onto established field-lined rootstocks of P. *padus*. P. *avium* seedlings or clonal roostocks will work [*Gardeners Chronicle* 1977 (20):22 (1972)].

Prunus pensylvanica Pin Cherry

SEED: The red fruit ripens July through August. A pound of cleaned seed averages about 14,000 seeds. Pin cherry requires both warm and cold stratification pretreatments with 2 months warm followed by 3 months cold recommended. A germinative capacity of 62% has been reported. Pin cherry can be early fall sown without pretreatment or spring sown after pretreatment (SWP-US). Seeds retained high viability after 10 years at 34 to 38°F under sealed conditions [*Amer. Nurseryman*, Nov. 15, p. 12 (1967)].

CUTTINGS: Pin cherry has been rooted at very low percentages (less than 10%) from semi-hardwoods collected in June and treated with 8000 ppm IBA [*The Plant Propagator* 30(4):7 (1984)].

Prunus persica Common Peach

SEED: Peach seed must be cold stratified (41°F) for 2 months. A 3 month treatment reduced germination and 1 month gave no germination. Raising or lowering the temperature to 50 or 36°F reduced germination from 75% to 10 and 5%, respectively.

CUTTINGS: Two-node stem cuttings of 23 to 79-day-old seedlings of peach rooted readily under mist in 3 weeks without hormone treatment. Cuttings of older seedlings rooted better with a 2000 ppm IBA dip [*HortScience* 19:249 (1984)].

Sprouted nodal cuttings with leaves partially expanded and made by undercutting sprouted buds 1/12" deep, from 1/5" above and 1/5" below each node, rooted 50% after 2 weeks. Rooting was greatest when the basal hardwood tissue was dipped for 10 seconds in 100 to 500 ppm IBA before sticking under mist in a peat:perlite medium [*HortScience* 15:579 (1980)].

Semi-hardwood (August) cuttings can be easily rooted [*HortScience* 15:41 (1980)]. Cultivars rooted included: 'Autumnglo', 'Loring', 'Redhaven', 'Jerseyqueen', 'M.A. Blake', and 'Rio Oso Gem'. Collect 8 to 10" long terminal cuttings in August, strip all leaves except for the top 3 to 5, wound on opposite sides and dip the basal 1½" for 5 seconds in 2500 ppm IBA solution. The cuttings are placed under mist in a peatmoss:perlite:vermiculite medium. Rooting is completed in about 4 weeks.

Hardwoods collected in March, dipped in 1000 ppm IBA for 10 seconds, and held at 41°F in a mixture of peat:perlite (1:1, v/v) with 65°F bottom heat, rooted well. Hardwoods rooted 52 to 100% when treated with 2000 to 5000 ppm IBA and sealed in plastic bags and kept at 50°F. Higher temperatures resulted in fungal diseases.

GRAFTING: T-budding or chip budding on peach seedlings are the most common with 'Elberta', 'Halford', 'Lovell', or 'Rutgers Red Leaf' usually used as understock. Apricot and almond seedlings are also used. Fall budding is preferred but spring budding is possible.

Two understocks, P. *besseyi* and P. *tomentosa*, have been utilized as dwarfing understocks for peach [PIPPS 26:304 (1976)]. Neither is completely satisfactory for dwarfing peach because of incompatibility problems. Not all budded plants fail. Some of the types are successful and they need to be propagated as clones. Both P. *besseyi* and P. *tomentosa* are reported to carry latent viruses, so any propagation with those plants should start with heat-treated material.

A combination of 'Lovell' roots with a 12" stempiece of P. *subcordata* with a peach scion has proven successful and produced dwarf trees [PIPPS 14:271 (1964)].

TISSUE CULTURE: Actively growing shoot tips of 'Nemaguard' peach roostock proliferated in vitro on a Murashige and Skoog medium supplemented with 50 mg/liter L-ascorbic acid, 20 ml/liter Staba vitamin mixture, 2.0 mg/liter BA and 0.1 mg/liter NAA. The pH was adjusted to 5.7. Actively growing shoot tip explants were surface-sterilized in 0.5% sodium hpyochlorite containing a few drops of detergent for 10 minutes; rinsed twice in sterilized water; soaked 5 minutes in 70% ethanol; and rinsed twice in sterile distilled water. Cultures were maintained at 81±2°F and continuous fluorescent light at 4.4 klx. An average of 6 lateral shoot tips was obtained after a 3-week culture period. Microcuttings rooted best in vitro in the presence of 0.1 mg/liter NAA. After 6 weeks, 95% rooting occurred. Rooted plants were transferred after 5 weeks to a sphagnum peatmoss and perlite medium and placed under mist [*HortScience* 17:194 (1982)].

Prunus pumila Sand Cherry

SEED: Cold stratification (41°) for 3 months produces good germination.

Prunus sargentii Sargent Cherry

SEED: Cold stratification for 3 to 6 months produces good germination. Untreated seeds do not germinate.

CUTTINGS: In one trial, mid-June cuttings, 8000 ppm IBA-talc + thiram, sand:perlite, mist, rooted 11%.

Prunus serotina Black Cherry

SEED: The black fruit ripens in August and September. A pound of cleaned seed contains approximately 4,200 seeds. Warm storage at a high moisture content for a few months is harmful to seed viability. Cold stratification for 4 months produced 86% germination. A warm period for 14 days prior to cold stratification improved germination slightly. In the wild, the plant is extremely weedy, and seedlings germinate readily the spring following dissemination.

CUTTINGS: Softwood from juvenile and mature (grafted) plants grown in pots and pruned to promote shoot formation, were rooted under mist in a sand:peat medium. IBA (8000 ppm in talc) stimulated rooting of cuttings propagated in spring and early summer. Wide tree-to-tree variation in rooting percentage (0-100%) was found [*Silvae Genetica* 23:104 (1974); PIPPS 19:30 (1969)].

GRAFTING: Black cherry cultivars can be grafted on seedlings of the species and P. *padus* using a whip or whip and tongue graft. Grafts of black cherry on P. *virginiana*, P. *besseyi*, P. *japonica*, and P. *cerasus* declined or died within 2 years [*The Plant Propagator* 16(3):3 (1970)].

Prunus serrula Paperbark Cherry

SEED: 2 months cold stratification.

CUTTINGS: Soft tip cuttings, late July, 8000 ppm IBA-talc, perlite:sand, mist, rooted 60%. Cuttings should not be disturbed after rooting and overwintered above freezing.

GRAFTING: *Prunus serrula* can be successfully grafted on P. *avium*.

Prunus serrulata Oriental Cherry

SEED: Three months cold stratification produces good germination.

CUTTINGS: The best rooting of the Japanese flowering cherry cultivars, 'Amanogawa' (18%), 'Kwanzan' (70%), 'Mikurumagaeshi' (89%), 'Roseo-plena' (90%), 'Shimidsu Sakura' ('Longipes') (20%), 'Tai-Haku' (100%), and 'Troiayo' (69%), occurred with cuttings taken in mid and late summer. Rooting conditions included wounding, 8000 ppm IBA-talc, mist, peat:sand (2:1, v/v), and 70 to 75°F bottom heat [*Gardeners Chronicle* 169(3):44 (1971); PIPPS 23:170 (1973)]. In another study, 'Shirofugen', late June, 1000 ppm IBA-talc, perlite:peat (3:1, v/v), mist, rooted 90%. 'Shirotae' is difficult to root. Commercially, 'Kwanzan' and others are rooted in summer using firm wood, about 8000 to 10,000 ppm IBA-talc or solution, sandy soil, mist, in white polytent outside. Rooting percentages are high.

GRAFTING: Japanese flowering cherry cultivars, such as 'Amanagawa', 'Fugenso', 'Hokusai', 'Kiku-Shidare', 'Kwanzan', 'Miyako', 'Shogetsu' and 'Takasago' are generally grafted on P. *avium* (mazzard) seedlings or cuttings. It has been reported that those budded or grafted at 3½ to 4′ developed a heavier head at digging than those budded or grafted at 6 feet. The weeping form is budded or grafted at 6′.

Bench grafting of Japanese flowering cherry in winter is possible [PIPPS 26:145 (1976)]. The bareroot stocks can either be seedling or clonal, such as 'Mazzard F 12/1'. A whip graft is used and after tying, the union and scion are dipped in paraffin wax. The grafts are plunged in moist peat and shaded with the air temperature maintained around 45°F. Regular fungicidal applications are followed to reduce losses from diseases, such as *Botrytis*. The containerization of the grafts is carried out after 6 weeks. In addition to being hardened off, some scion growth should occur before potting.

Prunus avium seedling rootstocks can be infected with virus diseases, such as prune dwarf virus, which can cause bud failure, graft incompatibility, or poor growth of many ornamental *Prunus* species. It is important that virus-tested scions be grafted onto virus-tested rootstocks [PIPPS 28:220 (1978)].

Prunus subhirtella Higan Cherry

SEED: A 3 month cold stratification (41°F) produces good germination; seeds sown at once did not germinate.

CUTTINGS: Higan cherry roots readily from softwoods in spring when large enough to handle [PIPPS 23:170 (1975)]. 8000 ppm IBA-talc, mist, peat:sand (2:1, v/v), 70 to 75°F bottom heat are ideal. Rooted cuttings will make additional growth the first year. Var. *autumnalis* from late June to July cuttings, 8000 ppm IBA-talc, rooted 40 to 60% in 4 to 5 weeks in one study and 79% in another. 'Fukubana' from mid-May cuttings only rooted 22% [*Gardeners Chronicle* 169(21):44 (1971)]. 'Pendula', mid-June (Pennsylvania), 8000 ppm IBA-quick dip, light wound, rooted 70%. 'Pendula' from cuttings must be staked to develop an upright form. 'Pendula Plena Rosea' softwoods, perlite:aged sawdust (6:4, v/v), mist, 8000 ppm IBA-talc, rooted 80%. July cuttings should be wounded. P. *subhirtella* may be difficult to bring through the first winter.

GRAFTING: Higan cherry cultivars are generally budded or grafted on P. *avium* (mazzard cherry) seedlings or rooted cuttings. Grafts at 3½ to 4′ develop a heavier head at digging time than those budded or grafted at 6′, the exception being P. *subhirtella* var. *pendula* [PIPPS 20:278 (1970)]. A drawing bud, a bud that is left on the understock at its highest point, improves success. Both T-budding and chip budding have been used successfully. Understock seedlings of P. *avium* are field planted in April to May (Oregon). For standards, the seedlings are cut back the following spring and a single straight trunk is developed for grafting in February and March [PIPPS 33:54 (1982)]. Bench grafting 'Pendula' on bareroot P. *subhirtella* seedlings is preferred by some because the normally thin twigs and tiny, thin-barked buds are difficult to field bud [PIPPS 32:569 (1982)].

Prunus tenella Dwarf Russian Almond

SEED: A cold stratification period of 3 months produces better germination than direct sowing.

CUTTINGS: Softwoods, mid-May, 4000 ppm IBA-talc, peat:perlite, mist, rooted 25%; early July, 8000 ppm IBA-talc, rooted 30%. Rooting is slow and regrowth is virtually nonexistent. Mound and French layering work well but the species and 'Fire Hill' are not prolific shoot producers [*Gardeners Chronicle* 166(15):18 (1969)].

GRAFTING: The dwarf Russian almond cultivar 'Firehill' can be grafted using a modified whip graft. The understock is 'Myrobalan B' and it is lifted from the cutting bed from early November until mid-March and held in cold storage. Scion wood is cut from mid-December to early January and used or held in cold storage. The grafting operation takes place anytime from January to mid-March. The understock is prepared by cutting the top off to leave a 3 to 4″ long stem. The top of the cut is trimmed to leave a slight slope. A cut is made ¾ to 1″ long from the top of the understock and deep enough to reveal the cambium layer. This leaves the understock with a flap.

The 4 to 6″ long scion is selected to match the understock and a cut is made to match the understock. A second cut not as long is made on the opposite side, and finally a third cut to form a short wedge at the base. After tying,the scion and union are waxed with low melting wax. Next the grafts are dipped in Benlate, packed into crates and placed in cold storage. The grafts are checked weekly and sprayed with Benlate. The grafts remain in cold storage for 6 to 10 weeks, after which they are potted and grown on in a greenhouse or polyhouse [PIPPS 29:202 (1979)].

Prunus tomentosa Nanking Cherry

SEED: The red fruits ripen from mid-July through August. The number of cleaned seeds per pound is approximately 4,700. Seeds stored for 21 months at room temperature did not lose viability. Nanking cherry requires 2 to 3 months cold stratification for germination. Fall sowing or spring sowing after stratification are successful. Seedlings produced from northern plains windbreak plantings are reportedly more vigorous than seedlings grown from imported Japanese seed [*Arnoldia* 24(9):81 (1964)].

CUTTINGS: Softwoods, mid-July, 8000 ppm IBA-talc plus thiram, sand, mist, rooted 100%. Cuttings were taken from the basal part of the plant. The addition of ethrel at 480 ppm improved rooting. Shoots taken from greenhouse-forced plants and placed in sand with bottom heat also rooted readily.

TISSUE CULTURE: Actively growing shoot tips proliferated in vitro on a Linsmaier and Skoog medium containing 2.5 mg BA/liter. Explants were surface-sterilized for 15 minutes in 0.5% sodium hypochlorite containing a few drops of detergent and rinsed twice in sterilized water. Cultures were maintained at 77±2°F and light was provided by Cool White fluorescent tubes at 4.4 klx for 16 hours per day. An average of 15 shoot tips was obtained after a 4-week culture period. Microcuttings were rooted in vitro in the presence of 2.5 mg IBA/liter. Rooting increased when cultures were incubated in the dark for 1 week prior to illuminated incubation [*HortScience* 17:190 (1982)].

Prunus triloba Flowering Almond

CUTTINGS: Flowering almond from softwoods (see 'Multiplex' below) is not difficult, however, problems with regrowth after the first winter may occur. The application of GA_3 to established rooted cuttings in late July is reported to increase both the number of shoots and survival of plants overwintered in cold frames [*The Plant Propagator* 39(1):12 (1983)]. 'Multiplex', mid-June, softwoods and basal cuttings, 3000 or 8000 ppm IBA-talc, rooted 60 or 100%, respectively. The higher hormone treatment produced excellent root systems. 'Multiplex' is reported to benefit from the incorporation of Captan in the rooting powder.

Rooting of hardwood cuttings has been moderately successful (33%). The hardwoods are collected in mid-January (Netherlands) and the rooting conditions include: 1% IBA, peat:sand (3:1, v/v), cool, frost-free greenhouse, bottom heat and a spray with antitranspirant (1:10 WiltPruf) [*The Plant Propagator* 30(2):3 (1984)].

Mound layering works [*Gardeners Chronicle* 166(15):8 (1969)].

GRAFTING: Whip and tongue grafting, using a 3 to 4 bud scion, on established field lined stock of 'Brompton' and 'St. Julien A' stocks in late winter is used in England.

Prunus tomentosa and P. *nigra* have been used to produce standards [PIPPS 17:303 (1967)].

Prunus virginiana Common Chokecherry

SEED: The purplish black fruits ripen August to October. A pound of cleaned seed averages approximately 4,800 seeds. Warm storage at a high moisture content for only a few months is harmful to seed viability. Chokecherry requires cold stratification for germination. In a controlled study, cleaned seed sown immediately did not germinate; 3 months cold stratification induced good germination, and 52% of the seeds germinating during the 6 month cold stratification period. The seeds can be fall sown without pretreatment or spring planted after pretreatment. 'Shubertii' (Canada red cherry) comes relatively true-to-type from seed and is produced this way by the nursery trade. The new growth is green and the older leaves gradually assume the reddish purple coloration.

CUTTINGS: Chokecherry has been rooted from terminal and basal cuttings collected in June using 8000 ppm IBA-talc, sand and mist [*The Plant Propagator* 30(4):7 (1984)].

Prunus × yedoensis Yoshino Cherry

SEED: Of hybrid origin and listed as a P. *serrulata* × P. *subhirtella* hybrid but also P. *speciosa* × P. *subhirtella*. Senior author used fresh seed collected in late June and induced radicle emergence during 3 months cold stratification. Average germination percentage was 85% [*Arnoldia* 41:162 (1981)].

CUTTINGS: Cuttings root better from July than May collections (25 vs. 85%); use 8000 ppm IBA-talc, peat:perlite (1:1, v/v), mist or high humidity chamber. Rooting occurs in 30 days. In another study, June cuttings (Tennessee) from vigorous field stock not older than 3 years, Benlate wash, 10,000 ppm IBA-quick dip, sterilized medium (pine bark 60%, top soil 30%, and sand 10%), mist, rooted well [PIPPS 33:552 (1983)]. The species and 'Perpendens' are rooted from firm-wood cuttings. Extremely soft cuttings often rot in the mist bench.

GRAFTING: Budding is accomplished in mid-summer on a root stock, such as P. *avium*. Several selections are known with 'Perpendens', a weeping type, quite popular in the south. It should be grafted on a standard or rooted and trained into a straight trunk.

Pseudocydonia sinensis (*Cydonia sinensis*) Chinese Quince

SEED: The seeds are contained in a large, 5 to 7" long, egg-shaped pome which is yellow when ripe and matures in October. Seeds require cold stratification at 41°F for 2 to 3 months. Seeds extracted from fruits which have been kept in a cooler during winter germinate immediately upon planting.

CUTTINGS: Firm, July-August, 3000 to 8000 ppm IBA-solution, peat:perlite, mist, root in 6 weeks. Reduce mist as soon as cuttings start to root.

Pseudolarix kaempferi (*amabilis*) Golden-larch

SEED: Seed is the usual method of propagation. Male and female cones are borne separately on the same tree. The cones are ready for collection when they turn a light golden-brown. A pound of cleaned seed contains about 30,000 seeds. Under room storage conditions the seed loses viability rapidly, but storage of dry seed in sealed, refrigerated containers prolongs life. Isolated trees cone heavily but it is difficult to obtain viable seed. It has been proposed that the trees are self-sterile. A fair percentage of gold-larch seeds germinate without pretreatment, but 1 to 2 months of cold stratification (40°F) improves and unifies germination. This species can be fall sown in the nursery [*Arnoldia* 37(1):69 (1979); *Amer. Nurseryman* Sept. 15, p. 10 (1967)].

CUTTINGS: Cutting propagation has not been successful.

Pseudotsuga menziesii (P. *douglasi*, P. *taxifolia*) Douglas-fir

SEED: The species has two varieties, *menziesii* and *glauca*. In both varieties, characteristics such as growth rate, cold hardiness, and pest resistance vary greatly. To help match seed source with planting site, seed collection zones have been defined and mapped by the Western Forest Tree Seed Council. Seed reaches maturity in August or early September and cones become brown at the same time with seed dispersal in September and October. A commonly used, practical index of sufficient maturity is the seed's appearance — seedcoat golden brown with wing of the same color which detaches intact from its bract. Cones are air-dried in the open, or kiln-dried at 90 to 110°F for 2 to 48 hours.

A pound of cleaned seed contains between 32,000 and 44,000 seeds. Viability declines rapidly at room temperature but storage at 0 or 32°F maintains viability for 10 to 20 years. Heit noted that seeds from central and southern Rocky Mountain sources are not dormant. They are not recommended for fall planting because of potential fall germination but can be spring sown without any pretreatment. However, seeds from the Pacific coast, northern Rocky Mountain and British Columbia sources have varying degrees of dormancy. Those seed sources can be fall sown. Pretreatment by first soaking in water for 24 hours, followed by cold stratification for up to 2 months, can substitute for fall planting. The length of cold stratification varies with seed lot (*Amer. Nurseryman* Jan. 15, p. 12 (1967) and Nov. 15, p. 12 (1968)].

CUTTINGS: Rooting conditions include: current season's growth from 10- to 12-year old sheared (hedged) plants, December and January, 10% Jiffy Grow (containing 5000 ppm IBA and 5000 ppm NAA), sand:peat (5:1, v/v), bottom heat (65 to 79°F), 50°F air temperature, and long day lighting. Wide clonal differences in rooting (0 to 100%) existed. In another study using branched terminals and first-order laterals, 1376 ppm NAA stimulated optimum rooting (67%) and bud break. See *The Plant Propagator* 24(2):9 (1978); *J. Amer. Soc. Hort. Sci.* 99:551 (1974) and PIPPS 28:32 (1978).

GRAFTING: Douglas-fir cultivars can be side-grafted or top-cleft grafted in winter on established understock of the same species. Problems of scion-stock incompatibility have complicated the grafting of Douglas-fir. Graft mortality from incompatibility is evident the first spring and mortality continues with delayed symptoms of incompatibility. Mortality in the range of 50% has been reported after 6 years. The degree of severity of loss is so great that the practicality of grafting is questionable [*The Plant Propagator* 20(1):4 (1974); PIPPS 17:130 (1967)]. The fastigiate and weeping forms must be grafted.

Ptelea trifoliata Hoptree

SEED: The fruit is an elm-like, reddish brown samara that ripens from June to November. Ripe seeds require a few days drying before storage. Hoptree fruits abundantly and the average number of samaras per pound is 12,000. When sown without stratification, germination is erratic. Three months stratification (41°F) unifies germination. The seed can be fall sown or spring sown after cold stratification [*Contrib. Boyce Thompson Instit.* 8:355 (1937)]. Seeds of 'Aurea' sown immediately germinated 47%; those provided with 2 to 3 months cold stratification germinated 100%. Interestingly, some of the seedlings had yellow foliage color.

CUTTINGS: 'Aurea', late July, 8000 ppm IBA-talc, perlite:sand, mist, rooted 100%. Reportedly can be reproduced by root cuttings [*The Plant Propagator* 14(4):4 (1968)].

Pterocarya fraxinifolia Caucasian Wingnut

The propagation methods described below for *Pterocarya fraxinifolia* also apply to *P. stenoptera*, Chinese Wingnut, *P.* × *rehderana*, Rehder Wingnut, *P. rhoifolia* and others.

SEED: Easily grown from seed, a winged nut that changes from green to brown when ripe in September or October. Seed stores satisfactorily, at least for short periods, when kept dry. Fresh seed germinates sporadically; 3 months cold stratification improves germination. See *Gardeners Chronicle* 170(12):22 (1971).

CUTTINGS: Summer cuttings from young shoots with a heel will sometimes root. Young plants provide the best cuttings. Probably best to collect cuttings in July and treat with high IBA concentration. Hardwoods in a Garner bin root well. The best cuttings come from vigorously pruned hedge plants. Basal shoots, late February or early March (England), 1000 ppm IBA-quick dip, basal heat of 70°F. Line out in fertile soil in a sheltered position. *P. rehderana* responds similarly.

This species shows a propensity to sucker and should be a candidate for root cuttings. Small quantities can be propagated by simple layering.

Pterostyrax corymbosus Shrubby Epaulettetree

SEED: Seeds germinate best after 3 months of cold stratification. When sown without pretreatment the seed started to germinate in 8 days and continued to do so for at least 150 days. Stratification for 3 months produced a uniform stand in 12 days.

CUTTINGS: Easily rooted, see *P. hispidus*.

Pterostyrax hispidus Fragrant Epaulettetree

SEED: Seeds may germinate when directly sown but 3 months cold stratification at 40°F is recommended to ensure more uniform germination. Several arboreta reported that the species achieved weed status and germinated in out-of-the-way places.

CUTTINGS: Softwoods (June-July) root easily [*The Plant Propagator* 6(2):5 (1960)] and the senior author had 80 to 100% success with August cuttings treated with 3000 to 8000 ppm IBA-quick dip. The root systems were profuse. Cuttings transplant readily and continue to grow.

Punica granatum Pomegranate

SEED: Fleshy fruits should be collected in late summer-fall, seeds removed, cleaned and stored dry under refrigeration or planted. Seeds germinate readily without pretreatment but the progeny are extremely variable.

CUTTINGS: Softwoods root readily; senior author has rooted 'Nana', June, July, August, 1000 ppm IBA-quick dip, peat:perlite, mist, in high percentages. Cuttings continue to grow after transplanting. Open field propagation of hardwoods is possible in warmer climates such as Alabama [PIPPS 27:255 (1977)]. Eight to 10″ long cuttings from current season's growth are used. Insert cuttings half their length and as close as possible together (about 1 to 1½″ apart). Soil is then packed as tightly as possible about the cuttings.

Pyracantha coccinea Scarlet Firethorn

SEED: Collect fruits in fall or winter, macerate, remove pulp, dry seeds and store under refrigeration. Seeds require cold stratification for 3 months. In a controlled study, fresh seed did not germinate; 3 months cold stratification produced 82% germination; radicles emerged during the 6 month cold stratification period.

CUTTINGS: *Pyracantha* cultivars, such as 'Mohave', 'Shawnee', 'Lalandei', can be propagated from softwood, semi-hardwood or hardwood cuttings. Best rooting occurs in June, July and August. Scarlet firethorn is a disease prone plant and spraying with a fungicide 24 hours before collecting cuttings was suggested. Rooting conditions: 5 to 6″ long cuttings, 1000 to 5000 ppm IBA-quick dip, perlite:peat (9:1, v/v), and mist. See PIPPS 27:102 (1977).

Scarlet firethorn can be propagated from semi-hardwood cuttings September through November. Wound and treat with 8000 ppm IBA-talc or solution, sand:peat, and mist. Cuttings will be rooted in 4 to 6 weeks. Pot the rooted cuttings and place in cold frame or greenhouse [PIPPS 24:388 (1974)].

Pyracantha can be grown from root cuttings but this has not proven economical or practical. *Pyracantha* can be propagated by layering.

TISSUE CULTURE: Actively growing shoot tips proliferated in vitro on a Linsmaier and Skoog medium containing 2.5 mg BA/liter. Actively growing shoot tip explants were surface-sterilized in 0.5% sodium hypochlorite containing a few drops of detergent for 15 minutes and rinsed twice in sterile water. Cultures were maintained at 77±2°F and light was provided by Cool White fluorescent tubes at 4.4 klx for 16 hours per day. An average of 11 shoot tips was obtained after a 4-week culture period. Microcuttings were rooted in vitro in the presence of 5 and 10 mg IBA/liter. Rooting increased when cultures were incubated in the dark for 1 week prior to illuminated incubation [*HortScience* 17(2):190 (1982)].

Pyracantha koidzumii Formosa Firethorn

SEED: See P. *coccinea*.

CUTTINGS: Root readily from July to October when treated with 3000 ppm IBA-quick dip. This species can be rooted by any technique described under P. *coccinea*.

Pyrus amygdaliformis Almond Pear

SEED: A water soak for 24 hours followed by cold stratification at 41°F for 22 days produced 90% germination [*Proc. Amer. Soc. Hort Sci.* 92:141 (1968)].

Pyrus betulifolia Birchleaf Pear

SEED: Refer to P. *calleryana* for harvest, cleaning and storage details. A pound of cleaned seed contains about 44,000 seeds. A 24 hour water soak followed by cold stratification at 41°F for 52 days resulted in 90% germination.

Pyrus calleryana Callery Pear

SEED: The 1/4 to 3/8" diameter, speckled, rounded fruits are harvested in late fall-winter and macerated to remove the pulp. A pound of seed contains about 9,000 seeds. Seeds should be dried to about 10% moisture content. The stored seed will maintain high germinative capacity for 2 to 3 years. Cold stratification (41°F) for 1 month will produce 90% germination. Presoaking for 24 hours prior to stratification also may help. Callery pear seeds can be fall sown if unstratified or, if stratified, spring sown. Senior author collected fruits of 'Bradford' in January (Georgia), removed the seeds, sowed immediately with high germination. Cold stratification took place in the fruits while on the tree.

CUTTINGS: Reports indicated that some cultivars, such as 'Bradford', can be rooted from cuttings. 'Bradford' pear cuttings are taken in early June and rooted in a greenhouse in the coarsest perlite. Terminal cuttings about 6 to 8" long are treated with 8000 ppm IBA-talc and placed under mist. A large commercial nursery takes 'Bradford' when the growth is firm, 10,000 ppm IBA + 5000 ppm NAA-solution, bark:sand in ground beds in polyhouses, 70% rooting with strong root systems. Alabama, late May, 6 to 7" firm-wooded tip and subterminal cuttings, 0, 1.0% and 2% IBA or K-IBA in various solvents, wound, sand:peat, mist, evaluated after 10 weeks. 1.0% IBA in potassium hydroxide (KOH) or 25% ethanol and KOH produced 95 and 93% rooting, respectively. The potassium salt of IBA produced comparable results at 2.0%. The authors theorized that the carriers (ethanol and KOH) were more effective than water and "carried" the IBA into the stem. The tip cuttings rooted in higher percentages and had greater root numbers than the subterminal cuttings [*Proc. SNA Res. Conf.* 29:222-223 (1984)]. Senior author has observed 'Bradford' pears on their own roots and they do not make a fibrous, well-branched root system. In fact, they lodge and are difficult to transplant because of the sparse root systems. Other reports have noted the production of good root systems.

GRAFTING: 'Aristocrat', 'Bradford', 'Capital', 'Chanticleer', 'Respire', 'Select', 'Whitehouse' and others should be budded on the species. Seedling understock, about pencil caliper with a good root system, is lined out in the spring. After planting, the seedlings are fertilized to get them established quickly. Budding is much easier if the plants are actively growing. It is important to select budwood that is not too green or hard. The buds should separate from the wood without pulling. Budding is possible in June, July or August. P. *calleryana* is reported to be incompatible on P. *communis* [*The Plant Propagator* 31(1):9 (1985)], however, in England it is recommended. Senior author has observed decline and incompatability with 'Bradford' on P. *communis*.

Bud as close to the ground as possible to avoid a long shank of exposed understock. The T-cut is made into the west side or prevailing wind side of the understock as this makes the bud much more wind resistant. The tightness of the tying is critical for success. In 2 weeks the buds will be growing if successful. Pears tend to grow out sideways, and some growers recommend using Gro-Straight Stakes (J. Frank Schmidt & Co.) to correct the problem.

Bench grafting has also been employed to propagate P. *calleryana* cultivars. The understocks are No. 1 grade seedlings. Scion wood is current season's dormant wood and is cut to 7". Grafts are made during January-February using the whip and tongue method. The graft is made at the crown area so that ½ of the union is in the root tissue. Grafts are wrapped, dipped in rose wax and then covered with talc to prevent sticking, packed with shingletoe, and placed in cold storage at 36 to 40°F until planting time when danger of frost is over. Cover the union with soil at planting.

TISSUE CULTURE: Several commercial laboratories have produced 'Bradford' and 'Redspire'.

Pyrus communis Common Pear

SEED: Refer to P. *calleryana* for details on fruit harvest, cleaning, and seed storage. Cold stratification is required to break internal dormancy of the seed. Stratification temperatures of 32 to 36°F for 3 months produced 97% germination [*The Plant Propagator* 12(4):10 (1966)]. Presoaking the seed in water for 24 hours may improve stratification results. The seed can be frozen at 23°F if planting has to be delayed until after stratification.

CUTTINGS: Softwoods (Bartlett pear), 8000 to 10,000 ppm IBA-quick dip, mist, rooted 80%. Hardwoods are possible but more difficult (only 47%), mid-November (California), 8″ by 1/4 to 5/16″, 24 hour soak in 150 ppm IBA, place in heated bin at 75°F for 31 days with cool tops, and then plant in prepared outdoor beds. 'Old Home' softwoods, early June (California), 6000 ppm IBA-quick dip, mist, perlite:vermiculite (1:1, v/v), rooted 98% in 6 weeks. 'Old Home' hardwoods, mid- to late October, 100 ppm IBA soak for 24 hours or 2000 ppm IBA-quick dip, store 3 weeks in heated bin at 70°F, plant out in early November. See *The Plant Propagator* 9(2):6 (1962) and *J. Amer. Soc. Hort. Sci.* 82:92 (1963).

GRAFTING: A wide range of techniques is employed with this important fruit species. Bench grafting using a whip and tongue method on seedling or clonal rootstocks is possible. After taping and waxing, the grafts are placed in moist, well drained storage. In Oregon, seedlings are lined out in rows 6 to 12″ apart and budded in late July and August. The following spring the understock is cut off just above the bud [*The Plant Propagator* 16(2):9 (1970)]. Rootstocks recommended for 'Bartlett' pear include: *P. betulifolia* seedlings, 'Old Home' rooted cuttings, and *P. communis* 'Winter Nelis' seedlings. *P. calleryana* is popular in the south but not in colder regions because of hardiness problems. Seedlings of the oriental pears, *P. pyrifolia* and *P. ussuriensis*, should not be used because they are susceptible to pear psylla and pear decline, and a grafting disorder develops.

Fall T-budding or chip-budding on seedling or rooted cuttings of quince, *Cydonia oblonga*, as a dwarfing understock, is practiced commercially. However, graft incompatibility problems exist with some pear cultivars and double-working with an intermediate stock such as 'Old Home' is required. Cultivars needing such a compatible interstock include: 'Bartlett', 'Bosc', 'Clairgeau', 'Clapp's Favorite', 'Easter', 'El Dorado', 'Farmingdale', 'Guyot', 'Seckel' and 'Winter Nelis'. The following pears appear to be compatible with quince: 'Anjou', 'Comice', 'Duchess', 'Flemish Beauty', 'Gorham', 'Hardy', 'Maxine', 'Old Home', and 'Packham's Triumph'.

TISSUE CULTURE: Proliferating cultures from nodal explants of pear cultivars have been established on Murashige and Skoog (MS) medium containing mixtures of 3 cytokinins. Zeatin and 2iP were both supplied at 10 uM and BA as follows: 'William's Bon Chretien', 6 uM; 'Packham's Triumph', 8 uM; 'Beurre Bosc', 10 uM. Microcuttings averaged 80% rooting in vitro with IBA and NAA both at 10 uM and approximately 50% of the rooted cuttings were established in containers [*Scientia Hortic.* 23:51 (1984)].

'Seckel' pear shoot tips were established on MS medium modified to include, 100 mg myo-inositol, 0.4 mg thiamine, 1 mg BA, and multiplication occurred with 2 mg BA/liter. Optimal shoot proliferation was obtained on media containing 0.3% agar. Cultures were maintained under 16 hours light provided by a combination of daylight and GroLux fluorescent lights (3.4 klx) at 77±2°F. See *HortScience* 19:227 (1984) and *Proc. Conf. on Nursery Production of Fruit Crops Through Tissue Culture*, Beltsville, MD. p. 59 (1980).

Pyrus elaeagrifolia Oleaster Pear

SEED: In one study, a 24 hour water soak followed by cold stratification for 3 months produced good germination (90%).

Pyrus fauriei

SEED: A water soak for 24 hours followed by cold (41°F) stratification for 35 days gives good germination (90%).

Pyrus kawakamii Evergreen Pear

SEED: Refer to *P. calleryana* for details.

GRAFTING: The grafting season begins in December (California). The understocks are dormant, bare root callery pear seedlings. Match the scion to the caliper of the understock. The 1/4 to 3/8″ scion and understock are joined by a cleft graft and tightly wrapped with 1/4″ masking tape. Grafting wax is applied to completely seal the top area of the graft. The completed graft is planted into 1-gallon containers and placed in unheated plastic houses in full sun. Deviations from this procedure lead to poor results [*The Plant Propagator* 21(2):2 (1975)].

Pyrus pashia Pashi Pear

SEED: Soak the seed in water for 24 hours and follow with 15 days cold (41°F) stratification.

Pyrus pyrifolia Chinese Sand Pear

SEED: A pound of cleaned seed contains about 18,000 seeds. Fresh seed germinated 7%, 72% after 60 days, and 90% after 75 days cold (41°F) stratification. Soak seeds in water 24 hours before stratification.

TISSUE CULTURE: Shoot tips from an adult, seedling plant were established on Murashige and Skoog (MS) medium supplemented with 1 mg/liter BA. Multiplication occurred on an MS medium supplemented with 1.5 mg/liter BA plus 0.02 mg/l NAA. Microcuttings rooted in vitro in the presence of 2 mg/liter NAA plus 162 mg/liter phloroglucinol. Five named cultivars have been established and multiplied but rooting has been poor. See *Scientia Hortic.* 23:247 (1984.)

Pyrus salicifolia Willowleaf Pear

SEED: Four months cold stratification (41°) produced the best germination.

GRAFTING: 'Pendula' is whip grafted on bareroot seedlings of *P. communis* or *P. salicifolia* [NGC 184(12):35 (1978)].

Pyrus ussuriensis Ussurian Pear

SEED: The fruit ripens in September. Soak the seed in water for 24 hours before cold stratification. Thirty to 60 days cold stratification produced 90% germination.

GRAFTING: In England, *P. communis*, common pear, from continental sources is recommended as an understock for budding and grafting. T-budding of the stock close to the ground is accomplished during late summer. Polyethylene strips are used and the bud is enclosed to eliminate infestation by the red bud borer. 'Quince A' was noted as being incompatible with *Pyrus ussuriensis*. Whip and tongue grafting on established field lined rootstock during late winter and early spring just prior to bud break was reported in England.

Quercus Oak

SEED: Oaks are the most important hardwoods in North America. Seventy species are native to the United States with about 400 species world-wide. The fruit, an acorn, is one seeded and ripens in one year for white oaks and two years with black oaks. Fruit ripening and seed dispersal occur August to early December. The acorn is generally green when immature and turns brown to brownish black when ripe. Seed selection from superior parents is always good practice. Unfortunately oaks are wind pollinated and highly heterozygous as well. There is a constant succession of natural hybrids which appear in large seedling populations. Seed bearing and seed production vary from year to year. In a given locality, heavy seed crops may be infrequent, ranging to ten or more years between good crops. Seeds of the white oak group should be collected immediately after falling to prevent early germination. Acorns are attacked by about 10 species of weevils, at least 3 moth species, and several species of gall wasps. Acorn weevils are common and destructive pests. The larva feeds on the kernel until the acorn falls; then the larva emerges and disappears into the soil. In years of light acorn production the acorns can be heavily infested with weevils and the collection of sound seed may be difficult. The amount of sound seed can be increased by removing hollow and defective acorns and killing weevil larvae within the seed. Flotation is used to remove defective seed. Weevils in seeds can be destroyed by immersing the acorns in hot water (120°F) for 40 minutes. Temperature control is important because a temperature of 125°F will kill the seeds. *Quercus alba*, which has a root already present, and possibly other members of white oak group, suffer root damage with this treatment. Fumigation with methyl bromide, carbon disulfide, or thiamine bisulfite is also possible. See SWP-US p. 692 and NGC 178(7):30 (1975).

Acorns in the white group are not dormant and emergence of the root occurs in the fall. The shoot appears the following spring. Acorns of the white group can be stored dry for only a short time. Seeds of the black oak group remain dormant through the winter and germinate in the spring. This group can be fall planted or stored. Best germination in the black oak group occurs when acorns are mixed in a moist medium and kept at low temperatures (33 to 41°F) for 1 to 3 months prior to spring planting. Seeds of the white oak group should not be stored, and it is not practical to store seeds of the black oak group for more than 6 months. Loss of viability occurs because life processes in oak seed are critically dependent on the quantity of water. For germination, moisture content for white oaks must not drop below 30 to 50%, and 20 to 30% for black oaks [PIPPS 12:166 (1962)]. A 15% weight loss reduced germination of Q. *virginiana* to 66%; and 20% loss to 4%.

Fall sowing is preferred. If spring sowing of black oaks is practiced, the seed should be given cold stratification. Better seedlings are produced when seeds are sown in rows in moist, well drained, humus-filled soil. Germination is usually high but will vary depending on the amount of weevil damage. Rodents can be destructive after planting and control measures are required.

CUTTINGS: Although oaks have been considered impossible to root on a commercial basis several researchers and nurserymen have had success with a limited number of species. The wealth of oak species and the wonderful variation that is present should be incentive enough for researchers to work with the genus. Selections for fine habit, foliage color, and other desirable traits have been made but propagation by grafting is anything but satisfactory.

Live oak (Q. *virginiana*) has been successfully propagated from cuttings and trees grown to maturity [PIPPS 29:113-115 (1979)]. The following observations/recommendations evolved from the above work. Cuttings taken from young trees root more readily than those from mature acorn producing (adult) trees. Cuttings taken during the early, warm days of October (Texas) rooted equally with those taken in the May to August period. Cuttings taken during the colder months of November to March failed to root. Hormone application was essential with 10,000 ppm K-IBA-solution most effective. Stem-tip cuttings (terminal) from outer branches rooted well; non-terminal cuttings also rooted. Cuttings should be semi-hardwood. Soft, green cuttings seldom root. Pencil size or slightly smaller cuttings are ideal. A well drained perlite:peat (3:1, v/v) medium and mist were best. Polytent was not satisfactory. Bottom heat in cooler weather is beneficial..Cuttings should remain under mist for 12 weeks with hardening taking place under reduced mist. Cuttings taken from the "first generation" rooted cuttings rooted better than those taken from the original plants. The juvenility factor is most important. Cuttings taken from 5 to 8-year-old trees rooted poorly, with the older ones producing the poorest rooting.

Senior author followed the above procedures with a selected clone of *Quercus phellos* **x** either Q. *rubra* or Q. *palustris*. The tree was 6 to 8-years-old. Collected late June, 10,000 ppm K-IBA-solution, 2 perlite:1 peat, mist, 50% shade, terminal, firm-wooded and subterminal (next stem unit)). Terminal cuttings rooted 73% with strong root systems; subterminal 81% with similar systems. The approach works with several species and might be extended to other oaks.

Cuttings that were transplanted produced excellent root growth and were overwintered in a double-layer polyhouse.

POT GRAFTING: A reliable method for the vegetative propagation of oaks is grafting upon potted seedlings of the same species [PIPPS 12:168 (1962)]. The grafted tree which results is often stunted for several years and normal growth resumes slowly. Small seedlings are usually potted in the spring prior to the season when grafting will take place. The understock is brought into the greenhouse in February and scions of the past summer's wood about 12″ long are collected. The scions are grafted with a veneer graft, tied with rubber strips, and the understock cut back to slightly less than the scion length. The grafts can be set upright on an open bench of a humid greenhouse and covered with moist peat. After the grafts have callused and the scions are showing growth, the understock is trimmed back, the grafts are placed into the humid greenhouse for one week, and then gradually hardened off. The graft union is very weak and losses can be prevented by staking for the first season. Equally successful grafting can be accomplished in a humid greenhouse in late August, using current season's scion wood and cutting back the scion leaves to 1/2 their length. See *The Plant Propagator* 10(1):5 (1964) and especially PIPPS 12:168 (1962) for a discussion of vegetative propagation of oak.

FIELD GRAFTING: A number of European nurseries practice open field grafting. The understock is lined out in rows and established. Vigorous scion wood is gathered in March and stored in peat under cold storage. Just as the buds of the understock are swelling, the understock is cut off squarely, the stubs are split and wedge grafted. Excellent growth of the scion occurs because of the extensive root system. This technique can be practiced only under the cool, moist spring conditions of the low countries of Europe, British Isles, and Washington and Oregon in the United States.

DORMANT BUDDING: Dormant field budding has generally been unsuccessful and commercial stands have not been obtained.

ROOT GRAFTING: Leiss reported the successful grafting of Q. *robur* var. *fastigiata* to the species root pieces [PIPPS 34:526 (1984)]. Root pieces were potted into light potting soil in rose pots, placed in a grafting house and covered with peat, except for the top, on December 15 (Canada). In addition a piece of opaque plastic was used as a cover. Temperature ranged from 54°F at the start to 65°F later. Two months later the scion pieces, about 1/4* thick with 3 to 4 buds were side veneer grafted. The whole root piece was covered with damp peat. The grafting case was covered with glass sash, shaded during sunny weather, and kept closed for the first 14 days. Thereafter, it was aired one hour daily to start, and increased daily until after 4 weeks when the sash was kept open. It is important to protect new growth against mildew. Similar trials with Q. *rubra* and Q. *palustris* were not successful.

SCION/UNDERSTOCK RELATIONSHIPS: Oak grafting should be confined to putting the species on its own species understock or a hybrid oak on seedlings of one of the parent species. Incompatibility can be a problem and has occurred with 'Sovereign' pin oak. The situation was considered rectified when 'Sovereign' was grafted onto seedlings grown from acorns produced by 'Sovereign'. A large West Coast nursery reported only 53% take of grafted 'Crown Rite'. Recently, senior author was advised that this nursery had increased success to around 75%.

Quercus acutissima Sawtooth Oak

SEED: The sawtooth oak is reported to have a very regular seed crop frequency with a good seed crop produced one year and off the next but always a few acorns. The number of cleaned seeds per pound averages 100. A germinative capacity of 98% is reported. Sawtooth oak is a member of the white oak group and should be fall planted. In Georgia, the species produces heavy crops on an alternate year basis but even off-years provide sufficient acorns.

Quercus alba White Oak

SEED: Intervals between good seed yields may vary from 4 to 10 years. Fruit size varies widely and the number of seeds per pound ranges from 70 to 210. The germinative capacity may also vary between 50 and 99%. Root emergence occurs very rapidly after the acorns fall and root damage can occur when treating with hot water to control weevils. The white oak is fall planted. In moist autumns, the root radicles emerge from the nut on open ground.

Quercus bicolor Swamp White Oak

SEED: There are 120 cleaned seeds per pound. A 3 to 5-year interval between seed crops has been reported. Seeds sown fresh or after 1, 2 or 3 months cold stratification germinated 96, 96, 93 and 88%, respectively. This oak is fall planted.

Quercus cerris Turkey Oak

SEED: Requires no pretreatment.

Quercus chrysolepis Canyon Live Oak

SEED: About 150 seeds per pound. Seeds should be sown at once.

Quercus coccinea Scarlet Oak

SEED: The scarlet oak produces a good seed crop every 3 to 5 years. A pound of cleaned seed averages about 235 seeds. Seed can be fall or spring sown; spring sown seed requires a cold stratification period of 1 to 2 months. The seed has a germinative capacity of 94 to 98%.

Moisture content is a key factor in viability retention of stored seed. Storage of scarlet oak seeds with a high moisture content (60 to 70%) in polyethylene bags at 34 to 35°F appears to be the best way of retaining viability for at least 3 years [*The Plant Propagator* 20(4):11 (1975)].

GRAFTING: *Quercus coccinea* 'Splendens' should be grafted on the species and not Q. *rubra*. Graft on 2-year-old seedlings, mid-August to early September (England), whip or cleft graft.

Quercus dentata Daimyo Oak

SEED: Requires no pretreatment.

Quercus falcata Southern Red Oak

SEED: Fall plant or provide cold stratification.

Quercus glandulifera Glandbearing Oak

SEED: Excellent germination when sown immediately after collection. All cold stratification treatments, 2, 3 or 4 months, significantly reduced germination.

Quercus glauca Blue Japanese Oak

SEED: Similar to Q. *myrsinifolia*; no pretreatment required.

Quercus imbricaria Shingle Oak

SEED: The interval between good seed crops varies from 2 to 4 years. Averages 415 seeds per pound. Wide variation in germinative capacity (28 to 66%) has been reported. If spring sowing is practiced, the seed should be given 1 to 2 months cold stratification. However, some reports indicated a longer period (4 months) is better.

Quercus lyrata Overcup Oak

SEED: Averages 140 seeds per pound. Although a member of the white oak group, fresh seed sown immediately germinated 63%; while seed from the same seed lot given 3 months cold stratification germinated 100%. Fruits every year on the Georgia campus but crops vary from light to heavy.

Quercus macrocarpa Bur Oak

SEED: The interval between good crops is 2 to 3 years with an average of 75 seeds per pound. A germinative capacity of approximately 45% has been reported. Seed should be sown as soon as collected. In one study, germination was improved by cold stratification.

Quercus marilandica Black Jack Oak

SEED: Freshly sown seed germinated 18%; while 3 months cold stratification produced 91%.

Quercus mongolica Mongolian Oak

SEED: Fresh seed germinated 48%; 3 months cold stratification produced 84% germination.

Quercus muehlenbergii Chinkapin Oak

SEED: A pound of cleaned seed contains approximately 395 seeds and has a germinative capacity of 98%. Fresh seed germinates well.

Quercus myrsinifolia Chinese Evergreen Oak

SEED: No pretreatment required; seeds sown immediately germinated in high percentages.

Quercus nigra Water Oak

SEED: A southern species which, although a member of the black oak group, will germinate without pretreatment, however, 2 months cold stratification improves germination.

Quercus palustris Pin Oak

SEED: The interval between good seed crops is 1 to 2 years and a pound of cleaned seed averages 410 seeds. Fall plant or provide a cold stratification period of 1 to 2 months if spring planted.

GRAFTING: The first step in budding pin oak cultivars is to line out 2-year-old seedlings in the early spring. The seedling is grown for one year. Early in the spring of the following year, the seedlings are cut back to ground level. After secondary shoots start to grow, all shoots except one are removed. In early July or August, depending on growing conditions, while the shoots are vigorously growing, budding is started. The bud wood should slip freely. It is important to have bud wood with plump buds. The following year, when new growth appears, the understock is cut back to the eye.

Another method is to bud young pin oak with a caliper of 1 to 1½″ and insert a top bud about 5 to 6′ above ground. Insert the bud on the back side of a crook. This procedure is accomplished in early June (Cincinnati). After about 3 weeks, the entire top of the tree is cut back. The bud will make 12 to 18″ growth the first year. Incompatibility problems have been reported even when grafted on the species. See PIPPS 32:569 (1982).

Quercus phellos Willow Oak

SEED: Good seed crops are produced virtually every year. Cleaned seed averages approximately 460 seeds per pound and has a germinative capacity of 65%. Fall plant or provide 1 to 3 months cold stratification before spring planting.

Quercus prinus (*Q. montana*) Chestnut Oak

SEED: A good seed crop is produced every 2 to 3 years. A pound of cleaned seed contains about 100 seeds. Fresh seed will germinate, but in a Georgia study, 2 months cold stratification improved germination.

Quercus robur English Oak

SEED: Good seed crops are produced every 2 to 4 years and a pound of cleaned seed contains approximately 130 seeds. The seed requires no pretreatment, should be sown immediately, and has a 70% germinative capacity. A water soak is recommended before sowing. There is a report of the successful storage of acorns in dry peat for 3 years. 'Fastigiata' comes relatively true-to-type and nurserymen rogue off-type seedlings. More fastigiate English oaks are seed-grown than grafted.

CUTTINGS: 'Fastigiata' rooted 50% from current season's growth while still soft in late June to early July (New Jersey), 2% IBA and mist beds or poly tunnel. The rooted cuttings were stored at abouty 38°F over winter and survived in high percentages. Many cuttings grew 10 to 12″ the first season. See PIPPS 12:168 (1962).

GRAFTING: 'Fastigiata' is grafted on the species using a side veneer or whip graft. Many cultivars have been selected in Europe and most are pot-grafted on seedlings of the species. Seedlings (2-year-old) are established in pots in spring and brought into the greenhouse in winter for grafting. Grafting takes place when root activity is observed. After grafting, place on open bench or in closed case.

TISSUE CULTURE: Shoot cultures have been established and multiplied in vitro using material from juvenile seedlings and stump sprouts of mature trees. Heller's medium was satisfactory with 1 mM $(NH_4)_2SO_4$. Shoot tips were surface sterilized by successive immersion in 70% alcohol for 30 seconds and 3 to 5% calcium hypochlorite for 5 minutes followed by 3 rinses in sterile distilled water. The explants were left in sterile water for 2 to 3 hours to reduce exudation. Benzyladenine (BA) at 1 mg/liter was used in culture initiation and 0.1 mg/liter during shoot multiplication. A rooting rate of 83% with juvenile shoots and 63% with shoots from stump sprouts were achieved by dipping the basal ends in 1 mg/liter IBA for 2 minutes followed by transfer to an auxin-free medium for root initiation and development [J. *Hort. Sci.* 60(1):99-106 (1985)].

Quercus rubra Northern Red Oak

SEED: A good seed crop is produced every 3 to 5 years. A pound of cleaned seed contains approximately 125 seeds and the seed has a germinative capacity of 58 to 100%. Fall sow or provide 2 to 3 months cold stratification if spring planted. In one study, 3 months cold stratification produced 95% germination. Seeds without pretreatment germinated in low percentages. Seedlings respond to long days with increased growth and this may be a method of shortening the time to produce salable plants.

Moisture content is a key factor in viability retention during seed storage. Storage of northern red oak seed with a high moisture content (60 to 70%) in polyethylene bags at 34 to 38°F was the most effective method of retaining viability for 2 years. See *The Plant Propagator* 20(4):11 (1975).

GRAFTING: Incompatibility problems are reported as the grafts mature, even when grafted on the species [PIPPS 32:569 (1982)].

Quercus × sargentii Sargent Oak

SEED: Seed should be sown as soon as harvested.

Quercus shumardii Shumard Oak

SEED: A good seed crop is produced every 2 to 3 years. A pound of cleaned seed contains about 100 seeds. Fresh seed germinated 36%, while 3 months cold stratification produced 84% germination.

TISSUE CULTURE: Single node stem sections of seedlings served as the explants. Woody Plant Medium with 2 mg/l BA produced greatest number of shoots. Shoots were rooted with 73% success in Jiffy-7 peat pellets after a basal dip in 0.5 mg/l IBA. Plantlets were acclimated successfully to greenhouse conditions [*HortScience* 21:1045 (1986)].

Quercus stellata Post Oak

SEED: No pretreatment required; sow as soon as possible after collection.

Quercus variabilis Oriental Oak

SEED: The interval between good seed crops is 2 years. A pound of cleaned seed averages 105 seeds. Fresh seed germinated 100% and only 80% after 3 months cold stratification.

Quercus velutina Black Oak

SEED: The interval between good seed crops is 2 to 3 years. A pound of cleaned seed contains about 245 seeds. In one experiment, fresh seed germinated 76%, 88% after 1 month cold stratification, 68% after 3 months cold stratification.

Quercus virginiana Live Oak

SEED: Acorns collected fresh from tree branches germinated better than those from the ground. Live oak is a white oak member and the seed should be planted immediately after harvest. Live oak reportedly produces a good crop of seed every year and a pound of cleaned seed contains approximately 350 seeds. Seeds failed to germinate when frozen or stored in dry peat moss [PIPPS 31:670 (1981)]. When stored in moist peat moss at 41°F or at 70 to 86°F, they germinated but were heavily infected with soil-borne fungal pathogens. Seeds dried at 93°F lost viability as they desiccated.

CUTTINGS: Taken from rooted propagules of a 2-year-old plant maintained in a greenhouse for 2 years rooted in greater numbers than did cuttings taken directly from the original tree 2 years later. Rooting cuttings from cutting propagated plants offer a possible method for vegetatively multiplying this plant and other oak species. See introduction for a discussion of Q. *virginiana* cutting propagation and *HortScience* 15:493 (1980).

Raphiolepis indica India Hawthorn

SEED: See R. *umbellata*.

CUTTINGS: Easy to root from June to August (Florida) cuttings [*The Plant Propagator* 30(1):2 (1984)]. One to 2% IBA-quick dip, wound, perlite:vermiculite (1:1, v/v), 47% shade and mist, produced 90 to 100% rooting. 'Springtime' from fairly mature shoot tips in November (California), 4000 ppm IBA-talc, mist, Sponge Rok:peat (3:2, v/v), bottom heat, and a greenhouse temperature of 70 to 75°F, rooted in 40 days [*The Plant Propagator* 17(1):6 (1971)]. 'Enchantress', 8000 ppm IBA + 2500 ppm NAA-solution, excellent rooting.

Raphiolepis umbellata Yeddo Hawthorn

SEED: Seed collected in mid-February and sown after removal of the pulp germinated in high percentages.

CUTTINGS: Semi-hardwood to hardwood can be rooted. 2500 ppm IBA + 2500 ppm NAA-solution and wounding have proven to be effective treatments.

Rhamnus alnifolia Alder Buckthorn

SEED: A 3 month cold stratification treatment produces excellent germination.

Rhamnus caroliniana Carolina Buckthorn

SEED: Requires cold stratification for best germination; 1 to 2 months proves ideal.

CUTTINGS: Mid-August (Georgia), 0, 1000, 5000 ppm IBA-solution, peat:perlite, mist, produced 27, 97 and 100% rooting in 8 weeks. Rooted cuttings averaged 23 roots/cutting.

Rhamnus cathartica Common Bucktorn

SEED: The fleshy berry-like drupe ripens in September-October and should be harvested about 2 weeks before it is fully ripe to prevent losses due to birds. The fruit is black when ripe and 5 pounds of fruit yield approximately one pound of seed (19,000 seeds). The seed can be stored for several years in sealed containers at temperatures around 41°F. Considerable variation exists within buckthorn species relative to pregermination treatments. Spring sown seed should probably receive 2 to 3 months cold stratification. Concentrated sulfuric acid for 20 minutes has been found to be harmful. Late September sowing is recommended [*Amer. Nurseryman* Aug. 15, p. 8 (1968)].

CUTTINGS: Cuttings are somewhat difficult to root. Late July cuttings, treated with 960 ppm Ethrel + 2500 ppm IBA-solution, and placed in peat:perlite under mist, rooted 80%. High IBA concentrations might prove beneficial.

Rhamnus davurica Dahurian Buckthorn

SEED: Treat like R. *cathartica*.

CUTTINGS: Mid-July, 8000 to 10,000 ppm IBA-talc, perlite:sand, mist, rooted 100% in 9 weeks.

Rhamnus frangula Glossy Buckthorn

SEED: The purple-black fruits ripen from July to October and should be harvested about 2 weeks before fully ripe to prevent losses from birds. Ten pounds of fruit yield approximately one pound of seed (27,000 seeds). Fall or spring sowing are possible. Spring sown seed should be stratified for 2 months. 'Tallhedge' comes partially true from seed. Hard-seededness can be a problem [*Amer. Nurseryman* May 15, p. 10 (1967)]. Rogue out non-upright types.

CUTTINGS: 'Tallhedge' ('Columnaris') is propagated by cuttings. Softwoods, June or July, 3000 ppm IBA-talc, peat:perlite, mist and bottom heat (75°F), rooted 85%. Cuttings of the species rooted 90% when stuck in early August. 'Asplenifolia', a lacy, fine-textured form, July, 8000 or 10,000 ppm IBA-talc, sand, mist, rooted 56 and 52%, respectively, in 4 weeks. See PIPPS 33:489 (1983).

Rhododendron Rhododendron and Azalea

The genus *Rhododendron* comprises over 900 species and infinite cultivars due to the ability of the various species to freely hybridize. They are cultivated for their beautiful flowers, and the evergreen forms for their foliage. Rhododendrons are native to many parts of the world with large concentrations in China, Japan, and the eastern United States. Azaleas, formally considered in their own genus, are now included with the rhododendrons. There are no clear cut lines for distinguishing all azaleas from all rhododendrons but: 1) True rhododendrons are usually evergreen but there are exceptions. 2) Azaleas are mostly deciduous. 3) True rhododendrons have 10 or more stamens and leaves are often scaly or with small dots on their undersurface. 4) Azalea flowers have mostly 5 stamens; leaves are never dotted with scales and are frequently pubescent. 5) Azalea flowers are largely funnel-form while rhododendron flowers tend to be bell-shaped.

Rhododendrons are normally propagated from seeds, cuttings, layers, grafting, and tissue culture.

SEED: Principally utilized for propagating wild species, raising understock and developing new hybrids. The perfect flowers appear from March to August and the fruit is an oblong, 5-valved, dehiscent capsule that generally ripens in the fall. Capsules are green and color yellow to brown when mature. Collection of capsules should begin as soon as they begin to turn brown. The fruit is dehiscent and if capsules are not collected before they open, most of the seed is lost. The capsules can be air or oven-dried for 12 to 24 hours at 95°F. Average number of cleaned seeds per pound is about 5,000,000 seeds for species such as R. *catawbiense*, R. *maximum* and R. *periclymenoides*. Seeds with a moisture content of 4 to 9% will remain viable about 2 years at room temperature. Long term storage should be at 20°F in sealed containers. Germination must be carried out under light and the seeds normally germinate completely without pretreatment.

Senior author has grown many rhododendron species from seeds. The process is simple and requires milled sphagnum or screened sphagnum, even moisture, light and patience. Seeds are sown on top of the medium, placed under mist or in a polybag, with germination taking place in 2 to 4 weeks. Senior author uses mist in the greenhouse but many hobbyists use Grolux or other "indoor" lights or simply a window sill. Seedlings can be transplanted in 8-10 weeks but must be handled carefully to prevent desiccation.

LARGE FLOWERED EVERGREEN TYPES

CUTTINGS: The best time to collect cuttings (New Jersey) is from early August to the end of November. Thin cuttings taken from side growths give better rooting than strong, vigorous terminal growths. Cuttings should be gathered early in the morning when the plant material is turgid. Cuttings 3 to 4″ long root better than longer types. The number of leaves is reduced to 4 to 5, and with large-leaf cultivars the leaf length may be reduced by half. Double wounding greatly increases the total percentage of well rooted plants. When using rooting hormones, it is important to consider cultivar differences. Most standard cultivars which root with relative ease respond to 8000 ppm IBA-talc, slightly more difficult types will need 10,000 ppm IBA, and the very difficult types (such as the red cultivars) require 20,000 ppm IBA. For the very difficult ones, 10,000 ppm of the potassium salt of IBA and 2500 ppm NAA have stimulated rooting. Wells recommended 70°F bottom heat for early cuttings. A rooting medium of peat:perlite (1:1, v/v) is satisfactory with mist or closed polytent. See GCHTJ 165(8):16 (1969) and *The Plant Propagator* 16(4):17 (1970).

LAYERING: For commercial propagation, deep beds are prepared in a lath house or natural half shade containing loam, peat moss and sand. Irrigation is essential. In late summer large stock plants are transplanted into the layering beds. The current season's growth that will be underground is stripped of leaves, wounded, placed into a trench, and bent upwards 6 to 8 inches from their tips. With the arrival of winter the beds are mulched, care being taken not to cover the protruding tips. The following growing season the beds are irrigated if rain is lacking, and in the fall the layers are severed. The layers are mulched again and transplanted the following spring to prepared beds. If rooting is slow, the layers are severed in mid-spring and transplanted.

GRAFTING: The first step in the production of grafted rhododendrons is selection of the understock. R. *ponticum* and R. *catawbiense* hybrids such as 'Roseum Elegans' are commonly used when about pencil thick. Refer to seed section for details on seed germination and growth. It takes 2 to 3 years to raise a good rootstock from seed. Understocks are lifted, transplanted into 3½″ pots, and maintained in a cool greenhouse at 60 to 65°F to stimulate new root growth. When new root growth begins, grafting can commence. Wells recommends the side graft. The completed grafts are placed in grafting cases containing well-moistened peat. The benches are kept as close to 70°F as possible. During the next 3 to 4 weeks, the grafts are aired each morning for ½ hour and then watered down before closing. If, after 3 or 4 weeks, union formation is slow, the peat is turned, watered and the plants placed back. Within 2 weeks callus formation is well developed and hardening begins. The grafts are next placed in a coolhouse maintained at 55 to 60°F, and this, together with frequent syringing, insures proper development. It is important to insure that the temperature never gets too high. In a few weeks the root stock is putting out new growth and the top of the understock can be removed. See *The Plant Propagator* 9(1):11 (1962).

DWARF RHODODENDRON SPECIES

CUTTINGS: Semi-ripe cuttings of R. 'Blue Tit', R. *cilipense*, R. *concatenans*, R. *impeditum*, R. *leucaspis*, R. *macrostemon*, R. *moupinense*, R. *mucronatum*, R. 'Seta', R. *tebotan*, and R. *tschonoskii* rooted better from July cuttings than in February (Ireland). However, plants rooted in February attained a greater size the first growing season. Treat with IBA (most standard types require 8000 ppm IBA-talc) and root under the warm bench (70°F) and plastic system or mist. [*Gardeners Chronicle* and *Horticultural Trade Journal*. 1979. 186(19):19]. For difficult types use 1.6% IBA-talc. Wounding and a fungicide, such as Captan or Benlate, at 5% is often recommended. Lamb et al. noted good rooting with February cuttings of the following: R. 'Baden-Baden', R. 'Blue Tit', R. 'Bountiful', R. *desypetalum*, R. *hippophaeoides*, R. *impeditum*, R. *leucaspis*, R. *moupinense*, R. *pemakoense*, R. *williamsianum*, and R. *yakusimanum*. (*Nursery Stock Manual*. Grower Books, London. 1975. p. 167)

DECIDUOUS AZALEAS

CUTTINGS: The Exbury, Ghent, Knap Hill and similar deciduous azaleas can be rooted, but there is extreme variability among species and clones. The stoloniferous North American species such as R. *atlanticum*, R. *arborescens*, R. *periclymenoides* (*nudiflorum*), and R. *viscosum* often root easier than the non-stoloniferous species. There are two problems in the rooting of deciduous azaleas: 1) rooting the cuttings, and 2) inducing new growth following rooting. Cuttings that fail to develop new shoot growth often have a low overwinter survival rate. During late fall (Virginia), stock plants are brought into a greenhouse that has not been covered. In early January, after the plants have gone through a cold spell, the greenhouse is covered. The plants are thawed out with temperatures just above freezing to 45°F over a two week period. The temperature is then raised to 70°F by the beginning of March. Cuttings are collected from the beginning of April until the end of April. The plants are fertilized every 10 to 14 days with liquid fertilizer. Suitable cuttings are about 6″ long, slightly firm, just about to snap when bent double, and still hairy. The cuttings must be kept turgid. The cuttings are stripped of all except four leaves, tip removed, wounded and treated with 4000 ppm IBA + Benlate, and placed in beds of pure peat moss (peat:perlite would work) kept at about 73°F under mist. Hormone requirements can vary from the above with different cultivars and some testing may be required.

In order to prevent *Botrytis* the cuttings are treated with a suitable fungicide. After 4 weeks, when rooting has started, the cuttings are fertilized with 1 tablespoon of 23-19-17 per 3 gallons of water. In 6 to 8 weeks the cuttings are rooted. Rooted cuttings are transplanted into 2-gallon containers and placed in a shaded house for 1 to 2 weeks. See the cutting chapter for details relative to extended photoperiod and overwintering of rooted cuttings. Also, see *Arnoldia* 20(1):1 (1960) and *Azaleas*, Timber Press. 1985.

ROOT CUTTINGS: North American azaleas can also be propagated by root cuttings. Root cuttings 3 to 4″ long and ¼ to ½″ in diameter can be used at all seasons of the year. Spring, however, is the best time. The root pieces are placed horizontally in a medium of equal parts of peat, perlite and shredded sphagnum moss, and covered with the same [*The Plant Propagator* 25(1):10 (1979)].

LAYERING: Layering is a reliable method for producing small numbers of plants. Stool beds must be prepared. A lath house is beneficial during the 18 months it takes to root the cuttings. Refer to large flowered rhododendrons for details.

Lamb *et. al.* recommend hard pruning in the spring and when the shoots are firm in July the layering operation is begun by pulling soil up around the shoots. The following spring the flower buds are removed from the layered shoots, and by autumn the rooted shoots can be harvested.

EVERGREEN AZALEAS

CUTTINGS: Cuttings are taken from vigorous plants in June-July and into fall. The cuttings are dipped in a fungicide containing Captan. Soft growth is pinched out, leaving 4 to 6 leaves and stripping off the rest. The entire cutting usually measures 4 to 6″ long and is placed in a shaded greenhouse, stuck 1½″ into a peat:perlite (1:1, v/v) medium. Mist or a plastic covering are supplied. Some propagators use no hormone while others use 8000 to 10,000 ppm IBA-talc. A weekly fungicide program is recommended and includes the use of Benlate and Captan, each being applied on a different week. Rooted cuttings are planted into flats. Senior author has observed tremendous root rot among the semi-evergreen to evergreen azaleas. Excessive moisture can prove disasterous. See PIPPS 34:444 (1984).

TISSUE CULTURE: In vitro culture has proven to be effective in the rapid clonal multiplication of the genus *Rhododendron*. It appears that most members of the genus *Rhododendron*, if not all, can be successfully micropropagated through the method of shoot tip culture. Wilbur Anderson (Washington State University) is responsible for developing the tissue culture medium that forms the basis for successful micropropagation. The most effective cytokinin is 2iP with N_6-benzyladenine shown to be quite toxic to rhododendron tissues. Kinetin is not effective. The essential compounds required for shoot development in the basal medium in mg/liter are: adenine sulfate dihydrate 80, 2iP 15, and IAA 4. A pH of 4.5 was found satisfactory [PIPPS 25:129 (1975)]. McCown and Lloyd examined the response of 7 genotypes of *Rhododendron* to culture conditions. The genotypes represented a broad genetic diversity in the genus and included: deciduous azaleas, subgenus Pentanthera; evergreen azaleas, subgenus Tsutsusi; elepidote evergreen rhododendrons, subgenus Hymananthes; and lepidote evergreen rhododendrons, subgenus Rhododendron. Actively growing 4/5 to 1 3/5″ long stem tips from stock plants were surface sterilized in 10% sodium hypochlorite with a wetting agent (0.05% Tween-20) added. After rinsing 3 times, the explants were placed in test tubes containing liquid medium supplemented with 8 microM 2iP. The shoots were transferred frequently at first and less often as the exudation from the explant decreased. After 1 to 4 months the shoot pieces were transferred to solid medium. Plants with strong episodic growth cycles required a longer period for establishment. The optimum 2iP level for shoot multiplication varied little with the genotype and levels of 4 to 16 microM proved optimal, depending on the specific species or cultivar. Adventitious shoot production was observed with some selections above 8 uM. Cultures were grown in rooms with 24 hours cool-white fluorescent lighting (20 uE m^{-2} sec^{-1}) and temperatures that averaged 82 to 86°F. Shoots are readily rooted out of culture in a peat:perlite (1:1, v/v) medium and high humidity chamber. See *Plant Cell Tissue Organ Cult.* 2:77 (1983).

Fordham, Stimart and Zimmerman examined the effect of 4 cytokinins: BA, kinetin, 2iP and zeatin, on axillary and adventitious shoot proliferation of Exbury azaleas. The greatest number of shoots was produced on medium containing zeatin, a lesser number on 2iP, and the least on BA and kinetin [*HortScience* 17:738 (1982)].

Flower pedicels and ovary bases of four R. *catawbiense* cultivars, 'Alba', 'Nova Zembla', 'Roseum Elegans', and 'Sefton' were cultured in vitro and proliferated granular masses of tissue on Anderson's medium supplemented with IAA (1 or 4 mg/liter) and 2iP (1 or 15 mg/liter) [*HortScience* 17:891 (1982)]. The granular masses formed numerous shoots when cultured on medium with lower hormone levels. The flower buds are removed from stock plants from October to April, and the outer resinous bud scales are removed until the white, papery covering of the florets is exposed. The buds are immersed in 10% Clorox and 0.1% Tween 20 for 20 to 30 minutes and then rinsed with sterile distilled water. The florets are excised with as much pedicel tissue as possible and pressed halfway into an agar culture medium.

Optimal 2iP concentration for selected *Rhododendron* species.

Species/cultivar	2iP Concentration (mg/liter)
R. arboreum	15 to 20
R. 'Boule de Neige'	1.6 to 3.2
R. canadense	1.6 to 3.2
R. chamae-thomsonii	15 to 20
R. chapmanii	10
R. 'Chikor'	15 to 20
R. 'Chinsayii'	15 to 20
R. dauricum	15 to 20
R. fastigiatum	15 to 20
R. forrestii	10 to 15
R. 'Gibraltar'	3.2 to 6.4
R. leucaspis	15 to 20
R. lutescens	15 to 20
R. 'P.J.M.'	0.8 to 3.2
R. 'Victor'	15 to 20
R. poukhanense	0.8 to 3.2
R. racemosum	10 to 15
R. schlippenbachii	0.8 to 3.2
R. 'Vuyks Rosy Red'	15 to 20
R. williamsianum	10 to 15

Rhodotypos scandens Black Jetbead

SEED: Fruit is a hard black drupe that can be collected from October through early winter. Extraction of the seed from the fruit may not be necessary. The number of cleaned seeds per pound averages about 5,200. Seeds can be air dried and sterilized at 34 to 50°F for 9 months without loss of viability. Seeds exhibit a double dormancy and require both warm and cold stratification. Warm (77 to 86°F) and 1 month cold stratification is recommended. Slightly green seeds germinate in 1 year if fall planted. In nursery practice, spring planted pretreated seeds would germinate the fastest. Work at the Arnold Arboretum indicated that fresh-collected seeds did not germinate; 1% germinated after 3 months cold stratification; 0% after 1 hour acid scarification; 12% after 1 hour acid and 3 months cold stratification. Obviously, the dormancy mechanism is quite complicated.

CUTTINGS: Stem cuttings can be rooted any time the plant has leaves but June and July are the best months. No hormone is required but is good insurance. Place under mist or in a shaded plastic tent. Hardwoods also root [*The Plant Propagator* 15(3):10 (1969)].

Black jetbead can be produced from root cuttings but the method is not economical or practical [PIPPS 27:402 (1977)].

GRAFTING: Black jetbead has been test grafted on *Kerria japonica*, a close relative, but the combination was incompatible.

Rhus aromatica Fragrant Sumac

SEED: Fragrant sumac germinates readily from seed which is the preferred method of propagation. The fuzzy red drupe ripens in August-September and persists into early winter. The dried fruit clusters can be easily broken into individual fruits by rubbing and screening. Pieces of remaining seed coat can be removed by running through a macerator with water. Fragrant sumac seed can be kept over winter and possibly for years without special treatment.

The seeds germinate poorly without pretreatment. Seed dormancy is caused by both a hard seed coat and dormant embryo. Scarification with sulfuric acid for approximately 30 minutes to one hour followed by cold stratification at 41°F for 1 to 3 months is recommended. There is some disagreement regarding the need for pretreatment if the seed is fall planted. Fall sowing after acid pretreatment is probably best. *Rhus trilobata*, skunkbush sumac, requires similar dormancy breaking treatments [*Amer. Nurseryman* 125(2):10 (1967)].

CUTTINGS: Softwood cuttings collected in July and treated with 1000 ppm IBA root readily. Stick in peat:perlite and place under mist [PIPPS 22:431 (1972)]. 'Gro-Low' is easily propagated in June or July when treated with 1000 ppm IBA [PIPPS 33:489 (1983)]. Reduce mist as soon as cuttings start to root. Fragrant sumac suckers freely from the roots and can be propagated from root cuttings. Late November, root pieces, sandy soil, greenhouse, produced shoots by late December; by mid January numerous shoots developed. Hardwood, early March, 4000 ppm IBA-talc + thiram, polytent, rooted 91% with strong root systems. If a few plants are desired both division and layering are satisfactory.

Rhus chinensis Chinese Sumac

SEED: Hot water soak for 24 hours followed by 3 months cold stratification produced good germination.

CUTTINGS: Root cuttings work well.

Rhus copallina Flameleaf or Shining Sumac

SEED: Fruit is crimson at maturity and ripens in September-October. A pound of cleaned seed contains approximately 57,000 seeds. The seeds germinate slowly and irregularly without pretreatment. Seed dormancy is caused by a hard seed coat. Seed coat dormancy problems can be overcome by soaking in concentrated sulfuric acid for 1 to 2 hours at room temperature. Spring sowing would be the preferred sowing time after acid treatment.

CUTTINGS: Root cuttings would be a valuable method for the production of flameleaf sumac [PIPPS 11:42 (1961)]. A method utilizing spring harvested root pieces can be found under *R. glabra*.

Rhus glabra Smooth Sumac

SEED: Fruit is scarlet when ripe and persists into winter. The dried fruit clusters are easily broken into individual fruits by rubbing and screening. Sound seed sinks in water. Seeds stored at room temperature for 10 years germinated over 60% which suggests that controlled storage conditions are not required. A pound of cleaned seed contains approximately 50,000 seeds.

The seed germinates slowly and irregularly without pretreatment. Seed dormancy is caused by a hard seed coat. Seed coat dormancy problems can be overcome by soaking in concentrated sulfuric acid for 1 to 3 hours at room temperature. Heit, however, recommended 6 hours acid. Pouring boiling water over the seed and allowing the seed to soak for 2½ days is also effective. Spring sowing is the preferred planting time after treatment. One report noted hot water treatment followed by 3 months cold stratification gave good germination.

CUTTINGS: Root cuttings offer a valuable method of propagation [PIPPS 11:49 (1961)]. There is controversy as to the best time to collect roots for propagation of *Rhus* species. Spring harvest is advocated by some propagators while other suggest fall. The technique utilizing spring harvested roots will be reported for *Rhus glabra* while a fall harvest method will be discussed for *Rhus typhina*. The root cutting method would typically be utilized with 'Laciniata' [*The Plant Propagator* 19(3):8 (1973)]. Roots are dug as soon as the frost leaves the ground in the spring. The ½ to 1″ diameter roots are cut into 2 to 3″ lengths, packed in boxes of almost dry sand, and stored in a cool place 3 weeks for callusing. When the soil has dried sufficiently a sandy spot is prepared. Care in marking the rows and subsequent weed control are required. After sprouting the cuttings are hilled up moderately.

Rhus typhina Staghorn Sumac

SEED: Fruits are crimson at maturity, ripen in October and persist on the plant through the winter. Seeds can be kept over winter and for several years at 32 to 41°F. The number of cleaned seeds per pound is 53,000. Seeds germinate poorly without pretreatment because of a hard coat. Treat the seeds with sulfuric acid. The length of acid treatment is variable (1 to 6 hours) and each lot should be tested before acid treatment. Spring sowing after pretreatment is the preferred nursery practice. Interestingly, seed sown immediately germinated 4%; 37% after a 24 hot water soak; 3 months cold stratification did not improve germination over freshly sown seed.

CUTTINGS: Staghorn sumac can be propagated by root cuttings and is the preferred method for 'Laciniata' [PIPPS 31:524 (1981)]. Root pieces for propagation are obtained from fall harvested plants. The roots pieces are between 1 and 2′ long and are stored in a cool place in moist sphagnum moss until worked into cuttings about 4″ long. The cuttings are treated with 3000 ppm IBA-talc and placed in moist sphagnum with the tops up. Root pieces (stored in boxes) are covered with poly, placed under cool conditions (50 to 60°F) for 10 days to initiate callus formation, and returned to cold storage at 35°F until planting time. Planting occurs the first part of May (Minnesota) with the cuttings planted 8 to 9″ apart. The cutting top should be even or slightly above the soil. Packing is important to insure adequate moisture. Cultivation is practiced to prevent soil crusting. Sprouts emerge in about a month with some taking 2 months.

Ribes alpinum Alpine Currant

SEED: The flowers on alpine currant are dioecious. Fruit is a scarlet berry that ripens in June-July. The fruits are macerated in water to remove the fleshy pulp. *Ribes* seeds maintain their viability for long periods when stored in sealed containers at a low moisture content.

Seeds germinate poorly without pretreatment because of a dormant embryo. Alpine currant seed requires a 3 month cold stratification treatment. Most *Ribes* seeds are fall sown, however, spring sowing after pretreatment is possible.

CUTTINGS: Softwoods taken in June and July root well when treated with IBA at 1000 ppm and placed under mist in a sand medium. Hardwood propagation is also possible with *Ribes*. Cuttings no smaller than a pencil from 1-year-old wood are cut into 6″ cuttings and stored in sand until early spring planting. 'Green Mound' can be propagated from softwood cuttings in sand under mist when treated with 1000 ppm IBA solution. Hardwood, late December, 8000 ppm IBA-talc + thiram, open bench (warm bottom/cool top), rooted 100%. See PIPPS 33:489 (1983).

Ribes missouriensis Missouri Gooseberry

SEED: Three months cold stratification induced good germination.

Ribes nigrum Black Currant

SEED: Refer to R. *alpinum*.

CUTTINGS: Readily propagated by single-bud hardwoods about ½″ long from the mid portion of the stem. Optimum rooting is obtained January through March after dormancy is broken.

Softwoods and semi-hardwoods root easily and can be harvested when the current season's shoots are 12 nodes long. When planted in the propagation frame with bottom heat, good rooting is obtained from April to June.

Ribes odoratum Clove Currant

SEED: Three months cold stratification should work.

CUTTINGS: Root pieces in late December produced shoots by late January. This species suckers and can be propagated by division.

Robinia fertilis
Robinia hispida Bristly Locust

SEED: Two similar species with R. *fertilis* producing abundant fruit and R. *hispida* little to none. They are easily produced from seed. The fruits ripen in September-October and dispersal occurs at that time although fruits persist into fall. The number of cleaned seeds per pound averages 28,000. See R. *pseudoacacia* for seed harvest, cleaning, storage and pregermination requirements. A 24 hour hot water soak prior to sowing improves germination. The hard seed coat inhibits germination and acid pretreatment with spring planting or fall planting result in good germination.

CUTTINGS: Root cuttings, 2 to 4″ long, mid-May (Indiana) [*The Plant Propagator* 19(3):8 (1973)]. Both species sucker prolifically and simple division produces a number of plants.

GRAFTING: Both species can be grafted on R. *pseudoacacia* understock in the early spring to produce bushy, pink-flowered small trees. See PIPPS 17:304 (1967).

Robinia pseudoacacia Black Locust

SEED: The fruit is a 2 to 4″ long, 4 to 10-seeded pod that matures in September-October. At maturity the brown-black pods split and shed the seed. Collect before opening and separate the seed by threshing. The number of cleaned seeds per pound is 24,000. Seeds can be stored in the open in a cool dry place or in sealed containers at 32 to 40°F for at least 10 years [*Amer. Nurseryman* Nov. 15, p. 12 (1967)]. The seeds germinate slowly and irregularly without pretreatment because of a hard seed coat. Mechanical, acid, and hot water scarification have been successful. A 24-hour soak prior to sowing is also recommended to speed up germination. 93% of the seeds given a hot water soak imbibed water.

CUTTINGS: Root cuttings are a valuable method for the production of black locust [*Gardeners Chronicle* 169(3):22 (1971); PIPPS 11:42 (1961)]. Some propagators suggest fall root harvest while others spring. The technique using spring root harvest will be presented, however, a fall harvest method would be similar to that described under *Rhus typhina*. Roots (½ to 1″ diameter) are dug as soon as possible in the spring, cut into 2 to 3″ lengths, packed in boxes of almost dry sand, and stored in a cool place for 3 weeks for callusing. Then the roots are planted in rows and soil is raked over them. After sprouting, the cuttings are hilled up moderately. The same approach could also be applied in a greenhouse.

GRAFTING: 'Bessoniana', 'Frisia' and the pink-flowered forms are bench grafted successfully in winter on bareroot seedlings of the species. The understock should be at least 3/8″ in diameter and headed back to 6″ before use. The scion wood is taken from the previous year's growth and cut 6 to 9″ long. A whip or side-veneer graft is used and, after tying, the union and scion are dipped in paraffin wax. Plunge the grafts into moist peat and shade so that the air temperature is maintained around 45 to 70°F. The grafts placed at the warmer temperatures are held for 8 to 10 days until well callused and the buds on the scions are just beginning to break. They are then removed and placed upright in a cool greenhouse until planted outside. A modification of the warm treatment system is to heat-treat the stocks just to bud break and then dry before grafting. Regular fungicidal applications are carried out to reduce losses from diseases, such as *Botrytis*. The containerization of the grafts is carried out after 6 weeks. It is very important that there should be minimal root growth before potting. This procedure applies to the numerous cultivars of R. *pseudoacacia* as well as the hybrids that do not reproduce from cuttings.

Summer grafting of 'Frisia' is practiced. Seedlings are sown in 3.4 × 6″ pots in the winter and allowed to grow until early summer. Scion preparation includes removing the leaves, the terminal, and 4 or 5 nodes. A top cleft or wedge graft is used and the scion has only one bud. See PIPPS 33:164 (1983).

Rosa Rose

The tremendous number of roses available makes it impossible to discuss propagation for each species and cultivar. In addition to the species there are numerous modern rose hybrids including floribunda, grandiflora, hybrid tea, climber and miniature types.

SEED: Seed propagation is used mainly for the production of roostocks, the production of certain species, and breeding new cultivars. When raising ornamental rose species from seed it is important that they be isolated to reduce the chances of hybridization. Flowers are bisexual (perfect) and are produced either singly or in clusters. The fruit is a fleshy hip containing hairy achenes that are erroneously called seeds. Depending on species and cultivar, mature fruit color ranges from orange-red to scarlet (other colors also) with ripening occurring in late summer and fall. Rose hips should be collected as the color changes from green to orange or red because seeds from the fully ripe stage have a higher dormancy level due to the accumulation of chemical inhibitors. The freshly harvested hips are macerated to remove the flesh, skin and empty seeds. Average numbers of cleaned seeds per pound range from 24,000 to 48,000. Seeds stored dry in sealed containers at 34 to 38°F have shown fair germination after 2 to 4 years. The seeds (achenes), in most cases, display embryo dormancy which first must be removed before germination can take place. Hybrid rose seed is generally planted in a sand and peat medium and placed directly in a greenhouse maintained at 60°F. Some crosses germinate immediately; other, however, germinate poorly, if at all, until they have received cold stratification. Most hybrid crosses fall into the 10 to 30% germination range. In a study [*Contrib. Boyce Thompson Instit.* 3:385 (1931)] of the after-ripening and germination of 8 rose species: R. *canina*, R. *carolina*, R. *fendleri*, R. *helenae*, R. *multiflora*, R. *rubrifolia*, R. *rugosa*, *and* R. *setigera*, seeds after-ripened at 41°F exhibited great variation in the time to complete germination. For example, R. *multiflora* germinated in 4 months, while R. *canina* required 27 months. The dormancies of hybrid rose seeds, and the subsequent germination response patterns, vary widely between years even for identical crosses, and low-temperature treatment is not an absolute prerequisite to good germination. A good correlation was found between germination and preharvest climate, particularly with the mean of the average daily temperatures for the 30-day period preceeding harvest [*Amer.* J. *Bot.* 43:7 (1956)]. It appears that higher temperatures favor lower dormancy. Aseptic culture of excised embryos from fresh seeds also has produced uniform germination.

All rose species have a hard seed coat to some degree and this may account for the wide germination differences reported in *Contrib. Boyce Thompson Instit.* 3:385. Acid scarification, followed by a period of warm stratification and then cold stratification, has proven satisfactory for achieving a high rate of germination with R. *dumetorum* 'Laxa', the most widely used understock in Great Britain. After the process of maceration the seed must be dried before acid scarification or embryo damage can occur. Because of the great variability in seed size it is graded 2mm plus and below 2mm seed size. Concentrated sulphuric acid is used to break down a third of the testa (seed coat). About 1 pound of seed is combined with 900 to 1000 ml of acid. The temperature should be maintained between 100 to 120°F. The average treatment time is 45 to 60 minutes per batch. A sample is removed after 35 minutes, neutralized, and cut in half to determine the state of seed coat digestion. Sampling is repeated until 1/3 of the seed coat has been degraded. Constant stirring is required to assure even digestion of the seed coats. Seed is drawn out of the acid and continually washed in water to remove all the acid. The seed is warm stratified for one month at about 76°F and then for 3 months at 40 to 42°F. To obtain maximum germination, sow immediately after treatment when soil temperatures are rising and late enough to avoid frosts. This procedure should be tried with other difficult to germinate roses. A safe general recommendation is 3 to 5 months warm/3 months cold or acid scarification and 3 months cold.

CUTTINGS: Softwoods root readily. Treat with 3000 to 5000 ppm IBA-talc or K-IBA in water, wound, peat:perlite and mist. Mention is made of fungicide dips and preventative sprays, such as Benlate, for disease control. See miniature rose rooting for additional details.

MINIATURE ROSE ROOTING: Cuttings are made with 2 to 3 nodes of rather soft to semi-hard wood, any time of the year material is available. Rooting of hard or mature wood is possible but much slower. The soft to semi-hard leafy cuttings will root in 3 to 4 weeks. Stock plants maintained in pots provide a good source of cutting wood. Cuttings are dipped into a fungicide solution and stuck directly into pots containing peat:perlite:fir bark (1:1:1, v/v/v) medium. Cutting bases are dipped in 3000 ppm IBA-talc and misted during the daylight hours [PIPPS 30:54 (1980)]. Some evidence points to tremendous alcohol sensitivity and IBA should be used in a talc formulation or as K-IBA in water.

HARDWOOD CUTTINGS: Hardwoods are often used for the propagation of rose rootstocks and may also be used with other vigorous types, such as climbers. In colder climates the hardwoods are made in late fall or early winter and stored at about 40°F until spring planting.

LAYERING: Layering is possible if only a few plants are needed.

GRAFTING: Many clonal roostocks used for grafting and budding are contaminated with viruses and only virus indexed stock should be used. Clonal rootstocks commonly used include: R. *canina*, commonly used in Europe and raised from seed; R. *chinensis* 'Gloire de Rosomanes' and 'Ragged Robbin', used in the western United States because of its resistance to heat and drought, however, susceptible to verticillium; R. *dumetorum* 'Laxa', the most widely used understock in Great Britain and grown from seed; R. *multiflora*, a widely used species for outdoor roses that is nematode resistant, adaptable to a wide range of soils, and propagates easily from cuttings or seed; R. *multiflora* 'De La Grifferaie', widely used for tree rose production, virgorous and very hardy; R. × *noisettiana* 'Manetti', widely used for grafting greenhouse roses; R. *odorata* (Odorata 224490), not cold hardy and widely used for greenhouse roses; R. *rugosa*, a long-lived species that is propagated from cuttings and seed; R. 'Dr. Huey' is widely used in Arizona and certain parts of California, but not hardy below 0°F and susceptible to verticillium; and IXL (Tausendschon/Veilchenblau), whose main use is in tree rose production, vigorous, no thorns, but not as hardy as 'Dr. Huey' [*American Rose Annual* 36:101 (1951)].

SUMMER BUDDING: Budding is the preferred method for the propagation of hybrid roses. Summer budding, as outlined by Davies for field rose production in east Texas, begins with field sticking of R. *multiflora* hardwood cuttings November through February. Hardwoods (about 8″ long) are disbudded except for the top 1 or 2 buds. Budwood, collected and stored in late fall under nondesiccating conditions at 30 to 32°F, is T-budded to actively growing understock from March through August, but mostly in the spring, with no budding during dry conditions without irrigation. Early budded plants will force some growth the first year. In December and January the scion wood forced during the previous season is cut back to about ½″ before cutting back the understock. Areas with a shorter growing season are summer budded and the rootstock is cut off in late winter or early spring. Between September and December the rose bushes are harvested [*HortScience* 15:817 (1980)]. It is important to use budwood free of verticillium.

BENCH CHIP BUDDING AND MINI-GRAFTING: Rose scions are chip budded onto dormant, 8″ long unrooted R. *multiflora* rootstocks. Parafilm tape is better than other wrapping materials. Budded rose cuttings are stored in plastic bags containing moist sphagnum in the dark at 80°F for 2 weeks, and then field planted. Mini-grafting is similar except the stock is an unrooted cutting with foliage, sliced almost to the middle so a scion can be inserted and tied. The budded cutting is placed in the propagation bench under mist. The cutting roots and the graft heals at the same time in little over one month [*The Plant Propagator* 27(2):11 (1981)].

TREE (STANDARD) ROSE PROPAGATION. R. *multiflora* is used as the root stock, and budded the first summer with the Grifferaie stock or IXL (*Tausendschon/Veilchenblau*). These are trained to a straight form and kept free of suckers. In the second summer, 3 to 4 buds of the desired cultivar are budded at about 3′. During the winter the understock above the inserted buds is removed. The cultivar buds develop the following summer along with sucker buds that must be removed. In the fall, the tree roses are ready for sale.

Miniature tree rose production is similar and rooted cuttings can be used for rootstock production. Sequoia Nursery uses 'Pink Clouds' which is nearly thornless and dark green in color. Cuttings are 16 to 17″ long and de-eyed with two leaves at the top. Cuttings are wounded with a Multi-Rooter tool (four, 1″ vertical slits) to stimulate a balanced root system, treated with 8000 ppm IBA-talc, placed in individual pots and misted. After rooting, they are transferred to 5″ pots to grow on. As soon as good growth is underway, the desired cultivar is budded. When the buds have taken, most of the 'Pink Clouds' top is removed to stimulate scion growth. As each new shoot becomes 2 to 3″, it is pinched to develop branching [PIPPS 30:54 (1980)].

TISSUE CULTURE: Rose species and cultivars can be readily propagated by tissue culture. Actively growing shoot tips of *Rosa* 'Improved Blaze' proliferated in vitro on Murashige and Skoog salts containing in mg/liter: thiamine 0.5, pyridoxine 0.5, nicotine 0.5, glycine 2.0, i-inositol 1000, sucrose 30,000, agar 8000, IAA 0.3 and BA 1.0, 3.0 or 10.0. Explants were surface-sterilized in 0.5% sodium hypochlorite containing a few drops of detergent for 15 minutes and rinsed twice in sterilized water. Cultures were maintained at 77±2°F and light was provided by Cool White fluorescent tubes at 4.4 klx for 16 hours per day. An average of 6 shoot tips was obtained after a 4-week culture period. Microcuttings were rooted in vitro with low MS salts + NAA 0.03 or 10.0 mg, or IAA 1.0 mg. A 90 to 100% establishment was obtained [*HortScience* 14:610 (1979)].

Khosh-khui and Sink observed variation in growth regulator requirements among species and cultivars. For multiplication, 2.0 mg/liter of BA was optimal for hybrid roses plus 0.05 and 0.10 mg/liter of NAA for 'Tropicana' and 'Bridal Pink.' Old world species required lower BA (1.0 mg/liter), and 0.10 NAA for R. *damascena* and 0.15 mg/liter for R. *canina* [J. *Hort. Sci.* 57:315 (1982)].

Rosmarinus officinalis Rosemary

SEED: No pretreatment required.

CUTTINGS: Easy to root as long as medium is well drained and mist is monitored. Excessive moisture in the rooting bench leads to decline. Cuttings should be firm, 2 to 3″ long, bottom one-half of the leaves removed, 1000 ppm IBA-talc or K-IBA solution, sand or well drained medium, mist, or open bench in fall and winter months.

Ruscus aculeatus Butcher's-broom

CUTTINGS: Butcher's-broom can be propagated in small numbers by division. Dicing the rhizomes into small pieces works. The rhizome is cut into 1 to 1½″ pieces, insuring that each piece has an active green leaf stalk and a prominant dormant bud. The cut surfaces are dusted with Captan and boxed up in peat and placed in a cold frame during February in England [PIPPS 21:257 (1971)].

Salix Willow

Willows comprise a large genus (200 species) complicated by their freedom of hybridization. The amenity value of willows is great. Many offer striking winter bark colors of yellow, orange, red, chestnut brown, purple or black; catkins are often colorful; foliage of several species is silvery or glaucous; and lastly the growth habits are unusual. See PIPPS 28:235 (1978). The propagation of all species is easy and similar. For the propagation of S. × *babylonica*, S. *elaeagnos*, S. × *elegantissima*, S. *gracilistyla*, S. *matsudana*, S. *melanostachys*, S. *pentandra*, S. *purpurea*, and S. *sachalinensis*, see below.

SEED: The seeds are borne in a dehiscent follicle and must be collected when the fruit changes from green to yellowish. Most willow fruits ripen in spring. Seed is viable for only a few days and the maximum storage period is 4 to 6 weeks, with germination rates dropping off fast after 10 days at room temperature. Moistened seed may be stored up to a month if refrigerated and sealed. Willow seeds have no dormancy and germinate within 12 to 24 hours after falling on moist ground. Seed dormancy has not been found in any species. In nursery practice seed is sown immediately after collection. Seed beds must be kept moist until seedlings are well established. Shading is often applied.

CUTTINGS: Most willows are readily propagated by softwood or hardwood cuttings, as they have preformed root initials. Softwoods, peat:perlite, under mist, root in high percentages.

Hardwood cuttings can be collected and prepared for insertion at any time after leaf fall, provided they are well ripened, from November through March. In practice there is no advantage from early insertion. Cuttings, 7 to 10″ long and ½ to 1″ thick, are initially stuck close and dug after one year. Even very large cuttings will root. Willows have a rooting percentage of 90 to 100% and root number is not promoted by rooting hormones [*The Plant Propagator* 15(3):10 (1969); PIPPS 20:342 (1970) and 34:281 (1984)].

Considerable research has been conducted on the "willow rooting cofactor" that is present in willow stems. When extracted, it has facilitated good rooting in other species. See PIPPS 35:509-518. Unfortunately, it does not perform on a consistent basis.

GRAFTING: Cultivars, such as S. *caprea* 'Pendula', are winter grafted to produce standards. Weeping forms can be rooted and trained to a single trunk before being allowed to flop over. See *The Plant Propagator* 26(3):5 (1980) and PIPPS 34:287 (1984).

Sambucus Elder

The elders include about 20 species of deciduous shrubs or small trees native to temperate and subtropical regions. Clusters of small white or cream flowers occur in spring or summer. The fruit is a berrylike drupe containing 3 to 5, one-seeded nutlets (seeds). When ripe the fruit varies from red to nearly black.

SEED: Fruits are collected as soon as ripe to reduce losses to birds. Macerate the fruit to remove the pulp and store the seed dry until used. Elder seeds can be stored dry at 41°F for several years. Seeds are difficult to germinate because of their hard seedcoat and dormant embryo. Good germination usually results after 2 to 3 months warm stratification followed by 3 months cold stratification. Many reports indicate that 3 to 6 months cold stratification produces good germination.

CUTTINGS: Elders are easy to root if collected when soft in July and August. Treat with 1000 to 3000 ppm IBA-talc, peat:perlite, sand, or any well drained medium and mist. Rooting takes 5 to 8 weeks. Hardwoods taken from December to March are treated with 4000 to 8000 ppm IBA-talc, sand, bottom heat, and cool air temperatures. Do not use a plastic tent as the buds break too soon and carbohydrates are shunted to the top instead of the root system.

Sambucus canadensis American Elder

SEED: Fruits should be harvested in August-September. A pound of cleaned seed contains approximately 232,000 seeds. Seeds can be stored dry at 41°F for 2 years. Nonstratified seeds do not germinate. Although response varies with seed lot, good germination usually results after 2 months warm (68°F) stratification followed by 3 to 5 months cold stratification (41°F). Heit recommends 10 to 20 minutes in concentrated sulfuric acid, followed by 2 months cold stratification (34 to 41°F). Southern seed material does not normally require acid treatment. In nursery practice the seed can be fall sown after acid treatment or spring sown after acid scarification and cold stratification. Often complete germination does not occur for 2 years. See *Amer. Nurseryman* 125(12):10 (1967) and *The Plant Propagator* 28(3):4 (1982).

CUTTINGS: Softwoods, mid-July, 3000 ppm IBA-talc, rooted 47% in 7 weeks. 'Acutiloba', mid-July, 4000 ppm IBA-talc + thiram, perlite:sand, mist, rooted 100% in 6 weeks.

Hardwoods are collected in October and November from stock hedges and cut to 6″ lengths with the cut made just below the node. The stems should be from one-year-old wood and no thinner than a pencil — the thicker the better. Heel cuttings have also been used to control basal rot. A fungicide, such as Captan, helps to prevent rotting, which is common with pithy cuttings. The prepared cuttings are placed in a suitable medium until planted. Late winter cuttings also will root satisfactorily. In England, the cuttings are stuck in February and March in sandy loam beds. A poorly drained bed will cause the cuttings to rot at the base. The rooting beds commonly contain 4 rows at 12″ with the cuttings 3″ apart. 'Aurea' hardwood cuttings, early February, 1000 ppm IBA-talc, rooted 80% after 15 weeks. 'Chlorocarpa' hardwoods, early March, 3000 ppm IBA-talc, sand, bottom heat, rooted 93%.

GRAFTING: Mature, one-year scion wood can be grafted in late winter or early spring on S. *nigra* and S. *racemosa* understock using a side graft.

Sambucus nigra European Elder

SEED: A pound of cleaned seed contains about 98,000 seeds. A 3 month cold stratification induces fair germination; 2 to 5 months warm stratification followed by 3 to 5 months cold stratification gives good (79%) germination. 'Laciniata', when isolated, comes partially true from seed [*The Plant Propagator* 14(3):18 (1968)].

CUTTINGS: Softwoods are treated with a hormone containing 1000 ppm NAA plus 2% Captan [PIPPS 28:192 (1978)]. Place under mist in a quick draining medium, such as sharp sand, and maintain the medium temperature at approximately 75°F. 'Pyramidalis', late January, 3000 ppm IBA-talc, open bench (cool top/warm bottom), rooted 88%.

GRAFTING: Mature, one-year scion wood can be grafted on S. *nigra* and S. *racemosa* understocks in winter or late spring using a side graft.

Sambucus pubens Red Elder

SEED: Fruits can be harvested before maturity (red) and will germinate. Seeds harvested from ripe fruits benefit from a 10 to 20 minute sulfuric acid treatment followed by 2 to 3 months cold stratification. Heit noted that fair to low germination with this species is associated with improperly developed seeds. Seed sown fresh germinated in low percentages; good germination after 3 months cold stratification. *Sambucus glauca*, blueberry elder, has similar germination requirements [*HortScience* 17:618 (1982)].

CUTTINGS: See S. *canadensis*.

Sambucus racemosa European Red Elder

SEED: Seed sown immediately after cleaning did not germinate; 6 months cold stratification produced excellent germination; and 5 months warm stratification and 3 months cold stratification produced good germination.

CUTTINGS: The authors found no published data for the species but suspect it should root similarly to S. *canadensis*. The cultivar 'Sutherland Gold', early June, 4000 ppm IBA-talc, Captan solution, mist, rooted well. Mist control is extremely important to prevent rotting and should be monitored to prevent excessive moisture in the medium [*The Plant Propagator* 28(1):13 (1982)].

Santolina chamaecyparissus Lavender Cotton
Santolina virens Green Santolina

SEED: Seeds germinate readily without pretreatment.

CUTTINGS: Softwood should be treated with 3000 ppm IBA-talc and semi-hardwoods with 8000 ppm IBA-talc. A rooting medium of peat and perlite (1:1, v/v) is satisfacotry. Apply a fungicide solution and place under mist. Keep the rooting medium on the dry side for best results. Cuttings deteriorate under high moisture regimes and should be removed from mist or transplanted as soon as rooting has taken place.

Sapindus drummondii Western Soapberry

SEED: The rounded, ½″ diameter, yellow-orange drupe ripens in October and persists into late winter. Supposedly, there is no difference in germination of seeds collected from September through March. Extraction of seed from dried fruits is facilitated by soaking the fruit before macerating to remove the pulp. Dry storage of the seed at low temperature is satisfactory. Cleaned seeds per pound range from 700 to 2000.

Germination of seed is highly variable. The major cause is embryo dormancy. Some seed lots also have a hard seed coat while others do not. A sulfuric acid treatment for 1 to 2 hours followed by 3 months at 41°F is recommended. The requirement for an acid pretreatment can be determined by soaking in water for 5 to 7 days and observing if the seed swells. Warm stratification at 70 to 85°F is reported to replace the acid treatment. In Georgia work, 1½ to 2 months cold stratification produced excellent germination (95%). Unstratified seeds germinated 3%. Texas work [*HortScience* 19:712-713 (1984)] noted that seeds collected in late autumn-early winter and provided 60 minutes acid plus 3 months cold germinated 85%; seeds collected in late winter required only 60 minutes acid. Untreated seeds germinated 45%.

CUTTINGS: Cuttings can be rooted when taken in May-June and treated with 1.6% IBA. Stick in peat:perlite (2:1, v/v) under mist. In another study, 1 to 3% IBA increased rooting equally with early June (Kansas) terminal cuttings of mature plants. NAA especially at higher concentrations was detrimental [*Amer. Nurseryman* 164(6):65-73 (1986)].

Sapium japonicum Japanese Tallow Tree

SEED: Sow immediately, no seed dormancy.

Sapium sebiferum Chinese Tallow Tree, Popcorn Tree

SEED: The 3-valved capsules ripen in October-November to expose the white waxy seeds. Seeds persist on the tree into winter. Seeds germinate readily without stratification or wax removal. Fresh seed with the waxy, white coat intact, germinated 90% in 4 weeks. Stratification for 2 months reportedly induces a secondary dormancy.

Sarcococca Sweet-box

Sarcococca comprises 16 species of Asian, evergreen shrubs allied to *Buxus*. *Sarcococca* differs in alternate leaves and female flowers borne below the male flowers. Flowers are produced in late winter or early spring and tend to be white or greenish and very fragrant. Fruit is a ¼″ diameter, rounded, black, purplish or red berry. The pulp should be removed by maceration. Seeds germinate without pretreatment [*The Plant Propagator* 21(2):4 (1975)]. Cuttings from current season's growth root readily in a cold frame or mist bench without bottom heat or hormone. A 3000 ppm IBA-quick dip, however, hastens and unifies rooting.

Sarcococca confusa

SEED: Refer to introduction.

CUTTINGS: This species is as easy to root as S. *hookerana* and roots slightly faster. Flowers may be slightly more fragrant, but are not as cold hardy. Senior author collected cuttings in late February, 1000 ppm IBA, peat:perlite, mist, with 90% rooting in 11 weeks. Rooted cuttings transplant readily and grow off vigorously.

Sarcococca hookerana Himalayan Sarcococca

SEED: Germinate without pretreatment. In one study, fresh seed germinated in high numbers; seeds given 3 months cold stratification did likewise.

CUTTINGS: September through December is the preferred rooting period although cuttings can be rooted successfully when taken through early spring. Growth should be firm. Treat the cuttings with 3000 ppm IBA-talc or quick dip, mist or polytent, peat:perlite (2:1, v/v), and bottom heat at 70°F. Rooting approaches 100%. Senior author collected cuttings of var. *humilis* and var. *digyna* in late February and treated with 1000 ppm IBA-quick dip, peat:perlite and mist with 100% rooting in 11 weeks. No bottom heat was used and the rooting process took longer.

Cuttings placed in a cold frame in October will be rooted by the following May. Select basal cuttings and treat with 8000 ppm IBA-talc. Easily grown from rhizome cuttings in July; set horizontally in a peat moss:perlite medium [PIPPS 24:423 (1974)].

Sarcococca ruscifolia Fragrant Sarcococca

SEED: Refer to introduction.

CUTTINGS: Late April, 4000 ppm IBA-talc + thiram, polytent, rooted 100% with heavy root systems.

Sassafras albidum Common Sassafras

SEED: The ½″ long, oval dark blue drupe ripens in September, quickly abscises or is consumed by birds. The pulp is removed by maceration. A pound of cleaned seed contains approximately 5,800 seeds. Seeds should be stored dry in sealed containers at 35 to 41°F. Seeds exhibit strong embryo dormancy and require 4 months stratification at 41°F. Fall planted seed should be sown as late as possible and spring sown seed should be stratified. The seeds germinate rather late in the spring. Obtaining viable seed may be a problem since the species is dioecious and birds literally ravage the fruits.

CUTTINGS: Root cuttings are a valuable method [PIPPS 11:42 (1961); *Gardeners Chronicle* 169(3):22 (1969)]. Roots are dug as soon as the frost leaves the ground. The ½ to 1″ thick roots are cut into 2 to 3″ lengths. Refer to *Rhus glabra* for details on spring root cuttings and R. *typhina* for details on fall root cuttings. Shoots from root cuttings can also be rooted. Sassafras can also be propagated by layering.

GRAFTING: Selected forms can be chip budded on the species in summer.

Schizophragma hydrangeoides Japanese Hydrangea-vine

SEED: Treat as described for *Hydrangea anomala* subsp. *petiolaris*.

CUTTINGS: Easier than climbing hydrangea. Late June cuttings, 3000 ppm IBA-talc, sand:peat, mist, rooted 56% in 3 months. Senior author had reasonable success (66%) with late July (Boston) cuttings, 4000 ppm IBA-quick dip, peat:perlite and mist. Results have been variable.

Sciadopitys verticillata Umbrella-pine

SEED: Male and female cones are borne on the same tree. Cones ripen the second season and are gray-brown when mature. The ripe cones are harvested and placed in a warm place to open. Number of cleaned seeds per pound averages 17,000. Seeds stored at a moisture content of 10% or less in sealed containers at 41°F will retain good viability for at least 2 years. Umbrella-pine seeds germinate erratically when sown fresh. Either a warm stratification at 63 to 70°F for 3 months or cold stratification for 3 months in moist peat at 32 to 50°F have been recommended. A combination of the two may be required. In one study, 3 months cold stratification produced an even, general germination; seed sown in mid-December without pretreatment germinated over a 7 month period. When sowing umbrella-pine seed, care must be taken to cover no more than ¼". Seedlings are extremely slow to grow-off and it may take 3 years to develop a 1' plant [*Arnoldia* 37(1):71 (1977)]. See *J. Env. Hort.* 4:145-148 (1986) for interesting work with germination and subsequent chilling of the seedlings to induce faster growth.

CUTTINGS: There is a wide range in rooting response among different clones. Waxman has shown that the highest levels of rooting occur in February-March and again in July-August. Cuttings are wounded, treated with 3000 ppm IBA-talc + Captan, and placed under mist in a peat moss:perlite medium (1:1, v/v). The removal of resin by soaking the bases of the cuttings in water for 24 to 48 hours significantly improved all aspects of rooting — percentage, number, and length. There is a significant negative correlation between resin exudate and rooting percentages, with the highest rooting percentages corresponding to the lowest internal levels. Maintain bottom heat at a minimum of 72°F. Rooting occurs in about 6 months (See PIPPS 10:128 and 28:546-550).

Securinega suffruticosa Securinega

SEED: In a controlled study, fresh seed germinated 41%, 67% after 2 months cold stratificaiton, and 86% after 3 months cold stratification.

CUTTINGS: This is a rather graceful shrub with rich grass green foliage and arching branches. Summer softwoods, mist, 8000 ppm IBA-talc, peat:sand, rooted 94 to 100% in 30 days [*The Plant Propagator* 17(4):3 (1971)]. If a few plants are needed, simple division would work.

Sequoia sempervirens Coast Redwood

SEED: Male and female cones are produced on the same tree. Cones mature at the end of the first season and seed is collected when cones turn greenish yellow in September and October. Cones dry in 10 to 14 days at 70 to 75°F or in 24 hours at 120°F. Seeds are separated by screening. Cleaned seeds average about 120,000 per pound with a soundness of about 23%. Seed is relatively short lived even under ideal storage conditions. Viability is maintained best at subfreezing temperatures (26 to 30°F) in sealed containers for 1 year. Redwood seed requires no pretreatment for germination. In nursery practice the seed is spring sown when the soil is warm and frost is unlikely. Seedling plants are variable in growth habit. When germinated under laboratory conditions seed requires artificial light and 68 to 86°F alternating temperatures.

CUTTINGS: Clones are variable in rooting ability with some rooting 90%: use terminal or lateral cuttings in March, 8000 ppm IBA, mist, peat:perlite and bottom heat. Cuttings may callus rapidly and this needs to be rubbed off. Stake the strongest shoot and trim back the others to develop a leader. In one study, 'Majestic Beauty'-3000 ppm IBA + 3000 ppm NAA rooted 65%; 'Santa Cruz'-16,000 ppm IBA-talc rooted 70%; and 'Soquel'-6000 ppm IBA + 6000 ppm NAA rooted 47%.

GRAFTING: Cultivars are grafted on species understock in winter using a whip and tongue graft.

Sequoiadendron giganteum Big Tree

SEED: Male and female flowers occur on the same tree. By late August of the second year following pollination the embryos are mature. Cones have been air dried at 85°F for 7 days and the seed extracted in a screened tumbler. A pound of cleaned seed averages 81,000 seeds and the percentage of properly filled seed is often low. Seed is stored at 0°F in polybags with good, long-term survival. An overnight soak in water followed by 2 months cold stratification produces good germination. In nursery practice the seeds are spring sown in fumigated beds after stratification.

CUTTINGS: Difficult, but possible with some clones of trees 20 to 100 years old. Use winter cuttings, 2000 to 4000 ppm IBA-quick dip, organic medium. Cuttings from young trees root in satisfactory numbers.

GRAFTING: Cultivars are winter grafted on the species using a whip and tongue graft.

Shepherdia argentea Silver Buffaloberry

SEED: The fruit is a drupe-like achene enclosed in a fleshy covering that is yellow to red at maturity. Fruit ripening occurs in June to December and a pound of cleaned seed averages about 40,000 seeds. Seeds with a moisture content of 13.1% germinated 97% after 3½ years in storage at 42°F. Seeds have a dormant embryo and germination is increased by cold moist stratification at 41°F for 2 to 3 months. Seed sown fresh germinated 10%; 36% after 2 to 3 months cold stratification. Ethrel at 50 ppm reportedly improves germination of non-cold treated seeds. Seeds should not be allowed to dry out during shipping or viability can be lost.

CUTTINGS: Silver buffaloberry can be propagated by digging up root sprouts. Root cuttings might work. Senior author had no success rooting this species. It appeared excess moisture in the rooting bed resulted in premature leaf drop and cutting death. Yellow fruited form was rooted 100% from late July cuttings, 8000 ppm IBA-talc + thiram, sand:perlite, mist.

Shepherdia canadensis Russet Buffaloberry

SEED: The yellowish red fruits mature in June and July. There are approximately 9,100 seeds per pound. Seeds stored dry at 41°F retain good viability for at least 3½ years. Seeds have a hard seed coat and scarification with sulfuric acid for 20 to 30 minutes followed by 2 to 3 months cold stratification produces good germination [*Amer. Nurseryman* May 15, p. 10 (1967)]. In nursery practice, cleaned seeds are fall planted without scarification.

CUTTINGS: July cuttings treated with 8000 ppm IBA-talc and mist rooted well. Rooting is not always consistent.

Sibiraea laevigata Altai Spirea

SEED: In late August to early September the yellow fruits turn brown and persist on the plant through most of the winter. Seeds germinate without any pretreatment. Store the seeds cool and dry until planting in mid-March [PIPPS 33:484 (1983)].

CUTTINGS: Early June, rooted 50% with 2000 ppm IBA-quick dip and mist.

Sinowilsonia henryi Henry Wilson Tree

SEED: Uniform germination after 3 months cold stratification.

CUTTINGS: Have been rooted in low percentages.

Skimmia japonica Japanese Skimmia

SEED: Evergreen species is dioecious (some perfect flowered selections), and the 3/8" diameter red drupe ripens in September and may persist into summer of the following year. Clean seeds of pulp and sow immediately. Seeds germinate like green beans. It was reported that seed coats cracked and germination had started in uncleaned fruits.

CUTTINGS: Can be rooted successfully from July to the following spring as long as the wood is firm. Rooting takes place in about 4 weeks. Synthetic hormones have been reported to inhibit root development, while other reports recommended the use of 3000 to 8000 ppm IBA-talc. Both authors have rooted the species using 1000 to 3000 ppm IBA-quick dip and 8000 ppm IBA-talc or solution, peat:perlite, and mist. Rooting averages 90% or greater. Rooted cuttings transplant readily and grow off well.

Skimmia reevesiana Reeves Skimmia

SEED: No pretreatment necessary. Senior author collected fruits from a large planting at Swarthmore College, removed pulp, sowed immediately and within 4 weeks germination occurred. Seedlings grew slowly. This is a perfect flowered species and every plant produces fruits. The rich crimson fruits persist into winter.

CUTTINGS: Easy to root; follow procedure described under S. *japonica*.

Sophora japonica Japanese Pagodatree

SEED: The fruit is a 2 to 4" long, 1 to 6-seeded pod that changes to yellow-brown when ripe — late August in Georgia to October in Pennsylvania. Cleaning the seed is difficult and messy because the fruit wall is sticky and gummy. Soak the fruit in warm water for 24 to 48 hours before cleaning. Cleaned seed can be kept at room temperature until planted. Seeds germinate readily without pretreatment. Dried seed may require a hot water soak or slight scarification for best germination. The best test is the finger nail method. If seed coat can be broken by the nail, no scarification is necessary. Some nurseries plant directly in the fall while others spring plant. Seedlings are very irregular and require staking if a straight trunk is desired.

CUTTINGS: Japanese pagodatree produces very poor, sparse root systems from cuttings. Hardwoods in the greenhouse or softwoods under mist root, but the root system after field planting is sparse and the plants must be root pruned. The Garner Bin method is used with hardwoods [PIPPS 32:569 (1982)]. No doubt, juvenility is extremely important and cuttings from young plants will root better.

GRAFTING: Budding — one year, pencil-size seedlings are lined out in the spring and T-budded in August. The following spring the understocks are cut back to the bud and the new growth staked. Grafting on S. *japonica* seedlings in winter or late spring using a side graft is successful. Scion wood is collected from the current season's growth (1-year-old wood) of the desirable cultivar.

Sophora microphylla Littleleaf Sophora

CUTTINGS: 'Earlygold', 'Goldie's Mantle' and 'Goldilocks' rooted 100% from semi-hardwood cuttings in winter using 8000 ppm IBA-talc, mist, "Fibremix" bark:pumice (1:1, v/v), bottom heat (68°F) and mist. The cuttings and trays were drenched with a 1 g/liter thiram solution after sticking and weekly thereafter.

Sorbaria sorbifolia Ural Falsespirea

(Also applies to S. *aitchisoni*, Kashmir Falsespirea; S. *arborea*, Tree Falsespirea; and S. *grandiflora*, Showy Falsespirea.

SEED: The fruit is a dehiscent follicle that ripens in August and later. Fruits are rubbed to break them up and then fanned to separate the seed. A pound of seed contains approximately 750,000 seeds. The seed can be stored dry in sealed containers at 41°F. Fresh seed will germinate without pretreatment. Some of the seeds may have an internal dormancy and, if a germination test shows dormancy, cold stratify for 1 to 2 months.

CUTTINGS: Easy to root from softwood, greenwood and hardwood cuttings in a sand medium under mist. Senior author used summer softwoods when the stems had firmed and a slight reddish green color developed, 1000 ppm IBA-solution, peat:perlite, mist, with 90% rooting in 3 to 4 weeks. Plants can also be divided in late winter or early spring if a few plants are desired. Suckering is common to all species but S. *sorbifolia* is the worst. Root cuttings will work.

Sorbus Mountainash

The mountainashes include more than 80 species of deciduous shrubs and trees distributed through the northern hemisphere. The flowers are white and borne in large, flattened corymbs from April until July depending on species and location. Fruits are 2 to 3-celled, berry-like pomes that are orange-red to bright red at maturity, and ripen from August until October. Although fruits may remain on plants until late winter, they should be harvested as soon as ripe to prevent losses to birds. A seed crop is usually produced every year. Seeds are extracted by maceration and floating off the pulp and empty seeds. Cleaned seeds have been stored successfully under cold, dry conditions for 2 to 8 years. For best results, store in sealed containers at 6 to 8% moisture content and temperatures of 34 to 38°F. Dry seeds may develop a hard seed coat [*Gardeners Chronicle* 170(27):24 (1971) and PIPPS 23:262 (1973)].

SEED: Species hybridize freely, and it is unreliable to grow any species from seed unless the tree is sufficiently isolated. Seeds of most *Sorbus* species have only dormant embryos. Those *Sorbus* species that require 3 months cold stratification include: × *arnoldiana*, *aucuparia*, *bakonyensis*, *cashmiriana*, *chamaemespilus*, *commixta*, *discolor*, *esserteauiana*, *folgneri*, *hupehensis*, *intermedia*, *pluripinnata*, *pohuashanensis*, *prattii*, × *pseudovertesensis*, *randiensis*, *reflexipetala*, *rehderiana*, *rufo-ferruginea*, *sargentiana*, *scopulina*, *sibirica*, *subsimilis*, *tianshanica* and *wilsoniana*. Seeds of S. *alnifolia* are variable and may be doubly dormant which would necessitate warm and cold stratification.

GRAFTING: *Sorbus* species can be divided into three groups: 1) the mountainashes (Aucuparia section) with pinnate leaves; 2) the whitebeams (Aria section) with simple or lobed leaves; and 3) Micromeles section. Well developed buds are taken in July or early August and T-budded as close to the ground on S. *aucuparia* for Aucuparia and Micromeles sections, or S. *aria* for whitebeams. *Sorbus intermedia* is often used as a rootstock for whitebeams, however, it does not produce a consistent and acceptable crop with 'Wilfrid Fox' and 'Mitchellii' and S. *aria* must be used with those cultivars. S. *latifolia* is sometimes used for whitebeams, but it is much less satisfactory [*The Plant Propagator* 25(4):12 (1979)].

Chip budding is also used. Cut the ties as soon as possible to prevent any constriction of the stem.

Bench grafting using the whip and tongue method is carried out on bareroot or potted understocks in winter (February).

SUITABLE ROOTSTOCK/SCION COMBINATIONS WITHIN THE GENUS SORBUS [*Gardeners Chronicle* 170(27):24 (1971)].

S. *aucuparia* can serve as a suitable rootstock for Aucuparia section: *americana, aucuparia* cultivars, Lombarts Hybrids, *cashmeriana, discolor, esserteauiana, hupehensis,* 'Joseph Rock', *matsumurana, pluripinnata, pohuashanensis, poteriifolia, prattii, rehderiana, rufo-ferruginea, sargentiana, scopulina,* × *thuringiaca, vilmorinii.* Micromeles section: *alnifolia, caloneura, folgneri, japonica, keissleri, meliosmifolia.* S. *aria* can serve as a rootstock for: *aria* cultivars, *cuspidata (vestita), hybrida, intermedia,* × *magnifica,* 'Mitchellii', *thomsonii,* 'Wilfrid Fox'. S. *intermedia* can serve as a roostock for: *aria* 'Lutescens', 'Magnifica', *cuspidata (vestita),* × *magnifica, mongeotii.* S. *latifolia* can serve as a rootstock for *aria* 'Lutescens', 'Magnifica', *domestica, torminalis.*

Sorbus alnifolia Korean Mountainash

SEED: In nursery practice, Korean mountainash is seed propagated. The fruit ripens in September-October and is pinkish red to scarlet. Korean mountainash is easily grown from seed which is collected in the autumn, cleaned of fleshy pulp and placed in outdoor seedbeds for germination the following spring. Seed germinates best when provided 3 to 4 months cold stratification. Seed from some trees is doubly dormant and requires warm stratification for 5 to 6 months. Best information comes from the Arnold Arboretum where in extensive tests, 3 to 4 months warm stratification and 3 months cold stratification produced the best germination (60%) compared to only 16% with 3 months cold stratification. See *Arnoldia* 14(5):25 (1954) and 36(5):159 (1978); also PIPPS 30:432 (1980).

CUTTINGS: Senior author has rooted a few cuttings. Early May, wound, 10,000 ppm IBA-talc, gave 20% rooting in 7 weeks. Late July, 4000 or 8000 ppm IBA-talc + thiram and perlite:sand produced 33 and 47% rooting, respectively. Root systems were rather fleshy.

Sorbus americana American Mountainash

SEED: The fruit is orange-red and ripens in August to October. Seeds dried to 6 to 8% moisture can be stored in sealed containers at 34 to 38°F for several years. A pound of cleaned seed contains approximately 160,000 seeds. 3 to 5 months warm stratification followed by 3 months cold stratification has produced good germination. In nursery practice if the seeds are sown in late fall they may not germinate well until the second spring; if sown in the summer they may germinate well the next spring [PIPPS 17:99 (1967)].

Sorbus aria Whitebeam Mountainash

SEED: The orange-red or scarlet fruits ripen September through October. A pound of cleaned seed contains about 16,500 seeds. Seeds can be stored under cool dry conditions without loss of viability. Whitebeam mountainash seed has a dormant embryo and requires 4 months cold stratification before sowing. It can be fall planted [PIPPS 29:205 (1979)]. In one study, 3 months warm stratificaiton and 3 months cold stratification produced good germination.

GRAFTING: The cultivars 'Lutescens' and 'Magnifica' can be grafted onto S. *intermedia* using a whip graft in February. The rootstocks are root- and stem-pruned as necessary before grafting. Following grafting the complete scion and union are dipped into liquid wax. The bare-root whip grafts are plunged into moist peat and covered with plastic. Maintain the air temperature as close to 45°F as possible. 'Lutescens' also reportedly can be grafted on S. *americana.* In England, S. *aria* was reported to grow better on the native thorn, *Crataegus laevigata,* than on S. *aria* seedlings. Use a side graft with one year mature scion wood. Summer budding of S. *aria* seedling understock using a T-bud is also possible with the cultivars. See PIPPS 17:389 (1967); 18:343 (1968); 26:145 (1976).

Sorbus aucuparia European Mountainash

SEED: The orange-red fruit ripens in late August and September. A pound of cleaned seed averages 125,000 seeds. Store in sealed containers at 6 to 8% moisture and temperatures of 34 to 38°F. Embryo dormancy occurs and germination of seeds is increased by cold stratification for 2 to 4 months. In one study, freshly cleaned seed sown at once failed to germinate; excellent germination occurred after 3 months cold stratification. With long cold stratification periods, germination (radicle emergence) often occurs during stratification. Dry seeds of this species may develop a hard seedcoat which can affect germination. Seeds from 'Pendula' reportedly produce 60% normal and 40% pendulous types.

CUTTINGS: Mid-May, 3000 ppm IBA-talc, sand:perlite, mist, rooted 65% in 12 weeks.

GRAFTING: *Sorbus aucuparia* is used at 6' for standards with 'Pendula'. *Sorbus intermedia* can also be used for 'Pendula'. 'Beissneri' is maintained by grafting onto the species. There are numerous cultivars of this species and all can be grafted or budded successfully on seedling understock.

Sorbus cashmiriana

SEED: 3 months cold stratification gave good germination with some germination during cold stratification.

Sorbus commixta

SEED: A cold stratification period of 3 months produced good germination.

Sorbus decora Showy Mountainash

SEED: Sulfuric acid for 20 minutes plus 3 months cold stratification produces good germination.

Sorbus discolor Snowberry Mountainash

SEED: Seed sown fresh failed to germinate; 3 months cold stratification produced very good germination with many germinated in the cold.

Sorbus dumosa Arizona Mountainash

SEED: Sown fresh failed to germinate; 3 months cold stratification produced good germination.

Sorbus esserteauiana

SEED: Cold stratification for 3 months produced very heavy germination.

Sorbus folgneri Folgner Mountainash

SEED: Cold stratification for 3 months produced good germination.

Sorbus intermedia Swedish Mountainash

SEED: Seed requires 3 months cold stratification.

CUTTINGS: Hardwoods (18″ long and as thick as possible), 2500 ppm IBA-quick dip, Garner Bin set a 70°F basal temperature for 2 weeks and then allowed to cool for 1 week; next placed 15″ deep in north facing frames containing peat:perlite [*The Plant Propagator* 26(1):6 (1980)].

Sorbus japonica Japanese Mountainash

SEED: Seeds have a double dormancy and require 3 months at 60 to 85°F followed by 3 months at 41°F [*Arnoldia* 14(5):25 (1954)].

Sorbus matsumurana Matsumura Mountainash

SEED: Seed sown fresh failed to germinate, however, 3 months cold stratification produced 75% germination.

Sorbus sargentiana Sargent Mountainash

SEED: A 5 month warm stratification followed by a 3 month cold stratification produced good germination.

Sorbus scopulina Greenes Mountainash

SEED: Cold stratification for 3 months produced good germination.

Sorbus sibirica Sibirian Mountainash

SEED: Cold stratification for 3 months produced good germination.

Sorbus tianshanica

SEED: Cold stratification for 3 months produced good germination.

Spiraea Spirea

A larges genus of spring or summer flowering shrubs with attractive white to rose and carmine flowers and delicate, clean foliage. Important species and cultivars include: S. *albiflora*, S. *arguta*, S. × *billiardii*, S. *bullata*, S. × *bumalda*, S. *cantoniensis* 'Lanceata', S. × *cinerea* 'Grefsheim', S. *japonica*, S. *nipponica* 'Snowmound', S. *prunifolia*, S. *thunbergii*, S. *trilobata*, and S. × *vanhouttei*.

SEED: The fruit is a dehiscent follicle and should be collected when it turns brown, air dried, and seeds removed by shaking. Fresh seed requires no pretreatment. Seeds, if allowed to dry, may require 1 to 2 months cold stratification.

CUTTINGS: Easily rooted anytime the plants are in leaf. Generally, the softer the wood the easier and faster the cuttings root. Senior author has rooted cuttings of many species from May to early September (Georgia). 1000 ppm IBA-talc or solution improves, hastens and unifies rooting. Any well drained medium like peat:perlite is suitable. Cuttings root in 2 to 4 weeks and can be transplanted immediately. An early rooted cutting will fill a one-gallon container by the end of the growing season.

Hardwood cuttings of most species except S. *thunbergii* can be rooted [*The Plant Propagator* 15(3):10 (1969); PIPPS 34:540 (1984)]. Take cuttings after leaf drop in the fall, place in cold storage, and plant out in early spring. Cut down one year later to develop bushy plants. Rooting should range from 80 to 100%.

TISSUE CULTURE: The micropropagation of S. × *bumalda* 'Froebelii' could serve as a model for other spirea species. Explants were selected from field-grown plants in active growth in May, surface sterilized in 0.5% sodium hypochlorite for 15 minutes, and rinsed in sterile distilled water. Linsmaier and Skoog medium was gelled with Difco Bacto agar. BA at 0.5 mg per liter was the best cytokinin rate. Temperature was maintained at 77±4°F and light was provided at 4.4 klx for 16 hours per day [J. *Hort. Science* 61:43 (1986); *HortScience* 17:190 (1982)].

The following information relative to seed and cutting propagation of spirea comes from the Arnold Arboretum.

Spiraea aemiliana

SEED: Good germination from fresh sown seed.

Spiraea alba Narrowleaf Meadowsweet

SEED: No pretreatment necessary; good germination upon sowing.

Spiraea albiflora Japanese White Spirea

SEED: No pretreatment necessary.

Spiraea amoena

SEED: No pretreatment necessary.

Spiraea arcuata

CUTTINGS: Early August, 3000 ppm IBA-talc, 100% rooting with profuse roots.

Spiraea arguta 'Compacta' Compact Garland Spirea

CUTTINGS: Mid June, 3000 ppm IBA-talc, sandy soil, outdoor frame with bottom heat, rooted 72% and were left to overwinter in place. Late July, 3000 ppm IBA-talc, 44% rooting.

Many of the fine-stemmed spireas are more difficult to root than the coarser stem types. They tend to rot and die during the rooting process. They are also subject to rapid desiccation.

Spiraea betulifolia Birchleaf Spirea

CUTTINGS: In outdoor frame as described above rooted 100%. Early August, 3000 ppm IBA-talc, sand, rooted 77%.

Spiraea × *billardii* Billiard Spirea

CUTTINGS: Late May, 3000 ppm IBA-talc, 40% rooting.

Spiraea bullata Crispleaf Spirea

CUTTINGS: Softwoods root well but remove from mist as soon as rooted. Mid-December hardwoods, 3000 ppm IBA-talc, open bench, rooted 47% in 11 weeks.

Spiraea × bumalda Bumald Spirea

A worthy group of summer flowering spireas with refined habit and environmental toughness. All flower on new growth of the season. In general, collect cuttings in May-June, 1000 ppm IBA-solution, peat:perlite, mist, with 80 to 100% rooting in 3 weeks. Transplant immediately, fertilize lightly and plants will continue to grow. Do not take cuttings with flower buds.

'Anthony Waterer' — Late July, 4000 ppm IBA-talc + thiram, mist, 100% rooting.

'Crispa' — Late July, 3000 ppm IBA-talc, sand, mist, 100% rooting; Early August, 3000 ppm IBA-talc, sand, mist, 73% rooting.

'Froebelii' — Late November, 3000 ppm IBA-talc, sand, 32% rooting.

'Goldflame' — Late March, 4000 ppm IBA-talc + thiram, polytent, 10% rooting; same treatment except open bench, 100% rooting.

'Norman' — Early February, 3000 ppm IBA-talc, sand:perlite, open bench, 50% rooting.

Spiraea cantoniensis Reeves Spirea

CUTTINGS: Mid-April, 3000 ppm IBA-talc, 33% rooting. Senior author has rooted 'Lanceata', a double flowered form, many times. The wood should be firm and green. As the stems mature they turn brown which is a sign that their peak rooting window is past. In Georgia, late May, June, early July are best. Cuttings taken in mid-August, 1.0% K-IBA-solution, rooted 70%.

Spiraea chamaedryfolia Germander Spirea

SEED: No pretreatment required. Fresh seed germinated well.

Spiraea densiflora Subalpine Spirea

SEED: No pretreatment required.

Spiraea × foxii Fox Spirea

SEED: No pretreatment required.

Spiraea hypericifolia

SEED: Excellent germination when sown immediately.

Spiraea japonica var. *acuminata* Taperleaf Japanese Spirea

SEED: No pretreatment required.

Spiraea japonica var. *alpina* Alpine Japanese Spirea

CUTTINGS: Early December, 3000 ppm IBA-talc, rooted 83%. 'Little Princess' is included here. In general, the cultivars of S. *japonica* should be handled like those of S. *× bumalda*.

'Atrosanguinea' — Late July, 4000 ppm IBA-talc + thiram, 100% rooting; late January, 4000 ppm IBA-talc + thiram, polytent or open bench, rooted 100%.

'Coccinea' — Late July, 4000 ppm IBA-talc + thiram, 91% rooting.

'Fortunei' — Late December, hardwoods, 8000 ppm IBA-talc + thiram, open bench, 66% rooting.

'Ruberrima' — Late July, 8000 ppm IBA-talc + thiram, mist, 100% rooting.

Spiraea × lemoinei Lemoine Spirea

CUTTINGS: Hardwood, early January, polytent and open bench, 4000 ppm IBA-talc + thiram, rooted 66 and 100%, respectively.

Spiraea menziesii Menzies Spirea

CUTTINGS: 'Triumphans', hardwood, early December, 3000 ppm IBA-talc, 92% rooting.

Spiraea miyabei var. *glabrata*

SEED: No pretreatment required.

Spiraea × multiflora Snowgarland Spirea

CUTTINGS: 'Compacta', hardwood, mid January, 4000 ppm IBA-talc + thiram, open bench, 72% rooting with excellent root systems.

Spiraea nipponica Nippon Spirea

SEED: Variety *tosaensis*, heavy germination when seed is sown fresh.

CUTTINGS: 'Rotundifolia', early August, 4000 ppm IBA-talc + thiram, 17% rooting. 'Snowmound' is a selection of S. *nipponica* and is easily rooted from June-July softwoods. Follow procedure under S. *cantoniensis*.

Spiraea pubescens

SEED: No pretreatment required.

Spiraea splendens

SEED: No pretreatment required.

Spiraea syringaeflora Lilac Spirea

CUTTINGS: Mid-September, soft tip cuttings, 3000 ppm IBA-talc, 90% rooting.

Spiraea ussuriensis

SEED: No pretreatment required.

Spiraea virgata

SEED: No pretreatment required.

Staphylea Bladdernut

A genus of about 10 large shrubby species indigenous to the Northern Hemisphere. The white or white-tinged pink flowers occur in panicles. Fruits are 1 to 4" long inflated capsules. The capsules should be collected when brown and the seeds extracted by threshing. Do not allow the seed to dry out before treatment. The seeds are extremely hard and require acid scarification or warm stratification to break down the seed coat and permit water imbibition. Seeds germinate readily if given 5 months warm stratification followed by 3 months cold stratification. Seeds should not be allowed to dry or viability will be lost [*Arnoldia* 40(2):26 (1980)].

Cuttings have been rooted in July using 8000 ppm IBA-talc, well drained medium and mist. Species such as *bolanderi*, *bumalda*, *×*

coulombieri, emodii, holocarpa, pinnata, and *trifolia* can be propagated by division, preferably in the spring.

Root cuttings also work. Senior author has moved S. *trifolia* several times and was always rewarded with new plants that developed from root pieces.

Staphylea bumalda Bumalda Bladdernut

SEED: A rather interesting study showed that seeds given a hot water soak did not germinate; hot water plus 3 months cold stratification yielded 0% germination; 1½ hours acid scarification produced 28% germination; 1½ hours acid plus 2 months cold stratification, 25% germination; and 1½ hours acid plus 3 months cold stratification, 72% germination.

Staphylea colchica Colchis Bladdernut

SEED: The seed is doubly dormant and requires 4 months warm stratification followed by 3 months cold stratification. No doubt acid treatment followed by cold stratification would work. See PIPPS 12:57 (1962).

CUTTINGS: July or any time the wood is firm, 8000 ppm IBA-talc, under mist. Might be best not to disturb cuttings after rooting.

Staphylea holocarpa Chinese Bladdernut

SEED: Refer to *Staphylea* introduction.

CUTTINGS: 'Rosea', early July, rooted 89% in 6 weeks with 3000 ppm IBA-talc, sand:perlite, mist, and bottom heat.

Staphylea pinnata European Bladdernut

SEED: Untreated seeds do not germinate; 5 months warm stratification and 3 months cold stratification produces reasonable germination.

CUTTINGS: Refer to *Staphylea* introduction.

Staphylea trifolia American Bladdernut

SEED: The fruit is an inflated capsule that changes to light brown when ripe. Each capsule contains several yellowish brown seeds. The seed is doubly dormant and requires 4 months warm stratification followed by 3 months of cold stratification. Interestingly, 3 months cold stratification produced no germination.

CUTTINGS: Cuttings rooted most successfully when taken in late July, treated with 8000 ppm IBA, and placed under mist. Division works quite well. Senior author divided plants in late winter with excellent success. Remaining root pieces produced shoots.

Stephanandra incisa Cutleaf Stephanandra

SEED: The fruit is a follicle. Seeds appear to require a warm/cold regime or perhaps light acid scarification (15 minutes) followed by 3 months cold stratification. Seeds sown immediately did not germinate but given 15 minutes acid scarification plus 3 months cold stratification germinated during the cold period; 3 months warm stratification followed by 3 months cold stratification produced 89% germination.

CUTTINGS: One of the easiest shrubs to root. A nurseryman was on a plant hunting expedition and collected cuttings of 'Crispa'. They were placed in plastic bags with moist paper towels and by the time he returned home they had rooted in the bags. Softwoods from May to August root well when treated with 1000 ppm IBA and placed under mist. Cuttings will continue to grow after transplanting. See PIPPS 33:489 (1983).

Stephanandra tanakae Tanaka Stephanandra

SEED: Germination requirements are similar to S. *incisa*. Seed sown fresh, 0% germination; 15 minutes acid plus 3 months cold stratification, 41% germination; 3 warm/3 cold or 5 warm/5 cold produced 80 and 82% germination, respectively.

CUTTINGS: Refer to S. *incisa*.

Stewartia Stewartia

A small genus of deciduous or evergreen shrubs and trees native to southeastern United States and eastern Asia with lovely white flowers, rich autumn color and ornamental bark.

SEED: The flowers are generally solitary with 5 petals (rarely more) and numerous stamens. *Stewartia* seeds are produced within 5-valved dehiscent woody capsules and each chamber, depending on species, contains potentially 2 to 4 seeds. Seed dispersal is by wind. About mid-September the capsules change from green to brown and, at this stage, the seeds are mature and the capsules can be collected. The capsules are allowed to dry and the seeds will separate from the capsules with shaking. Difficulty in separating the seeds from the capsules may be experienced only with S. *malacodendron* because the seeds are angular. Stored seed loses its viability quickly and should be sown or placed in pretreatment immediately. The taxa: S. *koreana*, S. *malacodendron*, S. *monadelpha*, S. *ovata* var. *grandiflora*, S. *pseudocamellia*, S. *pseudocamellia* 'Korean Splendor', S. *rostrata*, S. *serrata*, and S. *sinensis*, have doubly dormant seeds. Warm stratification for 4 months (3 to 5 months) followed by cold stratification for 3 months prepares the seed for germination. If the warm stratification is started in October the seeds are ready for sowing in May. Even with those manipulations, germination is seldom uniform. Senior author has worked extensively with *Stewartia monadelpha* and seeds given 5 months warm/3 months cold germinated over a 5 month period. Some seeds of S. *monadelpha* provided the above treatment have germinated during the cold stratification period.

Seeds can be sown as soon as collected in fall and will germinate in two years. For a commercial nurseryman this may be the best approach. The seed of *Stewartia* is extremely expensive and averages $49.00 per ounce. On the flip side, *Stewartia* seedlings sell for $1.25 each while common denominator plants sell for $0.25 to $0.50 each. More skill and patience are required to grown *Stewartia* seedlings but the payoffs are greater.

The capsules dehisce naturally but are often borne upright along the stems and usually contain numerous seeds as late as December. Senior author collected abundant seeds of *Stewartia monadelpha* as late as mid-December.

Stewartia species appear to be self-fertile for seed collected from an isolated plant germinated in high percentages. The most ornamental barked species include S. *koreana*, S. *monadelpha* and S. *pseudocamellia*. See *Arnoldia* 35(4):165 (1975); PIPPS 18:325 (1968) and 32:476 (1982).

CUTTINGS: At the Arnold Arboretum the best time to select cuttings is between late June and early July. Rooting conditions include: 8000 ppm IBA + 15% thiram or IBA + NAA (each at

2500 ppm) quick dip, sand:perlite (1:1, v/v) and mist. The main problem is survival during the winter. Do not transplant after rooting. After rooting and hardening off, transfer to cold storage at 34°F in November. In February or March the plants are returned to the greenhouse and when new growth starts they are transferred to containers. Taxa that have been propagated by the above procedure include: S. *henryae*, S. *malacodendron*, S. *monadelpha*, S. *ovata*, S. *ovata* var. *grandiflora*, S. *pseudocamellia*, S. *pseudocamellia* 'Korean Splendor', S. *rostrata*, S. *serrata*, and S. *sinensis*. Getting a flush of growth on the cuttings improves survival. The use of extended photoperiod (up to 18 hours) may induce a flush of growth after rooting.

Senior author has conducted numerous experiments with S. *monadelpha* to determine the best rooting and overwintering procedures. Rooting is relatively easy with June cuttings (Georgia), wood firm at base but still growing at apex (later firm wood cuttings also root), 5000 ppm to 1.0% IBA-solution, peat:perlite, direct stuck in pots or cells, mist, with 80% or greater rooting in 8 weeks. Unfortunately, a single rooted cutting has never been overwintered. Root growth is excellent and cuttings appear robust going into storage but generally bark split or simply die. There is considerable research that needs to be conducted with this genus. During the overwintering period, the medium should be kept as dry as possible; never overwater.

Layering is practiced with S. *monadelpha* and could be extended to other species. Unfortunately, this procedure supplies limited numbers of plants.

Stranvaesia davidiana Chinese Stranvaesia

SEED: The fruit is a bright red pome that persists into the early winter. Macerate the fruit to remove the pulp. The seeds display embryo dormancy and require 2 months cold stratification for germination. Some germination will occur with freshly planted seed but a cold period improves germination. In nursery practice the seed can be fall sown or spring sown after stratification.

CUTTINGS: September through November are the best months. Wound, treat with 8000 ppm IBA-talc and place under mist. This species is similar to *Photinia* × *fraseri* and could be propagated by similar procedures. 5000 to 10,000 ppm IBA-solution would prove effective. See PIPPS 24:338 (1974).

Styrax Snowbell

The fruit is an ovoid dry drupe, about ½" long and grayish in color. Fruit is effective in August and drops by November. At maturity the outer covering separates from the single shiny brown hard seed. *Styrax* seed is difficult to germinate and benefits from warm stratification, acid scarification or repeated warm and cold periods.

Styrax americanus American Snowbell

SEED: Cold stratification for 3 months has induced good germination. An Alabama nurseryman sows the seed outside as soon as collected and cleaned in the fall with excellent germination in spring.

CUTTINGS: This plant is very easy to root from softwoods: June-July and into August, 1000 ppm IBA solution, peat:perlite and mist result in 100% rooting in 4 weeks. Late July (Boston), 4000 ppm IBA-quick dip, peat:perlite, mist, 100% rooting in 3 to 4 weeks. Senior author has rooted this plant many times. Will continue to grow in rooting bed.

Styrax dasyanthus

SEED: Three months warm followed by 3 months cold stratification.

CUTTINGS: Treat as described under S. *americanus*.

Styrax grandifolius Bigleaf Snowbell

SEED: Authors suspect seed should germinate like S. *americanus*. It is a southern species found in the understory along streambanks.

CUTTINGS: Root as described under S. *americanus*.

Styrax hemsleyana Hemsley Snowbell

SEED: Treat like S. *obassia*.

Styrax japonicus Japanese Snowbell

SEED: Three to 5 months warm and 3 months cold stratification induces satisfactory germination. In a quantitative study, untreated seed did not germinate; seed given 3 or 4 months cold stratification did not germinate, however, seed provided 3 months warm and 3 months cold stratification or 3 months warm stratification and 4 months cold stratification germinated 64 and 76%, respectively [PIPPS 12:157 (1962)].

CUTTINGS: Softwoods are extremely easy to root: treat with 1000 to 3000 ppm IBA-quick dip, peat moss:perlite, and mist. Cuttings will root in 3 to 4 weeks. Rooted cuttings are readily transplanted and will continue to grow. Many new cultivars, including weeping and pink forms have been recently introduced into this country. These all root readily from cuttings. Senior author rooted cuttings with 100% success from a 100-year-old tree in the Arnold Arboretum with 4000 ppm IBA-solution, peat:perlite, mist. Rooted cuttings are easy to overwinter and in a single growing season may reach 3′.

Styrax obassia Fragrant Snowbell

SEED: Germinates best after 3 to 5 months warm stratification followed by 3 months cold stratification. A 3 month warm then 3 month cold stratification treatment induced 88% germination [PIPPS 12:157 (1962)]. Fresh seed is the best.

CUTTINGS: Softwood cuttings, May to July, are very easy to root: 1000 to 3000 ppm IBA quick dip, peat moss:perlite medium, mist. Cuttings overwinter without difficulty.

Symphoricarpos albus Common Snowberry

SEED: The fruit is a white, berry-like drupe that ripens in September through November. The fruit is macerated to remove the pulp and after drying the seed is ready for storage. The number of cleaned seeds per pound averages about 76,000. Seeds are extremely difficult to germinate because of hard, impermeable coats and partially developed embryos. Highest germination is usually obtained after warm stratification for 3 to 4 months followed by cold stratification for 4 to 6 months. Sulfuric acid scarification has been used but it has not been as successful as the warm stratification.

CUTTINGS: Cuttings are extremely easy to root. Softwoods and semi-hardwood cuttings from June through August root readily.

IBA-talc or solutions of 1000 to 3000 ppm have produced 90 to 100% rooting. Cuttings are easily over-wintered.

Hardwood cuttings of all *Symphoricarpos* species and cultivars root readily [PIPPS 34:540 (1984); *The Plant Propagator* 15:10 (1969)]. December-January cuttings, 3000 ppm IBA-talc, open bench with bottom heat and cool tops, root 90 to 100% in 4 to 6 weeks. The variety *laevigatus*, from early August cuttings, 4000 ppm IBA-talc, peat:perlite, mist, rooted 60%.

Symphoricarpos × chenaultii Chenault Coralberry

SEED: This hybrid species resulted from a cross between S. *orbiculatus* and S. *microphyllus*. The selection 'Hancock' with small leaves and reduced mildew susceptibility is superior. Fruits are produced and should be handled as described for S. *orbiculatus*.

CUTTINGS: Refer to S. *albus* for details. Late August (Boston), 8000 ppm IBA-quick dip, sand:perlite, mist, rooted 80% in 8 weeks. Ideally, June-July cuttings root in high percentages. Hardwoods are also possible [*The Plant Propagator* 15(3):10 (1969)].

Symphoricarpos orbiculatus Indiancurrant Coralberry

SEED: The fruit is a purplish red drupe that ripens in October and persists into winter. A pound of cleaned seed contains 140,000 seeds. The seeds are very difficult to germinate because of a hard seed coat and immature embryo. Warm stratification for 4 months followed by cold stratification for 4 months produced satisfactory germination. Sulfuric acid scarification for 30 minutes prior to warm stratification slightly improved germination.

CUTTINGS: Refer to S. *albus* for specifics on this easy to root species.

Symplocos paniculata Sapphireberry

SEED: The fruit is a single-seeded sapphire-blue drupe that ripens in September-October. Plants are relatively self-sterile and may not produce an abundance of fruit unless several different individuals are in proximity. The fruit is macerated after harvest to remove the pulp. Seed propagation is difficult. The mature seed contains a fully formed embryo, which requires 2 years to germinate. A long period of warm stratification is required to induce radicle emergence, followed by 3 months cold stratification. Stratification, scarification, and gibberellic acid treatment have failed to give consistent results. See *Amer. Nurseryman* 150(12):42-48 (1979) and PIPPS 33:487 (1983).

CUTTINGS: Softwoods and semi-hardwoods under mist are easy and the preferred method. Softwoods, stuck in early July are rooted by September. A hormone is not required but may improve uniformity. After rooting the cuttings should not be transplanted until growth commences the following spring. Potting after rooting will result in failure to grow out. Peter Del Tredici and the senior author conducted a rooting study at the Arnold Arboretum and found that 4000 and 8000 ppm IBA-quick dips were the best and produced profuse root systems. Even the control rooted 100%, however, all of the cuttings died during the winter [PIPPS 33:487 (1983)]. To authors' knowledge, no one has satisfactorily worked out the specifics for successful overwintering.

Symplocos tinctoria Sweetleaf or Horsesugar

SEED: The fruit is yellowish brown when ripe. Germination requirements have not been published.

CUTTINGS: The plants grow as colonies and the shoots grow from very thick, sparsely branched roots. Plants that reproduce in this manner are very difficult to propagate by division. Such plants are good candidates for propagation from root cuttings. One to 2″ long root pieces taken in December and January produce shoots within 1 month and roots within 2 months [*Arnoldia* 44:34 (1984)]. Root growth is slow and the cuttings probably should not be disturbed until the second spring. The rooting of leafy cuttings has not been studied, however, it may be possible since S. *paniculata* is so easy. The same overwintering problems that apply to S. *paniculata* may apply to this species.

Syringa Lilac

SEED: The perfect, often showy flowers are produced in April, May and June. The fruit is a dehiscent capsule that ripens in late summer or fall. Seeds can be stored up to 2 years if placed in bags in a dry environment. For long term storage, air dried seed should be kept in sealed containers at 34 to 38°F. There are conflicting reports on seed dormancy in lilacs. Dormancy varies with seed lot but is generally not very strong. If dormancy is present a cold period of 1 to 3 months can be used [*Arnoldia* 19(67):36 (1959)].

Syringa afghanica

CUTTINGS: Late June (Boston), 8000 ppm IBA-talc, sand:peat, mist, rooted 100% with heavy root systems in 9 weeks.

Syringa × chinensis Chinese Lilac

CUTTINGS: There appears to be a narrow period in late June and early July when good rooting will occur. Best rooting occurs just before terminal bud set with a marked decrease thereafter. Treat with 8000 ppm IBA-talc and mist. A 4000 ppm IBA-solution works well also. Wounding is not important [*The Plant Propagator* 22(3):8 (1976)].

Hardwood cuttings have been successful. One-year-old wood is taken from specially grown stock plants and cut to 8″. The cuttings are fall planted in sandy, well drained soil. Insert the cuttings 7″ into the soil. A 48% success rate has been obtained [PIPPS 8:94 (1958)].

Syringa henryi Henry Lilac

SEED: Untreated seed germinated 60%; 95% after one month cold stratification.

Syringa josikaea Hungarian Lilac

CUTTINGS: Early May (Boston), 3000 ppm IBA-talc, sand, outside under poly, rooted 90%.

Syringa laciniata Cutleaf Lilac

CUTTINGS: The best lilac for the south as it flowers profusely every year and does not develop mildew. Cuttings in early June (Boston), 8000 ppm IBA-talc, rooted 21% in 8 weeks; mid June, 8000 ppm IBA-talc, sand, polytent, rooted 78%. Excessive moisture appears to induce premature leaf defoliation. Also alcohol quick dips may be harmful and a K-IBA aqueous solution might prove better. Most literature indicated that cuttings under plastic or a high humidity chamber *without mist* rooted in higher percentages than under mist. Early July, 8000 ppm IBA-talc, mist or polytent, rooted 36% and 96%, respectively.

Syringa meyeri Meyer Lilac

CUTTINGS: Often difficult to root; timing is quite important. Young shoots collected from plants forced in the greenhouse rooted 80% in 3 weeks when treated with 3000 ppm IBA-talc and placed in sand. Many commercial nurserymen have no trouble rooting this species.

Syringa microphylla Littleleaf Lilac

CUTTINGS: Early May, 3000 ppm IBA-talc, rooted 50% while untreated cuttings rooted 26%. Cuttings collected later in early July rooted only 16% with 8000 ppm IBA-talc treatment. Again this illustrates the importance of softwood cuttings. 'Superba', mid-May, soft cuttings, 8000 ppm IBA-talc, rooted 97% in 7 weeks. Untreated cuttings only callused. Late July, 3000 ppm IBA-talc, rooted 46% in 8 weeks.

Littleleaf lilac can be propagated by layering.

Syringa oblata Late Lilac

CUTTINGS: Early May, wound, 10,000 ppm IBA-talc, rooted 65% in 5 to 6 weeks; 8000 ppm IBA-talc plus wound produced 55% rooting. The var. *dilatata* in early June with 8000 ppm IBA-talc plus thiram, polytent, rooted 50% with excellent root systems. The same treatment under mist produced 33% rooting and poor root systems.

Syringa patula (S. *palibiniana*) Korean Lilac

CUTTINGS: Cuttings from plants forced in a greenhouse in early May and treated with 8000 ppm IBA-talc rooted 100%. 'Miss Kim' is included under this species.

Syringa pekinensis Pekin Lilac

SEED: Handle as described under S. *reticulata*.

CUTTINGS: Timing is apparently quite critical; senior author was unable to root late July cuttings.

Syringa reflexa Nodding Lilac

SEED: Good germination occurred after 3 months cold stratification; seedlings appeared 12 days after planting.

Syringa reticulata (S. *amurensis* var. *japonica*) Japanese Tree Lilac

SEED: There are 12,000 seeds per pound. Refer to S. *vulgaris* for handling and storage details. Most reports indicated that the fruit matures in late summer or early fall and should be planted immediately, mulched and maintained until germination the following spring. Based on work at Morden Arboretum, 2 months warm stratification followed by 2 months cold stratification produced good germination when spring planted [PIPPS 23:359 and 515 (1973)]. Seeds of var. *mandshurica*, sown at once, germinated 7%; 56% after 3 months cold stratification.

CUTTINGS: The Japanese tree lilac is easily rooted from cuttings [PIPPS 23:515 (1973)]. Early July (Canada), 8 to 10″ long, rooted 87% by October after treatment with 8000 ppm IBA-talc in sand:perlite under mist. Propagation in a greenhouse is better because of the higher rooting medium temperature. The rooted cuttings were placed in cold storage and planted out the following spring. Senior author used 10,000 ppm IBA-quick dip on June (Illinois) cuttings with 90% rooting and strong root systems. Cuttings can be planted after rooting and over wintered successfully. It is surprising that this species is not rooted from softwood cuttings. Hardwoods are reported to root.

GRAFTING: Budding of cultivars, such as 'Ivory Silk', is practiced on seedlings of the species. Three-year-old seedlings are field planted at a spacing of 18 × 48″. T-budding is accomplished in July after the growth has hardened. The bud is placed in the direction of the prevailing wind. Rubber strips are used and are cut off after union formation. The following spring the stock is pruned back to 4″ above the bud and the shoot is tied to the understock as soon as it reaches 4″. The bud grows straight to about 6′ without staking. The second year any side branches are reduced to 8″ and a small head is produced. In late summer, all side branches are removed. The third year a good head develops.

Syringa × swegiflexa Swegiflexa Lilac

CUTTINGS: Mid-July, 8000 ppm IBA-talc, rooted 44% in 13 weeks; mid-August, wound, sand, mist, rooted 60% in 4 weeks.

Syringa sweginzowii Sweginzow or Chengtu Lilac

CUTTINGS: Hardwood, January, 8000 ppm IBA-talc + thiram, sand:perlite, open bench, rooted 50%. In a limited study, 100% rooting occurred with early June cuttings, 4000 ppm IBA-talc, bottom heat (80°F), peat:sand, and mist [*The Plant Propagator* 16(2):14 (1970)].

Syringa villosa Late Lilac

SEED: Refer to S. *vulgaris*. The number of cleaned seeds per pound averages 41,200. Cold stratification for 1 to 3 months has been used but may not be necessary for all lots of seed. In one study, fresh seed germinated 66%; 82% after 1 month cold stratification; and 90% after 2 months cold stratification. Sowing in October is recommended if seeds are not pretreated [*Amer. Nurseryman* Aug. 15, p. 8 (1967)].

CUTTINGS: Softwoods, late June to July (Canada), 3000 ppm IBA-talc, perlite:Turface (1:1, v/v), mist, rooted 90%. This procedure can be used for the Preston hybrids [*The Plant Propagator* 28(2):10 (1982)]. Other reports noted 36 to 50% rooting.

Syringa vulgaris Common Lilac

SEED: In practice only cultivars are offered for sale and seed propagation is used in breeding and to produce understock. The capsule ripens in August to October and should be collected in the fall, the later the better, and allowed to dry. Seeds will remain viable up to two years if kept dry and refrigerated. Number of cleaned seeds per pound averages 86,000 [*Amer. Nurseryman* Nov. 15, p. 12 (1967)]. Dormancy is variable. In one study, 2 months cold stratification induced 93% germination; no stratification was required in another. The optimum temperature for germination is 68°F.

CUTTINGS: Much discussion has occurred about the propagation of lilacs by shoot cuttings. There are as many methods as lilac propagators. The optimum timing has been related to a number of plant growth characteristics. Some reports relate the timing to condition and size of the cuttings, i.e., when new growth has reached 4 to 6″ and before stems become woody.

Other reports recommend taking cuttings when flowers first begin to open or shortly thereafter. Finally, another group suggests taking cuttings before the terminal bud is visible.

Spring softwood cuttings are preferred. Cuttings are collected just as the flowers are starting to open and until the end of the blooming season. Treat with IBA (1500 to 5000 ppm depending upon cultivar) and place in coarse sand or perlite under a white poly tunnel or mist.

Outdoor beds under mini-quonset structures were used in Oklahoma for rooting. The rooting medium, a light clay loam with peat moss added, is treated with methyl bromide for weed control. The beds are covered with 6 × 6″ concrete reinforcing wire and then with clear plastic. Cuttings are taken about April 15. Timing is critical as cuttings taken too late have a lower rooting percentage. The best length is 5 to 6″. The lower third of leaves are removed. Cuttings are dipped in a Captan solution followed by a 5-second quick dip in 1000 to 4000 ppm IBA. Use the lowest amount of mist that still keeps the cuttings turgid. Excessive mist is detrimental. Following sticking, a weekly spray program is started using Zineb, Captan, and Benlate in rotation. Banrot is used to combat soil borne diseases if needed. Callus development begins in 2 to 3 weeks and this is a critical time for mist control, as too much mist will result in excessive callus. Once 60 to 70% of the cuttings are showing roots, hardening-off is started. The top of the plastic is cut out, then the east side is let down in 3 days, and lastly the west side 3 days later. Misting is reduced gradually. Following rooting the cuttings are sheared to increase stem size and sprayed regularly for mildew control [PIPPS 28:348 (1978)].

Mezitt (Massachusetts) has used a sub-irrigation method with good results [PIPPS 28:494 (1978)]. Water is applied to the cuttings from beneath the rooting medium. The greenhouse is shaded with 60% saran cloth and white polyethylene. Metal pans (8′ × 3′ × 6″) are the only equipment needed. The pans are filled with 2″ of ¾″ diameter stone and the rest with horticultural grade perlite. The stones are separated from the perlite with a fine-sized plastic mesh screen. A 4″ perforated drain tile is placed in each corner. The pan is filled with water the day before the cuttings are stuck and drained prior to cutting insertion. The lilac cuttings are taken at flowering time. Cuttings are gathered early or late in the day and refrigerated overnight. A heavy wound is applied to one side before applying 8000 ppm IBA-talc + thiram. After sticking, the pan is again flooded with water until the cuttings are turgid. The pans are flooded again only if the cuttings begin to wilt. Cuttings are rooted in one month and potted in 6 weeks. Rooting percentages are high.

Layering is also possible for small numbers [*Arnoldia* 19 (6-7):36 (1959)]. Bend down shoots in the spring when the buds are swelling and wound the bases. Cuttings are ready for harvest the following spring.

Division is practiced for small numbers.

The possibility of using hardwood cuttings in cold frames was also suggested, but is not practiced to any degree.

Common lilac grown from root cuttings produces plants in half the time required by softwood stem cuttings. Dormant plants are fall dug before the ground freezes. They are either stored or cuttings are made. Heavy, fleshy roots are removed and cut into 4″ lengths with polarity being maintained. Place into 3″ clay pots (plastic or peat pots do not work as well) containing soil, peat moss and perlite. The roots are allowed to project 1″ above the soil line. Place the pots in a bench and cover with perlite to a depth of 1″. The pots are carried from December until March in a cool greenhouse at a night temperature of 33 to 36°F. Watering about once a month is required. From March through April the temperature is increased to 50°F and watering is increased to 2 to 3 times per month. By late April the plants will be 6 to 8″ tall and light soluble fertilizer can be applied. Water as needed and field plant in July. This method has its limitations as to quantities and varieties [PIPPS 27:402 (1977)].

GRAFTING: Bench grafting can provide a means for increasing the propagating season. Root pieces from one year old seedlings of green ash, *Fraxinus pennsylvanica*, are used as a nurse graft [PIPPS 6:75 (1956)]. The hybrid lilacs are bench grafted to the root pieces using a whip graft and secured with grafting thread which is better than tape. The completed grafts are placed in poly bags and then in refrigerated storage at 31 to 35°F to keep the buds dormant. The root will remain attached to the scion until roots emerge from the latter. At that time the lilac will reject the ash root. On a few of the white flowered varieties the ash root may remain attached. The grafts should not be planted out until the ground has warmed. Late planting, such as mid-June, will give poor establishment. Plant the grafts in a trench, almost completely covering with soil, and water well. At the end of the first growing season the grafts are cut back to force breaks.

Lilac cultivars can also be field budded in July and early August or bench grafted in winter to S. *vulgaris* and S. *tomentella*. *Syringa tomentella* appears to establish quicker after planting and has fewer suckers. S. *tomentella* also has a different leaf and suckers can be easily spotted and removed.

Lilacs grafted on species lilac or privet have a shorter useful life span because of graft incompatibility, suckering, stem borers and non-renewal of wood.

TISSUE CULTURE: Actively growing shoot tips from greenhouse-grown plants of 'Vesper' lilac initiated new shoots in vitro on a modified Murashige and Skoog revised medium supplemented with in mg/liter: myo-inositol 100, nicotinic acid 1, pyridoxine HCl 1, thiamine HCl 1, sucrose 30,000, agar 7,000, BA 0.1, and either 0.125, 0.25, or 0.5 IAA. Multiplication was achieved on the same medium but with 7.5 mg/liter BA and 0.1 mg/liter NAA. Cultures were maintained at 80 to 82°F with a light intensity of about 41 microE $m^{-2}sec^{-1}$ for 18 hours. In 5 to 6 weeks about 6 shoots, 1 to 1½″ long, were produced. Microcuttings treated with 1000 ppm IBA rooted best in vermiculite in a plastic covered flat. After 3 to 4 weeks, 81% rooting occurred [*HortScience* 18:432 (1983)].

Syringa wolfii Wolf Lilac

SEED: Cold stratification for 2 months produced good germination.

CUTTINGS: 100% rooting was obtained in 8 weeks from mid-July cuttings, 8000 ppm IBA-talc, sand and mist. Early September cuttings with soft tips, 8000 ppm IBA-talc, rooted 90% with excellent root systems.

Tamarix ramosissima (*pentandra*) Five-stamen Tamarix

The genus *Tamarix* contains about 54 species of shrubs and small trees; most of which are not used ornamentally. They are easy to propagate by seed, softwood and hardwood cuttings. The propagation information for T. *ramosissima* should also apply to other tamarix species, such as T. *anglica*, T. *gallica*, and T. *tetranda*.

SEED: The fruit is an inconsequential capsule. Seeds stored at 40°F for over one year retained their viability but seeds stored at room temperature remained viable for only a short time. No pretreatment is necessary for seed germination. Fresh seed germinates within 24 hours of imbibing water.

CUTTINGS: Cuttings will root during any season. Softwood cuttings placed in peat and perlite under mist root readily. The root systems may be sparse and the roots coarse and somewhat difficult to handle. Cuttings should be weaned from mist as soon as rooting starts since excess moisture results in decline. *Tamarix* roots readily from hardwood cuttings and, provided the material is thick enough (¾″), may approximate 100% [*The Plant Propagator* 15(3):10 (1969); PIPPS 20:238 (1970)].

Taxodium ascendens (T. *distichum* var. *nutans*) Pondcypress

SEED: Pondcypress is commonly propagated by seed. Male and female cones are produced on the same tree. The female cones turn from green to brownish purple as they ripen in October to December. Mature dry cones should be broken apart. Separation of cone fragments from seeds is difficult and both are often sown. The number of cleaned seeds per pound averages about 4000. Seeds can be stored dry at 41°F for at least one year. Pondcypress seeds respond well to 2 to 3 months cold stratification at 38°F. Fall sowing of untreated seeds or spring sowing of pretreated seeds are practiced in the nursery industry.

CUTTINGS: Refer to T. *distichum*.

GRAFTING: Refer to T. *distichum*.

Taxodium distichum Common Baldcypress

SEED: Refer to T. *ascendens* for information on seed harvest, extraction, and storage. About 50 pounds of seeds can be obtained from 100 pounds of fresh cones, and there are 2,600 to 3,500 cones per bushel. A pound of cleaned seed contains 5,200 seeds. Seeds exhibit an apparent internal dormancy that can be overcome by 3 months cold stratification, preceded by a 5-minute soak in ethyl alcohol. In nursery practice, fall sowing (in the north) or spring sowing after pretreatment are both practiced.

CUTTINGS: Cuttings root with difficulty and success varies with each clone. Very soft cuttings treated with 1000 ppm IBA and placed under mist reportedly root in low numbers. Cuttings from 2 to 3-year-old seedlings root readily when placed under mist in a sand:peat medium. This indicates the importance of juvenility [PIPPS 20:282 (1970)].

Hardwood cuttings (England), 4 to 5″ long, treated with 8000 ppm IBA-talc, and placed in a well shaded and dampened cold frame have been reported to root. In April, the rooted cuttings are lifted and lined out into another cold frame before lining out into nursery beds in the fall.

GRAFTING: Grafting on the species is the most practical method for 'Monarch of Illinois', 'Pendula' and 'Shawnee Brave'. Field graft with dormant scions in the spring or in late summer with well-matured scion wood. In England, bench grafting is practiced in March. Two to 3-year-old seedlings, whip and tongue, 55 to 60°F during callusing [*The Plant Propagator* 9(1):11 (1962); PIPPS 17:376 (1967)].

Taxus Yew

SEED: Male and female flowers are usually borne in the leaf axils of the previous season's wood on separate plants (dioecious). Male flowers are rounded clusters of stamens, while the female flower contains a solitary ovule enclosed in a fleshy aril that ripens the first year. The fleshy aril is red when mature and ripens in late summer or autumn. Under the aril, a hard coat covers the oily, white endosperm and immature embryo. Seeds should be harvested when ripe and the pulp removed by macerating the fleshy arils in water. A pound of cleaned seed contains 7,500 seeds (T. *cuspidata*). Properly cleaned seeds can be stored for a few years in sealed containers under refrigeration. The oily texture of the seed coat may appear to repel immediate water penetration, but it does not permanently prevent water absorption. Seed production is fraught with difficulty, particularly with seeds purchased on the open market. The key to success is fresh, high-quality seeds [*Amer. Nurseryman* Jan. 15, p. 10 (1969)].

Taxus seeds should be tested for viability because much of the seed shipped into the country is of poor quality, immature or dead. Heit devised a viability test which has proven satisfactory and reliable. Seeds are placed at room temperature (65 to 85°F) for 1 to 2 months under moist conditions. During this time the seeds develop or break down internally according to their vitality and viability. Viability is next tested by cutting the seeds in half longitudinally to determine color and firmness. Strong, healthy seeds remain firm, plump and pure white. Dead or unfit seeds break down, decay or discolor to varying degrees from gray to dark brown.

Yew seeds are very slow to germinate and several factors contribute to this condition. Germination may take 1, 2 or 3 years. No germination can be secured with *Taxus* seeds unless they are given both a warm and a cold treatment in the proper sequence and at the optimum temperatures and lengths of time. Heit noted that 5 to 7 months warm stratification at 60 to 65°F followed by 2 to 4 months cold stratification at 34 to 40°F produced good results.

For successful seedling production in the nursery, and maximum germination of all viable seeds the first spring, it is recommended that seeds be sown no later than July of the preceding year. It is essential to mulch the seedling beds to maintain a constant moisture supply and prevent excessive heat buildup. Shading should be utilized. A large Canadian nursery mixes seed with moist sand and peat in January, treats with fungicidal drench, packs in shallow boxes and places outside and leaves until the spring two years following. The necessary cycle is cold, warm, cold and then plant.

Taxus baccata English Yew

SEED: A pound of cleaned seed contains 6,400 seeds. The viability can be maintained for 5 to 6 years by drying after harvest for 1 to 2 weeks, and then storing in sealed containers at 34 to 36°F. Warm stratification at 68°F for 4 months followed by cold stratification at 34 to 40°F for 2 to 4 months is recommended. Untreated seed, or seed given 3 months cold stratification, did not germinate. Seed given 4 months warm stratification/3 months cold stratification germinated 26%; 55% after 5 months warm/3 months cold.

CUTTINGS: Yews are relatively easy to propagate from cuttings and methods vary widely from grower to grower. *Taxus* cuttings are highly topophytic and maintain the growth habit they exhibited on the parent plant. Collect from September through December after a hard frost in the midwest and northeast and, if necessary, store in a cooler at 38 to 45°F until used. Cuttings taken after a frost root more readily than cuttings without frost. Strip the needles from the lower part of the cutting. 8000 ppm IBA-talc plus a fungicide is commonly used for all cultivars. 'Repandens' is considered difficult by some propagators but treatment with 8000 ppm IBA + thiram has yielded 100% rooting [PIPPS 16:190 (1966)]. The cuttings are stuck in greenhouse benches and remain there all winter with occasional fungicidal sprays and hand watering or misting. Cuttings can be placed in flats and literally positioned anywhere in the structure until rooting takes place.

Media varying from straight sharp sand to sand and perlite mixtures are used. Temperature at the base of the cuttings is maintained at 68 to 75°F with the tops cool. Watering is done in the morning and all foliage should be dry by late afternoon because wet foliage in a closed, heated house is an invitation to diseases during the cool, low light winter days. The cuttings will have well branched roots in 8 to 9 weeks. As soon as the cuttings are well rooted, the the bottom heat is lowered and watering is reduced to harden the cuttings. No top ventilation is given until late February or early March when roots are established. In early June, after danger of frost has passed, the cuttings are prepared for the transplant beds [PIPPS 4:76 (1954) and 29:71 (1979)].

Rooted cuttings of *Taxus* reportedly flush better the first year (3 to 4" vs. ½ to 1") if given a cold treatment at 34°F for 2 months prior to planting out [PIPPS 32:609 (1982)].

'Adpressa Aurea', 'Semperaurea' and other dwarf cultivars are taken with a heel, December to February, wound, 8000 ppm IBA-talc, 2% IBA-talc or 2500 ppm IBA-quick dip for difficult cultivars, peat:perlite or sand (2:1, v/v) [NGC 188(3):15 (1980) and 180(7):26 (1976)].

GRAFTING: 'Adpressa Aurea', 'Adpressa Fowle', 'Repandens', 'Silver Green', 'Semperaurea', 'Standishii', 'Variegata', and 'Washingtonii' have been grafted on T. *baccata*, T. *cuspidata* or T. × *media* 'Hicksii' but could be grafted on any *Taxus* cultivar that roots [NGC 188(21):31 (1980)]. Use a side-veneer graft in winter, bind with rubber tie, wax, and place in closed bench with shading. No basal heat is used.

Taxus brevifolia Pacific Yew

SEED: This native yew is not listed as requiring a pregermination treatment. Refer to *Taxus* introduction for general germination conditions if problems develop.

CUTTINGS: Follow the methods outlined under T. *baccata*. 'Nana' reportedly has a low survival rate the first winter in New England unless it is potted in June and kept in a frost free frame the first winter. Stem splitting during the first winter is a problem.

Taxus canadensis Canadian Yew

SEED: The seed ripens from July to September depending on location. This species has proven difficult from seed and 3 months warm stratification/3 months cold stratification produced no germination. The same seed given another cold period germinated sporadically upon sowing. Refer to *Taxus* introduction for additional details.

CUTTINGS: Cuttings are treated with 5000 to 10,000 ppm IBA-quick dip or 8000 ppm IBA-talc and placed in sand or sand:peat. Refer to T. *baccata* and T. *cuspidata* for additional information.

Taxus cuspidata Japanese Yew

SEED: A pound of clean seed (US collected) contains about 7,400 seeds. Imported seeds from Japan often contain nearly twice as many seeds per pound and this is partly due to the immature or empty seeds sometimes more prevalent in those seed lots. Seed viability can be maintained for 5 to 6 years by drying after harvest for 1 to 2 weeks and storing in sealed containers at 34 to 36°F. Warm stratification for 4 months followed by cold stratification for 4 months has been recommended. Also, 8 months to 1 year warm stratification followed by 3 months cold stratification gave good germination. Imported seed is often reported to germinate poorly and locally collected seed is preferred. In nursery practice, T. *cuspidata* var. *capitata* is often propagated from seed since it comes relatively true-to-type.

CUTTINGS: Wells (New Jersey) described a method for the propagation of *Taxus* in cold frames. Shoots are collected in September and trimmed to 8 to 10" with a small portion (2 to 3") of old wood at the cutting base. Wound, 8000 ppm IBA-talc, and place close together in a cold frame using a mixture of sandy top soil and sand (1:2, v/v). Syringe and ventilate as needed. Syringe not less than once a day in the early fall and reduce as the weather cools. When the weather is cold, close the frames for the winter.

Critical conditions can arise when a prolonged cold spell occurs with bright sun light. Protect the cuttings against direct sunlight and rapid temperature fluctuations. When the weather moderates, the cuttings can be thawed gradually by opening the cold frame gradually. After the weather begins to warm up, the frames are opened, dead or diseased cuttings are removed, and the medium is soaked. If disease is evident, fungicide sprays should be applied. At this time the shade on the glass is checked and replaced as needed. Follow normal watering and ventilation practices until early June when the cuttings should be rooted. Carry the cuttings through the summer in the cold frames with special attention to shading and watering. By September the sash can be completely removed [PIPPS 4:76 (1954)].

The greenhouse propagation of cuttings is described under T. *baccata*.

GRAFTING: Grafting is not commercially practiced with this species.

Taxus × media Anglojap Yew

SEED: A hybrid species first raised by T. D. Hatfield of the Hunnewell Pinetum, Wellesley, Massachusetts about 1900. Since that time numerous selections, such as 'Andersonii', 'Brownii', 'Hatfieldii', 'Hicksii', 'Kelseyii', 'Wellesleyana', et. al., have been selected and often it is difficult to tell if one is looking at T. *cuspidata*, T. *baccata* or T. × *media* types. The majority of the seedlings produced from 'Hicksii' seed reportedly exhibit a fairly uniform, dense, columnar growth habit typical of the cultivar. Other cultivars, however, do not produce consistently similar types. The T. × *media* group of cultivars is propagated by cuttings and not seed.

CUTTINGS: Refer to T. *cuspidata* and T. *baccata* for details on cutting propagation methods.

Ternstroemia gymnanthera Ternstroemia

SEED: Refer to *Cleyera japonica*.

Teucrium chamaedrys Wall Germander

SEED: Seed requires no pretreatment and will germinate upon sowing.

CUTTINGS: Easily rooted from softwoods in June through August provided there is not excessive moisture. A low level of hormone, 1000 to 3000 ppm IBA, improves rooting. In one study, late August cuttings rooted 94% with 3000 ppm IBA-talc in 10 days.

Thuja koraiensis Korean Arborvitae

SEED: The Korean arborvitae is closely allied to T. *standishii*, but the leaf sprays are flatter and the undersides of the leaves are usually glaucous to almost silvery. Refer to T. *occidentalis* for seed germination information.

CUTTINGS: Late May, rooted 62% in 8 weeks when treated with 8000 ppm IBA-talc + thiram, sand:peat and mist. Although this is not the ideal time to take arborvitae cuttings, it indicates the wide latitude in timing. Late December cuttings treated with 8000 ppm IBA-talc rooted 100% in 12 weeks. Mid-January cuttings rooted 100% while late January cuttings rooted 92%.

Thuja occidentalis Eastern or American Arborvitae

SEED: The 1/3″ long cones are composed of 5 to 20 scales, each bearing 2 to 5 winged seeds. Cones ripen in early autumn and should be harvested immediately after turning from yellow-green to light brown to prevent loss due to seed dispersal. For eastern arborvitae the time between cone ripening and opening is only 7 to 10 days. Large quantities can be kiln dried at 110°F. The seeds should not be dewinged because they are easily damaged. A pound of cleaned seed contains 346,000 seeds. Seed loses viability quickly at room temperature and should be stored in sealed containers at 32 to 38°F.

Some lots of seeds germinate as soon as ripe while others require pretreatment to overcome dormancy. To be assured of uniform germination all seed should receive 2 months cold stratification. In nursery practice, T. *occidentalis* is often fall planted. Half shade over the beds the first year is recommended. When germinated under laboratory conditions, seed requires artificial light and 68 to 86°F alternating temperatures [*Arnoldia* 37(1):75 (1977)].

CUTTINGS: Hardwoods of the species and cultivars, such as 'Densa', 'Goldcrest', 'Globosa', 'Holmstrup', Nigra', 'Pyramidalis', and 'Woodwardii', root well from September to March. November to December cuttings will be rooted by March. Bottom needles are removed and the base may be given a fresh cut before treatment with 3000 to 8000 ppm IBA-talc or solution; open greenhouse bench or polytent; sand, sand:peat, or peat:perlite media. Temperature at the basal zone of the cutting is maintained at 66 to 68°F with tops at 45 to 50°F. Rooting occurs in 2½ to 3 months. Cuttings can be handled like *Taxus*, with syringing only when necessary during the rooting process.

GRAFTING: Cultivars such as 'Pendula' can be grafted on the species. Use a side-veneer graft. See pine introduction for details.

Thuja orientalis (*Platycladus orientalis*) Oriental Arborvitae

SEED: Cones will open in 24 to 36 hours at a temperature of 90°F. A pound of cleaned seed averages between 20,000 and 25,000 seeds. Seeds have been stored in sealed containers at 32 to 38°F for over 5 years with little loss of viability. Cold stratification for 1 to 1½ months may promote germination in certain seed lots. The optimum temperature for seed germination is 68°F. In nursery practice, this species is generally spring planted.

CUTTINGS: Oriental arborvitae cultivars are more difficult to root than T. *occidentalis*. June to August cuttings are best although October to December are used in England with good results: wound, 8000 ppm IBA-talc + Benlate (8:1, v/v), peat:perlite (2:1, v/v), mist, bottom heat [NGC 188(3):15 (1980); PIPPS 25:81 (1975)]. See T. *occidentalis* for additional information.

GRAFTING: Cultivars can be grafted on species understock. Use side-veneer graft in February to early March and place in polytent.

Thuja plicata Giant or Western Arborvitae

SEED: The small female cones are composed of 8 to 10 scales. Cones ripen in early September and opening occurs in late September. The cones should be harvested as soon as they turn from green to brown to prevent losses due to seed dispersal. Peak rate of seed dispersal occurs 4 to 6 weeks after the first cones have opened. Cones of T. *plicata* open in 24 to 36 hours at a temperature of 90°F. The percentage of empty seeds in cones of *Thuja* is frequently very high; but a large portion of those empty seeds can be separated in an air stream. A pound of cleaned seed contains 414,000 seeds. Seeds usually do not require stratification. However, dormant seed lots have been encountered and cold stratification at 34 to 41°F for 1 to 2 months may be required. When germinated under laboratory conditions, seed requires artificial light and 68 to 86°F alternating temperatures.

CUTTINGS: T. *plicata* and its cultivars are easily rooted from cuttings. Early October, rooted 91% in 8 weeks with 3000 ppm IBA-talc, sand:peat (1:2, v/v), and bottom heat (65°F). Mid-November cuttings rooted 80% under similar conditions. 'Atrovirens' rooted 80% from mid-November cuttings and 60% from late December cuttings. Treat with 3000 to 8000 ppm IBA-talc + thiram, sand:perlite, and polytent.

Thuja standishii Japanese Arborvitae

SEED: Fresh seed germinates sporadically and 2 weeks cold stratification improves germination.

CUTTINGS: Late October cuttings rooted 33% using 8000 ppm IBA-talc + thiram, sand:perlite and polytent.

Thujopsis dolobrata Hiba Arborvitae

SEED: The small, almost globular cone is composed of 6 to 8 woody scales, with the fertile scales (usually six) bearing 3 to 5 winged seeds. Plants raised from seed of this monotypic Japanese species show considerable variation in habit and grow slowly for a period of years. No good information is available on seed propagation, but one report said seed sown immediately after harvesting and cleaning will germinate.

CUTTINGS: This is one of the easiest conifers to root. Hardwood cuttings rooted 100% when taken November through January, using 4000 ppm IBA-talc plus a fungicide, peat:perlite, and mist. Even cuttings from mature trees are easy to root.

'Nana' and 'Variegata' are easily rooted using the same methods. Senior author has rooted the two cultivars with 90% success.

Tilia Linden, Basswood, Lime

SEED: The fruit is a grayish nut-like structure (drupe) that ripens in the fall. Seeds have a hard pericarp that can be removed by running through a grinder or sulfuric acid treatment. All can be propagated from seeds, which provide understocks for budding. Linden seeds are among the toughest for obtaining good germination and seedling production. Seeds show delayed germination because of a tough pericarp, impermeable coat, and dormant embryo [*Contrib. Boyce Thompson Instit.* 6:69 (1936); *Amer. Nurseryman* 125(2):10 (1967)]. Seed may remain in the ground several years and never produce a good stand of seedlings. The seed is considered to have a triple dormancy. Seed treatments that consistently result in good germination have not been developed. Several reports suggest that best germination results when fruits are collected as early as possible and treated promptly. The ideal time to harvest is when the pericarp has turned grayish brown. However, some propagators have harvested early but obtained inconsistent results, with the seeds sometimes turning to mush. Flemer recommends mixing the seed with damp sand, putting the seed mix in a box, and burying it in a stratification bin out-of-doors in the winter. The boxes are dug up the following fall and the seeds are sown. Seeds are sown shallow and lath screens improve seedling stands. An alternate method is to remove the pericarp immediately and fall sow or treat with concentrated sulfuric acid for 10 to 15 minutes followed by 3 months cold stratification. Since all linden species are variable in their percentage and degree of hard-seededness, it is best to soak the seeds for 1 to 2 days to determine the degree of hard-seededness.

CUTTINGS: Stem cuttings of T. *americana*, T. *cordata*, and T. × *euchlora* have been rooted in high numbers in mist beds. Leaf abscision in the mist bed is a problem. Subsequent growth is much slower than budded trees.

GRAFTING: Lindens are very slow and difficult to propagate by bench grafting [PIPPS 30:333 (1980)]. Even when stands are acceptable, the grafts grow slowly and take 2 years longer to reach 6 to 8′ than budded trees. Stands are better if dormant scions are grafted in February or March on seedlings established in pots. Initial field growth is slow because of the small root systems.

Lindens are among the easier trees to bud successfully. Incompatibility problems are extremely rare if proper understock-scion combinations are used. If improper combinations occur, the stand will look good through the first winter, but growth the first summer will be slow and irregular.

In budding, the T-bud method is used, and the scion bud is dewooded for best results prior to insertion. Some English growers use chip budding. Best stands in the eastern U.S. are obtained by budding onto vigorous growing understocks about mid-August. Earlier budding can be unsuccessful because the bark may heal over the bud, and later buddings may be unsuccessful because of reduced bark slippage. The tying rubbers should be removed 2 to 3 weeks after budding. Understocks are cut back in January or February. Linden seedlings often sucker abundantly and the sucker growth must be removed several times the first year. Linden buds tend to grow horizontally for a few inches before growing up, thus causing an unsightly bow. Tying or staking using Schmidt's "Grow Straight" irons produces straight trees.

Stock-scion compatibility between linden species is critical. T. *cordata* will grow on T. *cordata* or T. × *europaea* and vice versa. T. × *euchlora* will grow on T. *cordata*, T. *platyphyllos*, or T. *americana*. T. *americana*, T. *platyphyllos*, T. × *euchlora* 'Redmond', T. *tomentosa*, and T. *petiolaris* will grow on seedlings of T. *americana*, T. *platyphyllos*, and T. *tomentosa*. T. *cordata* and T. *platyphyllos* will hybridize with each other but will not graft. T. × *euchlora* 'Redmond' is compatible on both T. *americana* and T. *cordata* but is more dwarf on T. *cordata*. There is some question as to whether 'Redmond' is really T. × *euchlora* or, more likely, a selection of T. *americana*.

Tilia americana American Linden, Basswood

SEED: Can be stored dry for 2 years at room temperature; longer storage requires lower temperatures. A pound of cleaned seed contains about 3,000 seeds. Seeds exhibit great variation in germination from year to year for no explainable reason. When poor stands occur, the seedlings should be removed carefully so as not to disturb the bed. Additional germination often occurs the second year after planting. According to a Canadian nurseryman [PIPPS 35:495-499 (1985)], seeds will germinate if picked before the seed coat and wing turn from gray to brown; sow in early September.

CUTTINGS: Layering is practiced on a limited scale in Holland. It works best in cool summers and in soil with abundant moisture. Mound layering is the preferred method. The stock plants are cut back to the crown each year and the bases of the new shoots are covered with a mound of soil when they are 6 to 10″ high. The soil is removed the following spring and the rooted shoots cut off before new growth starts. Layered shoots are usually so lightly rooted that they are cut back and bedded-out for one year.

Still reported the successful rooting (100% in 30 days) of softwoods and semi-hardwoods from mature plants when taken from mid-May to late June. Wound, 2 to 3% IBA-quick dip, perlite:peat (7:3, v/v), and mist. Young shoots from plants forced in the greenhouse have been rooted (37%); treat with 8000 ppm IBA-talc and wound. Linden stem cuttings are prone to leaf abscision after which they will not root. Rooted cuttings also may prove difficult to overwinter (*Scientia Hortic.* 11:391). See also *The Plant Propagator* 24(2):15 (1978).

GRAFTING: Refer to *Tilia* introduction for details.

Tilia cordata Littleleaf Linden

SEED: Seeds with 10 to 12 percent moisture content will keep satisfactorily for 2 to 3 years under ordinary dry storage conditions. A pound of cleaned seed contains approximately 13,800 seeds. Seeds are doubly dormant and it is often difficult to produce an economic stand. Pick before seed coat turns brown, keep seed moist and plant by middle of October (Canada). Will germinate the following spring [PIPPS 35:495-499 (1985)].

CUTTINGS: Apical cuttings as short as 3¼ to 4″ can be rooted between mid-June and early July. Some clones are easier to root than others. Best results occur with 6″, pencil-size cuttings, 8000 ppm IBA-talc or quick dip, a very porous medium and mist. Bottom heat increases rooting. The cuttings are overwintered in the rooting flats, under cold, but not freezing conditions.

Cuttings grow slowly after planting and should be cut back the following winter and trained to a single stem. Cutting-grown plants are particularly prone to wind-throw in wet weather and it is probable that bedding-out rooted cuttings for one year followed by careful pruning of the roots before transplanting is necessary. Still (Kansas) reported best rooting (100% in 30 days) in mid-May to mid-June with softwood and semi-hardwood cuttings from mature plants. Wound, 2 to 3% IBA-quick dip, perlite:peat (7:3, v/v), and mist (*Scientia Hortic.* 11:391).

This species can be layered like other lindens. It will be fully rooted in 2 years. Refer to T. *americana* for details.

GRAFTING: Linden seedlings, especially those of T. *cordata*, sucker abundantly from the base, and the young budded trees should have suckers removed several times during the first summer after cutting back. Refer to *Tilia* introduction for details. Most of the T. *cordata* selections are budded during the summer.

Tilia dasystyla Caucasian Linden

SEED: Five months warm/3 months cold produced low germination; seed flats were returned to cold and then to warm where germination improved.

Tilia × euchlora Crimean Linden

SEED: A hybrid between T. *cordata* and T. *dasystyla* and not seed propagated.

CUTTINGS: Hardwood cuttings were rooted in heated bins using the same technology as that for *Malus* rootstocks. However, survival upon transplanting was very poor [PIPPS 19:223 (1969)].

Still reported best rooting (100% in 30 days) in mid-May to early August with softwood and semi-hardwood cuttings from mature plants. Wound, treat with 2 to 3% IBA-quick dip, perlite:peat (7:3, v/v), and mist. Same after-care required as described under T. *cordata*.

GRAFTING: Refer to *Tilia* introduction for details on budding and grafting.

Tilia × europaea European Linden

SEED: A hybrid between T. *cordata* and T. *platyphyllos* and not propagated by seed.

CUTTINGS: European linden can be layered and will root in one year. Refer to T. *americana* for details. Shoots taken in late June and early July will root. Refer to *Tilia cordata* for details.

GRAFTING: European linden can be grafted and budded. The compatible understocks include T. *cordata* and probably T. *platyphyllos*. Refer to *Tilia* introduction for details.

Tilia flavescens Yellow Linden

CUTTINGS: 'Spaethii', late August, wound, 8000 ppm IBA-talc, 90% rooting in 11 weeks; mid-August, wound, 8000 ppm IBA-talc, 88% rooting but all failed to overwinter.

Tilia mongolica Mongolian Linden

SEED: Refer to *Tilia* introduction for information.

CUTTINGS: Senior author tried this species when at the Arnold Arboretum with 20% success using late July cuttings, 4000 ppm IBA-quick dip, peat:perlite and mist. Late May, 4000 ppm IBA-talc + thiram, sand:perlite, mist, rooted 66% but did not survive the overwintering.

Mongolian linden can be propagated by layering.

Tilia petiolaris Pendent Silver Linden

GRAFTING: See under *Tilia* introduction.

Tilia platyphyllos Bigleaf Linden

SEED: Requires a 3 to 5 month warm/3 month cold period and even this does not guarantee high germination.

CUTTINGS: Hardwood cuttings have been rooted in a heated bin but heavy losses after overwintering occurred. Refer to T. × *euchlora* for details. Hardwoods were rooted in a peat:sand medium at 75°F. Bigleaf linden and several of its cultivars and varieties can be propagated by layering.

GRAFTING: Grafting, T-budding and chip budding work successfully with this species. In chip budding, the polyethylene strip should entirely cover the cut surface but avoid the bud. Remove the strips as soon as possible (usually about 5 weeks) to prevent any "sleepy" bud effects. Refer to *Tilia* introduction for details on budding and grafting.

Tilia tomentosa Silver Linden

SEED: A difficult species to obtain good germination. See under *Tilia* introduction.

GRAFTING: See under *Tilia* introduction.

Torreya

A group of 6 evergreen species native to North America and eastern Asia, of which the oriental species are hardier. The species include: T. *californica*, T. *fargesii*, T. *grandis*, T. *jackii*, T. *nucifera*, T. *taxifolia*. Male and female flowers are borne on separate trees. Seeds ripen August-September of the second season and can be harvested in September-November. Seed propagation is the preferred method. *Torreya* seed germinates slowly and probably needs a cold stratification.

Torreya nucifera Japanese Torreya

SEED: Seed germinates slowly and probably needs a cold period.

CUTTINGS: Refer to *Torreya taxifolia* for a possible method.

Torreya taxifolia Florida Torreya

SEED: Seed germinates without stratification, but does so slowly, and possibly would benefit from cold stratification. Seeds of T. *californica* require a long after-ripening period before germination. Even after 3 months of stratification, seeds germinated 6 to 8 months after sowing. A pound of seed contains 225 seeds.

CUTTINGS: Late October, 8000 ppm IBA-talc, rooted 83% in 16 weeks. Summer cuttings from short side shoots have also been rooted.

Trachelospermum asiaticum Asiatic Jasmine

CUTTINGS: Nodal tip cuttings in July and August, 3000 ppm IBA, peat:sand (1:1, v/v), mist, root well. Rooting appears to be inhibited by excessive latex if taken too early in the season. Generally considered easy to root anytime wood is firm. See NGC 185 (21):14 (1979) and *Gardeners Chronicle* 180(1):17 (1976).

Trachelospermum jasminoides Star Jasmine, Confederate Jasmine or Chinese Jasmine

CUTTINGS: The species is easily rooted from softwoods in June or July but could be rooted about anytime of the year as long as the wood is firm; use 1000 to 3000 ppm IBA-talc or quick dip, perlite:vermiculite (1:1, v/v), and mist. Senior author has rooted this species in summer and fall. 'Variegatum' also roots readily, use terminal cuttings and 3000 ppm IBA-talc.

Tripterygium regelii Regel's Threewingnut

SEED: The flowers give way to lime green, bladder-like, winged fruits which ripen to a tan color. Fresh seed germinates at a low percentage. Cold stratification for 3 months induces heavy germination [PIPPS 34:606 (1984)].

CUTTINGS: This plant can be increased vegetatively by softwood and hardwood cuttings and root suckers. Early July, 4000 and 8000 ppm IBA-talc + thiram, mist, rooted 100%. Propagation from root suckers indicates that it may be a candidate for increase by root cuttings. It can also be propagated by layering.

Trochodendron aralioides Wheel Tree

SEED: Good germination when seed is sown without pretreatment.

Tsuga canadensis Canadian (Eastern) Hemlock

SEED: Cones ripen September and October of their first year. The brown cones should be harvested as soon as they mature to prevent losses due to the release of the winged seeds. Canadian hemlock cones when collected green are difficult to open, but can be opened by exposure to repeated cycles of moistening followed by drying at 100°F. Cones collected just as they turn tan will open readily upon drying. T. *canadensis* seeds may have low viability due to the difficulty of separating out poor seed. A pound of cleaned seed contains about 187,000 seeds.

Moisture content of *Tsuga* seeds should be maintained between 6 and 9%. Seed stored dry at 41°F in sealed containers remains viable for 4 years.

Dormancy is variable in hemlock seed with some seed lots requiring cold stratification. Stratification accelerates and improves total germination and, unless seeds are known not to require pretreatment, stratification at 33 to 41°F for 1 to 4 months is recommended.

Hemlock seedlings are difficult to grow. They are easily damaged in hot sun, and their small size the first year makes them susceptible to frost heaving. In nursery practice T. *canadensis* is often spring sown after stratification. However, there is no difference in growth between fall and spring sown seed. Seeds germinated under laboratory conditions require 21 to 28 days cold stratification followed by artificial light and 68 to 86°F for 14 to 28 days.

CUTTINGS: The preferred method for propagating hemlock cultivars, however, success is variable. Cuttings harvested from January to mid-February and placed over bottom heat (75°F) have been successful with most cultivars [*The Plant Propagator* 17(1):5 (1971), PIPPS 25:81 (1975) and 35:565 (1985)]. In other cases, cuttings consisting of two or more growth flushes and taken in October and November have rooted [PIPPS 21:470 (1971)]. A 5-second dip using NAA and IBA at 5000 ppm each has been most successful. The cuttings can be placed in an open bench or polyethylene chamber. Softwoods taken in July and placed under mist have also been successful. Wounding the cuttings may be beneficial. Treat the cutting base with 8000 ppm IBA-talc and Benlate (8:1, v/v).

Considerable variation also exists in rooting "witches' brooms". In one study, cuttings with 2 to 3-year-old wood at the base from six, 5-year-old "witches' broom" clones, early January, 2% IBA-quick dip, peat:perlite (1:1, v/v), rooted 40 to 100% in 4 months [PIPPS 35:555 (1985)].

GRAFTING: In the past it was customary to propagate cultivars by grafting. However, instances of incompatibility have developed. Root girdling, brought about by circling roots which arose on understocks established in small pots, can lead to failure shortly after grafting or in subsequent years.

Two-year-old stocks potted in 2¼" deep pots are used. In early December understocks are brought into a greenhouse and kept at 65°F for 6 weeks. Scions are collected daily and only when the temperature is above 35°F. A side-tongue graft can be used. A 1¼" long cut is made on the stock to a depth of ¼ the diameter and as close to the soil as possible. Match the scion to the understock and then cut a wedge on the base. A wedge is then made on two opposite sides to fit the opening on the understock. Bind the grafts, making sure to keep the cambial layers aligned and cover the union with wax. Keep the wax below 140°F [PIPPS 24:401 (1974)].

Tsuga caroliniana Carolina Hemlock

SEED: The female cone is about 1 to 1½" long at maturity. Cones mature in August and September of their first year, when they turn brown. The cones should be harvested as soon as they mature to prevent losses due to the release of the winged seeds. There are few problems in extracting seeds from *Tsuga* cones. Kiln drying temperatures range from 87 to 100°F for 48 hours. A pound of cleaned seed averages about 85,000 seeds. Carolina hemlock seed does not require fall sowing. Refer to T. *canadensis* for additional information on seed germination conditions [*Amer. Nurseryman* Sept. 15, p. 10 (1967)].

CUTTINGS: Information on the cutting propagation of this species is not available. Refer to T. *canadensis* for possible suggestions.

GRAFTING: Refer to T. *canadensis*.

Tsuga heterophylla Western Hemlock

SEED: Cones are ready for harvest in August and September when mature. The cones should be harvested as soon as they mature to prevent losses due to the release of the winged seeds. A pound of cleaned seed averages 260,000 seeds.

Dormancy is variable with some seed lots requiring cold stratification. Stratification accelerates and improves total germination and, unless seeds are known not to require pretreatment, cold stratification at 41°F for 3 weeks to 3 months is recommended. At some nurseries seed is soaked for 24 to 36 hours prior to stratification.

Ulex europaeus Common Gorse

SEED: A close relative of *Cytisus* and the authors suggest that the germination requirements are similar to that species. Either a brief acid scarification or hot water soak should suffice.

CUTTINGS: The species and 'Plenus' root from hardwood cuttings taken from October to January. Nodal cuttings, wounding, fungicidal soak prior to hormone treatment, 8000 ppm IBA-talc, peat:perlite (2:1, v/v), and mist produce good results [NGC 188(2):33 (1980)]. Heel cuttings also root well. The plant is difficult to transplant and should be direct stuck in containers or transplanted immediately after rooting.

Ulmus Elm

SEED: The fruit is a winged samara. Under natural conditions elm seeds that mature in the spring usually germinate in the same growing season; seeds ripening in the fall germinate the following spring. Generally, elm seeds require no presowing treatments; exceptions being U. *americana*, U. *crassifolia*, U. *parvifolia*, U. *rubra*, and U. *serotina* which benefit from 1 to 3 months stratification at 41°F. Presoaking for 24 hours in water prior to sowing increases seedling stands. Seeds of most species can be kept viable for 1 year under refrigerated storage (*Contrib. Boyce Thompson Instit.* 10:221).

CUTTINGS: The majority of species, hybrids and cultivars can be vegetatively propagated [*The Plant Propagator* 20(4):14 (1975)]. Layering is an oldtime method which can be practiced when limited numbers are desired.

Elms can be propagated from root cuttings. Results are the same on roots taken from adult or juvenile plants. Roots are lifted during December-January and cut into 2 to 3″ long sections. The roots are dusted with Captan and placed into a medium to their complete depth in individual containers. Individual containers are used because the roots are brittle. Proper polarity is important.

Soft shoots produced from root cuttings can also be rooted. Roots are prepared as above except that the uppermost end is left protruding about ½″. After sticking, the trays are covered with poly and provided 68°F bottom heat. A green callus forms at the cut surface and differentiates into adventitious shoots. Eight to 10 weeks after insertion the young shoots contain 6 to 8 leaves and the stems are woody enough to root. Treat with 8000 ppm IBA-talc and root under mist.

A technique of propagating elms by softwood cuttings has been developed. Stock plants are encouraged to produce soft growth. It is important to ensure that the cuttings do not dry out or are not removed too early from the mist. Root in individual pots because of brittle roots. Elm rooted cuttings are often difficult to overwinter.

A number of elm species can be propagated by layering.

GRAFTING: The more unusual species and cultivars are often propagated by grafting. In England the rootstock used is U. *glabra* because it does not sucker. The normal grafting methods include twig and bench grafting, and chip budding [*The Plant Propagator* 20(4):14 (1975)].

Twig grafting is practiced with elm species that have very thin branches. This technique should be done between March and May in a greenhouse. The normal T-cut is made as for budding, but a one-year-old, 2 to 5″ long dormant twig is used.

Bench grafting is accomplished in January on rootstocks which have been dried. Drying should last at least one week because a vigorous sap flow will inhibit union formation. A splice graft is often used and, after tying and waxing, the grafts are healed in a cool frame or placed in a heated grafting case. Union should take place in 3 to 4 weeks.

Chip budding can take place in the same year as the rootstocks are lined out in the nursery, between April and September. It is essential that the bud be tied with polyethylene strips and that the strip should be both above and below the bud, but not on the bud. Many losses can result from improper tying.

Ulmus alata Winged Elm

SEED: A pound of seed contains about 111,500 seeds. Freshly collected fruit should be air dried a few days before storage. Seed can be stored for one year at 39°F. Seeds are usually sown or stored with the wings attached. This species does not require any pretreatment before germination, but may benefit from cold stratification for 1 month.

CUTTINGS: The winged elm has been rooted from spring and summer cuttings like U. *parvifolia* [PIPPS 26:37 (1976)]. Three to 6″ long cuttings are made when the new shoots are 6 to 12″ long. If the tip growth is very soft, it is removed. Very sharp shears are recommended for making elm cuttings because of the tendency for bark to peel off if shears are dull. Cuttings are harvested in the early morning to prevent wilting. Treat with 4000 ppm alpha-naphthylacetamide-talc, perlite:vermiculite (1:1, v/v), and mist or high humidity chamber. Other reports have suggested the use of 2 to 3% IBA-quick dip levels or 8000 ppm IBA-talc. Cuttings will root in 6 to 10 weeks.

Ulmus americana American Elm

SEED: Flowering occurs from February to March and fruit ripening from late February to June. Seed dispersal occurs from mid-March to mid-June. The fruit is greenish brown at maturity. Seeds at 3 to 4% moisture have been stored for 15 years at a temperature of 25°F. Seeds stored at 40°F remained viable for 2 years. A pound of seed contains 80,000 seeds. Although the seed of some lots requires no presowing treatment, practically all the seeds in some lots remain dormant until the second season. Dormant seed lots should receive cold stratification for 2 to 3 months.

CUTTINGS: American elm can be rooted from semi-lignified cuttings and clonal differences have been reported. In one report the cuttings were collected from grafted stool plants growing in a greenhouse [PIPPS 28:490 (1978)]. The cuttings were pruned 2 to 4″ long with one or more leaves, stuck in perlite:peat (1:1, v/v), and placed in a plastic case. Occasional misting was necessary. A medium temperature of 75 to 85°F was best. Cutting survival with elms depends on adequate moisture, especially during the critical period immediately after insertion. A wilted cutting often does not recover. Air temperatures should be kept as close to 68°F as possible. Elm cuttings can be stored in a cold room at 37°F for several days if lightly packed. Early June, 8000 ppm IBA-talc, peat moss:perlite, mist, rooted 94%. See *HortScience* 10:615 (1975).

GRAFTING: Chip budding has been recommended as an alternative to T-budding for elm species. Chip budding is possible any time of the year [PIPPS 27:357 (1977)]. In practice it is confined to the period of April to September with the best time July and August.

Understocks are bench grafted in February or March, using one or two-year-old scion wood. A side or splice graft can be used. Heel in the grafts after tying and waxing in a shaded cold frame. As the buds begin to break in the spring, plant out in nursery beds. Refer to *Ulmus* introduction for additional details.

Ulmus carpinifolia Smoothleaf Elm

SEED: Freshly collected seeds germinate immediately. If the seed shows some dormancy, cold stratification at 41°F for 2 to 3 months before sowing should improve germination.

CUTTINGS: On a limited basis this elm can be propagated by digging suckers. Five inch terminal softwoods from young trees, 1000 ppm IBA, under mist, have rooted [PIPPS 14:36 (1964)].

GRAFTING: Refer to *Ulmus* introduction for details.

Ulmus crassifolia Cedar Elm

SEED: A fall flowering and fruiting species. The ripe fruit color is green and a pound of seed contains about 67,000 seeds. Stratification at 41°F for 2 to 3 months improves germination. See *Ulmus* introduction for additional details.

Ulmus glabra Scotch Elm

SEED: The fruits ripen in May and June with dispersal occurring at the same time. Fruit color is yellowish brown when ripe. The seed is collected in May-June when the green pigment has disappeared from the seed wing; once the seed is fully ripe it will be dispersed by wind, which makes collection difficult. Fruits are not dewinged because of potential damage to the seed. A pound of cleaned seed contains 40,000 seeds. Air-dried seed stored at 34 to 50°F had only a 6-month viable period. Scotch elm seed has no pregermination requirements. After collection, the seed is sown the same day if possible; if stored, keep cool and avoid drying. In a controlled study, fresh seed germinated 98%; 88% after 2 months cold stratification.

CUTTINGS: This elm does not produce suckers naturally. Considerable clonal variation in rooting exists. Refer to *Ulmus* introduction for details.

GRAFTING: *Ulmus glabra* 'Pendula' and 'Umbraculifera' are spring grafted in the field at 6 to 7′. Graft just as the frost is going out of the ground but not during frosty weather [PIPPS 17:303 (1967)].

Ulmus x hollandica Dutch Elm

SEED: A series of clones which resulted from crossing U. *carpinifolia* and U. *glabra*.

CUTTINGS: Softwoods in mid-summer, 8000 ppm IBA-talc, mist, coarse sand:peat moss (1:1, v/v) or vermiculite:sphagnum (3:1, v/v), rooted in about 6 weeks. Refer to U. *procera* for further details. 'Commelin' hardwoods, December, heated bin, 1500 ppm IBA-quick dip, 60°F bottom heat, rooted in 6 weeks. Propagation by root cuttings and cuttings taken from root cuttings is possible; refer to *Ulmus* introduction for details. Seasonal variability exists with hardwoods [*The Plant Propagator* 21(3):4 (1975)].

GRAFTING: For bench grafting and chip budding details refer to *Ulmus* introduction.

Ulmus Hybrids

CUTTINGS: Dutch Elm disease resistant hybrids through breeding programs: Hardwoods of 'Dodoens', 'Lobel', 'Plantyn', 'Sapporo Autumn Gold', and 'Urban' rooted 32, 32, 8, 10, and 30%, respectively. Conditions included basal cuttings 10″ long, 1000 ppm IBA-quick dip, peat:pumice (1:1, v/v), and cold frame with bottom heat. Cuttings were watered and ventilated daily. After 6 weeks the heat was turned off and the cuttings were transplanted [PIPPS 33:332 (1983)].

Softwoods, 5000 ppm IBA-talc, peat:pumice (1:1, v/v), and mist were successful.

Root cuttings 0.2 to 0.6″ diameter and 2 to 6″ long were dipped in thiram solution and the ends sealed with Shell Grafting Matrix (a petroleum-based product containing captafol). The root cuttings were placed in pumice and covered with sphagnum moss on a heated bed under mist. Resulting softwoods with a small piece of root were treated with 3000 ppm IBA-talc and rooted under mist. Rooting percentages were 'Dodoens' - 96%, 'Groeneveld' - 46%, 'Lobel' - 54% and 'Plantyn' - 72%.

GRAFTING: 'Sapporo Autumn Gold' and 'Urban' were both successfully (100%) grafted using a whip and tongue or cleft graft on U. *glabra* stock.

TISSUE CULTURE: 'Homestead' protoplasts were isolated enzymatically from friable callus cultures. Complete plants were not regenerated [*Plant Science* 41:117-120 (1985)]. 'Pioneer' produced shoots on an MS medium with 10 micro-molar BA. The entire process was complicated and anyone interested should see *Plant Cell, Tissue and Organ Culture* 7:237-245 (1986).

Ulmus japonica Japanese Elm

SEED: This species flowers in April-May and the fruits ripen in June. A pound of seed contains 5,800 seeds. Requires no pretreatment.

Ulmus laevis Russian Elm

SEED: This species flowers and fruits in May and June. The ripe fruit is yellow brown and a pound of seed contains 63,000 seeds. Seed has no pregermination requirements.

Ulmus parvifolia Chinese or Lacebark Elm

SEED: The Chinese elm flowers in August to September and the fruits ripen in September to October. If the seed is collected fresh (green to greenish brown stage) it will germinate. If allowed to dry, it may require cold stratification. Even fresh seeds germinate sporadically. One to 2 months cold stratification is recommended to hasten and unify germination. See PIPPS 30:428 (1980).

CUTTINGS: The Chinese elm and catlin elm (U. *parvifolia* 'Catlin') can be propagated from spring and summer cuttings [PIPPS 26:37 (1976)]. Cuttings 6 to 12″ long are trimmed to 3 to 6″ and the bottom 2 to 3 leaves removed leaving a minimum of 5 leaves at the top. Cuttings are prepared in the early morning to prevent wilting, treated with 4000 ppm alpha-naphthyl-acetamide-talc, placed in perlite:vermiculite (1:1, v/v) under mist or high humidity. Other reports have suggested the use of very high IBA levels (8000 ppm in talc or 2 to 3% quick dip). Rooted cuttings can be potted up in 6 to 10 weeks or lined out in the field or cold frame. Senior author rooted early July cutting (48%) from 10 to 12-year-old plants with 8000 ppm IBA-quick dip, peat:perlite (1:2, v/v) and mist [*The Plant Propagator* 29(2):10 (1983)]. Large clonal differences in rooting occur and from13 clones — 3 did not root, 5 rooted less than 25%, and 5 others — 68, 73, 81, 86, and 96%. Cuttings were taken in early May (Oklahoma). See *Proc.* SNA *Res. Conf.* 28:221-222.

GRAFTING: The cultivars can be grafted or budded on the species. Chip or T-budding in late summer are both used. Bench grafting in winter is possible. Refer to *Ulmus* introduction for details.

Ulmus procera English Elm

SEED: The English elm rarely produces seed in England and is vegetatively propagated. Fresh seed, when available, will germinate in high percentages. Fresh seed germinated 100%; the same after 2 months cold stratification.

CUTTINGS: Softwoods have been rooted in mid-summer in unheated benches. The cuttings should be about 6″ long and, after removal of the lower leaves and treatment with hormone, inserted to a depth of 2″. Soil temperatures of 70 to 75°F favor callus development and rooting. Place in a sand:peat (1:1, v/v) medium, treat with 8000 ppm IBA-talc and use mist. Cutting survival depends on adequate watering, especially during the critical period immediately after insertion. Variation in rooting exists among different clones. *Ulmus procera* 'Van Houttei' softwood cuttings can be rooted about 75% from late spring to early summer. Particular care is required not to damage the leaves [PIPPS 30:174 (1980)]. Standard practice involves 6″ cuttings, Captan dip, wound, 3700 ppm NAA + 8000 ppm IBA-talc mixture, sawdust:peat:sand (3:2:1, v/v/v) medium, and mist. The leaves are kept moist at all times because they dehydrate rapidly. In the propagation house it is important to maintain high mist, no air movement and heavy shade. Bottom heat is maintained between 68 and 77°F. A preventative fungicide program is practiced. In 6 to 8 weeks, 70% rooting occurs. Gradually reduce the mist when root initials first become visible. When well rooted, the cuttings are potted and returned to mist or high humidity conditions for 4 weeks. Gradually harden off and shift to the shade. It is necessary to force new growth on the plants to insure overwinter survival.

In New Zealand, hardwood cutting propagation of 'Van Houttei' is practiced [PIPPS 30:174 (1980)]. Soil preparation includes treatment with chloropicrin. Wood from specially maintained stock blocks, between ½ and ¾″ diameter, is collected from the previous season's growth in mid-winter. Cuttings are pruned to 8 to 10″, wounded, dipped in 8000 ppm IBA-talc and placed in poly-covered beds. Place 4¾ by 4¾″ apart and insert 3″ into the soil. Cover with temporary shade structure to protect spring growth. A rooting percentage of about 50% is obtained.

GRAFTING: *Ulmus procera* 'Van Houttei' has been root grafted using the whip and tongue method onto roots of 2-year-old *U. parvifolia* seedlings. After tying, the grafts are planted in the field, with the union below the soil level to prevent dehydration. T-budding on rooted cuttings is possible [PIPPS 25:278 (1975)].

Ulmus pumila Siberian Elm

SEED: The Siberian elm flowers in March and April with fruit ripening in April and May. Seed dispersal occurs when ripe. Excessive drying and dewinging should be avoided as they reduce viability. Siberian elm seeds with 3 to 8% moisture can be stored at 36 to 40°F in sealed containers for 8 years. A pound of cleaned seed contains 72,000 seeds. Seed does not require a pretreatment and should be spring sown immediately after harvest. In one test, fresh seeds germinated 96%; cold stratification did not improve germination. This is a very weedy species because of the high seed viability and the fact that germination takes place when moisture conditions are good for seedling survival [*Amer. Nurseryman* Nov. 15, p. 12 (1967)].

CUTTINGS: Siberian elm can be rooted from semi-lignified (firm) cuttings harvested from greenhouse-forced plants. Untreated cuttings, mid-March, rooted 80%; those treated with 8000 ppm IBA-talc rooted 84%.

Ulmus rubra Slippery Elm

SEED: The ripe fruit is green and a pound of seed contains 41,000 seeds. Northern sources show some dormancy and benefit from 2 to 3 months cold stratification. Fresh seed germinated 70%; 57% after 2 months cold stratification.

Ulmus serotina September Elm

SEED: The September elm flowers in September and fruit ripening and dispersal occur in November. The ripe fruit is light green to brownish and a pound of seed contains 149,000 seeds. Stratification at 41°F for 2 to 3 months before sowing improves germination.

Ulmus thomasii Rock Elm

SEED: The rock elm flowers from March to May and fruit ripening and dispersal occur in May to June. Mature fruit is yellow or brown and a pound of seed contains 7,000 seeds. The seeds have no pregermination requirements [*The Plant Propagator* 6(4):7 (1960)].

Ulmus × *vegeta* Huntingdon Elm

CUTTINGS: Root cuttings, 1½ to 2″ long, November, perlite, full light. Leave in medium until both top and root growth occur. Extra softwood cuttings from root cuttings can be rooted. Softwoods from greenhouse grown plants root 100%.

GRAFTING: 'Commelin' is chip budded in spring and summer with 75% success. 'Christine Buisman' has been budded on Chinese and Scotch elms. Refer to *Ulmus* introduction for additional grafting and budding information.

Vaccinium Blueberry, Cranberry, Huckleberry and others

A large group of deciduous and evergreen shrubs with rather delicate pink to white flowers and usually a fleshy berry containing several seeds. Fruits are relished by man and animal with *V. angustifolium*, *V. ashei*, *V. corymbosum*, *V. macrocarpon* and several others commercially significant. Fruits should be collected when ripe, macerated and the pulp and debris removed. The seeds should be dried before storage. Based on work at the Arnold Arboretum, seed dormancy is shallow and many species germinate without pretreatments. If in doubt, a 1 to 3 month cold stratification might prove beneficial. In a controlled study with *V. macrocarpon*, fresh seed germinated 4%; 56% after one month cold stratification and 100% after 3 months.

The following species were researched at the Arnold Arboretum and the results are presented in tabular form.

	Treatment inducing germination
Vaccinium alaskaense	none required
Vaccinium angustifolium	(see main entry)
Vaccinium arboreum	3 months cold stratification
Vaccinium arctostaphylos	none required
Vaccinium axillare	none required

	Treatment inducing germination
Vaccinium bracteatum	none required
Vaccinium canadense	none required
Vaccinium caespitosum	none required
Vaccinium cylindricum	none required
Vaccinium deliciosa	none required
Vaccinium erythrinum	none required
Vaccinium gaultherioides	none required
Vaccinium glauco-album	none required
Vaccinium hirsutum	none required
Vaccinium japonicum	none required
Vaccinium macrocarpon	(mentioned above)
Vaccinium membranaceum	none required
Vaccinium myrtilloides	none required
Vaccinium myrtillus	none required
Vaccinium oldhamii	some germination with no pretreatment; 3 months cold produced the best germination
Vaccinium ovalifolium	none required
Vaccinium oxycoccos	none required
Vaccinium patifolium	none required
Vaccinium scoparium	none required
Vaccinium simulatum	none required
Vaccinium smallii	none required
Vaccinium uliginosum	none required
Vaccinium vacillans	none required
Vaccinium versicolor	none required
Vaccinium vitis-idaea	none required
Vaccinium vitis-idaea var. ***majus***	none required
Vaccinium vitis-idaea var. ***minus***	none required

Vaccinium angustifolium Lowbush Blueberry

SEED: The fruit is a blue-black berry that ripens in July and August. After harvest the fruit is macerated with the pulp and unsound seed floated off. Dried seed can be stored under normal refrigeration for up to 12 years. A pound of cleaned seed contains 2,000,000 seeds. Seeds of some blueberry species are apparently not dormant while others are. Large seeds are more viable than small seeds and germination varies with clone. In one study, lowbush blueberry, when given 83 days of cold moist stratification, germinated 52% after 169 days [*HortScience* 15:587 (1980)].

CUTTINGS: The lowbush blueberry, unlike the highbush blueberry, is rhizomatous and rhizome cuttings offer a useful method of propagating this species. Optimum times for taking rhizome cuttings were found to be either early spring or late summer and autumn. Rhizomes are dug, cut into 4″ lengths and placed in vermiculite at a constant temperature of 70°F. The application of gibberellic acid and kinetin failed to increase new shoot development. IBA strongly inhibited shoot development [*The Plant Propagator* 15(2):3 (1969)].

TISSUE CULTURE: See *HortScience* 13:698 (1978) and *Can. J. Plant. Sci.* 63:467 (1983).

Vaccinium ashei Rabbiteye Blueberry

SEED: No pretreatment required. See *J. Amer. Soc. Hort. Science* 103:530 (1978).

CUTTINGS: Terminal softwoods ('Tifblue' and 'Woodward') harvested mid-May (Georgia) rooted best in media containing milled pine bark or bark:perlite (1:1, v/v) under mist with partial shade (25%). For hardwoods refer to *V. corymbosum* [*HortScience* 17:640 (1982)].

TISSUE CULTURE: Actively growing shoot tips from seedlings and mature cultivars proliferated in vitro on a modified Knops medium supplemented with in mg/liter:casein hydrolysate 300, glycine 2, myo-inositol 100, nicotinic acid 0.5, pyridoxin HCl 0.5, thiamine HCl 0.1, sucrose 30,000, 2iP 15, and agar 9,000. The pH was adjusted to 5.7. Actively growing juvenile shoot tip explants were obtained from sterilized seeds growing on water agar. Nonjuvenile shoot tips were difficult to establish. Shoot tips from mature cultivars blackened and died on agar medium and were first established on liquid medium for 2 to 4 weeks. Mature shoots produced fine filamentous shoots which were as amenable to subculture as juvenile shoots. In some cases mature leaves produced juvenile-type shoots adventitiously from leaves of the original explant when they touched the medium. A 50-fold multiplication rate every 4 months was obtained with some clones. Micro-cuttings rooted best in vitro in the presence of 0.1 mg/liter NAA. After 6 weeks, 95% rooting occurred. Rooted plants were transferred after 5 weeks to a sphagnum peat moss and perlite medium and placed under mist. Cultures were maintained at 72 to 90°F with illumination of about 1 klx for 16 hours [*HortScience* 15:80 (1980)].

Vaccinium corymbosum Highbush Blueberry

SEED: This species is important horticulturally for its edible fruit, and a number of asexually propagated cultivars are available. The fruit is a ¼ to ½″ diameter bloomy blue-black berry that ripens in late July and August. Collect, macerate and float the pulp and unsound seed off. A pound of cleaned seed contains 1,000,000 seeds. After drying for 48 hours at 60 to 70°F, the seed can be stored under normal refrigeration for up to 12 years. The seed does not require a pretreatment for germination.

CUTTINGS: Softwoods root best from May cuttings (Illinois) with decreased success in late summer [*The Plant Propagator* 18(2):9 (1972)]. Use 8000 ppm IBA-talc + thiram or 4000 ppm naphthaleneacetamide, peat:perlite, and mist. Heel cuttings from plants that have just completed their primary growth produced bushier plants. The rooted cuttings can be overwintered outside if a protective mulch is applied.

Hardwoods in March, sphagnum peatmoss:sand (1:1, v/v), root about 80%. Although no hormone is required, 8000 ppm IBA-talc + thiram improves rooting with some cultivars like 'Stanley' and 'Blue Crop'. Hardwoods collected in early dormancy and stored at 35 to 36°F in moistened polyethylene bags rooted well and avoided winter injury problems [*HortScience* 16:316 (1981)].

GRAFTING: *V. corymbosum* can be grafted on *V. ashei* and *V. arboreum* to take advantage of different soil types. A cleft, whip or side graft can be used. Mid-August T-budding on the species is possible [*J. Amer. Soc. Hort. Sci.* 26:294 (1971)].

TISSUE CULTURE: The following in vitro technique has been used with the cultivars Atlantic, Jersey, Dixie, Stanley, Burlington, Darrow, Berkeley, Ivanhoe, Blueray, Bluecrop and Earliblue. Conditions in the culture room include 79°F temperature with illumination supplied for 16 hours by 40-watt, cool-white fluorescent tubes mounted 8″ above the shelf. The culture medium was ½ strength Murashige and Skoog minerals with full strength Linsmaier and Skoog vitamins (10 mg/l myo-inositol, 0.4 mg/l thiamine-HCl), sucrose (30 g/l), and IPA (iospentenyl adenine, 5 mg/l). IPA, when added at 15 mg/l, increased shoot production but also induced adventitious shoots, and is less desirable. The medium was solidified with 6 g/l agar and sterilized for 15 minutes at 15 psi (250°F). Stock plants were grown in a greenhouse and shoots were collected at the end of a growth flush, surface sterilized in 0.6% sodium hypochlorite with 0.1% Multifilm X77 (a non-ionic detergent) for 30 minutes followed by 3 rinses in sterile distilled water. The shoots were cut into single-node sections and implanted. After 4 to 6 weeks shoots were produced and those were cut into 1 to 3 nodes and replanted into the same medium. The procedure can be repeated at 6 to 8 week intervals and gives a 5-fold proliferation rate per culture. In vitro shoots were rooted in a pumice:peat moss (1:1, v/v) medium with high humidity and shade. Treat the base with 3000 ppm IBA-talc. After two weeks about 90 to 95% of the shoots are rooted. See *HortScience* 18:703 (1983) and *Proc. Conf. Nursery Prod. Fruit Plants Through Tissue Culture*. p. 44-47. USDA SEA, Agr. Res. Results ARR-NE-11.

Vaccinium macrocarpon American Cranberry

SEED: In one study, fresh seeds germinated 4%, 56% after 1 month cold stratification, and 100% after 3 months cold stratification.

CUTTINGS: This is one of the easiest ericaceous plants to root. Treat softwoods with 1000 ppm IBA-talc and place under mist in a peat moss:perlite medium. Rooting should approach 100%. Ideally, softwood and semi-hardwood root easily, however, fall and winter hardwoods can be rooted.

Vaccinium vitis-idaea Cowberry

CUTTINGS: Current and 1-year-old wood, collected in autumn or spring just before bud break, 3000 ppm IBA, mist, peat, will root.

TISSUE CULTURE: Shoot proliferation was achieved from shoot tips excised from 1-year-old stock plants cultured on a low-salt medium (see deciduous azalea tissue culture propagation) supplemented with 20 mg/liter 2iP. The cultures were incubated at 72°F under 16 hours cool-white fluorescent light at 86 umol $s^{-1}m^{-2}$ at explant height. Rooting was obtained without auxin application by placing microshoots in a peat:vermiculite medium in a high humidity chamber. After 6 weeks the rooted cuttings were transplanted into a peat-lite mix and acclimated to greenhouse conditions. Within 9 months the plants developed a rhizomatous growth habit, in contrast to the non-rhizomatous habit of conventionally rooted cuttings [*HortScience* 20:364 (1985)].

Viburnum Viburnum

Superb shrubs or small trees (in a few species) offering excellent flower, fruit and fall color. Flowers occur in compound cymes or panicles. The drupaceous fruit ranges in color from yellow, orange, red, blue, purple to black. Fruits should be collected as mature color change occurs, macerated, and the cleaned seeds dried and stored under refrigeration or planted. Viburnum seeds are often double dormant and require extended warm stratification of 3 to 9 month duration followed by 3 months cold. A classic paper by Barton [*Contrib. Boyce Thompson Instit.* 9:79 (1937)] describes seed propagation of viburnums.

Viburnum acerifolium Mapleleaf Viburnum

SEED: The fruit is a one seeded drupe with a soft pulp and thin stone that is generally black at maturity. Fruit ripening occurs from July to October and seed dispersal in the fall. Viability of air-dried seeds has been maintained for 10 years by storing in sealed containers at 34 to 38°F. A pound of cleaned seed contains 13,000 seeds. Based on studies with arrowwood, nannyberry, and blackhaw viburnums, viburnum seeds can be stored for at least 10 years either at 34 to 38°F or as low as 0°F if properly dried and sealed in a moisture proof container. Mapleleaf viburnum requires a warm stratification of 6 to 17 months followed by a 2 to 4 months cold stratification at 34 to 41°F. Some growers collect the seed early in the fall and plant immediately. This circumvents the warm period.

CUTTINGS: Late June, 8000 ppm IBA-talc + thiram, sand:perlite, mist, rooted 100%. Control cuttings rooted about 50% [*The Plant Propagator* 18(1):7 (1972)]. Probably best to leave rooted cuttings in place and transplant after new growth occurs in spring.

Viburnum alnifolium Hobblebush

SEED: The purple-black fruit ripens during August and September. A pound of cleaned seed contains 11,500 seeds. The seed requires a warm stratification of 5 months at 68 to 86°F followed by 2½ months cold stratification at 34 to 41°F.

Viburnum betulifolium

SEED: 5 months cold/3 months warm.

Viburnum bitchiuense Bitchiu Viburnum

SEED: Should be handled as described for V. *carlesii*.

CUTTINGS: Late May, 8000 ppm IBA-talc + thiram, perlite:sand, mist, rooted 63%.

Viburnum × bodnantense

SEED: A hybrid between V. *farreri* and V. *grandiflorum*, represented in the trade by the cultivars 'Dawn', 'Charles Lamont' and 'Deben' which are vegetatively propagated.

CUTTINGS: Softwoods produced from the first flush of growth (late May and June) are the best. Treat with 8000 ppm IBA-talc and place under mist in a peat moss and perlite medium. Late July cuttings treated with 3000 ppm IBA-talc rooted 100% in 7 weeks, while cuttings in polytents rooted 100% with 4000 ppm IBA-talc + thiram. In Denmark, low white polyethylene tunnels are used in the propagation of V. × *bodnantense*. The cuttings are inserted in August, potted up in March and placed under heated glass. Hardwood cuttings are an economical method of producing this species [PIPPS 20:378 (1970)]. Refer to V. *dentatum* for hardwood propagation details.

Small quantities can be produced by French layering. The shoots are laid horizontal to the ground early in the year so that the lateral buds on the stems break into growth and grow upward. When the new shoots are a few inches long they are covered by soil and again in early summer. After leaf-drop the layers are harvested.

Viburnum bracteatum Bracted Viburnum

SEED: Should respond like V. *dentatum*. 5 months warm/3 months cold produced poor germination. A handsome southern arrowwood with lustrous dark green leaves and a more compact habit.

CUTTINGS: Refer to V. *dentatum* for cutting propagation details.

Viburnum buddleifolium Woolly Viburnum

CUTTINGS: Early July, 4000 ppm IBA-talc + thiram, sand:perlite, mist, rooted 75% with heavy root systems in 11 weeks.

Viburnum burejacticum

CUTTINGS: Easy to root, late July, 4000 ppm IBA-talc + thiram, rooted 90% with excellent root systems.

Viburnum × burkwoodii Burkwood Viburnum

SEED: This species is a hybrid between V. *carlesii* and V. *utile* and is therefore asexually propagated to maintain its unique characteristics. Fruits are often evident and the resultant seedlings might prove interesting. Handle as described under V. *carlesii*.

CUTTINGS: The rooting of stem cuttings is not a problem, especially under mist. Established plants that are flowering produce cuttings with a lower capacity to regenerate, whereas stock plants which are pruned regularly produce shoots with high rooting capacity. Timing extends into July-August for this species, however, softwood cuttings in June are better. The earlier the cuttings are taken the better because a subsequent growth flush often occurs. The cuttings are inserted into a well drained medium under mist after treating with a hormone such as a 1000 ppm IBA-quick dip. Mid August cuttings treated with 8000 ppm IBA-talc + thiram rooted 100%. Late August cuttings treated with 8000 ppm IBA-quick dip also rooted 100% in 8 weeks. If possible, do not disturb the cuttings after they have rooted. 'Mohawk' can be easily propagated only when growth is very soft.

In Denmark, low white polyethylene tunnels are used in the propagation of V. × *burkwoodii*. The cuttings are inserted in August and potted in March under heated glass.

Any viburnum species or cultivar containing V. *carlesii* as a parent is most logically rooted early and left in place. Some nurserymen actually leave the cuttings in the rooting beds (outside) for two seasons. Senior author roots V. × *burkwoodii*, 'Chesapeake', 'Eskimo', etc., in individual containers in early summer. Root systems are not disturbed and the rooted cuttings are overwintered in a double-layer inflated polyhouse. Survival is excellent.

GRAFTING: Burkwood viburnum can be readily propagated by budding on V. *lantana* or V. *opulus* rootstocks. Well-established field-lined rootstocks are budded in August using a T-bud method. The rootstocks are headed back in February after which continued attention to removal of suckers is necessary. This technique is not desirable because of suckering and incompatibility problems and has largely been superseded by cuttings [PIPPS 20:378 (1970)].

The traditional grafting system is bench grafting during August and September. Seedling rootstocks of V. *lantana* are potted the previous autumn into 3″ diameter pots. The grafting operation is conducted on understocks headed back to within 1/5″ of the soil and grafted with a two bud scion using a whip graft. A good union should form in 4 to 5 weeks. The use of rootstocks produced from cuttings which have had the lower buds removed will eliminate the suckering problem.

Viburnum × carlcephalum Carlcephalum or Fragrant Viburnum

SEED: A hybrid between V. *carlesii* and V. *macrocephalum* var. *keteleeri* which is asexually propagated.

CUTTINGS: Relatively easy to root from softwood cuttings. Refer to V. × *burkwoodii* for details. 'Cayuga' can be propagated only when growth is very soft and before the terminal flower buds have been initiated. Late June (Boston) cuttings treated with 8000 ppm IBA-talc + thiram under mist rooted 84% in 4 weeks. Late August (Boston) treated similarly rooted 80% in 8 weeks. Again this hybrid species can prove difficult to overwinter and might be handled like V. × *burkwoodii*.

Viburnum carlesii Koreanspice Viburnum

SEED: Some nurseries propagate Koreanspice viburnum from seed because seedlings grow faster than cuttings. Collect fruits from large flowered, fruited and leaved forms. Two months warm (68 to 86°F) stratification followed by 2 months cold (41°F) stratification induces good germination. This is one of the easiest viburnums to grow from seeds. Collect seeds as they color in August and before coats are soft and black, clean and sow at this time. Many seeds germinate the following spring.

CUTTINGS: Softwoods up to mid-July, 8000 ppm IBA-talc, peat:perlite, mist, rooted about 80%. Root in flats in the fall and move to a cold frame, poly or pit house maintained at 35°F. A large Tennessee grower roots cuttings from summer softwoods using 1% IBA-quick dip, a sandy soil and mist in a shaded poly tunnel outside. Rooted cuttings are kept in the beds 2 years before lifting. The grower noted that all cultivars or hybrids with V. *carlesii* parentage should not be lifted until the second growing season. 'Compactum' rooting conditions include, softwoods, 2000 ppm IBA + 5000 ppm NAA, and mist [J. *Amer. Soc. Hort. Sci.* 93:699 (1968)]. June cuttings (Rhode Island) are best and the cuttings are overwintered in a cold frame without disturbing. Fall cuttings are dormant and fail to root.

Propagation by layering is also possible.

GRAFTING: Koreanspice viburnum can be grafted onto V. *lantana* or V. *dentatum*. This is not a preferred method because both understocks sucker. When V. *dentatum* is used as an understock, 1-year seedlings are used because they have fewer sucker buds. Pot the seedlings in the winter and graft by the end of August under double glass. Any sucker buds left will be forced out by the grafting operation and can be cut out. Overwinter the grafts in a cold frame and plant out the following spring [PIPPS 21:384 (1971)]. Refer to V. × *burkwoodii* for details when using V. *lantana*.

Viburnum cassinoides Withe-rod Viburnum

SEED: The fruit is a drupe that turns from green to pink to red to blue and finally black at maturity. Fruits ripen in September and October. Extraction of the seed from the pulp is recommended by running through a macerator. A pound of cleaned seed contains 27,600 seeds. Seeds of withe-rod viburnum are doubly dormant and require warm stratification at 68 to 86°F for 2 months followed by 3 months cold stratification at 34 to 41°F.

CUTTINGS: Easy to root from June and July cuttings. Treat with 3000 to 5000 ppm IBA-quick dip and place under mist. Overwintering can be a problem. Follow the procedures described under V. × *burkwoodii*. 'Nanum' rooted 100% in 6 weeks from late June cuttings when treated with 8000 ppm IBA-talc and placed in sand under mist.

Viburnum × chenaultii Chenault Viburnum

CUTTINGS: Closely allied to V. × *burkwoodii* and considered a seedling selection of the same parentage. Easily rooted from June and July cuttings; late July cuttings, treated with 3000 ppm IBA-talc, mist, rooted 90%. Overwinter as described under V. × *burkwoodii*. Chenault viburnum can be propagated by layering.

Viburnum cylindricum Tubeflower Viburnum

SEED: Three months cold stratification produced good germination.

Viburnum davidii David Viburnum

SEED: Seven months warm stratification followed by 3 months cold stratification induces germination.

CUTTINGS: Very soft shoot tips taken early in the season and placed under mist root well. Later cuttings root with 8000 ppm IBA + 500 ppm NAA solution. A liquid formulation hastens rooting and induces a heavier root system. Hardwoods with flower buds removed, 8000 ppm IBA-talc, a suitable medium, and mist will root [PIPPS 22:123 (1972) and 31:109 (1981)]. Overwintering may be a problem and procedures described under V. × *burkwoodii* should be followed.

Viburnum dentatum Arrowwood Viburnum

SEED: The blue-black drupe ripens in September and October. A pound of cleaned seed contains 20,400 seeds. This species is widely distributed from New England to Florida and its germination requirements are equally varied. With seeds from the southern part of the range, temperature is not critical. Northern seed sources may require 12 to 17 months of warm stratification at 68 to 86°F followed by 15 to 30 days cold stratification at 41°F. It has been reported that if the fruit is picked just prior to ripening and sown directly in August, germination will occur the following spring for northern sources [*Amer. Nurseryman* 86(5):58 (1947); PIPPS 8:126 (1958)]. Refer to V. *acerifolium* for additional details on seed germination.

CUTTINGS: Arrowwood can be propagated from softwood, hardwood or root cuttings. Softwoods are preferred because the cuttings root easily and establish quickly. A wide range of hormone treatments, varying from 1000 ppm IBA + 500 ppm NAA solution to 8000 ppm IBA-talc, have been used successfully. Place in a suitable medium such as peat moss:perlite or sand, and mist. Rooting percentages of 94 to 100% are reported. Hardwoods selected from flowering plants are not suitable. Plants cut back hard produce suitable cuttings. At leaf fall the canes are cut into 5 to 6″ long nodal cuttings, treated with 8000 ppm IBA-talc, and placed in a cold frame protected from frost. Basal sections of the shoots should be used, as there is a demonstrable positional effect on regeneration. Peat moss:sand (2:1, v/v) is used and the cuttings are inserted so that only the top ½″ is exposed. As the cutting will remain and develop in place until the following autumn, sufficient space must be allowed (3 × 4″). In the spring, harden off gradually as the cuttings leaf out.

Viburnum dilatatum Linden Viburnum

SEED: The red fruit ripens in September and October and often persists into winter with the appearance of "dried red raisins". The seeds have a complex dormancy and are difficult to germinate. Five to 7 months warm stratification followed by 3 to 4 months cold stratification is recommended. Germination response is variable and 5 months warm with 3 months cold produced only 8% germination. The same treatment produced heavy germination of 'Xanthocarpum' (yellow fruited form).

CUTTINGS: Softwoods under mist in early summer with 8000 ppm IBA-talc or solution, peat moss:perlite, sand or suitable medium, mist. Rooting will average 80 to 100% [*The Plant Propagator* 17(4):3 (1971)]. Early August (Boston) cuttings, treated with 4000 ppm IBA quick dip, rooted 60%. If the cuttings are rooted early and kept under long days (18 hours) to induce a flush of growth, they will overwinter in a cold frame. Later rooted cuttings that do not make a new growth flush must be overwintered in a cool greenhouse. In general, this species is not as difficult to overwinter as V. *carlesii* but should not be disturbed after rooting.

Viburnum erosum Beech Viburnum

SEED: 3 to 5 months warm/3 months cold.

CUTTINGS: Early August, 3000 ppm IBA-talc, rooted 80%.

Viburnum erubescens

CUTTINGS: Roots readily from softwoods; follow procedure described under V. *farreri*.

Viburnum farreri (V. *fragrans*) Fragrant Viburnum

SEED: Seed is sparsely produced because the very early flowering habit results in late frost damage. Also, the absence of different clones or seedlings for cross pollination probably affects fruit set.

CUTTINGS: Softwoods root easily. The earlier the cuttings are selected the better for ultimate survival the first winter. Senior author has rooted the species in June using 1000 ppm IBA-solution, peat:perlite and mist. 'Nanum' rooted 80% in early August when treated with 4000 ppm IBA-quick dip.

Hardwoods from stock hedges are an economical means of producing this species. Cut into 5 to 6″ lengths, 8000 ppm IBA-talc, insert in a cold frame and protect from frost.

Small quantities can be produced by French layering [PIPPS 20:378 (1970)]. This technique is described under V. × *bodnantense*.

Viburnum furcatum Forked Viburnum

CUTTINGS: Late June, 8000 ppm IBA-talc, rooted 21%.

Viburnum hupehense Hupeh Viburnum

SEED: Seeds are doubly dormant and require 5 months at 60 to 85°F followed by 3 months at 41°F [*Arnoldia* 14(5):25 (1954)].

CUTTINGS: July cuttings root readily.

Viburnum japonicum Japanese Viburnum

CUTTINGS: Readily rooted from June, July and August cuttings. Treat with 3000 to 8000 ppm IBA-talc or solution and place in a well drained medium under mist. Leave in place and transplant in spring. This is a large-leaved evergreen form that can also be rooted in October through March from hardwood cuttings. Bottom heat coupled with the above treatments would be beneficial.

Viburnum × juddii Judd Viburnum

SEED: A hybrid between *V. carlesii* and *V. bitchiuense*. Occasionally fruits are produced and the resultant seedling populations might prove interesting. Treat as described for *V. carlesii*.

CUTTINGS: This hybrid species roots more easily than *V. carlesii*. Cuttings taken in late June, treated with 3000 ppm IBA, placed in sand:peat moss medium with mist, rooted about 90% in 12 weeks. Overwintering the rooted cuttings can be a problem. For additional details on rooting and overwintering refer to *V. × burkwoodii*.

Judd viburnum can be propagated by layering.

GRAFTING: Judd viburnum can be grafted on *V. lantana* and *V. opulus*. Refer to *V. × burkwoodii* for details. This is not a preferred method because of the suckering problem.

Viburnum lantana Wayfaringtree Viburnum

SEED: This species is commercially propagated from seed. The fruit is a drupe that changes from yellow to red to black as it matures in August to late September and may persist into winter. A pound of cleaned seed averages 8,700 seeds. Two months cold stratification at 40°F is adequate to stimulate germination. Embryos extracted from dormant seeds germinate readily which indicates that the dormancy problem does not reside in the embryo itself [PIPPS 12:150 (1962)]. Collect seed before it turns black in August, clean, sow and many seedlings will germinate the following spring.

CUTTINGS: Softwoods root easily [PIPPS 30:398 (1980)]. Treat with 3000 to 5000 ppm IBA-talc or quick dip, peat:perlite (1:1, v/v) or suitable medium, and mist. Refer to *V. × burkwoodii* for additional information on the rooting of viburnums from softwood cuttings. In one study 'Mohican' rooted 89% from semi-hardwoods, 5000 ppm IBA, peat:perlite, mist [*The Plant Propagator* 26(3):5 (1980)].

Viburnum lentago Nannyberry Viburnum

SEED: This species is primarily grown from seed. The blue-black drupe ripens in September to October. A pound of cleaned seed averages 5,900 seeds. The seed requires 5 to 9 months warm stratification at 68 to 86°F fluctuating temperatures followed by 2 to 4 months cold stratification at 41°F.

CUTTINGS: Nannyberry viburnum roots easily from softwood cuttings. Treat with 2000 to 8000 ppm IBA-talc or solution and place under mist or in a plastic tent. Overwintering may be a problem.

Viburnum lobophyllum

SEED: Seeds are doubly dormant and require 5 months at 60 to 85°F followed by 3 months at 41°F [*Arnoldia* 14(5):25 (1954)].

Viburnum macrocephalum Chinese Snowball Viburnum

SEED: Sterile form and sets no fruit.

CUTTINGS: Can be rooted from summer softwood cuttings when the wood has firmed at the base. Soft succulent cuttings should be avoided and this applies to all viburnums. If tips are soft they should be removed. Treat with 5000 to 10,000 ppm IBA-talc or solution, and place under mist in a well drained medium. A rooting percentage of 90 to 100% can be expected. Treat rooted cuttings as described for *V. carlesii*. This species resists moving after rooting. A Tennessee nurseryman related that this is a difficult species to root. Senior author collected mid-August (Georgia) cuttings, 1.0% K-IBA solution, rooting plugs, mist, with 30% rooting in 9 to 10 weeks.

Viburnum nudum Smooth Withe-rod

SEED: This is a southern coastal plain species and requires no pretreatment for germination.

CUTTINGS: Easily rooted as described under *V. carlesii*. No problem overwintering this species.

Smooth withe-rod viburnum can be propagated by layering.

Viburnum opulus European Cranberry Viburnum

SEED: The red drupe ripens September to October and persists into winter. A pound of cleaned seed contains 13,600 seeds. The seed requires warm and cold stratification to germinate. A 2 to 3 month warm stratification at 68 to 86°F followed by 1½ months cold stratification at 41°F is most effective [PIPPS 8:126 (1958)]. One reference noted 5 months warm and 3 months cold was satisfactory.

CUTTINGS: *Viburnum opulus* and its cultivars are easy to propagate. Soft summer cuttings root well. Treat with 1000 ppm IBA and place under mist in peat moss and perlite or suitable medium. If the cuttings are taken from the first flush of growth, the long growing season promotes the development of a well-established plant by autumn. Overwintering is the main concern as with many other viburnums. Supplemental light has been used to induce new growth on *V. opulus* 'Nanum' rooted cuttings and improve survival. Senior author has observed "after-rooting" decline and overwinter survival problems. Unless cuttings are rooted early and induced to grow, they should be handled like *V. × burkwoodii* [PIPPS 20:378 (1970)].

Although no information is available, it is probable that this plant would propagate satisfactorily from hardwood cuttings if stooling was practiced. Make sure the cuttings are nodal, as this plant has a large pith. Refer to *V. dentatum* for hardwood cutting methods.

For small quantities simple layering or French layering (refer to *V. × bodnantense* for method) are possible. It is necessary to stool the plants so vigorous shoots are available.

Viburnum plicatum Japanese Snowball Viburnum

Viburnum plicatum var. *tomentosum* Doublefile Viburnum

SEED: The variety is fertile and sets viable seed which ripens in June and July. Might be cleaned and sown immediately with germination taking place the following spring.

CUTTINGS: Overwintering rooted cuttings is a problem although it is not as serious as with *V. carlesii*. Softwoods, mist, 1000 ppm IBA solution, peat:perlite, root easily [*The Plant Propagator* 18(1):7 (1972)]. It is important to encourage new vegetative growth after rooting and not transplant until new growth starts. The best cuttings are produced from pruned stock plants that develop vigorous young growth. Cuttings rooted early are more easily overwintered than those rooted in August. The rooted cuttings can be left in place after rooting for 2 years.

Hardwood cuttings root provided they are in a vigorous vegetative condition. The technique is the same as that described for V. *farreri* and V. *dentatum*.

A reliable method for the propagation of small numbers is layering, such as French layering. The plants are stooled back to produce a low crown and many vigorous shoots. Ten to 12 of the most vigorous shoots are selected and pegged down early in the new year. The timing is important to allow sufficient time fo apical dominance to be dissipated before bud break. In the spring, after the buds have grown out 4 to 6″, the layer leads are dropped 4″ into the soil and then covered with earth to their tips. Further attention is not required until autumn harvest.

Viburnum × pragense Prague Viburnum

This hybrid between V. *utile* and V. *rhytidophyllum* was raised in Prague and is very hardy. At the Arnold Arboretum it survived -6°F without leaf damage and has displayed excellent heat tolerance in the south. It is an attractive evergreen shrub with lustrous dark green leaves. The flowers are pink in bud, opening to white.

CUTTINGS: Firm wood cuttings in June, July and August (Georgia) and later are rooted 90-100% with 1000 to 3000 ppm IBA-quick dip. Place in peat moss:perlite or suitable medium under mist. Rooted cuttings transplant easily and continue to grow if fertilized. Probably can be rooted year-round except from extremely soft growth.

Viburnum prunifolium Blackhaw Viburnum

SEED: The black fruit (drupe) ripens September to October and persists into early winter. A warm stratification period of 5 to 9 months followed by 1 to 2 months cold stratification is required.

CUTTINGS: Although information is not available on cutting propagation, June cuttings should root if treated with 3000 to 8000 ppm IBA in talc or solution and placed in a peat moss:perlite or suitable medium. Overwintering may be a problem.

Viburnum pubescens Downy Viburnum

SEED: 4 months warm/3 months cold induced good germination.

Viburnum × rhytidophylloides Lantanaphyllum Viburnum

SEED: A hybrid of V. *lantana* and V. *rhytidophyllum* and must be asexually propagated although prodigious quantities of fruits are often set if another seedling or related species are in the vicinity.

CUTTINGS: Cultivars of this hybrid are easy to root from softwoods using 1000 ppm IBA, peat moss:perlite, and mist. Rooted cuttings present no overwintering problems. 'Alleghany' in one study rooted above 90% with semi-hardwoods in June, 5000 ppm IBA, peat:perlite, and mist [*The Plant Propagator* 26(3):5 (1980)].

Layering is also possible for a limited number of plants.

Viburnum rhytidophyllum Leatherleaf Viburnum

SEED: Possibly similar to V. *lantana* in germination requirements.

CUTTINGS: Propagation of this evergreen species is easy. It can be propagated throughout summer into fall and winter. Ideally, 1000 to 3000 ppm IBA-solution, well drained medium, mist, produce excellent results. Hardwoods taken in March and April, 8000 ppm IBA, peat:perlite, 65°F bottom heat, mist, root readily. Wounding is helpful with hard cuttings [PIPPS 24:207 (1974)].

Summer (June, July, August) cuttings, fairly small in size, can be rooted very easily with no apparent seasonal decline in rooting capacity.

Rooting has been reported under thermo-blankets (refer to *Pachysandra terminalis* for details).

Viburnum rufidulum Southern or Rusty Blackhaw

SEED: The blue-black fruit (drupe) ripens in September and October with dispersal at the same time. A pound of cleaned seed contains 5,500 seeds. Southern blackhaw requires a warm stratification of 6 to 17 months followed by 3 to 4 months of cold stratification.

CUTTINGS: Cuttings root easily when taken as soft or maturing wood from late May through August and treated with 3000 to 8000 ppm IBA-talc or quick dip and placed under mist. Cuttings taken in late May, 4000 ppm IBA-talc + thiram, mist, rooted 71%. Growth ceases after the cutting is taken and overwintering may be a problem. Refer to V. × *burkwoodii* or V. *carlesii* for additional information on overwintering rooted cuttings.

Viburnum sargentii Sargent Viburnum

SEED: The fruit is a one-seeded red drupe with a soft pulp and thin stone. The fruit ripens from August through October. Seeds possess a double dormancy similar to V. *opulus*. A warm stratification of 2 to 3 months followed by cold stratification for 1 to 2 months promotes germination. In one test, 5 months warm/3 months cold produced excellent germination. 'Flavum' shows radical emergence after 3 months warm stratification.

CUTTINGS: It is important to induce a flush of growth for successful overwintering. Cuttings rooted early and kept under long days until a flush of growth has occurred will overwinter. Treat with 8000 ppm IBA, peat moss:perlite or perlite:vermiculite and mist. Cuttings can be rooted in place and left for 2 years before digging. 'Ononadaga' rooted 78% from semi-hardwoods in June using 5000 ppm IBA, peat:perlite, and mist [*The Plant Propagator* 26(3):5 (1980)].

Hardwood cuttings can be rooted from cuttings produced in stool beds. Cuttings have a large pith and should be nodal.

For small quantities layering is a useful method, either simple layering of individual stems or French layering. In either case it is necessary to stool the plants.

Viburnum setigerum Tea Viburnum

SEED: A 3 to 5 month warm period followed by 3 months cold produces the best results.

CUTTINGS: Very easy to root. Although the species can be rooted into September, the earlier the cuttings are taken the better. Treat with 8000 ppm IBA-talc or solution and place in a peat:perlite medium under mist. The major problem is the overwintering of the rooted cuttings. Stimulating a flush of growth is important. Refer to V. × *burkwoodii* and V. *carlesii* for details.

Propagation from hardwood cuttings and layering are additional possibilities, although neither are reported.

Viburnum sieboldii Siebold Viburnum

SEED: Cold stratification for 3 months produced 36% germination. Perhaps a brief warm period (1 to 2 months) prior to cold stratification would improve germination.

CUTTINGS: Summer cuttings from early June to August are the best; the earlier the better to overcome the overwintering problem. Treat with 8000 ppm IBA-talc and place in a peat moss:perlite medium under mist. Early August (Boston), 4000 ppm IBA-quick dip, rooted 100% with a profuse root system.

Viburnum suspensum Sadankwa Viburnum

CUTTINGS: Treat like V. *tinus*.

Viburnum tinus Laurustinus

CUTTINGS: This species and its cultivars are very easy to root from stem cuttings and it is merely a question of when to fit it into the propagation schedule. Softwoods, mid-summer, 8000 ppm IBA-talc, peat moss:perlite, mist, are the best. Cuttings will be well rooted in 3 to 4 weeks. June cuttings in low plastic tunnels rooted satisfactorily when treated with 8000 ppm IBA-talc. Autumn and winter hardwoods with a wound under mist root in high percentages.

Viburnum trilobum American Cranberry Viburnum

SEED: Four months warm/3 months cold has produced good germination.

CUTTINGS: Handle as described under V. *opulus*.

Viburnum wrightii Wright Viburnum

SEED: Treat as described under V. *dilatatum*. This species can best be described as a nonpubescent V. *dilatatum*.

CUTTINGS: Softwoods, 2000 ppm IBA + 500 ppm NAA solution, mist, root. See V. *dilatatum* for additional information.

Vinca major Large Periwinkle

CUTTINGS: Easily rooted year-round except when growth is extremely soft.

Vinca minor Common Periwinkle

SEED: Not commercially propagated from seed to the authors' knowledge. The fruit (follicle) is seldom set under cultivation. The only study authors noted was with seed of 'Gertrude Jekyll'. Seeds treated with 1000 ppm gibberellic acid + 3 months cold stratification germinated 70% in 1 month. Neither gibberellic acid nor cold alone induced germination.

CUTTINGS: Division of the clumps and rooted runners in spring is possible. Roots readily from softwood cuttings which often have preformed root initials at the nodes. Treat with 1000 to 3000 ppm IBA-talc or quick dip and place under mist in a peat:perlite, sand or suitable medium. IBA-talc at 8000 ppm can be used with mature cuttings. Periwinkle cuttings that wilt reportedly do not recover, so care during preparation is necessary. Fall hardwood cuttings in a cold frame are also possible. Appears to be a strong nodal rooter so collect multiple node cuttings.

TISSUE CULTURE: Junior author has propagated this species by tissue culture. Shoot tips from actively growing plants were disinfected with 10% Clorox for 10 minutes with continuous agitation. After rinsing 3 times with sterile distilled water the shoot tips were trimmed to 5 to 10 mm. Stage 1 establishment medium consisted of Woody Plant Medium (WPM) salts, Murashige and Skoog vitamins, 30 grams/liter sucrose, 7 g/liter Difco Bacto-agar, 8.9 microM benzyladenine (BA), and 0.107 microM NAA. The cultures were maintained at 80°F with a light intensity of 14 microM $s^{-1}m^{-2}$ for 24 hours. In the multiplication medium the maximum production of shoots 5 mm and longer occurred in a medium containing 64 microM BA and 0.1 microM NAA, whereas 64 microM BA plus 1.0 microM NAA produced the most shoots. Proliferated shoots were rooted and established in a soil-less medium under high humidity.

Vitex Chaste Tree

Summer flowering shrubs and trees with purple, lilac, rose and white flowers that occur on the current season's growth. Although killed back to the ground in northern climates, regrowth approximates 6' by the end of summer. Fruit is a small drupe that ripens in September and October. The entire infructescence should be collected and dried. Seeds can be shaken from the dried structures and stored dry for one year. Seeds do not require a pretreatment.

Vitex agnus-castus Lilac Chaste Tree

SEED: Heavy germination without pretreatment.

CUTTINGS: Softwoods root readily (100%). Collect in May through July before the inflorescence appears in August. Treat with 1000 to 4000 ppm IBA-quick dip, peat moss:perlite and mist. Reduce misting as soon as rooting commences because *Vitex* declines rapidly under excessive moisture conditions. Cuttings will continue to grow after rooting.

Hardwoods taken before cold weather will root.

Vitex negundo Chaste Tree

SEED: No pretreatment required.

CUTTINGS: This species and the cutleaf forms, 'Heterophylla' and 'Incisa', are hardier than V. *agnus-castus*. All are easily rooted as described under V. *agnus-castus*. 'Heterophylla', taken in late July (Boston) and placed in a peat moss:perlite medium after treatment with a 4000 ppm IBA-quick dip, rooted 100%. Cuttings decline in the mist bed like the previous species and mist should be reduced after rooting (about 2 to 3 weeks). Cuttings are easily transplanted and will continue to grow.

Weigela florida Old Fashioned Weigela

SEED: Seeds of *Weigela* species have no dormancy and can be sown directly. This plant is not commercially propagated by seed. Seeds of *Weigela grandiflora*, W. *maximowiczii*, W. *middendorfiana* and W. *wagneri* germinate readily without pretreatment.

CUTTINGS: A very easy plant to root from cuttings about any time of the year. June and July is the best time to take softwood cuttings [*The Plant Propagator* 30(2):11 (1984)]. Softwoods in June and July, treated with 8000 ppm IBA-talc and placed under mist root readily. Actually, 1000 to 3000 ppm IBA solution proves ideal. The medium selected is not that important as long as it is well drained. Cuttings should be well rooted in 3 to 5 weeks and will continue to grow after rooting. Either polyethylene tents outdoors or mist can be used. If polytents are used the cuttings are completely covered with white poly. Lath or shade cloth providing about 75% shade is used to cover the tent. The tent is left unattended for 5 weeks after which both ends are opened to harden off the rooted cuttings for 2 to 3 days. After 3 days the plastic is removed and the 75% shade is left in place [PIPPS 19:290 (1969)].

In England, hardwood cuttings are used. In the fall, after leaf drop, the cuttings are collected from stock hedges, cut to 6″ lengths, tied into bundles, and treated with 8000 ppm IBA-talc. The material should be one-year-old wood and no thinner than a pencil. The bundles of cuttings are placed in sand pits and covered with sand. In February or March the cuttings which have callused are planted into 12″ rows with cuttings 3″ apart. Cuttings are inserted with only one inch above the soil surface [PIPPS 27:64 (1977)].

Old fashioned weigela can be propagated by layering.

Wisteria Wisteria

The wisteria species, W. *floribunda*, W. *frutescens*, W. *sinensis*, and W. *venusta* can all basically be propagated using the same techniques. The fruits are green bean shaped pods that persist into winter. The pods dehisce slowly and often not at all. Collect pods, extract seeds, and store dry under refrigeration. A 24 hour water soak followed by planting results in excellent germination.

Wisteria floribunda Japanese Wisteria

SEED: Seeds germinate readily without treatment. If the seeds are dry, a 24 hour soak in warm water is recommended. Do not sow too deeply or decay may be a problem. Seeds should not be used to propagate plants for sale because seedling produced plants are variable and are slow to flower. Seedlings should be used only for grafting understock production or in breeding work.

CUTTINGS: June through August; defoliation may be a problem with cuttings taken too early. A rooting hormone is not necessary although a weak hormone increases the number of roots. Place in a peat moss:perlite medium under mist. Hardwood cuttings root readily. Collect in December and store until spring. They may be rooted by the time they are planted.

Wisterias are well adapted to propagation by layering. Use the serpentine method. Select the long shoots in the spring and lay on the ground. These shoots are buried at intervals in a snake-like fashion. Roots will form on the buried part and shoots on the above ground part.

Root cuttings 1 to 2″ long placed in bottom heat will develop shoots in 4 to 5 weeks.

GRAFTING: Wisteria cultivars are often grafted. Sow the seeds in prepared beds in the spring. About early December, when the leaves have fallen, lift the understocks. If is important to lift the understock with fibrous roots intact. Store the seedlings between layers of moist peat under refrigeration. In January prepare a heated bench with a 50/50 peat:sand mix, to a depth of 5″. Insure that the peat is well moistened. Heat to 55°F (gentle bottom heat). Cover with a polytent.

Take the scion material from stock plants and dip in Captan solution. Cut the scions into 6″ sections containing 3 buds. The scion should have ½ to 1″ of stem below the bottom bud. Head the rootstock below the cotyledons (to remove all eyes) and trim the roots. It is possible to use sections of root when understock is short. Use a wedge, whip and tongue, or reverse veneer graft; insure that the cambial layers meet. Use rubber ties and string from the bottom. Leave 1/8″ gap between ties for callus expansion. Place grafts in trenches in the prepared bench. Cover the union to 1/2″. The poly should be 6″ above the grafts and they should be left for 6 weeks. Ventilate once a week for first 3 weeks. When root development starts, pot into 50/50 peat:sand mix, making sure the union is below soil level. It is important to shade the grafts.

Xanthoceras sorbifolium Yellowhorn

SEED: Seeds require no pretreatment for germination, however, cold stratification for 2 to 3 months at 41°F hastens and unifies germination.

CUTTINGS: Root cuttings with moderate bottom heat can also be used [*Gardeners Chronicle* 169(3):22 (1971)].

Xanthorhiza simplicissima Yellow-root

SEED: A follicle that ripens in late summer. Fall planting should work.

CUTTINGS: Late June cuttings, 4000 or 8000 ppm IBA-talc + thiram, mist, rooted 97 and 93%, respectively, in 4 weeks.

Clumps can be divided in late winter with excellent results.

Yucca filamentosa Adam's-needle Yucca

SEED: The fruit is a dehiscent capsule containing a large number of seeds (over 100). The capsules ripen from August to September and seed dispersal occurs at that time. Since the capsules dehisce they should be gathered before or at the time of ripening. Number of cleaned seeds per pound is not known but averages about 22,000 seeds for other yucca species. Yucca seeds have been satisfactorily stored dry at room temperature. A pretreatment is not needed for germination, however, there may be some hard-seededness. The speed of germination can be increased by soaking in water for 24 hours at room temperature or mechanical scarification. This procedure applies to *Yucca glauca*, Y. *flaccida*, Y. *gloriosa*, Y. *smalliana*, and other species.

CUTTINGS: The species and variegated cultivars can be propagated by removing the small toes (swollen buds) around the base of the plant when potting in the spring. They are cut off with a piece of root, boxed up and kept frost free until they show signs of activity and then put outside to grow. Root (rhizome) cuttings work well. Collect in late winter, divide into 3″ long pieces, and place horizontally in sand or suitable medium with bottom heat. Shoots develop in 6 to 8 weeks and can be potted [PIPPS 28:212 (1978)]. Senior author produced numerous plants of Y. *filamentosa* 'Bright Edge' by root cuttings.

If small numbers are wanted, division in the spring is possible.

Zanthoxylum americanum Common Prickly-ash

SEED: *Zanthoxylum* fruits are single-seeded capsules which ripen in August and later. The capsule opens when ripe to reveal a black seed. Capsules are easily removed when ripe and unopened capsules discharge their seeds after several days of air drying. A pound of cleaned seed contains 25,600 seeds. *Zanthoxylum* seeds exhibit strong dormancy. Stratification in moist sand for 4 months at 41°F is recommended. Scarification may also be necessary. Senior author has germinated *Z. simulans* successfully using 3 months cold stratification. *Zanthoxylum piperitum* and *Z. schinifolium* should be treated similarly.

CUTTINGS: On a small scale suckers can be dug up. The plant is a good candidate for propagation by root cuttings and this is the preferred vegetative method of propagation [*Gardeners Chronicle* 169(3):22 (1971)].

Zelkova serrata Japanese Zelkova

SEED: Seed does not require a pretreatment for germination, however, percentage germination is better after cold stratification (41°F) for 2 months. Seeds should not be allowed to dry out or viability will be lost.

CUTTINGS: Senior author rooted early July cuttings on which the end bud had formed. The cuttings were 5″ long and treated with 0, 0.8, 1.6, and 3.2% IBA-quick dip, and placed in perlite:peat (2:1, v/v) medium under mist. Control cuttings rooted 32% with the IBA treated cuttings rooting 48, 62, and 54%, respectively. Clonal differences may exist because results from another study failed to show any rooting with a 12-year-old tree. Senior author followed up on early work [*The Plant Propagator* 29(2):10 (1983)] and has successfully rooted cuttings every year. Unfortunately, no cuttings have been successfully overwintered. Extended photoperiods induced new growth but cuttings still died. Considerable research should be conducted with this species. In a study using ethanol (control), 5000 ppm P-ITB, 5000 ppm IBA, and 2.0% P-ITB (all in solution), early June, peat:perlite, mist, cuttings rooted 47, 77, 70, and 80% in 6 weeks, respectively. Again, all failed to survive after transplanting.

GRAFTING: *Ulmus* reportedly can serve as an understock, however, zelkova seedlings are a better choice. Use a side graft.

Zenobia pulverulenta Dusty Zenobia

SEED: Fruit is a dehiscent capsule that should be collected in the yellow brown stage, dried inside and the seeds shaken loose. Fresh seeds require no pretreatment and will germinate immediately after sowing. Handle like *Calluna* or *Rhododendron*.

CUTTINGS: Difficult. Senior author had 20% success with late August cuttings, 8000 ppm IBA-solution, peat:perlite, mist. Timing may be important for this species. Excessive moisture leads to defoliation.

Ziziphus jujuba Chinese Date

SEED: No good information available but 3 months cold stratification produced 20% germination; 3 months warm/3 months cold, 40%.

CUTTINGS: Root cuttings have been utilized successfully.

GRAFTING: Cultivars are whip grafted on seedling understock.

Scientific Name Index

Abelia 80
'Edward Goucher' 80
× grandiflora 80
Abeliophyllum distichum 80
Abies 80
balsamea 80
concolor 81
fraseri 81
koreana 81
Acanthopanax 81
henryi 81
sieboldianus 81
Acer 81
barbatum 81
buergeranum 81
campestre 81
capillipes 82
carpinifolium 82
circinatum 82
cissifolium 82
davidii 82
floridanum 81
ginnala 82
griseum 82
henryi 83
japonicum 83
leucoderme 83
mandshuricum 83
miyabei 84
negundo 84
nigrum 84
nikoense 84
palmatum 84
pensylvanicum 85
platanoides 85
pseudoplatanus 86
pseudosieboldianum 86
rubrum 86
rufinerve 86
saccharinum 86
saccharum 87
sieboldianum 87
spicatum 87
tataricum 87
tegmentosum 87
triflorum 87
truncatum 87
subsp. mayrii 87
subsp. mono 87
Actinidia 88
arguta 88
chinensis 88
kolomikta 88
polygama 88
Aesculus 88
arguta 89
californica 89
× carnea 89
discolor 89
flava 89
georgiana 89
glabra 89
hippocastanum 89
indica 89
octandra 89
parviflora 89
pavia 89
splendens 89
sylvatica 89
turbinata 89
Ailanthus altissima 90
Akebia 90
× pentaphylla 90
quinata 90
trifoliata 90
Albizia julibrissin 90
Alnus 90
cordata 90
crispa 90
firma 90
glutinosa 91
hirsuta 91
incana 91
rhombifolia 91
rubra 91
rugosa 91
serrulata 91
Amelanchier 91
alnifolia 91
arborea 91
asiatica 92
canadensis 92
florida 92
× grandiflora 92
laevis 92
spicata 92
Amorpha 92
brachycarpa 92
canescens 92
fruticosa 92
nana 92
Ampelopsis 92
aconitifolia 92
brevipedunculata 92
humulifolia 92
Andrachne colchica 92
Andromeda 93
glaucophylla 93
polifolia 93
Anisostichus capreolatus 93
Aralia 93
elata 93
spinosa 93
Arctostaphylos uva-ursi 93
Ardisia japonica 93
Aristolochia durior 93
Aronia 93
arbutifolia 94
melanocarpa 94
prunifolia 94
Asimina triloba 94
Aspidistra elatior 94
Aucuba japonica 94
Baccharis 94
angustifolia 94
halimifolia 94
Berberis 94
buxifolia 94
candidula 94
× chenaultii 94
circumserrata 95
darwinii 95
gagnepainii 95
gilgiana 95
× gladwynensis 'William Penn' 95
julianae 95
koreana 95
× mentorensis 95
ottawensis 95
sargentiana 95
stenophylla 95
thunbergii 95
triacanthophora 95
vernae 95
verruculosa 95
vulgaris 96
Betula 96
albo-sinensis 96
alleghaniensis 96
ermanii 96

grossa . . . 96
jacquemontii . . . 96
lenta . . . 96
maximowicziana . . . 96
nigra . . . 96
papyrifera . . . 96
pendula . . . 97
platyphylla . . . 97
populifolia . . . 97
Bignonia capreolata . . . 93
Broussonetia papyrifera . . . 97
Buddleia . . . 97
alternifolia . . . 97
davidii . . . 97
Buxus . . . 98
microphylla . . . 98
sempervirens . . . 98
Callicarpa . . . 98
americana . . . 98
bodinieri . . . 98
dichotoma . . . 98
japonica . . . 98
Calluna vulgaris . . . 98
Calocedrus decurrens . . . 98
Calycanthus floridus . . . 99
Camellia . . . 99
japonica . . . 99
sasanqua . . . 99
sinensis . . . 99
Campsis . . . 99
radicans . . . 99
× tagliabuana . . . 100
Caragana . . . 100
arborescens . . . 100
aurantiaca . . . 100
frutex . . . 100
maximowicziana . . . 100
microphylla . . . 100
pygmaea . . . 100
sinica . . . 100
Carpinus . . . 100
betulus . . . 100
caroliniana . . . 100
cordata . . . 101
japonica . . . 101
laxiflora . . . 101
orientalis . . . 101
turczaninovii . . . 101
Carya . . . 101
cordiformis . . . 101
glabra . . . 101
illinoensis . . . 101
laciniosa . . . 101
ovalis . . . 101
ovata . . . 101
tomentosa . . . 101
Caryopteris . . . 102
× clandonensis . . . 102
incana . . . 102
Castanea . . . 102
dentata . . . 102
mollissima . . . 102
sativa . . . 102
Catalpa . . . 102
bignonioides . . . 102
bungei . . . 102
fargesii . . . 102
ovata . . . 102
speciosa . . . 102
Ceanothus . . . 102
americanus . . . 102
ovatus . . . 102
Cedrela sinensis . . . 103
Cedrus . . . 103
atlantica . . . 103
deodara . . . 103
libani . . . 103
Celastrus . . . 103
loeseneri . . . 103
orbiculatus . . . 103
scandens . . . 104
Celtis . . . 104
jessoensis . . . 104
laevigata . . . 104
occidentalis . . . 104
Cephalanthus occidentalis . . . 104
Cephalotaxus . . . 104
harringtonia . . . 104
Ceratostigma plumbaginoides . . . 104
Cercidiphyllum japonicum . . . 104
Cercis . . . 105
canadensis . . . 105
chinensis . . . 105
occidentalis . . . 105
reniformis . . . 105
siliquastrum . . . 105
Chaenomeles . . . 105
japonica . . . 105
speciosa . . . 105
× superba . . . 105
Chamaecyparis . . . 106
lawsoniana . . . 106
nootkatensis . . . 106
obtusa . . . 106
pisifera . . . 106
thyoides . . . 107
Chamaedaphne calyculata . . . 107
Chimonanthus praecox . . . 107
Chionanthus . . . 107
retusus . . . 107
virginicus . . . 107
Cladrastis lutea . . . 107
Clematis . . . 107
microphylla . . . 108
orientalis . . . 108
paniculata . . . 108
tangutica . . . 108
vitalba . . . 108
viticella . . . 108
Clerodendron trichotomum . . . 108
Clethra . . . 108
acuminata . . . 108
alnifolia . . . 108
barbinervis . . . 108
Cleyera japonica . . . 108
Colutea . . . 108
arborescens . . . 108
× media . . . 108
orientalis . . . 108
Comptonia peregrina . . . 108
Cornus . . . 109
alba . . . 109
alternifolia . . . 109
amomum . . . 109
canadensis . . . 109
controversa . . . 109
florida . . . 109
kousa . . . 110
macrophylla . . . 110
mas . . . 110
officinalis . . . 110
racemosa . . . 110
sanguinea . . . 111
sericea . . . 111
stolonifera . . . 111
walteri . . . 111
Corylopsis . . . 111
glabrescens . . . 111
pauciflora . . . 111
sinensis . . . 111
spicata . . . 111
willmottiae . . . 111
Corylus . . . 111
americana . . . 111
avellana . . . 111
colurna . . . 112
cornuta . . . 112
maxima var. purpurea . . . 112
Cotinus . . . 112
americanus . . . 113
coggygria . . . 112
obovatus . . . 113
Cotoneaster . . . 113
acutifolius . . . 113
adpressus . . . 113
apiculatus . . . 113
congestus . . . 113
conspicuus . . . 113
dammeri . . . 113
divaricatus . . . 113
glaucophyllus . . . 113
horizontalis . . . 114
× hybridus . . . 114
lacteus . . . 114
lucidus . . . 114
microphyllus . . . 114
multiflorus . . . 114
parneyi . . . 114
racemiflorus var. soongoricus . . . 114
salicifolius . . . 114
Crataegus . . . 114
arnoldiana . . . 114
cordata . . . 115
crusgalli . . . 114
laevigata . . . 115
mollis . . . 115
monogyna . . . 115
phaenopyrum . . . 115
prunifolia . . . 115

punctata 115
succulenta 115
Croton alabamensis 115
Cryptomeria japonica 115
Cunninghamia lanceolata 115
× Cupressocyparis leylandii 115
cupressus 116
arizonica 116
macrocarpa 116
sempervirens 116
Cyrilla racemiflora 116
Cytisus 116
beanii 117
decumbens 117
hirsutus 117
× praecox 117
scoparius 117
Danae racemosa 117
Daphne 117
× burkwoodii 117
cneorum 117
genkwa 117
giraldii 117
mezereum 118
odora 118
retusa 118
Davidia involucrata 118
Deutzia 118
crenata 118
gracilis 118
× lemoinei 118
rosea 118
scabra 118
Diervilla 119
lonicera 119
rivularis 119
sessilifolia 119
Diospyros 119
kaki 119
virginiana 119
texana 119
Dipelta floribunda 119
Dirca palustris 119
Disanthus cercidifolius 119
Distylium racemosum 119
Elaeagnus 119
angustifolia 119
commutata 120
× ebbingii 120
macrophylla 120
multiflora 120
pungens 120
umbellata 120
Elliottia racemosa 120
Elsholtzia stauntonii 120
Enkianthus 120
campanulatus 120
cernuus 121
deflexus 121
perulatus 121
Epigea repens 121
Erica 121
carnea 121
ciliaris 121
cinerea 121
tetralix 121
vagans 121
Eriobotrya japonica 121
Eucommia ulmoides 121
Euonymus 121
alatus 122
americanus 122
atropurpureus 122
bungeanus 122
europaeus 122
fortunei 122
hamiltonianus var. sieboldianus . . . 122
japonicus 122
kiautschovicus 123
nanus var. turkestanicus 123
oxyphyllus 123
sachalinensis 123
Eurya japonica 123
Evodia 123
daniellii 123
hupehensis 123
Exochorda 123
giraldii 123
korolkowii 123
× macrantha 123
racemosa 123
Fagus 123
grandifolia 123
sylvatica 123
× Fatshedera lizei 124
Fatsia japonica 124
Feijoa sellowiana 124
Ficus 124
carica 124
pumila 125
Firmiana simplex 125
Fontanesia fortunei 125
Forestiera neomexicana 125
Forsythia 125
europaea 125
× intermedia 125
mandschurica 125
ovata 125
suspensa 125
var. atrocaulis 125
var. fortunei 125
var. sieboldii 125
viridissima 125
Forthergilla 126
gardenii 126
major 126
monticola 126
Franklinia alatamaha 126
Fraxinus 126
americana 126
angustifolia 127
excelsior 127
holotricha 127
nigra 127
ornus 127
oxycarpa 127
pennsylvanica 127
quadrangulata 127
uhdei 127
Gardenia jasminoides 127
Gaultheria procumbens 127
Gaylussacia brachycera 127
Gelsemium sempervirens 128
Genista 128
germanica 128
hispanica 128
lydia 128
pilosa 128
tinctoria 128
Ginkgo biloba 128
Gleditsia triacanthos 128
var. inermis 128
Gordonia lasianthus 129
Gymnocladus dioicus 129
Halesia 129
carolina 129
diptera 129
monticola 129
parviflora 129
Hamamelis 129
× intermedia 130
japonica 130
mollis 130
vernalis 130
virginiana 131
Hedera 131
canariensis 131
colchica 131
helix 131
Helleborus orientalis 131
Hibiscus syriacus 132
Hippophae rhamnoides 132
Hovenia dulcis 132
Hydrangea 132
anomala 132
anomala subsp. petiolaris 132
arborescens 132
macrophylla 133
paniculata 133
quercifolia 133
serrata 133
Hypericum 133
androsaemum 133
calycinum 133
frondosum 133
kalmianum 133
× moseranum 134
olympicum 134
patulum 134
prolificum 133
Iberis sempervirens 134
Idesia polycarpa 134
Ilex 134
× altaclarensis 134
aquifolium 134
× aquipernyi 135
× attenuata 135
cassine 135
ciliospinosa 135

cornuta 135
crenata 135
decidua 135
glabra 136
integra 136
× koehneana 136
latifolia 136
× meserveae 136
'Nellie R. Stevens' 136
opaca 136
pedunculosa 136
pernyi 137
rugosa 137
serrata 137
sugerokii 137
verticillata 137
vomitoria 137
yunnanensis 137
Illicium 137
anisatum 137
floridanum 137
parviflorum 137
Indigofera 137
amblyantha 137
gerardiana 138
incarnata 138
kirilowii 138
Itea 138
ilicifolia 138
japonica 138
virginica 138
Jasminum 138
floridum 138
mesnyi 138
nudiflorum 138
officinale 138
Juglans 138
cinerea 138
nigra 138
regia 139
Juniperus 139
chinensis 140
communis 141
conferta 141
davurica 141
excelsa 141
horizontalis 141
procumbens 141
sabina 141
scopulorum 141
squamata 142
virginiana 142
Kalmia 142
angustifolia 142
cuneata 142
hirsuta 142
latifolia 142
polifolia 143
Kalmiopsis leachiana 143
Kalopanax pictus 143
Kerria japonica 143
Koelreuteria 143
bipinnata 144
paniculata 144
Kolkwitzia amabilis 144
+ Laburnocytisus adamii 144
Laburnum 144
alpinum 144
anagyroides 144
× watereri 144
Lagerstroemia indica 144
Larix 145
decidua 145
× eurolepis 145
kaempferi 145
laricina 145
sibirica 145
Lavandula angustifolia 145
Ledum groenlandicum 146
Leiophyllum buxifolium 146
Leitneria floridana 146
Lespedeza 146
bicolor 146
thunbergii 146
Leucothoe 146
axillaris 146
fontanesiana 146
grayana 146
keiskei 146
populifolia 146
Libocedrus decurrens 98
Ligustrum 146
amurense 147
× ibolium 147
japonicum 147
lucidum 147
obtusifolium 147
ovalifolium 147
sinense 147
× vicaryi 147
vulgare 147
Lindera 147
benzoin 147
obtusiloba 147
Liquidambar 147
formosana 147
styraciflua 148
Liriodendron tulipifera 148
Lonicera 148
alpigena 148
× bella 148
× brownii 148
caerulea 148
fragrantissima 148
× heckrottii 148
japonica 148
korolkowii 148
maackii 148
maximowiczii 148
morrowii 148
nitida 148
pileata 148
sempervirens 148
tatarica 148
xylosteum 148
Loropetalum chinense 148
Lycium chinense 149
Lyonia 149
mariana 149
ovalifolia 149
Maackia 149
amurensis 149
chinensis 149
× Macludrania hybrida 149
Maclura pomifera 149
Magnolia 149
acuminata 150
ashei 150
cordata 150
cylindrica 150
fraseri 150
grandiflora 150
heptapeta 151
hypoleuca 151
kobus 151
× loebneri 152
macrophylla 152
officinalis 152
quinquepeta 152
salicifolia 152
sieboldii 152
× soulangiana 152
stellata 152
tripetala 153
virginiana 153
× Mahoberberis 153
aquicandidula 153
aquisargentiae 153
neubertii 153
Mahonia 153
aquifolium 153
bealei 153
japonica 154
lomariifolia 154
nervosa 154
pinnata 154
repens 154
wagneri 154
Malus 154
× atrosanguinea 154
baccata 154
communis 154
florentina 154
floribunda 154
'Hopa' 154
hupehensis 154
pumila 154
purpurea 154
sargentii 154
sikkimensis 154
toringoides 154
tschonoskii 154
Melia azedarach 156
Menispermum canadense 156
Metasequoia glyptostroboides 156
Michelia figo 156
Mitchella repens 156
Morus 156
alba 156

rubra . . . 157
Myrica . . . 157
cerifera . . . 157
gale . . . 157
pensylvanica . . . 157
Nandina domestica . . . 157
Neillia sinensis . . . 158
Nemopanthus mucronatus . . . 158
Nerium oleander . . . 158
Neviusia alabamensis . . . 158
Nyssa . . . 158
aquatica . . . 158
sinensis . . . 158
sylvatica . . . 158
Orixa japonica . . . 158
Osmanthus . . . 159
americanus . . . 159
delavayi . . . 159
× fortunei . . . 159
fragrans . . . 159
heterophyllus . . . 159
Ostrya . . . 159
carpinifolia . . . 159
virginiana . . . 159
Oxydendrum arboreum . . . 159
Pachysandra . . . 160
axillaris . . . 160
procumbens . . . 160
terminalis . . . 160
Paeonia suffruticosa . . . 160
Parrotia persica . . . 160
Parrotiopsis jacquemontia . . . 161
Parthenocissus . . . 161
henryana . . . 161
quinquefolia . . . 161
tricuspidata . . . 161
Paulownia tomentosa . . . 161
Paxistima . . . 161
canbyi . . . 161
myrsinites . . . 161
Perovskia atriplicifolia . . . 161
Phellodendron . . . 162
amurense . . . 162
chinense . . . 162
japonicum . . . 162
lavallei . . . 162
sachalinense . . . 162
Philadephus . . . 162
coronarius . . . 162
× lemoinei . . . 162
lewisi . . . 162
microphyllus . . . 162
× nivalis . . . 162
× virginalis . . . 162
Photinia . . . 162
× fraseri . . . 162
glabra . . . 162
koreana . . . 162
parvifolia . . . 162
serrulata . . . 162
villosa . . . 162
Phyllodoce glandulifera . . . 162
Physocarpus opulifolius . . . 162
Picea . . . 163
abies . . . 163
brewerana . . . 164
engelmannii . . . 165
glauca . . . 165
jezoensis . . . 165
koyamai . . . 165
mariana . . . 165
omorika . . . 165
orientalis . . . 165
pungens . . . 165
rubens . . . 166
sitchensis . . . 166
smithiana . . . 166
Picrasma quassioides . . . 166
Pieris . . . 166
floribunda . . . 166
'Forest Flame' . . . 166
formosa . . . 166
japonica . . . 166
phillyreifolia . . . 167
Pinus . . . 168
albicaulis . . . 168
aristata . . . 168
armandii . . . 168
balfouriana . . . 168
banksiana . . . 168
bungeana . . . 168
cembra . . . 168
cembroides . . . 167
contorta . . . 169
densiflora . . . 169
echinata . . . 169
flexilis . . . 169
halapensis . . . 169
koraiensis . . . 169
leucodermis . . . 169
monophylla . . . 169
monticola . . . 169
mugo . . . 169
nigra . . . 169
parviflora . . . 170
peuce . . . 170
ponderosa . . . 170
pumila . . . 170
resinosa . . . 170
rigida . . . 170
strobus . . . 170
sylvestris . . . 170
tabulaeformis . . . 171
taeda . . . 171
thunbergiana . . . 171
virginiana . . . 171
wallichiana . . . 171
Pistacia chinensis . . . 171
Pittosporum . . . 171
tenuifolium . . . 171
tobira . . . 171
Platanus . . . 171
× acerifolia . . . 171
occidentalis . . . 172
orientalis . . . 172
wrightii . . . 172
Platycladus orientalis . . . 172
Podocarpus macrophyllus . . . 172
Polygonum . . . 172
aubertii . . . 172
baldschuanicum . . . 172
cuspidatum . . . 172
Poncirus trifoliata . . . 172
Populus . . . 173
acuminata . . . 173
alba . . . 173
× canadensis . . . 173
× canescens . . . 173
deltoides . . . 173
× euroamericana . . . 173
grandidentata . . . 173
heterophylla . . . 173
nigra . . . 173
songarica . . . 173
tremula . . . 173
tremuloides . . . 173
trichocarpa . . . 173
yunnanensis . . . 174
Potentilla fruticosa . . . 174
Prinsepia . . . 174
sinensis . . . 174
uniflora . . . 174
Prunus . . . 174
americana . . . 174
amygdalis . . . 176
armeniaca . . . 174
avium . . . 175
besseyi . . . 175
× blireiana . . . 175
campanulata . . . 175
caroliniana . . . 175
cerasifera . . . 175
cerasus . . . 176
× cistena . . . 176
cyclamina . . . 176
dulcis . . . 176
glandulosa . . . 176
'Hally Jolivette' . . . 177
× hilleri . . . 177
× incam . . . 177
incisa . . . 177
laurocerasus . . . 177
lusitanica . . . 177
maackii . . . 177
maritima . . . 177
nigra . . . 177
padus . . . 178
pensylvanica . . . 178
persica . . . 178
pumila . . . 178
sargentii . . . 178
serotina . . . 178
serrula . . . 179
serrulata . . . 179
subhirtella . . . 179
tenella . . . 179
tomentosa . . . 179
triloba . . . 180
virginiana . . . 180

× yedoensis 180
Pseudocydonia sinensis 180
Pseudolarix kaempferi 180
Pseudotsuga menziesii 180
Ptelea trifoliata 181
Pterocarya 181
fraxinifolia 181
× rehderana 181
stenoptera 181
Pterostryrax 181
corymbosus 181
hispidus 181
Punica granatum 181
Pyracantha 181
coccinea 181
koidzumii 182
Pyrus 182
amygdaliformis 182
betulifolia 182
calleryana 182
communis 182
elaeagrifolia 183
fauriei 183
kawakamii 183
pashia 183
pyrifolia 183
salicifolia 183
ussuriensis 183
Quercus 183
acutissima 185
alba 185
bicolor 185
cerris 185
chrysolepsis 185
coccinea 185
dentata 185
falcata 185
glandulifera 185
glauca 185
imbricaria 185
lyrata 185
macrocarpa 185
marilandica 185
mongolica 185
muehlenbergii 185
myrsinifolia 186
nigra 186
palustris 186
phellos 186
prinus 186
robur 186
rubra 186
× sargentii 186
shumardii 186
stellata 186
variabilis 187
velutina 187
virginiana 187
Raphiolepis 187
indica 187
umbellata 187
Rhamnus 187
alnifolia 187
caroliniana 187
cathartica 187
davurica 187
frangula 187
Rhododendron 187
arborescens 188
arboreum 189
atlanticum 188
canadense 189
chamae-thomsonii 189
chapmanii 189
catawbiense 188
cilipense 188
concatenans 188
dauricum 189
desypetalum 188
fastigiatum 189
forrestii 189
hippophaeoides 188
impeditum 188
leucapsis 188
mactrostemon 188
maximum 188
moupinense 188
mucronatum 188
nudiflorum 188
pemakoense 188
periclymenoides 188
ponticum 188
poukhanense 189
racemosum 189
schlippenbachii 189
tebotan 188
tschonoskii 188
viscosum 188
williamsianum 188
yakusimanum 188
Rhodotypos scandens 189
Rhus 190
aromatica 190
chinensis 190
copallina 190
glabra 190
typhina 190
Ribes 191
alpinum 191
missouriensis 191
nigrum 191
odoratum 191
Robinia 191
fertilis 191
hispida 191
pseudoacacia 191
Rosa 192
canina 192
carolina 192
chinensis 192
damascena 193
dumentorum 192
fendleri 192
helenae 192
multiflora 192
× noisettiana 192
odorata 192
rubrifolia 192
rugosa 192
setigera 192
Rosmarinus officinalis 193
Ruscus aculeatus 193
Salix 193
caprea 194
× babylonica 193
elaeagnos 193
× elegantissima 193
gracilistyla 193
matsudana 193
melanostachys 193
pentandra 193
purpurea 193
sachalinensis 193
Sambucus 194
canadensis 194
nigra 194
pubens 194
racemosa 194
Santolina 194
chamaecyparissus 194
virens 194
Sapindus drummondii 195
Sapium 195
japonicum 195
sebiferum 195
Sarcococca 195
confusa 195
hookerana 195
ruscifolia 195
Sassafras albidum 195
Schizophragma hydrangeoides 195
Sciadopitys verticillata 196
Securinega suffruticosa 196
Sequoia sempervirens 196
Sequoiadendron giganteum 196
Shepherdia 196
argentea 196
canadensis 196
Sibiraea laevigata 196
Sinowilsonia henryi 197
Skimmia 197
japonica 197
reevesiana 197
Sophora 197
japonica 197
microphylla 197
Sorbaria sorbifolia 197
Sorbus 197
alnifolia 198
americana 198
aria 198
aucuparia 198
cashmiriana 198
commixta 198
decora 198
discolor 198
dumosa 199
esserteauiana 199
folgneri 199

intermedia . . . 199
japonica . . . 199
matsumurana . . . 199
sargentiana . . . 199
scopulina . . . 199
sibirica . . . 199
tianshanica . . . 199
Spiraea . . . 199
aemiliana . . . 199
alba . . . 199
albiflora . . . 199
amoena . . . 199
arcuata . . . 199
arguta . . . 199
betulifolia . . . 199
× billardii . . . 199
bullata . . . 199
× bumalda . . . 200
cantoniensis . . . 200
chamaedryfolia . . . 200
densiflora . . . 200
× foxii . . . 200
hypericifolia . . . 200
japonicum . . . 200
× lemoinei . . . 200
menziesii . . . 200
miyabei . . . 200
× multiflora . . . 200
nipponica . . . 200
pubescens . . . 200
splendens . . . 200
syringaeflora . . . 200
ussuriensis . . . 200
virgata . . . 200
Staphylea . . . 200
bumalda . . . 201
colchica . . . 201
holocarpa . . . 201
pinnata . . . 201
trifoliata . . . 201
Stephanandra . . . 201
incisa . . . 201
tanakae . . . 201
Stewartia . . . 201
henryae . . . 202
koreana . . . 201
malacodendron . . . 201
monadelpha . . . 201
ovata . . . 201
pseudocamellia . . . 201
rostrata . . . 201
serrata . . . 201
sinensis . . . 201
Stranvaesia davidiana . . . 202
Styrax . . . 202
americanus . . . 202
dasyanthus . . . 202
grandifolius . . . 202
hemsleyana . . . 202
japonicus . . . 202
obassia . . . 202
Symphoricarpos . . . 202
albus . . . 202
× chenaultii . . . 203
orbiculatus . . . 203
Symplocos . . . 203
paniculata . . . 203
tinctoria . . . 203
Syringa . . . 203
afghanica . . . 203
× chinensis . . . 203
henryi . . . 203
josikaea . . . 203
laciniata . . . 203
meyeri . . . 204
microphylla . . . 204
oblata . . . 204
palibiniana . . . 204
patula . . . 204
pekinensis . . . 204
reflexa . . . 204
reticulata . . . 204
× swegiflexa . . . 204
sweginzowii . . . 204
villosa . . . 204
vulgaris . . . 204
wolfii . . . 205
Tamarix ramosissima . . . 205
Taxodium . . . 206
ascendens . . . 206
distichum . . . 206
Taxus . . . 206
baccata . . . 206
brevifolia . . . 207
canadensis . . . 207
cuspidata . . . 207
× media . . . 207
Ternstroemia gymnanthera . . . 208
Teucrium chamaedrys . . . 208
Thuja . . . 208
koraiensis . . . 208
occidentalis . . . 208
orientalis . . . 208
plicata . . . 208
standishii . . . 208
Thujopsis dolobrata . . . 208
Tilia . . . 209
americana . . . 209
cordata . . . 209
dasystyla . . . 210
× euchlora . . . 210
× europaea . . . 210
flavescens . . . 210
mongolica . . . 210
petiolaris . . . 210
platyphyllos . . . 210
tomentosa . . . 210
Torreya . . . 210
nucifera . . . 210
taxifolia . . . 210
Trachelospermum . . . 210
asiaticum . . . 210
jasminoides . . . 211
Tripterygium regelii . . . 211
Trochodendron aralioides . . . 211
Tsuga . . . 211
canadensis . . . 211
carolinana . . . 211
heterophylla . . . 211
Ulex europaeus . . . 211
Ulmus . . . 212
alata . . . 212
americana . . . 212
carpinifolia . . . 213
crassifolia . . . 213
glabra . . . 213
× hollandica . . . 213
hybrids . . . 213
japonica . . . 213
laevis . . . 213
parvifolia . . . 213
procera . . . 214
pumila . . . 214
rubra . . . 214
serotina . . . 214
thomasii . . . 214
× vegeta . . . 214
Vaccinium . . . 214
alaskaense . . . 214
angustifolium . . . 215
arboreum . . . 214
arctostaphylos . . . 214
ashei . . . 215
axillare . . . 214
bracteatum . . . 215
canadense . . . 215
caespitosum . . . 215
corymbosum . . . 215
cylindricum . . . 215
deliciosa . . . 215
erythrinum . . . 215
gaultherioides . . . 215
glauco-album . . . 215
hirsutum . . . 215
japonicum . . . 215
macrocarpon . . . 216
membranaceum . . . 215
myrtilloides . . . 215
myrtillus . . . 215
olhamii . . . 215
ovalifolium . . . 215
oxycoccos . . . 215
patifollum . . . 215
scoparium . . . 215
simulatum . . . 215
smallii . . . 215
uliginosum . . . 215
vacillans . . . 215
versicolor . . . 215
vitis-idaea . . . 216
Viburnum . . . 216
acerifolium . . . 216
alnifolium . . . 216
betulifolium . . . 216
bitchiuense . . . 216
× bodnantense . . . 216
bracteatum . . . 216
buddleifolium . . . 217
burejacticum . . . 217

× burkwoodii . . . 217
× carlcephalum . . . 217
carlesii . . . 217
cassinoides . . . 217
× chenaultii . . . 218
cylindricum . . . 218
davidii . . . 218
dentatum . . . 218
dilatatum . . . 218
erosum . . . 218
erubescens . . . 218
farreri . . . 218
fragrans . . . 218
furcatum . . . 218
hupehense . . . 218
japonicum . . . 218
× juddii . . . 219
lantana . . . 219
lentago . . . 219
lobophyllum . . . 219
macrocephalum . . . 219
nudum . . . 219
opulus . . . 219
plicatum . . . 219
plicatum var. tomentosum . . . 219
× pragense . . . 220
prunifolium . . . 220
pubescens . . . 220
× rhytidophylloides . . . 220
rhytidophyllum . . . 220
rufidulum . . . 220
sargentii . . . 220
setigerum . . . 220
sieboldii . . . 221
suspensum . . . 221
tinus . . . 221
trilobum . . . 221
wrightii . . . 221
Vinca . . . 221
major . . . 221
minor . . . 221
Vitex . . . 221
agnus-castus . . . 221
negundo . . . 221
Weigela florida . . . 221
Wisteria . . . 222
floribunda . . . 222
frutescens . . . 222
sinensis . . . 222
venusta . . . 222
Xanthoceras sorbifolium . . . 222
Xanthorhiza simplicissima . . . 222
Yucca . . . 222
filimentosa . . . 222
flaccida . . . 222
glauca . . . 222
gloriosa . . . 222
smalliana . . . 222
Zanthoxylum . . . 222
americanum . . . 222
piperitum . . . 223
schinifolium . . . 223
simulans . . . 223
Zelkova serrata . . . 223
Zenobia pulverulenta . . . 223
Ziziphus jujuba . . . 223

Common Name Index

Abelia 80
Glossy 80
Abelialeaf 80
Korean 80
Actinidia 88
Bower 88
Kolomikta 88
Akebia 90
Fiveleaf 90
Three-leaflet 90
Alder 90
American Green 90
European 91
Gray 91
Italian 90
Japanese Green 90
Manchurian 91
Red 91
Sierra 91
Speckled 91
Tag 91
Alexandrian-laurel 117
Allspice 99
Carolina 99
Almond 176
Dwarf Russian 179
Dwarf Flowering 176
Flowering 180
Amorpha 92
Indigobush 92
Leadplant 92
Ampelopsis 92
Hops 92
Porcelain 92
Andromeda 93
Downy 93
Angelica-tree 93
Japanese 93
Anise-tree 137
Florida 137
Japanese 137
Small 137
Apple 154
Apricot 174
Aralia 81
Fiveleaf 81
Henry's 81
Arborvitae 208
American 208
Eastern 208
Giant 208
Hiba 208
Japanese 208
Korean 208
Oriental 208
Western 208
Arbutus 121
Trailing 121
Ardisia 93
Japanese 93
Ash 126
Black 127
Blue 127
European 127
Flowering 127
Green 127
Manna 127
Narrowleaf 127
'Raywood' 127
Shamel 127
White 126
Aspen 173
European 173
Quaking 173
Aucuba 94
Japanese 94
Autumn-olive 120
Azalea 187
Deciduous 188
Evergreen 189
Baccharis 94
Eastern 94
Narrowleaf 94
Baldcypress 206
Common 206
Bamboo 157
Heavenly 157
Banana Shrub 156
Barberry 94
Black 95
Boxleaf 94
Chenault 94
Common 96
Cutleaf 95
Darwin 95
Japanese 95
Korean 95
Mentor 95
Ottawa 95
Paleleaf 94
Rosemary 95
Sargent 95
Threespine 95
Verna 95
Warty 95
Wildfire 95
'William Penn' 95
Wintergreen 95
Basswood 209
Bayberry 157
Northern 157
Bearberry 93
Beautyberry 98
American 98
Bodinier 98
Japanese 98
Purple 98
Beautybush 144
Beech 123
American 123
European 123
Big Tree 196
Birch 96
Chinese Paper 96
Erman 96
European White 97
Gray 97
Jacquemont 96
Japanese Cherry 96
Japanese White 97
Monarch 96
Paper 96
River 96
Sweet 96
Szechwan White 97
Yellow 96
Bittersweet 103
American 104
Loesener 103
Oriental 103
Bladdernut 200
American 201
Bumalda 201
Chinese 201
Colchis 201
Bladder-senna 108
Bluebeard 102
Common 102
Blueberry 214
Highbush 215

Lowbush . . . 215
Rabbiteye . . . 215
Bluemist Shrub . . . 102
Bog-rosemary . . . 93
Boxelder . . . 84
Boxwood . . . 98
Common . . . 98
Harland . . . 98
Littleleaf . . . 98
Broom . . . 116
Buckeye . . . 88
Bottlebrush . . . 89
California . . . 89
Ohio . . . 89
Painted . . . 89
Red . . . 89
Texas . . . 89
Yellow . . . 89
Buckthorn . . . 187
Alder . . . 187
Carolina . . . 187
Common . . . 187
Dahurian . . . 187
Glossy . . . 187
Sea . . . 132
Buffaloberry . . . 196
Russet . . . 196
Silver . . . 196
Bunchberry . . . 109
Bushclover . . . 146
Japanese . . . 146
Shrub . . . 146
Bush-honeysuckle . . . 119
Georgia . . . 119
Southern . . . 119
Dwarf . . . 119
Butcher's-broom . . . 193
Butterfly-bush . . . 97
Alternateleaf . . . 97
Orange-eye . . . 97
Butternut . . . 138
Buttonbush . . . 104
Camellia . . . 99
Japanese . . . 99
Sasanqua . . . 99
Candytuft . . . 134
Castor-aralia . . . 143
Catalpa . . . 102
Bunge . . . 102
Chinese . . . 102
Common . . . 102
Farges . . . 102
Southern . . . 102
Western . . . 102
Ceanothus . . . 102
Inland . . . 102
Cedar . . . 103
Atlantic White . . . 107
Atlas . . . 103
Deodar . . . 103
Cedar of Lebanon . . . 103
Cedrela . . . 103
Chinese . . . 103
Ceratostigma . . . 104
Blue . . . 104
Chaste Tree . . . 221
Lilac . . . 221
Checkerberry . . . 127
Cherry . . . 174
Black . . . 178
Cyclamen . . . 176
European Bird . . . 178
'Fugi' . . . 177
'Hally Jolivette' . . . 177
Higan . . . 179
Nanking . . . 179
'Okame' . . . 177
Oriental . . . 179
Paperbark . . . 179
Pin . . . 178
Purpleleaf Sand . . . 176
Sand . . . 175
Sargent . . . 178
Sour . . . 176
Sweet . . . 175
Taiwan . . . 175
Yoshino . . . 180
Cherrylaurel . . . 175
Carolina . . . 175
Common . . . 177
Portuguese . . . 177
Chestnut . . . 102
American . . . 102
Chinese . . . 102
Sweet . . . 102
Chinaberry . . . 156
China-fir . . . 115
Chokeberry . . . 93
Black . . . 93
Purple-fruited . . . 93
Red . . . 93
Chokecherry . . . 180
Amur . . . 177
Common . . . 180
Cinquefoil . . . 174
Bush . . . 174
Clematis . . . 107
Clethra . . . 108
Cinnamon . . . 108
Japanese . . . 108
Summersweet . . . 108
Cleyera . . . 108
Japanese . . . 108
Coffeetree . . . 129
Kentucky . . . 129
Coralberry . . . 203
Chenault . . . 203
Indiancurrant . . . 203
Corktree . . . 162
Amur . . . 162
Chinese . . . 162
Japanese . . . 162
Lavalle . . . 162
Sakhalin . . . 162
Corkwood . . . 146
Cotoneaster . . . 113
Bearberry . . . 113
Brightbead . . . 113
Cranberry . . . 113
Creeping . . . 113
Hedge . . . 114
Littleleaf . . . 114
Many-flowered . . . 114
Peking . . . 113
'Pendula' . . . 114
Pyrenees . . . 113
Rockspray . . . 114
Spreading . . . 113
Sungari Redbead . . . 114
Willowleaf . . . 114
Wintergreen . . . 113
Cotton . . . 194
Lavender . . . 194
Cottonwood . . . 173
Cowberry . . . 215
Crabapple . . . 154
Cranberry . . . 214
American . . . 215
Crapemyrtle . . . 144
Common . . . 145
Creeper . . . 161
Japanese . . . 161
Silvervein . . . 161
Virginia . . . 161
Cross-vine . . . 93
Cryptomeria . . . 115
Japanese . . . 115
Cucumber Tree . . . 150
Amercian . . . 150
Currant . . . 191
Alpine . . . 191
Black . . . 191
Clove . . . 191
Cypress . . . 116
Arizona . . . 116
Common Bald . . . 206
Italian . . . 116
Leyland . . . 115
Monterey . . . 116
Pond . . . 206
Cyrilla . . . 116
Swamp . . . 116
Dahoon . . . 135
Daphne . . . 117
Burkwood . . . 117
February . . . 118
Fragrant . . . 118
Giraldi . . . 117
Lilac . . . 117
Rose . . . 117
Winter . . . 118
Date . . . 223
Chinese . . . 223
Dawn Redwood . . . 156
Deutzia . . . 118
Fuzzy . . . 118
Lemoine . . . 118
Slender . . . 118
Devil's-walkingstick . . . 93

Devil-wood 159
Dogwood 109
  Bigleaf 110
  Bloodtwig 111
  Corneliancherry 110
  Flowering 109
  Giant 109
  Gray 110
  Japanese Cornel 110
  Kousa 110
  Pagoda 109
  Red-osier 111
  Silky 109
  Tatarian 109
  Walter 111
Douglas-fir 180
Dove-tree 118
Dutchman's Pipe 93
Elaeagnus 119
  Cherry 120
  Thorny 120
Elder 194
  American 194
  European 194
  European Red 194
  Red 194
Elliottia 120
Elm 212
  American 212
  Cedar 213
  Chinense 213
  Dutch 213
  English 214
  Huntingdon 214
  Hybrids 213
  Japanese 213
  Lacebark 213
  Rock 214
  Russian 213
  Scotch 213
  September 214
  Siberian 214
  Slippery 214
  Smoothleaf 213
  Winged 212
Elsholtzia 120
  Staunton 120
Enkianthus 120
  Bent 121
  Redvein 120
  White 121
Epaulettetree 181
  Fragrant 181
  Shrubby 181
Euonymus 121
  Dwarf 123
  European 122
  Japanese 122
  Sakhalin 123
  Spreading 123
  Winged 122
  Winterberry 122
  Wintercreeper 122
  Yeddo 122
Eurya 123
  Japanese 123
Evodia 123
  Hupeh 123
  Korean 123
False Willow 94
Falsecypress 106
  Hinoki 106
  Japanese 106
  Lawson's 106
  Nootka 106
Falsepirea 197
  Ural 197
Fatsia 124
  Japanese 124
Fig 124
  Common 124
  Creeping 125
Filbert 111
  European 111
  Purple Giant 112
  Turkish 112
Fir 80
  Balsam 80
  Concolor 80
  Douglas 180
  Fraser 80
  Korean 80
  White 80
Firethorn 181
  Formosan 182
  Scarlet 181
Floweringquince 105
  Common 105
  Japanese 105
Fontanesia 125
  Fortune 125
Forestiera 125
  New Mexican 125
Forsythia 125
  Albanian 125
  Early 125
  Greenstem 125
  'Vermont Sun' 125
Fothergilla 126
  Dwarf 126
  Large 126
Franklinia 126
Fringetree 107
  Chinese 107
  White 107
Gale 157
  Sweet 157
Genista 128
Georgia-plume 120
Germander 208
  Wall 208
Ginkgo 128
Glory-bower 108
Goldenchain 144
Golden-larch 180
Goldenraintree 143
  Bipinnate 144
  Panicled 144
Gooseberry 191
  Missouri 191
Gorse 211
  Common 211
Grancy Graybeard 107
Grapeholly 153
  Oregon 153
Graybeard 107
  Grancy 107
Guava 124
  Pineapple 124
Hackberry 104
  Common 104
  Jesso 104
  Sugar 104
Hardy-orange 172
Hawthorn 114
  Arnold 114
  Cockspur 114
  Dotted 115
  Downy 115
  English 115
  Fleshy 115
  India 187
  Plumleaf 115
  Singleseed 115
  Washington 115
  Yeddo 187
Hazelnut 111
  American 111
Heath 121
Heather 98
Hemlock 211
  Canadian 211
  Carolina 211
  Eastern 211
  Western 211
Henry Wilson Tree 197
Hickory 101
  Bitternut 101
  Mockernut 101
  Pignut 101
  Red 101
  Shagbark 101
  Shellbark 101
Hobblebush 216
Holly 134
  Altaclara 134
  American 136
  Blue 136
  Chinese 135
  English 134
  Foster's 135
  Japanese 135
  Longstalk 136
  Lusterleaf 136
  Merserve 136
  Mountain 158
  'Nellie Stevens' 136
  Perny 137
  Prostrate 137

Sugeroki . . . 137
'Wirt L. Winn' . . . 136
Yunnan . . . 137
Honeylocust . . . 128
Thornless Common . . . 128
Honeysuckle . . . 148
Hophornbeam . . . 159
American . . . 159
European . . . 159
Hoptree . . . 181
Hornbeam . . . 100
American . . . 100
European . . . 100
Heartleaf . . . 101
Japanese . . . 101
Loose-flower . . . 101
Oriental . . . 101
Horsechestnut . . . 88
Common . . . 89
Indian . . . 89
Japanese . . . 89
Red . . . 89
Horsesugar . . . 203
Huckleberry . . . 214
Box . . . 127
Hydrangea . . . 132
Bigleaf . . . 133
Climbing . . . 132
Oakleaf . . . 133
Panicle . . . 133
Smooth . . . 132
Hydrangea-vine . . . 195
Japanese . . . 195
Incensecedar . . . 98
Indigo . . . 137
Chinese . . . 138
Himalayan . . . 138
Kirilow . . . 138
Pink . . . 137
Indigobush . . . 92
Fragrant False . . . 92
Inkberry . . . 136
Ivy . . . 131
Algerian . . . 131
Boston . . . 161
Colchis . . . 131
English . . . 131
Jasmine . . . 127, 138
Asiatic . . . 210
Cape . . . 127
Chinese . . . 211
Common White . . . 138
Confederate . . . 211
Primrose . . . 138
Showy . . . 138
Star . . . 211
Winter . . . 138
Jessamine . . . 128
Carolina Yellow . . . 128
Jetbead . . . 189
Black . . . 189
Judas-tree . . . 105
Juniper . . . 139
Chinese . . . 140
Common . . . 141
Creeping . . . 141
Dahurian . . . 141
Japgarden . . . 141
Meyer's Singleseed . . . 142
Rocky Mountain . . . 141
Savin . . . 141
Shore . . . 141
Spiny Greek . . . 141
Kalmia . . . 142
Bog . . . 143
Sandhill . . . 142
Whitewicky . . . 142
Kalmiopsis . . . 143
Oregon . . . 143
Katsuratree . . . 104
Kerria . . . 143
Japanese . . . 143
Kiwi . . . 88
Labrador-tea . . . 146
Laburnum . . . 144
Common . . . 144
Scotch . . . 144
Waterer . . . 144
Lamb-kill . . . 142
Larch . . . 145
Dunkfeld . . . 145
Golden . . . 180
Laurel . . . 142
Mountain . . . 142
Laurustinus . . . 221
Lavender . . . 145
Common . . . 145
Leatherleaf . . . 107
Leatherwood . . . 116
Atlantic . . . 119
Lenten Rose . . . 131
Leucothoe . . . 146
Coast . . . 146
Drooping . . . 146
Florida . . . 146
Keisk's . . . 146
Lilac . . . 203
Chengtu . . . 204
Chinese . . . 203
Common . . . 204
Cutleaf . . . 203
Henry . . . 203
Hungarian . . . 203
Japanese Tree . . . 204
Korean . . . 204
Late . . . 204
Littleleaf . . . 204
Meyer . . . 204
Nodding . . . 204
Peking . . . 204
Swegiflexa . . . 204
Sweginzow . . . 204
Wolf . . . 205
Lime . . . 209
Linden . . . 209
American . . . 209
Bigleaf . . . 210
Caucasian . . . 210
Crimean . . . 210
European . . . 210
Littleleaf . . . 209
Mongolian . . . 210
Pendent Silver . . . 210
Silver . . . 210
Yellow . . . 210
Loblolly-bay . . . 129
Locust . . . 191
Black . . . 191
Bristly . . . 191
Loquat . . . 121
Loropetalum . . . 148
Chinese . . . 148
Lyonia . . . 149
Staggerbush . . . 149
Tibet . . . 149
Maackia . . . 149
Amur . . . 149
Magnolia . . . 149
Anise . . . 152
Ashe . . . 150
Big Leaf . . . 152
Cucumber Tree . . . 150
Fraser's . . . 150
Kobus . . . 151
Lily . . . 152
Loebner . . . 152
Saucer . . . 152
Siebold . . . 152
Southern . . . 150
Star . . . 152
Sweetbay . . . 153
Umbrella . . . 153
Whiteleaf Japanese . . . 151
Yulan . . . 151
Mahonia . . . 153
Cascades . . . 154
Cluster . . . 154
Creeping . . . 154
Japanese . . . 154
Leatherleaf . . . 153
Maple . . . 81
Amur . . . 82
Black . . . 84
Chalkbark . . . 83
David . . . 82
Florida . . . 81
Fullmoon . . . 83
Hedge . . . 81
Henry . . . 83
Hornbeam . . . 82
Ivy-leaved . . . 82
Japanese . . . 84
Manchurian . . . 83
Manchu-striped . . . 87
Miyabe . . . 83
Nikko . . . 84
Norway . . . 85
Oregon Vine . . . 82
Painted . . . 87

Paperbark 82
Purplebloom 86
Purpleblow 87
Red 86
Redvein 86
Siebold 87
Silver 86
Striped 85
Sugar 87
Tatarian 87
Threeflower 87
Trident 81
Meadowsweet 199
Narrowleaf 199
Mimosa 90
Mockorange 162
Monk's Hood Vine 92
Moonseed 156
Common 156
Mountainash 197
American 198
Arizona 199
European 198
Folgner 199
Greenes 199
Japanese 199
Korean 198
Matsumura 199
Sargent 199
Showy 198
Siberian 199
Snowberry 199
Swedish 199
Whitebeam 198
Mountainheath 162
Cream 162
Mountain-holly 158
Mountain-laurel 142
Mulberry 156
French 98
Paper 97
Red 157
White 156
Nectarine 174
Neillia 158
Chinese 158
Ninebark 162
Common 162
Oak 183
Black 187
Black Jack 185
Blue Japanese 185
Bur 185
Canyon Live 185
Chestnut 186
Chinese Evergreen 186
Chinkapin 185
Daimyo 185
English 186
Glandbearing 185
Live 187
Mongolian 185
Northern Red 186
Oriental 187
Overcup 185
Pin 186
Post 186
Sargent 186
Sawtooth 185
Scarlet 185
Shingle 185
Shumard 186
Southern Red 185
Swamp White 185
Turkey 185
Water 186
White 185
Willow 186
Oleander 158
Olive 159
American Tea 159
Autumn 120
Delavay Tea 159
Fortune's Tea 159
Fragrant Tea 159
Russian 119
Orange 162, 172
Hardy 172
Mock 162
Osage 162
Orixa 158
Japanese 158
Osage-orange 149
Osmanthus 159
Holly 159
Pachysandra 160
Allegheny 160
Chinaspurge 160
Japanese 160
Pagodatree 197
Japanese 197
Parasol Tree 125
Chinese 125
Parrotia 160
Persian 160
Patridgeberry 156
Paulownia 161
Royal 161
Pawpaw 94
Paxistima 161
Canby 161
Myrtle 161
Peach 174
Common 178
Pear 182
Almond 182
Birchleaf 182
Callery 182
Common 182
Evergreen 183
Oleaster 183
Pashi 183
Ussurian 183
Willowleaf 183
Pearlbush 123
Common 123
Redbud 123
Turkestan 123
Peashrub 100
Siberian 100
Pecan 101
Peony 160
Japanese Tree 160
Periwinkle 221
Common 221
Large 221
Persimmon 119
Common 119
Japanese 119
Texas 119
Photinia 162
Fraser 162
Korean 162
Littleleaf 162
Red-tip 162
Smooth Oriental 162
Veinyleaf Oriental 162
Pieris 166
Chinese 166
'Forest Flame' 166
Himalayan 166
Japanese 166
Mountain 166
Pine 167
Aleppo 169
Armand 168
Austrian 169
Balkan 169
Bristlecone 168
California Mountain 169
Chinese 171
Eastern White 170
Foxtail 168
Himalayan 171
Jack 168
Japanese Black 171
Japanese Red 169
Japanese Stone 170
Japanese White 170
Korean 169
Lacebark 168
Limber 169
Loblolly 171
Lodgepole 169
Mugo 169
Pitch 170
Ponderosa 170
Red 170
Scotch 170
Shortleaf 168
Singleleaf Pinon 169
Swiss Stone 168
Umbrella 196
Virginia 171
Western White 169
Whitebark 168
Pistache 171
Chinese 171
Pittosporum 171

Japanese . . . 171
Tawhiwhi . . . 171
Planetree . . . 86, 171
American . . . 172
Arizona . . . 172
London . . . 171
Oriental . . . 172
Plum . . . 174
American . . . 175
Beach . . . 177
Blireiana . . . 175
Canada . . . 177
Myrobalan . . . 175
Plum-yew . . . 104
Harrington . . . 104
Podocarpus . . . 172
Yew . . . 172
Polygonum . . . 172
Pomegranate . . . 181
Pondcypress . . . 206
Popcorn Tree . . . 195
Poplar . . . 173
Balsam . . . 173
Black . . . 173
Gray . . . 173
White . . . 173
Yellow . . . 148
Porcelain-vine . . . 92
Possumhaw . . . 135
Prickly-ash . . . 222
Common . . . 222
Prinsepia . . . 174
Cherry . . . 174
Privet . . . 146
Amur . . . 147
Border . . . 147
California . . . 147
Chinese . . . 147
European . . . 147
Glossy . . . 147
Golden Vicary . . . 147
Ibolium . . . 147
Japanese . . . 147
Regel's Border . . . 147
Waxleaf . . . 147
Quassia-wood . . . 166
India . . . 166
Quince . . . 105, 180
Chinese . . . 180
Common Flowering . . . 105
Japanese Flowering . . . 105
Raisintree . . . 132
Japanese . . . 132
Redbud . . . 105
Chinese . . . 105
Eastern . . . 105
'Oklahoma' . . . 105
Western . . . 105
Redcedar . . . 142
Eastern . . . 142
Redwood . . . 156, 196
Coast . . . 196
Dawn . . . 156
Rhododendron . . . 187
Dwarf . . . 188
Rose . . . 192
Lenten . . . 132
Rosemary . . . 193
Rose-of-Sharon . . . 132
Rubber Tree . . . 121
Russian-olive . . . 119
Russian-sage . . . 161
Sage . . . 161
Russian . . . 161
Sandmyrtle . . . 146
Box . . . 146
Santolina . . . 194
Green . . . 194
Sapphireberry . . . 203
Sarcococca . . . 195
Fragrant . . . 195
Himalayan . . . 195
Saskatoon . . . 91
Sassafras . . . 195
Common . . . 195
Scotch Broom . . . 116
Securinega . . . 196
Serviceberry . . . 91
Allegheny . . . 92
Apple . . . 92
Asian . . . 92
Downy . . . 91
Dwarf . . . 92
Pacific . . . 92
Shadblow . . . 92
Silktree . . . 90
Silverbell . . . 129
Carolina . . . 129
Small . . . 129
Two-wing . . . 129
Silverberry . . . 120
Silver-vine . . . 88
Skimmia . . . 197
Japanese . . . 197
Reeves . . . 197
Smoketree . . . 112
American . . . 113
Common . . . 112
Snowbell . . . 202
Bigleaf . . . 202
Fragrant . . . 202
Hemsley . . . 202
Japanese . . . 202
Snowberry . . . 202
Common . . . 202
Snow-wreath . . . 158
Soapberry . . . 195
Western . . . 195
Sophora . . . 197
Littleleaf . . . 197
Sourwood . . . 159
Spicebush . . . 147
Common . . . 147
Japanese . . . 147
Narrowleaf . . . 147
Spindle Tree . . . 122
Spirea . . . 196, 199
Alpine Japanese . . . 200
Altai . . . 196
Billiard . . . 199
Birchleaf . . . 199
Bumald . . . 200
Compact Garland . . . 199
Crispleaf . . . 199
Fox . . . 200
Germander . . . 200
Japanese White . . . 199
Lemoine . . . 200
Lilac . . . 200
Menzies . . . 200
Nippon . . . 200
Reeves . . . 200
Snowgarland . . . 200
Subalpine . . . 200
Taperleaf Japanese . . . 200
Spruce . . . 163
Black . . . 165
Brewer . . . 164
Colorado . . . 165
Engelmann . . . 165
Himalayan . . . 166
Koyama . . . 165
Norway . . . 163
Oriental . . . 165
Red . . . 166
Serbian . . . 165
Sitka . . . 166
White . . . 165
Yeddo . . . 165
St. Johnswort . . . 133
Aaronsbeard . . . 133
Golden . . . 133
'Hidcote' . . . 134
Kalm . . . 133
Moser's . . . 134
Olympic . . . 134
Shrubby . . . 133
Stephanandra . . . 201
Cutleaf . . . 201
Tanaka . . . 201
Stewartia . . . 201
Stranvaesia . . . 202
Chinese . . . 202
Strawberry-bush . . . 122
Sugarberry . . . 104
Sumac . . . 190
Chinese . . . 190
Flameleaf . . . 190
Fragrant . . . 190
Shining . . . 190
Smooth . . . 190
Staghorn . . . 190
Sweet-box . . . 195
Sweetfern . . . 108
Sweetgum . . . 147
American . . . 148
Formosan . . . 147
Sweetleaf . . . 203
Sweetshrub . . . 99

Schmitt

Sweetspire . . . 138
Hollyleaf . . . 138
Japanese . . . 138
Virginia . . . 138
Sycamore . . . 86, 172
Tallow Tree . . . 195
Chinese . . . 195
Japanese . . . 195
Tamarix . . . 205
Five-stamen . . . 205
Tea . . . 99
Labrador . . . 146
New Jersey . . . 102
Tea-olive . . . 159
American . . . 159
Delavay . . . 159
Fortune's . . . 159
Fragrant . . . 159
Ternstroemia . . . 208
Threewingnut . . . 211
Regel's . . . 211
Toon . . . 103
Torreya . . . 210
Florida . . . 210
Japanese . . . 210
Tree of Heaven . . . 90
Trumpetcreeper . . . 99
Common . . . 99
'Mme. Galen' . . . 100
Tuliptree . . . 148
Tupelo . . . 158
Black . . . 158
Chinese . . . 158
Water . . . 158
Tutsan . . . 133
Twinberry . . . 156
Umbrella-pine . . . 196
Viburnum . . . 216
American Cranberry . . . 221
Arrowwood . . . 218
Beech . . . 218
Bitchiu . . . 216
Blackhaw . . . 220
Bracted . . . 216
Burkwood . . . 217
Carlcephalum . . . 217
Chenault . . . 218
David . . . 218
Doublefile . . . 219
Downy . . . 220
European Cranberry . . . 219
Forked . . . 218
Fragrant . . . 218
Hupeh . . . 219
Japanese Snowball . . . 219
Japanese . . . 218
Judd . . . 219
Koreanspice . . . 217
Lantanaphyllum . . . 220
Leatherleaf . . . 220
Linden . . . 218
Mapleleaf . . . 216
Nannyberry . . . 219
Prague . . . 220
Rusty Blackhaw . . . 220
Sadankwa . . . 221
Sargent . . . 220
Siebold . . . 221
Smooth Withe-rod . . . 219
Southern . . . 220
Tea . . . 220
Tubeflower . . . 218
Wayfaringtree . . . 219
Withe-rod . . . 217
Woolly . . . 217
Wright . . . 221
Wahoo . . . 122
Eastern . . . 122
Walkingstick . . . 93
Walnut . . . 138
Black . . . 138
English . . . 139
Waxmyrtle . . . 157
Southern . . . 157
Weigela . . . 221
Old Fashioned . . . 221
Wheel Tree . . . 211
Willow . . . 193
Wingnut . . . 181
Caucasian . . . 181
Winterberry . . . 137
Common . . . 137
Japanese . . . 137
Winterhazel . . . 111
Buttercup . . . 111
Chinese . . . 111
Fragrant . . . 111
Spike . . . 111
Willmott . . . 111
Wintersweet . . . 107
Fragrant . . . 107
Wisteria . . . 222
Japanese . . . 222
Witch-hazel . . . 129
Chinese . . . 130
Common . . . 131
Japanese . . . 130
Vernal . . . 130
Withe-rod . . . 219
Smooth . . . 219
Woadwaxen . . . 128
Dyer's . . . 128
German . . . 128
Silkyleaf . . . 128
Spanish . . . 128
Wolfberry . . . 149
Chinese . . . 149
Woodbine . . . 161
Yaupon . . . 137
Yellowhorn . . . 222
Yellowwood . . . 107
American . . . 107
Yellow-root . . . 222
Yew . . . 206
Anglojap . . . 207
Canadian . . . 207
English . . . 206
Japanese . . . 207
Pacific . . . 207
Yucca . . . 222
Adam's-needle . . . 222
Zelkova . . . 222
Japanese . . . 222
Zenobia . . . 222
Dusty . . . 222

St. Louis Community College
at Meramec
Library